T0324140

Sequencing Technologies in Microbial Food Safety and Quality

Food Analysis & Properties

Series Editor
Leo M. L. Nollet
University College Ghent, Belgium

This CRC series **Food Analysis and Properties** is designed to provide a state-of-art coverage on topics to the understanding of physical, chemical, and functional properties of foods: including (1) recent analysis techniques of a choice of food components; (2) developments and evolutions in analysis techniques related to food; (3) recent trends in analysis techniques of specific food components and/or a group of related food components.

Fingerprinting Techniques in Food Authenticity and Traceability
Edited by K. S. Siddiqi and Leo M. L. Nollet

Hyperspectral Imaging Analysis and Applications for Food Quality
Edited by Nrusingha Charan Basantia, Leo M. L. Nollet, and Mohammed Kamruzzaman

Ambient Mass Spectroscopy Techniques in Food and the Environment
Edited by Leo M. L. Nollet and Basil K. Munjanja

Food Aroma Evolution: During Food Processing, Cooking and Aging
Edited by Matteo Bordiga and Leo M. L. Nollet

Mass Spectrometry Imaging in Food Analysis
Edited by Leo M. L. Nollet

Proteomics for Food Authentication
Edited by Leo M. L. Nollet and Semih Otles

Analysis of Nanoplastics and Microplastics in Food
Edited by Leo M. L. Nollet and Khwaja Salahuddin Siddiqi

Chiral Organic Pollutants
Monitoring and Characterization in Food and the Environment
Edited by Edmond Sanganyado, Basil Munjanja, and Leo M. L. Nollet

Sequencing Technologies in Microbial Food Safety and Quality
Edited by Devarajan Thangadurai, Leo M. L. Nollet, Saher Islam, and Jeyabalan Sangeetha

For more information, please visit the Series Page: https://www.crcpress.com/Food-Analysis--Properties/book-series/CRCFOODANPRO

Sequencing Technologies in Microbial Food Safety and Quality

Edited by
Devarajan Thangadurai
Leo M. L. Nollet
Saher Islam
Jeyabalan Sangeetha

CRC Press
Taylor & Francis Group
Boca Raton London New York

CRC Press is an imprint of the
Taylor & Francis Group, an **informa** business

First Edition published 2021
by CRC Press
6000 Broken Sound Parkway NW, Suite 300, Boca Raton, FL 33487-2742

and by CRC Press
2 Park Square, Milton Park, Abingdon, Oxon, OX14 4RN

© 2021 Taylor & Francis Group, LLC

CRC Press is an imprint of Taylor & Francis Group, LLC

Library of Congress Cataloging-in-Publication Data

Names: Thangardurai, Devarajan, editor. | Nollet, Leo M. L., 1948- editor.
| Islam, Saher, editor. | Sangeetha, Jeyabalan, editor.
Title: Sequencing technologies in microbial food safety and quality /
edited by Devarajan Thangadurai, Leo M.L. Nollet, Saher Islam,
Jeyabalan Sangeetha.
Description: First edition. | Boca Raton : CRC Press, 2021. | Series: Food
analysis & properties | Includes bibliographical references and index.
Identifiers: LCCN 2020055202 (print) | LCCN 2020055203 (ebook) | ISBN
9780367351182 (hardback) | ISBN 9780429329869 (ebook)
Subjects: LCSH: Food--Analysis. | Food--Microbiology. | Microbial genomics.
| Nucleotide sequence. | Molecular epidemiology--Methodology. |
Food--Safety measures. | Food--Quality.
Classification: LCC TX541 .S467 2021 (print) | LCC TX541 (ebook) | DDC
641.3001/579--dc23
LC record available at https://lccn.loc.gov/2020055202
LC ebook record available at https://lccn.loc.gov/2020055203

ISBN: 978-0-367-35118-2 (hbk)
ISBN: 978-0-367-76053-3 (pbk)
ISBN: 978-0-429-32986-9 (ebk)

Typeset in Sabon
by Deanta Global Publishing Services, Chennai, India

Contents

PART II   Generation of Sequencing Technologies

PART III High-Throughput Sequencing Technology Applications for Food Safety and Quality

Series Preface

There will always be a need to analyze food components and their properties. Current trends in analyzing methods include automation, increasing the speed of analyses, and miniaturization. Over the years, the unit of detection has evolved from micrograms to picograms.

A classical pathway of analysis is sampling: sample preparation, cleanup, derivatization, separation, and detection. At every step, researchers are working and developing new methodologies. A large number of papers are published every year on all facets of analysis. So, there is a need for books that gather information on one kind of analysis technique or on the analysis methods for a specific group of food components.

The scope of the CRC Series on Food Analysis & Properties aims to present a range of books edited by distinguished scientists and researchers who have significant experience in scientific pursuits and critical analysis. This series is designed to provide state-of-the-art coverage on topics such as:

1. Recent analysis techniques on a range of food components
2. Developments and evolution in analysis techniques related to food
3. Recent trends in analysis techniques for specific food components and/or a group of related food components
4. The understanding of physical, chemical, and functional properties of foods

The book *Sequencing Technologies in Microbial Food Safety and Quality* is volume number 14 of this series.

I am happy to be a series editor of such books for the following reasons:

- I am able to pass on my experience in editing high-quality books related to food.
- I get to know colleagues from all over the world more personally.
- I continue to learn about interesting developments in food analysis.

Much work is involved in the preparation of a book. I have been assisted and supported by a number of people, all of whom I would like to thank. I would especially like to thank the team at CRC Press/Taylor & Francis Group, with a special word of thanks to Steve Zollo, senior editor.

Many, many thanks to all the editors and authors of this volume and future volumes. I very much appreciate all their effort, time, and willingness to do a great job.

I dedicate this series to:

- My wife, for her patience with me (and all the time I spend on my computer).
- All patients suffering from prostate cancer; knowing what this means, I am hoping they will have some relief.

Dr Leo M. L. Nollet (Retired)
University College Ghent
Ghent, Belgium

Preface

Safety of food products is a worldwide concern and all consumers have the right to have nutritious and safe food in their lives. In many countries, considerable efforts are being taken to improve food quality. Though progress is being made, the global burden on food through foodborne diseases and foodborne infections and economic loss in the food sector are unacceptably high. Food health and economy are achieved only through confirming food safety and security, which can be affected by unsafe food processes and practices. In this perspective, global food sectors are making efforts to build food control systems from legislative perspective to surveillance and monitoring programs through advanced analytical tools in laboratories.

Whole-genome sequencing (WGS) enables better understanding of differences both within and between species through revealing the complete DNA sequence of an organism. It is allowing researchers to differentiate the organisms accurately, where other technologies fail to do so. It is a newly emerged advanced tool and has greater potential to manage food security issues, food inspection, foodborne disease surveillance, outbreak detection and investigation, eliminating unwanted uncertainty in microbiological food safety issues, and food technology developments. This tool attracts the food sector to confirm food safety and management by its rapidly decreasing costs and increasing simplicity in usage.

Next-generation sequencing (NGS) has encouraged the development of high-throughput technologies like genomics, proteomics, transcriptomics, metagenomics, metaproteomics, and metatranscriptomics. NGS is offering a deeper understanding of the microbial genomes in an ecosystem compared to culture-based detection methods. The emergence of NGS technology reduced the cost of sequencing DNA in high magnitude and has given a power to scientists and investigators in various fields and industries for genetic material sequencing. NGS is predominantly applied in two different ways–the shotgun method, sequencing the total nucleic acids of microbes, and targeted sequencing, like gene-specific sequencing. The NGS tool is applied widely in various sectors as it is cheaply available and quickly generates draft WGS data. Food microbial ecology is well explored through NGS technology, in addition to the genome sequencing of foodborne pathogens. The NGS tool, combined with bioinformatics approaches, is reforming the field of food microbiology.

The present book deals with recent developments and applications of genomics in the various fields of food sectors in three sections. Part I of this book discusses the importance of food safety and security in food industries, epidemiology of foodborne pathogens, traditional molecular approaches in food safety science, and the impact of sequencing methods in the area of food microbiology. Part II discusses in detail the various sequencing generations, advancements in sequencing tools, library preparations, single-cell sequencing, and whole-genome sequencing applications in food safety and quality. Finally, Part III deals in detail with the various applications of omics technologies in food technology and the significance of high-throughput sequencing methods to improve food quality and food surveillance.

We humbly thank the keen interest, active participation, and potential contribution of relevant chapters by the contributing authors. We also sincerely thank Steve Zollo, Senior Editor with CRC Press/Taylor & Francis Group, for his opinion, support, suggestions, and guidance in the fruitful completion of this book publishing program.

Devarajan Thangadurai

Leo M. L. Nollet

Saher Islam

Jeyabalan Sangeetha

Editors

Devarajan Thangadurai is Assistant Professor at Karnatak University, Dharwad, India, and did his postdoctoral research at the University of Madeira, Portugal; University of Delhi, India; and ICAR National Research Centre for Banana, India. He is the recipient of the Best Young Scientist Award with Gold Medal from Acharya Nagarjuna University, India, and the VGST-SMYSR Young Scientist Award of the Government of Karnataka, India. He has keen interest and expertise in the fields of biodiversity and biotechnology, genetics and genomics of food crops and beneficial microbes for crop productivity, and food safety towards sustainable agricultural development. He has authored and edited more than twenty books including *Genetic Resources and Biotechnology* (3 Vols.), *Genes, Genomes and Genomics* (2 Vols.), *Genomics and Proteomics*, and *Biotechnology of Microorganisms* with international scientific publishers in the USA, Canada, Switzerland, and India. He has extensively traveled to many universities and institutes in Africa, Europe, and Asia for academic works, scientific meetings, and international collaborations. He is also reviewer for several journals including *Comprehensive Reviews in Food Science and Food Safety*, *Food Science and Technology International*, *Journal of Agricultural and Food Chemistry*, *Journal of Food Science*, *Journal of the Science of Food and Agriculture*, and *International Journal of Plant Production*.

Leo M. L. Nollet earned an MS (1973) and PhD (1978) in biology from the Katholieke Universiteit Leuven, Belgium. He is editor and associate editor of numerous books. He edited for M. Dekker, New York—now CRC Press of Taylor & Francis Publishing Group—the first, second, and third editions of *Food Analysis by HPLC* and *Handbook of Food Analysis*. The last edition is a two-volume book. He also edited the *Handbook of Water Analysis* (first, second, and third editions) and *Chromatographic Analysis of the Environment* (third and fourth editions; CRC Press). With F. Toldrá, he coedited two books published in 2006, 2007, and 2017: *Advanced Technologies for Meat Processing* (CRC Press) and *Advances in Food Diagnostics* (Blackwell Publishing—now Wiley). With M. Poschl, he coedited the book *Radionuclide Concentrations in Foods and the Environment*, also published in 2006 (CRC Press). He has also coedited with Y. H. Hui and other colleagues on several books: *Handbook of Food Product Manufacturing* (Wiley, 2007), *Handbook of Food Science, Technology, and Engineering* (CRC Press, 2005), *Food Biochemistry and Food Processing* (first and second editions; Blackwell Publishing—now Wiley—2006 and 2012), and the *Handbook of Fruits and Vegetable Flavors* (Wiley, 2010). In addition, he edited the *Handbook of Meat, Poultry, and Seafood Quality* (first and second editions; Blackwell Publishing—now Wiley—2007 and 2012). From 2008 to 2011, he published five volumes on animal product-related books with F. Toldrá: *Handbook of Muscle Foods Analysis*, *Handbook of Processed Meats and Poultry Analysis*, *Handbook of Seafood and Seafood Products Analysis*, *Handbook of*

Dairy Foods Analysis, and *Handbook of Analysis of Edible Animal By-Products*. Also, in 2011, with F. Toldrá, he coedited two volumes for CRC Press: *Safety Analysis of Foods of Animal Origin* and *Sensory Analysis of Foods of Animal Origin*. In 2012, they published the *Handbook of Analysis of Active Compounds in Functional Foods*. In a coedition with Hamir Rathore, *Handbook of Pesticides: Methods of Pesticides Residues Analysis* was marketed in 2009; *Pesticides: Evaluation of Environmental Pollution* in 2012; *Biopesticides Handbook* in 2015; and *Green Pesticides Handbook: Essential Oils for Pest Control* in 2017. Other finished book projects include *Food Allergens: Analysis, Instrumentation, and Methods* (with A. van Hengel; CRC Press, 2011) and *Analysis of Endocrine Compounds in Food* (Wiley-Blackwell, 2011). His recent projects include *Proteomics in Foods* with F. Toldrá (Springer, 2013) and *Transformation Products of Emerging Contaminants in the Environment: Analysis, Processes, Occurrence, Effects, and Risks* with D. Lambropoulou (Wiley, 2014). In the series Food Analysis & Properties, he edited (with C. Ruiz-Capillas) *Flow Injection Analysis of Food Additives* (CRC Press, 2015) and *Marine Microorganisms: Extraction and Analysis of Bioactive Compounds* (CRC Press, 2016). With A. S. Franca, he coedited *Spectroscopic Methods in Food Analysis* (CRC Press, 2017), and with Horacio Heinzen and Amadeo R. Fernandez-Alba he coedited *Multiresidue Methods for the Analysis of Pesticide Residues in Food* (CRC Press, 2017). Further volumes in the series Food Analysis & Properties are *Phenolic Compounds in Food: Characterization and Analysis* (with Janet Alejandra Gutierrez-Uribe, 2018), *Testing and Analysis of GMO-containing Foods and Feed* (with Salah E. O. Mahgoub, 2018), *Fingerprinting Techniques in Food Authentication and Traceability* (with K. S. Siddiqi, 2018), *Hyperspectral Imaging Analysis and Applications for Food Quality* (with N. C. Basantia and Mohammed Kamruzzaman, 2018), *Ambient Mass Spectroscopy Techniques in Food and the Environment* (with Basil K. Munjanja, 2019), *Food Aroma Evolution: During Food Processing, Cooking, and Aging* (with M. Bordiga, 2019), *Mass Spectrometry Imaging in Food Analysis* (2020), *Proteomics in Food Authentication* (with S. Ötleş, 2020) and *Analysis of Nanoplastics and Microplastics in Food* (with K. S. Siddiqi, 2020).

Saher Islam is Visiting Lecturer at the Department of Biotechnology of Lahore College for Women University, Pakistan, and is a Higher Education Commission (HEC) Scholar of the Islamic Republic of Pakistan at the University of Veterinary and Animal Sciences, Lahore, where she received her BS, MPhil, and PhD in Molecular Biology, Biotechnology, and Bioinformatics. She was an IRSIP Scholar at Cornell University, New York, and Visiting Scholar at West Virginia State University, West Virginia. She has keen research interests in genetics, molecular biology, biotechnology, and bioinformatics, and has ample hands-on experience in molecular marker analysis, whole genome sequencing, and RNA sequencing. She has visited the USA, UK, Singapore, Germany, Italy, and Russia for academic and scientific trainings, courses, and meetings. She is the recipient of the 2016 Boehringer Ingelheim Fonds Travel Grant from European Molecular Biology Laboratory, Germany.

Jeyabalan Sangeetha is Assistant Professor at Central University of Kerala, Kasaragod, India. She earned her BSc in Microbiology (2001) and PhD in Environmental Sciences (2010) from Bharathidasan University, Tiruchirappalli, Tamil Nadu, India. She holds an MSc in Environmental Sciences (2003) from Bharathiar University, Coimbatore, Tamil Nadu, India. Between 2004 and 2008, she was the recipient of Tamil Nadu Government Scholarship and the Rajiv Gandhi National Fellowship of University Grants Commission,

Government of India for doctoral studies. She served as Dr. D. S. Kothari Postdoctoral Fellow and UGC Postdoctoral Fellow at Karnatak University, Dharwad, South India, during 2012–2016 with funding from University Grants Commission, Government of India, New Delhi. She has carried out her doctoral and postdoctoral research in the fields of microbiology and metagenomics, respectively. She has authored a textbook, *Fundamentals of Molecular Mycology*, and edited seven books including *Genomics and Proteomics, Industrial Biotechnology* and *Biotechnology of Microorganisms*. She was trained in next generation sequencing and data analysis at Malaysia Genome Institute, Malaysia.

Contributors

Arowora Kayode Adebisi
Department of Biochemistry
Federal University Wukari
Taraba State, Nigeria

Charles Oluwaseun Adetunji
Department of Microbiology
Edo University
Iyamho, Nigeria

Juliana Bunmi Adetunji
Department of Biochemistry
Osun State University
Osogbo, Nigeria

Mahvish Ajaz
Department of Eastern Medicine
Government College University
Faisalabad, Pakistan

Kashif Akram
Department of Food Sciences
Cholistan University of Veterinary and
    Animal Sciences
Bahawalpur, Pakistan

Muhammad Akram
Department of Eastern Medicine
Government College University
Faisalabad, Pakistan

Shanmugarathinam Alagarsamy
Department of Pharmaceutical Technology
Anna University
Tiruchirappalli, India

Jorianne Thyeska Castro Alves
State University of Pará (UEPA)
Pará, Brazil

Saadia Andleeb
Atta-ur-Rahman School of Applied
    Biosciences
National University of Sciences and
    Technology
Islamabad, Pakistan

Usha Antony
Department of Biotechnology
Anna University
Chennai, India

Marie Arockianathan
PG and Research Department of
    Biochemistry
St. Joseph's College of Arts and Science
    (Autonomous)
Cuddalore, India

Ayodele Eugene Ayeni
Department of Biological Sciences
Landmark University
Omu-Aran, Nigeria

**Sayed Muhammad Ata Ullah
Shah Bukhari**
Department of Microbiology
Abdul Wali Khan University
Mardan, Pakistan

Muhammad Baqir
Department of Food Science and
  Technology
MNS-University of Agriculture
Multan, Pakistan

Ramachandran Chelliah
Department of Food Science and
  Biotechnology
Kangwon National University
Chuncheon, South Korea

Eric Banan-Mwine Daliri
Department of Food Science and
  Biotechnology
Kangwon National University
Chuncheon, South Korea

Fazle Elahi
Department of Food Science and
  Biotechnology
Kangwon National University
Chuncheon, South Korea

Umar Farooq
Department of Food Science and
  Technology
MNS-University of Agriculture
Multan, Pakistan

Amala Geevarghese
School of Biosciences
Mahatma Gandhi University
Kottayam, India

Khizar Hayat
Department of Food Science and
  Technology
MNS-University of Agriculture
Multan, Pakistan

Zafar Hayat
Department of Food Sciences
Cholistan University of Veterinary and
  Animal Sciences
Bahawalpur, Pakistan

Daniel Ingo Hefft
University of Birmingham
Birmingham, United Kingdom

Adil Hussain
Department of Agriculture
Abdul Wali Khan University
Mardan, Pakistan

Amjad Iqbal
Department of Food Science and
  Technology
Abdul Wali Khan University
Mardan, Pakistan

Mariam Iqbal
Department of Food Science and
  Technology
MNS-University of Agriculture
Multan, Pakistan

Muhsin Jamal
Department of Microbiology
Abdul Wali Khan University
Mardan, Pakistan

Fahad Said Khan
Department of Eastern Medicine
Government College University Faisalabad
Faisalabad, Pakistan

Imran Khan
Department of Food Science and
  Biotechnology
Kangwon National University
Chuncheon, South Korea

and

Department of Biotechnology
University of Malakand
Khyber Pakhtunkhwa, Pakistan

Muhammad Zaki Khan
Department of Food Science and
  Technology
MNS-University of Agriculture
Multan, Pakistan

Radhakrishnan Edayileveettil Krishnankutty
School of Biosciences
Mahatma Gandhi University
Kottayam, India

Arvind Kumar
Department of Dairy Science and Food Technology
Banaras Hindu University
Varanasi, India

V. J. Rejish Kumar
Department of Aquaculture
Kerala University of Fisheries and Ocean Studies
Panangad, India

Umme Laila
Department of Eastern Medicine
Government College University Faisalabad
Faisalabad, Pakistan

Inamul Hasan Madar
Department of Biochemistry
Bharathidasan University
Thiruchirappalli, India

Shanthala Mallikarjunaiah
Department of Zoology
Bangalore University
Bangalore, India

Amisha Mathew
School of Biosciences
Mahatma Gandhi University
Kottayam, India

Olugbenga Samuel Michael
Department of Physiology
Bowen University
Iwo, Nigeria

Sadhna Mishra
Department of Dairy Science and Food Technology
Banaras Hindu University
Varanasi, India

Sumaira Miskeen
Department of Food Science and Biotechnology
Kangwon National University
Chuncheon, South Korea

and

Department of Biotechnology
University of Malakand
Khyber Pakhtunkhwa, Pakistan

Debadarshee Das Mohapatra
BITS-Pilani
Hyderabad, India

Gislenne Moia
Faculty of Computer Engineering
Federal University of Pará
Tucuruí, Brazil

Muhammad Asif Nawaz
Department of Biotechnology
Shaheed Benazir Bhutto University
Sheringal, Pakistan

Deog Hwan Oh
Department of Food Science and Biotechnology
Kangwon National University
Chuncheon, South Korea

Olugbemi Tope Olaniyan
Department of Physiology
Edo University
Iyahmo, Nigeria

Mônica Silva de Oliveira
Graduate Program in Applied Computing
Federal University of Pará
Tucuruí, Brazil

Rosyely da Silva Oliveira
Federal Rural University of Amazonia (UFRA)
Pará, Brazil

Wadzani Dauda Palnam
Department of Agronomy
Federal University
Gashua, Nigeria

Shikha Pandhi
Department of Dairy Science and Food
  Technology
Banaras Hindu University
Varanasi, India

Sunil Pareek
Department of Agriculture and
  Environmental Sciences
National Institute of Food Technology
  Entrepreneurship and Management
  (NIFTEM)
Sonepat, India

Mahesh Pattabhiramaiah
Department of Zoology
Bangalore University
Bengaluru, India

Smaranika Pattnaik
Department of Biotechnology and
  Bioinformatics
Sambalpur University
Sambalpur, India

Sidra Pervez
Department of Biochemistry
Shaheed Benazir Bhutto Women
  University
Peshawar, Pakistan

Radhakrishnan Preetha
Department of Food Process Engineering
SRM Institute of Science and Technology
Kattankulathur, India

Nayana Aluparambil Radhakrishnan
School of Biosciences
Mahatma Gandhi University
Kottayam, India

S. Prasanna Raghavender
Department of Food Process Engineering
SRM Institute of Science and Technology
Kattankulathur, India

Dinesh Chandra Rai
Department of Dairy Science and Food
  Technology
Banaras Hindu University
Varanasi, India

Sana Raza
Department of Microbiology
Abdul Wali Khan University
Mardan, Pakistan

Redaina
Department of Microbiology
Abdul Wali Khan University
Mardan, Pakistan

Pablo Henrique Caracciolo Gomes de Sá
Federal Rural University of
  Amazonia (UFRA)
Pará, Brazil

Akshaya Chekkara Thandayan Santhosh
School of Biosciences
Mahatma Gandhi University
Kottayam, India

Ronilson Santos dos Santos
Federal Rural University of
  Amazonia (UFRA)
Pará, Brazil

Kandasamy Saravanakumar
Department of Medical Biotechnology
Kangwon National University
Chuncheon, South Korea

Sebastain Korattiparambil Sebastian
Department of Zoology
Government College
Kottayam, India

Afshan Shafi
Department of Food Science and
  Technology
MNS-University of Agriculture
Multan, Pakistan

Liloma Shah
Department of Microbiology
Abdul Wali Khan University
Mardan, Pakistan

Muhammad Shahbaz
Department of Food Science and
  Technology
MNS-University of Agriculture
Multan, Pakistan

Sreejith Sreekumaran
School of Biosciences
Mahatma Gandhi University
Kottayam, India

Ghazala Sultan
Department of Computer Science
Aligarh Muslim University
Aligarh, Uttar Pradesh, India

Ajayi Kolawole Temidayo
Department of Microbiology
University of Ilorin
Ilorin, Nigeria

Mary Theresa
School of Biosciences
Mahatma Gandhi University
Kottayam, India

Benjamin Ewa Ubi
Department of Biotechnology
Ebonyi State University
Abakaliki, Nigeria

Aparna Sankaran Unni
School of Biosciences
Mahatma Gandhi University
Kottayam, India

Benssan K. Varghese
Department of Food Process Engineering
SRM Institute of Science and Technology
Kattankulathur, India

Thirumalai Vasan
Department of Biotechnology
Srimad Andavan Arts and Science College
Tiruchirapalli, India

Adonney Allan de Oliveira Veras
Graduate Program in Applied Computing
Federal University of Pará
Tucuruí, Brazil

Myeong-Hyeon Wang
Department of Medical Biotechnology
Kangwon National University
Chuncheon, South Korea

Shuai Wei
Guangdong Provincial Key Laboratory of
  Aquatic Product Processing and Safety
Guangdong Ocean University
Zhanjiang, China

Su-Jung Yeon
Department of Food Science and
  Biotechnology
Kangwon National University
Chuncheon, South Korea

C. Anoint Yochabedh
Department of Food Process Engineering
SRM Institute of Science and Technology
Kattankulathur, India

PART I

Introduction, Genomics, Proteomics, and Bioinformatics

CHAPTER 1

Food Quality and Food Safety
An Introduction

Umar Farooq, Afshan Shafi, Muhammad Shahbaz,
Muhammad Zaki Khan, Khizar Hayat,
Muhammad Baqir, and Mariam Iqbal

CONTENTS

1.1 Introduction 3
1.2 Food Quality 5
1.3 Food Safety 7
    1.3.1 Foodborne Illnesses 8
    1.3.2 Traceability 10
    1.3.3 Food Biotechnology 11
    1.3.4 GMO Food Regulations 14
1.4 Food Quality and Safety Standards 15
    1.4.1 Food Quality Standard (ISO 9001) 16
    1.4.2 Food Safety Standards 16
        1.4.2.1 British Retail Consortium 17
        1.4.2.2 International Food Standard 17
        1.4.2.3 Hazard Analysis Critical Control Point 17
        1.4.2.4 ISO 22000 18
        1.4.2.5 Food Distribution Systems 19
1.5 Conclusion 19
References 20

1.1 INTRODUCTION

Food and Agriculture Organizations (FAOs) characterize food quality as being "a dynamic aspect of food, which defines its significance or acceptability by consumers." While the nutritional value as well as the physiological and sensory attributes of a food corresponds to its presumed quality, another factor is the safety of a food. A safe food is one devoid of contaminants that may jeopardize a person's health. The hierarchy of responsibility for food safety was quite limited during prehistoric times, when human groups were primarily composed of hunter-gatherers and their relatives. But, as communities began to grow larger and more diverse, and as trade routes broadened and food was transported across long distances, the burden to provide safe food became more prominent. In recent decades, food safety control programs have been established, such as Hazard

Analysis Critical Control Points (HACCP), Good Manufacturing Practice (GMP), and Good Hygiene Practice (GHP). "Food safety" is a confirmation that food is fit for consumption. Characterizing two other terms–"toxicity" and "hazard"–can enhance our understanding of food safety. *Toxicity* is a substance's potential under any circumstances to cause damage or injury of some sort. *Hazard* is the associative risk of damage or injury if a product is not used in a controlled way. There are three main categories of hazards i.e. physical, chemical, and biological as shown in Table 1.1.

TABLE 1.1   Potential Hazards in Food (FDA 2016)

Hazard Category	Hazard Sub-Category	Examples
Biological	Bacteria	*Bacillus cereus* (*B. cereus*)
		Campylobacter jejuni (*C. jejuni*)
		Clostridium botulinum (*C. botulinum*)
		Clostridium perfringens (*C. perfringens*)
		Shiga-toxin producing *Escherichia coli* such as O157:H7 (*E. coli* O157:H7)
		Listeria monocytogenes (*L. monocytogenes*)
		Salmonella spp.
		Shigella spp.
		Staphylococcus aureus (*S. aureus*)
	Protozoa and parasites	*Cryptosporidium parvum*
		Cyclospora cayetanensis
		Giardia lamblia (*G. intestinalis*)
		Trichinella spiralis
	Viruses	Norovirus
		Hepatitis A
		Rotavirus
Chemical	Pesticide residues	Organophosphates
		Carbamates
		Chlorinated hydrocarbons
		Pyrethroids
	Heavy metals	Lead
		Arsenic
		Cadmium
		Mercury
	Drug residues (veterinary antibiotics)	Chloramphenicol
		Beta-lactams
	Industrial chemicals	Ammonia
	Environmental contaminants	Dioxins
	Mycotoxins	Aflatoxin
		Patulin
		Ochratoxin
		Fumonisin
		Deoxynivalenol
	Allergens	Milk, eggs, fish, crustacean shellfish, tree nuts, peanuts, wheat, and soybeans (commonly called "the Big 8")

(Continued)

TABLE 1.1 (CONTINUED)   Potential Hazards in Food (FDA 2016)

Hazard Category	Hazard Sub-Category	Examples
	Unapproved colors and additives	FD&C Red #4
		Melamine
	Substances associated with a food intolerance or food disorder	Lactose
		Yellow #5
		Sulfites
		Carmine and cochineal
		Gluten
	Radionuclides	Radium 226 and 228
		Uranium 235 and 238
		Strontium 90
		Cesium 137
		Iodine 131
Physical	—	Metal
		Glass
		Hard plastic

1.2 FOOD QUALITY

Food quality encompasses environmental factors including physical appearance (size, gloss, color, consistency, and shape), flavor, and microbial, chemical, nutritional, and physical aspects (Perez-Gago et al. 2006). It applies not only to the physical characteristics of food but also to the manner in which the end consumer perceives the product (Grunert 2005). This includes the microbial, textural, and flavor dimensions. Quality assurance is responsible for the production, auditing, distribution, certification, and expense of food as well as the technical advancements that generate greater productivity and reduce costs. New technological advancements include time-temperature integrators, measures that are used to enhance temperature control across the distribution network (Giannakourou and Taoukis 2003) thereby improving the shelf life of food items (Dalgaard et al. 2002; Raab et al. 2008). The Global Food Safety Initiative (GFSI), British Retail Consortium (BRC), International Organization for Standardization (ISO), Safe Quality Food (SQF), and International Food Standard (IFS), are by far the most significant quality control systems in the food industry.

Food quality is an essential prerequisite for food production, since food consumers are vulnerable to any sort of infection which may arise even during the production process. Most customers also depend on quality and processing standards to identify ingredients related to dietary (kosher, halal, vegetarian), nutritional needs, or medical circumstances (diabetes and allergic reactions).

In addition to product consistency, sanitation criteria often apply. It is critical to ensure that the food manufacturing system is just as healthy as possible so that the customer receives the safest food possible.

Food safety also coincides with product traceability, tracking the suppliers of ingredients in case of a product recall. This also involves concerns related to labeling to ascertain that listed ingredients are accurate and that complete details related to nutrition are included. Food quality seems to have become a dominant focus in agriculture and food processing, and a challenging technical and scientific discourse on the concept of quality is currently under way. Dramatic incidents of food contaminated with toxins from dioxin, hormones, pesticides, and BSE, and also the rising number of pollution-related cancers, concentrate the focus of consumers and industry on food safety.

Respect for the past is gaining momentum through a new perception of efficiency, the modern definition of which is summed up in a product's "price profile" which involves multiple aspects of its quality. Infact, manufacturers and consumers are now making proposals for reliable warranty and quality. Performance requires a profound restructuring of farming and technology for development. Control structures, quality requirements, and the entire management of the company must be reviewed according to this new perspective.

Scientific organizations may assist producers by establishing novel technologies (biological, operational and computerized), based on quality and safety considerations. Evaluation of food composition and emerging food processing techniques are the primary objectives of this approach, and they have resulted in significant improvements in the production and distribution of some food groups (organic foods and light products) over the past few decades.

Farmers continue to work to achieve strategic targets intended to meet both the requirements of industrial processes and the safety regulations in order to enhance their productivity in the global marketplace. The development of crop variants including the use of modern cultivation and food preservation techniques (integrated pest control, conservation of cereal crops by cold storage) are some of the examples.

With regard to food processing and preservation strategies, manufacturing is progressively tailored to the use of moderate technology consisting of highly targeted and restricted treatments that will decrease thermo-mechanical destruction and chemical or biological pollution. This method can be applied widely in the development of fatty foods especially comprising of cholesterol and saturated fatty acids.

Composition adjustment has predominantly been practiced in the meat production field although milk products are a category of foods wherein variations in configuration are becoming extremely prevalent. Low-fat or low-calorie foods are increasingly popular items in this field. To meet demand, changes to the law were required in order to deliver the necessary information to consumers.

There is a growing demand in developed countries for "functional foods" and herbal products, items that, when added to a regular diet, are deemed to maintain nutrition and health.

Functional foods are characterized as foods wherein essential nutrients are incorporated to enhance their nutritional benefits innate to them. This group covers a wide variety of commodities that alter the basic framework of a food by incorporating vitamins, minerals, and other bioactive components. This category also encompasses probiotics, bioactive peptides, prebiotics, fibers, vegetable extracts, and symbiotics. A specific product type is *dietary supplements*, consisting of distilled multi-micronutrients administered as tablets or other pharmacological formulations. This classification also comprises a number of herbal products.

Similarly, market share in cereals and "whole-meal" food products (biscuits, crackers) is growing substantially in response to customer preference for increased dietary fiber to obtain physiological and metabolic advantages.

The correlation between nutrition and health was historically related in terms of macronutrient content. Today, minor compounds such as antioxidants, vitamins, fiber, and other bioactive substances are now given prominence. The importance of bioactive compounds is universally acknowledged in reducing chronic metabolic syndromes including heart diseases.

1.3 FOOD SAFETY

Food safety encompasses how food is handled, prepared, and stored in order to avoid potential foodborne illness. As it has become a public health concern of increased significance, governments around the world are ramping up efforts to improve food safety (FAO 2002; Food Safety Authority of Ireland 2005), providing remedies to food safety issues and addressing customer concerns.

The food distribution system is shown in Figure 1.1.

Food safety basically means the prevention of illness caused by the ingestion of contaminated food.

By 2000, the occurrence of foodborne illnesses had decreased consumer confidence and customer trust throughout Europe, forcing the European Union (EU) to change its attitudes toward food safety.

The EU managed to launch food safety laws for all member countries, requiring rigorous scientific analysis of the various hazards. The United States review of peanut butter attributed to the contribution of salmonella is a particular example of a food safety crisis. This was the largest foodborne illness outbreak ever seen in the nation's history, impacting more than two hundred offshore food suppliers and more than 2100 products (Terreri 2009). Among the most significant things for food safety defense was the labeling of products that are very essential to ensure consumers' interests in the market. Food labeling comprises the origin of the food, preservation, preparation methods, and shelf life.

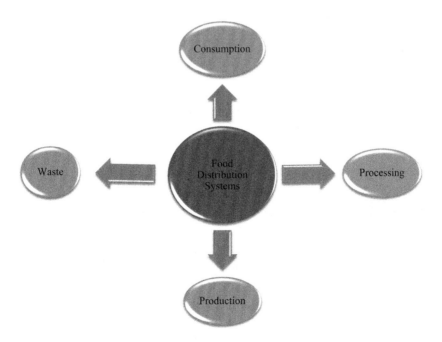

FIGURE 1.1 Food distribution system.

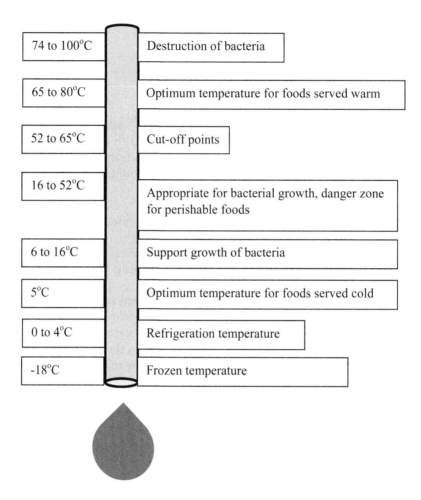

FIGURE 1.2   Food safety temperatures.

Different parameters are represented by food which could indicate that food is unfit for consumption. Avoiding unsafe food can limit the chances of microbial exposure to consumers and hence, prevent public outbreaks of food borne illnesses. If a person notices any signs of unsafe food, that food should not be chosen for consumption. After the food has been purchased, it should be kept safe by thoroughly washing raw foods and storing perishable foods at appropriate temperatures. Temperature ranges for foods are described in Figure 1.2.

1.3.1 Foodborne Illnesses

Foodborne disorders are ailments, generally infectious or toxic, induced through substances that access the body via food intake. The cause may be due to microscopic organisms or other factors including chemicals. They constitute substantial and wide-ranging national health concerns for both industrialized and developing economies (National Center for Food and Agricultural Policy 2005).

It is impossible to determine the worldwide occurrence of foodborne illnesses, but it has been documented that around 1.8 million individuals died from diarrhea in 2005 (FAO 2007). A substantial percentage of such instances can indeed be directly linked to food and potable water contamination. The proportion of individuals contracting foodborne illnesses was more in 30 developed countries. Though less extensively reported, developing nations are suffering from the consequences of the crisis due to the massive emergence of wide spectrum foodborne illnesses which also include parasite-induced diseases. In several developing countries, the higher incidence of diarrhea indicates significant inherent food safety concerns.

Even though most of the foodborne illnesses are erratic and not always confirmed, episodes of foodborne disease can infect large numbers of people. Few foodborne diseases seem to be quite understood and yet are doomed to emerge as they have become increasingly prevalent (Sudhakar et al. 1988). For instance, salmonellosis epidemics have been documented for centuries, however, the frequency of this illness has increased on almost every continent over the last 25 years. *Salmonella enteritidis* (SE) has been the prevailing strain in the Western Hemisphere and Europe. SE epidemic reports suggest its occurrence is primarily attributed to the consumption of contaminated poultry and eggs.

Although cholera has troubled most of Asia and Africa for decades, its appearance in 1991 in South America, for perhaps the first time in nearly a century, renders it yet another instance of a well-identified and developing infectious disease. Though cholera is mostly spread through contaminated water, foodstuffs spread infection as well. The frozen, unprocessed, or under-processed seafoods in Latin America are effective epidemiological vectors for the transmission of cholera.

Many foodborne pathogens are presumed to be emerging since they are novel microscopic organisms or because the transmission through food has only recently been addressed. Illness caused by *Escherichia coli* had first been identified as a pathogen in 1982. It has since quickly broken out as a significant cause of acute renal failure and bloody diarrhea. Quite often, the contamination is lethal especially among children. Incidences are generally linked with beef and have occurred in various European countries, Canada, South Africa, Australia, the USA, and Japan. Other food products including alfalfa sprouts, cheese curds, unpasteurized fruit juice, and spinach are also known sources of infection.

Listeria monocytogenes is the pathogen responsible for listeriosis. Infection is often the result of eating contaminated processed meats and cheeses or ingesting foods that have not been properly chilled. Listeriosis eruptions have also been documented in different countries including the USA, France, Australia, and Switzerland.

Foodborne trematodes are also appearing as a substantial public health hazard, particularly in Latin America and Southeast Asia, due to unhygienic conditions in fisheries and aquiculture and the use of unprocessed or slightly processed freshwater fish and their products. Foodborne trematodes could lead to acute liver disease and hepatic cancers.

Bovine spongiform encephalopathy (BSE), a catastrophic, communicable, neurodegenerative disease associated with cattle, was first reported in the UK in 1985. Almost nineteen countries have experienced endemic outbreaks of BSE since then, as the epidemic is no longer confined to the European Community: the first report of BSE in Japan's cattle herd was documented in 2001.

New risks of foodborne illness emerge for a multitude of reasons, including the growth of international commerce and travel, microbial modification and improvements in the food supply chain, and individual trends and behaviors. Foodborne illnesses present a significant risk to human health and to people, communities, and national economies.

Their regulation calls for effective trials on behalf of three partners including consumers, government, and the food industry.

The latest developments in worldwide agricultural production, manufacturing, and distribution are generating a growing demand for food safety studies to ensure a more reliable worldwide supply of safe food. Much of the food safety research has concentrated on the identification and understanding of foodborne hazards and managing them (European Commission 2000; Beier and Pillai 2007).

The "Farm-to-Fork" strategy was introduced by the European Commission (EC) to define and concentrate efforts on certain points in the food supply chain food contamination has been known to occur. The report is attempting to develop and improve national food safety frameworks to better control their supply of food and to minimize the risk of foodborne illnesses.

Thousands of research projects are produced in collaboration with suppliers and other organizations in order to enhance food safety (Buzby and Roberts 1996). The research priorities require applicable field related studies including chemical, biological, and physical such as with the production and assessment of current alternatives for analyzing the agricultural environments, individual animals, and herds in order to detect the classic and novel pathogens. Epidemiological reports are aimed at identifying critical control points where pathogens could be prohibited from accessing the food supply chain; assessing the efficacy of quality management systems and HACCP (hazard analysis and critical control points) programs to decrease the likelihood of microbial contamination of food items; introducing innovative diagnostic techniques to be used in farm environments; developing new approaches for the prevention, monitoring, reduction, or removal of microorganisms in food animals; and conducting risk evaluation and cost benefit analyses to evaluate the efficacy and appropriateness of farm-level preventive measures.

It is evident that engineering must be an essential aspect of the research on food safety. The development of physico-chemical processes for the identification of chemical and microbial hazards in food production requires engineering. All foods undergo multiple procedures from the farm level to ingestion. Each cycle has an impact on food safety and quality. Moreover, engineering helps calculate essential aspects of controlled processing and the effect of processing on food quality and safety.

1.3.2 Traceability

Food safety drives the need for traceability, as recent food safety reports indicate that approximately seven million people are impacted by foodborne disease each year. The solution is an effective monitoring system that traces a forthcoming product and successfully investigates the cause of the problem. Traceability is the ability to study in detail, step-by-step, the history of a process and associated activities.

Traceability can also be characterized as the background of a food product in relation to the increasing properties of that product and/or the properties correlated with it until such products have also been submitted to specific processes of added value using linked factors of production and under associated environmental factors (Thompson 2003). Knowledge of source relationships could be used vertically in the supply chain (for example, in the purchasing phase, to identify the specifications of an ordered product) or downwards (for example, in distribution processes to determine product characteristics) (Miotrag 2001). In addition, the details may be used in the supply chain or by private entities for analytic purposes (UNI 2001; UNI 2002; European Commission 2007).

In 1994 the ISO (ISO 2010) established a focused and more comprehensive definition of the food supply as ISO standard 8402:1994 and upheld by EC Regulation 178/2002 (European Parliament 2002). It describes 'traceability' as the potential to trace and monitor a plant, feed, or food manufactured from animals or different combinations of ingredients through all production and distribution stages (USDA 2002).

An appropriate and productive process which transmits accurate, comprehensive, and relevant product details via the supply chain could substantially minimize operational expenses and maximize productivity. At around the same time, any such program includes other elements of product safety: it ensures customers' safety by offering specific information of where an object originates, what are its elements, or what it contains and also provides background knowledge of processing.

Genetically Modified Organism (GMO) is an essential related matter. Genetic modification is used to create new and advantageous features, such as improved shelf life or greater tolerance to pests (Genetic Modification Advisory Committee 2002). Yet little is documented about the long-term impact of GMOs on health and the environment.

No proper global conclusion could be drawn on either GMO standards or research methods and, therefore, safety assessment (WHO 2002). As an effect, consumers need sufficient details to indicate not whether a food contains GMO ingredients, which could only be ensured by an effective tracing program.

In 2007, the ISO announced a novel standard: ISO 22005:2007–Traceability in the food and feed network–General standards and fundamental process design and execution criteria. The application of the food management systems outlined in the ISO 22000 series is expected to further guarantee the safety of food products and address consumer concerns.

According to ISO, the new guidelines and standards for the development and implementation of traceability for food and feed will help companies functioning anywhere in the food chain to:

- Trace content movement (food, its ingredients, and packaging)
- Identify documents and monitoring required for each development process
- Guarantee strong communication between the different parties concerned
- Allow at least each member to be notified about its consumer and direct suppliers

Because safety hazards may access the food chain at any point, proper control and coordination are necessary across the entire process, thus an ISO statement is added. A poor connection in the production process could lead to unsafe products that can pose a significant risk to customers and have expensive consequences for suppliers. Consequently, food safety is the collective concern of all associated actors.

ISO continued to notice how the diversity of retail and private quality systems in the food processing industry creates unequal standards of safety, uncertainty about specifications, and higher costs and complexities for suppliers forced to comply with numerous programs. ISO 2200 asserts, "Provides a unique alternative for good practices globally and therefore leads to the reduction of restrictions on trade."

1.3.3 Food Biotechnology

According to the Codex Alimentarius Commission's definition, *biotechnology* is defined as "the direct injection of nucleic acid (genetic material) into the host (organelles or cells), studying the deoxyribonucleic acid (DNA) and union of cells outside the taxonomic family which amplifies traditional reproductive or recombinant obstacles and classical breeding was not supported by this technique".

The introduction into the food system of genetically modified organisms (GMOs) is responsible for enhanced crop yields and added nutritional value that ultimately affects public health. GM crops sometimes support public health and safety indirectly by reducing the environmental effects of food production (American Medical Association Council 2000; Institute of Food Technologists 2000).

Modern biotechnology uses numerous research techniques to generate desired traits in plants, animals, and microorganisms (Yun-Hwa et al. 2007). Biotechnology has been used to explore novel approaches to productivity improvement ever since its implementation to agricultural processes at the start of the 1990s.

In 2005, approximately 21 countries cultivated a total of 222 million acres of biotech crops. Such crops include papaya, corn, canola, soybeans, squash, and cotton which are better versions of conventional varieties. Furthermore, fast-rise yeast and an enzyme used in the production of cheese were developed by exploiting the techniques of biotechnology (ISAAA 2005).

Biotechnology is a broad category for using live creatures and includes methods from basic to comprehensive (Zhao and McDaniel 2005). Individuals have interbred associated animal or plant species for decades to create valuable hybrids or newly developed varieties with improved characteristics, for example higher production and taste. Conventional interbreeding induces modifications in the genomes of plants and animals. The process can be quite time-consuming because it requires numerous generations to achieve the desired characteristics and eliminate undesirable ones.

Current agricultural biotechnology procedures, like genetic engineering, permits for a far more realistic production of different varieties of crops and animal varieties. The genomes which express desirable characteristics directly, including the agricultural output, are much more easily recognized. The genetic composition of food-producing crops and animals can thus be further enhanced. Genetic engineering not only offers the potential to choose favorable traits but also enables more accurate transmission of genes to achieve those traits. There are three important goals in the production of biotechnologically enhanced crops:

- Improved input characteristics like environmental stress, herbicide resistance, and insect and virus defense
- Various production features, like value addition, including certain corn enriched with lysine using the animal feed, or vegetable oils enriched with omega-3 fatty acids
- Crops generating pharmaceuticals or developing bio-based fuel processing; today crops are mainly those with enhanced input characteristics

Today the use of genetic knowledge to enhance animal selection and reproduction, known as *animal genomics*, is an important resource in agriculture (American Medical Association Council 2000). Genomics aids in assessing optimum animal nutritional needs and in producing reliably high-quality food, eggs, and milk. Cloning is yet another advanced technology which makes it easier to breed the healthiest and most efficient livestock. In this process, the animal's genetic structure is not altered in any way. Rather, this method of supported reproduction permits the breeding of livestock to develop twins of the best genetic makeup and facilitates breeding of the next generations. Foods developed from cloning of animals or their descendants have been available in the marketplace since about 2005.

Genetic engineering is also another useful strategy that has been explored in food animal-breeding programs. Significant benefits include animals which grow faster or have improved nutritional attributes, such as pork that contains higher levels of omega-3 fatty acids. Another highly anticipated commodity presently under regulatory consideration is a species of salmon which grows faster than its non-biotech equivalents. The advent of novel biotechnology to agricultural production poses new possibilities for human health and problems (Hsich and Ofori 2007).

The future benefits to public health include improved nutritional content of foods, decreased allergy-causing potential, and better quality of food processing processes. In contrast to this, the possible impacts of genetic manipulation of food on public health need to be studied carefully. The latest biotechnology will have to be examined extensively if it does facilitate further improvements in food production.

In the food production industry, the utilization of biotechnology covers four major groups, notably: (1) foods comprised of or possessing viable organisms; (2) food products produced from or carrying components acquired from GMOs; (3) food items usually containing single components or additives generated by genetically modified microorganisms (GMMs); and (4) foods containing enzyme-processed materials through (GMMs). GMMs have created a range of gums, amino acids, enzymes, and other additives used throughout food production in conjunction with large amounts of bio-fermentation at a reasonable cost but with little environmental effect. Enzymes extracted from plants, animals, and microorganisms are being used in the field of enzyme technology. Although natural enzymes cannot endure the processing to which the commodity is exposed, enzymes extracted from GMMs can be engineered to improve their heat stability and therefore allow them to resist extreme processing conditions. GMMs are often used for the processing of amino acids and vitamins for the development of dietary supplements. For instance, in order to combat vitamin A deficiency, carotenoids are formulated as a precursor for developing vitamin A supplements.

Current food biotechnology offers several benefits (American Dietetic Association 2006). One benefit is better crops with higher yields. To the farmer, this may mean lower production costs. An illustration is the improvement of some corn varieties to contain a common soil bacterium called *Bacillus thuringiensis* (BT). It encourages the corn to defend itself against insects that could kill plants, thereby reducing the use of insecticides. Other crops have been formed to tolerate plant viruses and other diseases. Plants can of course generate compounds to defend against invasive species. Multiple foods contain natural toxins, and researchers could determine certain genes that generate natural toxins. A further advantage of genetically modified crops is the tolerance to climatic conditions. This will allow certain plants to tolerate extreme weather conditions, spreading the growing season and emerging regions to enable year-round availability of more fruits, vegetables, and grains. Regions with poor soil or inadequate weather could grow into productive farmland. This could also reduce crop loss. Food biotechnology also provides strategies for farming on a limited landscape.

Transmitting different genetic traits into plants will generate fruits and vegetables with maturing qualities that enable them to be exported farther and longer without contamination and with better flavor on the arrival of fresh products. Products could also be cultivated to resist mold. Modern agricultural technology would make food safer by detecting bacteria and viruses within the food. This could lead to a reduction in food-borne diseases.

Vegetable oils have been modified to have fewer saturated fatty acids and more mono-unsaturated. It has improved soybeans, canola, and other oil-seeds that tend to have fewer saturated fats and far more oleic acid (an advantageous fatty acid). Current food biotechnology will encourage sophisticated crossbreeding of foods, generating novel food varieties. In several areas of the world, food biotechnology is progressing and provides tremendous potential for supporting the planet. Food biotechnology has many positive aspects, and as science advances even further new opportunities will be explored. Food biotechnology is one method for creating a worldwide high-quality, plentiful, nutritious, and less costly food supply and for supporting environmental protection. Nutritionally improved crops could overcome the consequences of malnutrition all around the world.

Different GM species contain numerous genes which are incorporated in various ways. That implies single GM foods and their safety must be evaluated on a case-by-case basis, although specific statements on the safety of all GM foods could not be generated. GM foods that are on the global market have exceeded risk evaluations and are unlikely to create public health threats (Society of Toxicology 2002).

Moreover, the intake of these foods by the general public in the areas where they were licensed seems to have no effect on public health. The consistent use of risk analysis focused on the Codex guidelines would form the foundation for evaluating the safety of GM foods, including post-marketing analysis wherever applicable.

The safety assessment of genetically modified crops typically examines: (a) immediate health effects (toxic effects), (b) inclinations to cause allergy (allergenicity); (c) particular elements that are assumed to have nutritional or hazardous characteristics; (d) the sustainability of the transgene; (e) nutritional impacts attributed to genetic alteration; and (f) any unintentional consequences that may result from the inserted gene.

1.3.4 GMO Food Regulations

The GM Foodstuffs Regulation-EC No. 1829/2003 and the Traceability and Labeling Regulation-EC No. 1830/2003 (Grujic and Blesic 2007) legislate the official permission and labeling of GM products inside the EU. A genetically modified food should not negatively impact human or animal health and the environment; should not deceive the customer; and should not be different from the food that it is supposed to bring adverse health effects through its consumption. The registration procedure for genetically modified crops is documented in Articles 5 to 7 of the GM Foodstuffs Regulation (EC No 1829/2003); details to facilitate the processing of the implementation document is provided in Regulation EC No 641/2004.

Articles 12 and 13 of the genetically modified Food and Feed Regulation describe the labeling laws for food and food ingredients which contain or are derived from GMOs. Limited GM labeling is mandated where more than 0.9 percent of a product or ingredient is or incorporates a GMO and this provision does not depend on the occurrence of GM DNA or protein. Though if the GM content was no more than 0.9 percent of the product or ingredient and its existence is beneficial or practically inevitable, strict GM labeling is not needed. The ingredient level is applicable when single ingredients constitute the food product.

To benefit from this exemption, manufacturers should be ready to demonstrate that they had undertaken sufficient measures to prevent the use of GM ingredients. For all foods, genetically modified foods are also subject to legal guidelines on food labeling (2000/13 / EC, SI No. 92 of 2000), premised on the requirement that labeling does not deceive the consumer.

At present, the varieties of crops that are genetically modified and approved for food consumption in the EU include 1 GM soya beans, 12 GM maize, 6 GM varieties of oil-seed rape, and 5 GM varieties of cotton. A few of these crops were genetically engineered to withstand the use of particular herbicides or avoid invasion by pests.

There is no constitutional meaning of the words "GM free," "Non-GM," or "Constructed without GM ingredients," etc. Nevertheless, customers tend to take these claims to mean that no GM products are present in a product bearing any of those labels. Some actors in the food industry knowingly add such labelling to products without genetically modified ingredients.

The general guideline on food labeling explicitly notes that labeling should not confuse people by aligning various characteristics with such a product if all other comparable foods contain the same features. For instance, labeling milk as "GM-free" may deceive customers to understand that GM milk is available in the market when no such product is available yet.

1.4 FOOD QUALITY AND SAFETY STANDARDS

Far more quality systems and food safety management systems are now being accredited according to the latest practices and specifications which require third-party audits. The demands of customers, the rapidly evolving legal and regulatory structures, and the increasing concentration of markets and purchasing power have laid the foundation for establishing private principles. Although both food safety and quality standards are seen as necessary elements for safeguarding and protecting companies from civil liability, extra specifications including workers', environmental, and animal wellbeing are getting more attention in an effort to encourage consumer trust and protect market share. Regional retailers tend to regard food safety harmonization as a stepping stone toward a worldwide food management system (Fulponi 2006).

The food industry presently uses a variety of audit modes, which include guidelines that have not been designed to endorse a specific group or community over the other–including the International Organization for Standardization (ISO) 9001, 22000, and 14001. The specifications are being released by ISO. These regulations provide a structure which could be used to design and execute programs which fulfill some requisites related to quality, food safety, and environmental protection. There are indeed industry standards which have been documented and advanced by some actors in the food industry that are required to address a particular need. Within this section the GFSI lies. GFSI is a partnership of large retail chains comprising Tesco, Walmart, Sainsbury's, Carrefour, and others whose mission is to standardize their suppliers' quality, safety, and ethical standards. Some of the requirements accepted by GFSI are the BRC, Dutch HACCP, the IFS, and SQF as well as the FSSC 22000. Such standards consist of ISO 22000, Food Safety Management Systems–specifications for every company in the food supply chain and ISO/TS 22004 technical specification–Food Safety Management Systems–instruction on the implementation of ISO 22000:2005, which are generally recognized as ensuring high standards of food safety (ISO/TS 22004:2005) (Stier 2009). Independent, third party resources could participate to maintain a great degree of conformity with food safety and food law. Accreditation bodies, including the National Sanitation Foundation International (NSF), use their knowledge and skills to benefit the food industry and management by offering cost-minimizing alternative facilities. Third-party certification guarantees unbiased and transparent guidance and even regulatory bodies are shifting their focus to certified third parties for help (Tanner 2000).

1.4.1 Food Quality Standard (ISO 9001)

The standards family ISO 9000 takes into consideration a number of dimensions of quality control and integrates some of the most well-recognized standards of ISO. Such standards assist and encourage businesses and organizations by offering them guidance that is coherent and consistent with the needs of their customers and enhances organizational quality characteristics. ISO standards consist of the following:

- ISO 9001:2015 responsible for the requisites of the quality management system
- ISO 9000:2015 designs for fundamentals and languages
- ISO 9004:2009 controls the efficiency of the quality management system
- ISO 19011:2011 provides instructions for overall audits related to the quality management system

As per the British Standards Institution (BSI 2014), the International Organization for Standards 9001:2015 establishes the requirements for a quality management system and it is the only community standard that can be accreditational. Any company, big or small, may use it irrespective to their activities. There have been presently over a hundred thousand companies and associations accredited to ISO 9001 across approximately 170 countries (ISO 9000, 2015). This standard was then replaced by ISO 9001:2008 which contains guidelines for developing, executing, maintaining, and enhancing quality management systems for all companies irrespective of their nature.

Systematic quality systems are sometimes misperceived as onerous (as per an administrative perspective), expensive, and, consequently, not required. Of course, that is not valid, due to proper implementation of the process. Systematic certification of an agency to an ISO quality system is vital in stressing the commitment of the organization to obtain client satisfaction. The accredited organizations are able to prosper not just in internal audits but also in third-party audits through accomplishing their agreements, while being capable to elaborate that productively (Atkin and Brooks 2009).

Poksinska et al. (2006) summarized the findings of audits conducted in Sweden at 269 ISO 9001:2000-certified companies. The organizations claimed, depending on the outcomes, that accreditation audits helped them enhance their quality management systems and boost performance and work quality. Auditors decided to audit organizations accredited to this level which will not only grant a credential but also have their own observations and recommendations for enhancing the processes. Audit results providing professional advice are not an unusual activity amongst auditors. The results indicated which significant variations occurred with respect to the requirement, which depends primarily on the audit committee as well as on the accreditation bodies. Eventually, an enhancement in the audit process was stated to have been found after the ISO 9000 revision in 2000.

1.4.2 Food Safety Standards

Due to the increasing consumer demands for food safety, numerous public and private guidelines are being introduced across the world during the last few decades. Though a great deal of work was already conducted from all perspectives of food safety, this current study was not very extensively customized (Hammoudia et al. 2009). The BRC, the IFS, the Dutch HACCP, the SQF 2000 Level 2, and ISO 22000:2005 are the most common standards. The adoption of the BRC specification took place in 1998, in an attempt to satisfy the criteria of food safety in the United Kingdom among manufacturers and retailers. After that it was recognized worldwide, particularly in North America and Europe (Mensah and Julien 2011).

1.4.2.1 British Retail Consortium

Popular standards for inspecting food suppliers have been identified by large UK retailers, such as Tesco and Sainsbury's, in cooperation with the BRC in 1998. The inspections are carried out by accredited organizations. The retailers used to carry out their own checks before BRC was adopted; they soon discovered that joint inspections are cost-effective. Other European retailers then decided to be audited by BRC and deemed to have their accreditation results. HACCP specifications are also included in BRC, but more focus is put in this specification on recording, industry and its services, product and process monitoring, and staff (Trienekens and Zuurbier 2008). Numerous revisions have been made to the BRC authorities to ensure that it incorporates all new food safety criteria and developments, and the standard has now been achieved global acceptance. BRC establishes a methodology to help food producers guarantee that the products they manufacture are safe, and aids them in controlling product quality to ensure the customer's expectations. Most supermarkets, food service firms, and manufacturers across the world who seek to evaluate their suppliers' efficiency embrace the BRC certification (BRC 2015). This phenomenon, a real opportunity for retailers, inevitably leads to the expectation that the retail industry will become more engaged in the reconfiguration of the supply chain as well as in other activities which have previously been performed by producers (Arfini and Mancini 2004).

1.4.2.2 International Food Standard

German and French wholesale and retail organizations developed the IFS, a standardized assessment framework for all retailer-branded food distribution organizations (Gawron and Theuvsen 2009). Food Marketing Inst owns the SQF project. This norm provides a blend of food safety and quality management accreditation for all the processors. Interest in food safety in the UK was a result of the increase in the number of food poisoning cases at the time of the *Escherichia coli* 0157 outbreaks. As a consequence, a great deal of attention was being directed to the administrative measures that the UK government introduced regarding the quality of food supply networks that are used to monitor food hazards (Henson et al. 1999).

1.4.2.3 Hazard Analysis Critical Control Point

HACCP is a cost-effective continual food safety program that allows product quality to be implemented, maintained, and retained. It is used by food companies to develop a greater level of food safety, which could never be accomplished simply by following fundamental hygiene practices. Via HACCP, all phases related to the processing and preparation of food are periodically evaluated and all critical points are controlled for food safety (Gaze 2009). The program is a seven-step process aimed at providing a high standard of food safety and promoting a two-way partnership between consumers and industry. It also helps businesses to easily recognize any regulatory issues relevant to a company's internal quality standards of an organization (Tzouros and Arvanitoyannis

2000). A study which documented the HACCP and ISO 9001 expertise of nine Canadian food producers found that the respondents had different trends. First, to assure product quality, all the organizations interrogated enforced HACCP in conjunction with several internal controls including analytical examining and supplier assessment (Nguyen et al. 2004). As per Mortimore (2000), the outcome of the evaluation is just to have documented that the supplier has: introduced a reliable HACCP program, the requisite expertise and experience to sustain it, and the required support. This is generally compliance to the Good Manufacturing Practices (GMPs). From the producer's point of view, there may be other reasons for evaluation especially when undertaking internal analysis. These could encompass assessing if the advantages of applying HACCP have been noticed in exercise, reducing costs resulting in a greater focus on strategic areas, or assuring that supplier audits concentrate on crucial areas.

HACCP should be supported by robust prerequisite programs (PRPs) comprised of analytical control programs, supplier assessment, traceability and recall systems, certification, good manufacturing practices (GMPs), water and air quality control, and internal audit schedules and inspections. HACCP verification requires attestation of observations, PRPs, and personnel assigned to monitor the critical control points and review the critical control points and other equipment calibration (Sperber 1998). Vela and Fernandez (2003) addressed a study performed in Madrid focused at defining the key challenges encountered by businesses when implementing HACCP programs. The findings revealed that lack of awareness and poor judgment contributed to inadequate detection of hazards—an issue which could not be fixed easily by approaching external advisors (Vela and Fernandez 2003). Gilling et al. (2001) address these concerns by analyzing processes and features that might pose challenges during HACCP implementation and maintenance. Data was gathered from compliance designs to medical clinical standards and practical knowledge of the issues faced when HACCP systems were introduced. Emphasis was placed on the implementation of behavior design to food safety management through telephonic interviews. The design is sometimes used as a method to identify and locate issues and to support the subsequent action required to address them. It was proposed that the challenges involved in complying with the HACCP principles will be further explored with a view to finding the appropriate solutions and hence strengthening the health systems. HACCP has two primary components: Hazard Analysis (HA) and Critical Control Points (CCPs). Hazard Analysis (HA) is a risk assessment that measures the rates of such hazards as well as how they impact human health at any point during production and distribution. CCPs in the food supply chain are all those areas that allow for total threat prevention, threat removal, or minimizing the threat to a tolerable level. Principles are shown in Figure 1.3.

1.4.2.4 ISO 22000

ISO 22000:2005 specifies the criteria for an efficient food safety and management system which illustrates the capacity of an entity to manage hazards to food safety. It is implemented by companies of varying sizes participating in every part of the supply chain (ISO 22000:2005). Implementing ISO 22000 produces enormous benefits, such as reduced waste, less production time loss, and higher productivity, particularly for small- and medium-sized organizations. Another benefit is customer satisfaction. Consumer requirements were shown to be one of several chief factors why catering companies are getting ISO accreditation. Companies regard ISO certification as an instrument that provides them with a competitive edge. ISO 22000 allows organizations to incorporate a food chain framework for creating, implementing, and enhancing the efficiency of their food safety and management system. ISO 22000 allows the company to understand the backward as well as forward impacts of the food web (ISO/TS22004:2005). There are several

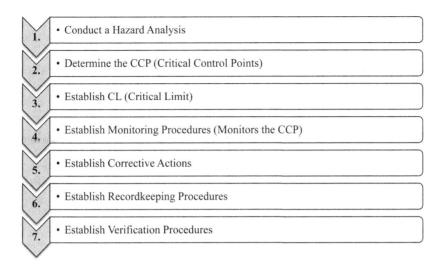

1. • Conduct a Hazard Analysis

2. • Determine the CCP (Critical Control Points)

3. • Establish CL (Critical Limit)

4. • Establish Monitoring Procedures (Monitors the CCP)

5. • Establish Corrective Actions

6. • Establish Recordkeeping Procedures

7. • Establish Verification Procedures

FIGURE 1.3   Principles of a HACCP plan.

ways by which ISO 22000 improves the HACCP plan. This is a principle in management and thus has many parallels with other principles of the management system. It covers, for instance, regulation, implementation, and organizational criteria, performance evaluation, and reviews of management. ISO 22000 does not facilitate the installation of a preventive activity protocol because HACCP (an ISO 22000 prerequisite) is structured to avoid food safety hazards. Conversely, ISO 22000 acknowledges the development of new hazards and the implementation of novel monitoring systems and technologies to regulate them. Hence, it entails a framework approach for their prevention (Surak 2007). ISO 22000 can be applied in conjunction with ISO 9001. While ISO 9001 emphasizes enforcement of efficacious quality management systems needed to fulfill consumers' needs, ISO 22000 assures that an effective food safety and management system is present to serve customer needs (ISO/TS 22004:2005).

1.4.2.5 Food Distribution Systems

Physical movements and product storage from production to the end user is called the distribution management process (Rushton et al. 2006). Food producers manufacture these before any kind of delivery and after the acquisition of agricultural starting materials. The processing stage covers everything from basic packaging of fresh product to substantial preservation techniques, such as cooking. The basic requirements for good sanitary practice in transportation as per HACCP Manual are described as follows:

- Transportation methods must be designed in such a way so as to allow for effective cleaning and disinfection; they should be physically isolated to avoid contamination
- The proper temperature control must be ensured at a specific temperature level

1.5 CONCLUSION

Food quality is basically a degree of acceptance and rejection by the consumer for a specific food product/service. It is an essential prerequisite for food production, since food consumers are vulnerable toward any sort of infection which may arise even during the production process. Customers depend on quality and processing standards, honest and understandable labeling, and consistency of product. In addition, sanitation criteria often apply. It is critical to ensure that the food manufacturing system is just as healthy as possible such that the customer receives the safest food possible. Because safe food is now a parameter of food quality, food safety is now an important discipline. It deals with the health of the consumer by the provision of safe food to the consumers. Various standards and systems like ISO:22000 and HACCP, specify the criteria for efficient food safety and management systems which illustrate the capacity of an entity to manage hazards to food safety. These standardization systems allow organizations to incorporate a food chain framework for creating, implementing, and enhancing the efficiency of their food safety and management system. They also allow the food industries to understand the backward as well as forward impacts of the food web.

REFERENCES

American Dietetic Association (ADA). 2006. Agricultural and food biotechnology. http://www.eatright.org/cps/rde/xchg/ada/hs.xsl/advocacy_abiotechnology_ENU_HTML.htm.

American Medical Association Council on Scientific Affairs Report. 2000. Genetically modified crops and foods. http://www.amaassn.org/ama/pub/category/13595.html.

Arfini, F. and Mancini, M. C. 2004. British Retail Consortium (BRC) standard: A new challenge for firms involved in the food chain. In: Schiefer, G., and Rickert, U., eds., *Analysis of economic and managerial aspects*. Parma, Italy: Department of Economic and Quantitative Studies, Institute of Agrifood Economics, Parma University, 1–9.

Atkin, B. and Brooks, A. 2009. Service specifications, service level agreements and performance. In: Atkin, B., and Brooks, A., eds., *Total facilities management*. West Sussex: Wiley-Blackwell, 101–8.

Beier, R. C. and Pillai, S. D. 2007. Future directions in food safety. In: Shabbir, S., ed., *Foodborne diseases*. Totowa, NJ: Humana Press, 511–30.

BRC. 2015. *Global standard food safety*. London: British Retail Consortium 7:1–24.

BSI. 2014. Moving from ISO 9001: 2008 to ISO 9001: 2015. *The new international standard for quality management systems*. ISO 9001 transition guide. London: BSI Group, 1–12.

Buzby, J. C. and Roberts, T. 1996. ERS updates US foodborne disease costs for seven pathogens. *Food Review* 19(3):20–5.

Dalgaard, P., Buch, P. and Silberg, S. 2002. Seafood spoilage predictor development and distribution of a product application software. *International Journal of Food Microbiology* 73(2–3):343–9.

European Commission. 2000. The European white paper on food safety. *COM* (1999) 719 Final, Brussel.

European Commission. 2007. Health and customer protection - Food traceability. Directorate-General for Health and Consumers, Brussel.

European Parliament. 2002. Regulation (EC) 178/2002 of the European Parliament and of the Council. *Official Journal of the European Communities* L31(1):24.

FAO. 2007. *Countries urged to be more vigilant about food safety*. Rome: FAO.

FAO/WHO. 2002. PAN-European conference on food safety and quality. Budapest.

FDA. 2016. Hazard analysis and risk-based preventive controls for human food. In: *Draft guidance for industry*. USA: Center of Food Safety and Applied Nutrition.

Food Safety Authority of Ireland. 2005. GM food survey. Dublin. www.fsai.ie.

Fulponi, L. 2006. Private voluntary standards in the food system: The perspective of major food retailers in OECD countries. *Food Policy* 31:1–13.

Gawron, J. C. and Theuvsen, L. 2009. The international food standard: Bureaucratic burden or helpful management instrument in global markets? Empirical results from the German food industry. *Journal of International Food and Agribusiness Marketing* 21:239–52.

Gaze, R. 2009. *HACCP: A practical guide* (4th ed.). Chipping Campden: Campden BRI, 1–88.

Genetic Modification Advisory Committee (GMAC). 2002. Biosafety guidelines. www. gmac.gov.sg.

Giannakourou, M. C. and Taoukis, P. S. 2003. Application of a TTI-based distribution management system for quality optimization of frozen vegetables at the customer end. *Journal of Food Science* 68(1):201–9.

Gilling, S. J., Taylor, E. A., Kane, K. and Taylor, J. Z. 2001. Successful hazard analysis critical control point implementation in the United Kingdom: Understanding the barriers through the use of a behavioral adherence model. *Journal of Food Protection* 5:599–743.

Grujic, S. and Blesic, M. 2007. *Food regulations*. Banja Luka: Faculty of Technology.

Grunert, K. G. 2005. Food quality and safety: Consumer perception and demand. *European Review of Agricultural Economics* 32(3):369–91.

Hammoudi, A., Hoffmann, R. and Surry, Y. 2009. Food safety standards and agri-food supply chains: An introductory overview. *European Review of Agricultural Economics* 36: 469–78.

Henson, S., Holt, G. and Northen, J. 1999. Costs and benefits of implementing HACCP in the UK dairy processing sector. *Food Control* 10:99–106.

Hsich, Y. H. P. and Ofori, J. A. 2007. Innovations in food technology for health. *Asia Pacific Journal of Clinical Nutrition* 16(1):65–73.

Institute of Food Technologists. 2000. IFT expert report on biotechnology and foods. http://members.ift.org/IFT/Research/IFTExpertReports/.

International Service for the Acquisition of Agri-Biotech Applications (ISAAA). 2005. Global status of commercialized biotech/GM crops. www.isaaa.org.

International Organization of Standardization (ISO). 2010. http://www.iso.org.

ISO 9000. 2015. Quality management. http://www. iso.org/iso/iso_9000.

ISO 22000. 2005. International standard. ISO 22000. *Food safety management systems - Requirements for any organization in the food chain*. Belgium: Comite Europeen De Normalisation, 1–32.

ISO/TS 22004. 2005. *Technical specification. Food safety management systems - Guidance on the application of ISO 22000:2005*. Geneva, Switzerland: International Organization for Standardization, 1–13.

Italian Standards Institute (UNI). 2001. Traceability system in agricultural food chain – General principles for design and development. Norma UNI 10939:2001. www. uni.com.

Italian Standards Institute (UNI). 2002. Traceability system in agrofood industries – Principles and requirements for development. Norma UNI 11020:2002. www.uni.com.

Mensah, L. D. and Julien, D. 2011. Implementation of food safety management systems in the UK. *Food Control* 22:1216–25.

Miotrag, M. 2001. Food safety – Using technology to improve traceability. In: *Proceedings of CIES convention*, Amsterdam, 21–34.

Mortimore, S. 2000. An example of some procedures used to assess HACCP systems within the food manufacturing industry. *Food Control* 11:403–13.

National Center for Food and Agricultural Policy (NCFAP). 2005. Plant biotechnology: Current and potential impact for improving pest management in U.S. agriculture. http://www.ncfap.org/whatwedo/pdf/2004biotechimpacts.pdf.

Nguyen, T., Wilcock, A. and Aung, M. 2004. Food safety and quality systems in Canada. An exploratory study. *International Journal of Quality and Reliability Management* 21:655–71.

Perez-Gago, M. B., Serra, M. and del Rio, M. A. 2006. Color change of fresh-cut apples coated with whey protein concentrate-based edible coatings. *Postharvest Biology and Technology* 39:84–92.

Poksinska, B., Dahlgaard, J. J. and Eklund, J. A. E. 2006. From compliance to value-added auditing - Experiences from Swedish ISO 9001:2000 certified organizations. *Total Quality Management and Business* 17:879–92.

Raab, V., Bruckner, S., Beierle, E., Kampmann, Y., Petersen, B. and Kreyenschmidt, J. 2008. Generic model for the prediction of remaining shelf life in support of cold chain management in pork and poultry supply chains. *Journal of Chain Network Science* 8(1):59–73.

Rushton, A., Croucher, P. and Baker, P. 2006. *The handbook of logistics and distribution management*, 3rd ed. London: Kogan Page.

Society of Toxicology Position Paper. 2002. The safety of genetically modified foods produced through biotechnology. *Toxicological Science* 71:2–8.

Sperber, W. H. 1998. Auditing and verification of food safety and HACCP. *Food Control* 9:157–62.

Stier, R. F. 2009. Third-party audits: What the food industry really needs. http://www.foodsafetymagazine.com/magazine-archive1/octobernovember2009/third-party-audits-what-the-food-industry-reallyneeds/.

Sudhakar, P., Nageswara, R. R., Ramesh, B. and Gupta, C. P. 1988. The economic impact of a foodborne disease outbreak due to *Staphylococcus aureus*. *Journal of Food Protection* 51:11.

Surak, J. G. 2007. A recipe for safe food: ISO 22000 and HACCP. *Quality Progress* 2007:21–7.

Tanner, B. 2000. Independent assessment by third-party certification bodies. *Food Control* 11:415–27.

Terreri, A. 2009. Preventing the next product recall. *Food Logistics* 111:20–5.

Thompson, A. K. 2003. Preharvest factors on postharvest life. In: *Fruit and vegetables*. Ames, IA: Blackwell Publishing Ltd., 1–8.

Trienekens, J. and Zuurbier, P. 2008. Quality and safety standards in the food industry, developments and challenges. *International Journal of Production Economics* 113:107–22.

Tzouros, N. E. and Arvanitoyannis, I. S. 2000. Implementation of hazard analysis critical control point (HACCP) system to the fish/seafood industry: A review. *Food Review International* 16(4):273–325.

USDA (United States Department of Agriculture). 2002. Traceability for food marketing and food safety: What's the next step? *Agricultural Outlook* 288:21–5.

Vela, A. R. and Fernandez, J. M. 2003. Barriers for the developing and implementation of HACCP plans: Results from a Spanish regional survey. *Food Control* 14:333–7.

WHO. 2002. Food safety programme. World Health Organization. http://www.who.int/fsf. www.ific.org/food/biotechnology/index.

Yun-Hwa, P. H. and Jack, A. O. 2007. Innovations in food technology for health. *Asia Pacific Journal Clinical Nutrition* 16(1):65–73.

Zhao, Y. and McDaniel, M. 2005. Sensory quality of foods associated with edible film and coating systems and shelf-life extension. In: Han, J. H., ed., *Innovations in food packaging*. San Diego, CA: Elsevier Academic Press, 434–53.

CHAPTER 2

Foodborne Pathogens and Food-Related Microorganisms

Mary Theresa, Aparna Sankaran Unni, Amala Geevarghese,
Sebastain Korattiparambil Sebastian, Sunil Pareek,
and Radhakrishnan Edayileveettil Krishnankutty

CONTENTS

2.1 INTRODUCTION

Food is a nutrient-rich medium and favors microbial growth. Microorganisms play a significant role in the formation of many food products and, at the same time, in the introduction of pathogenic organisms into the food, posing a significant threat to human health. Microorganisms associated with foodborne diseases are considered an important risk factor in public health and have adverse an impact on food security and trade. Frequent occurrences of foodborne outbreaks have been reported from all countries causing considerable morbidity and mortality (Kirk et al. 2015). According to the World Health Organization (WHO), about 600 million cases and 420,000 deaths from foodborne illnesses are reported worldwide each year, and 30% of all deaths are children. Identifying the pathogens behind these outbreaks is essential to understand the spread and pathogenicity and to develop effective preventive measures. This chapter deals with foodborne pathogens, their pathogenic mechanisms, and preventive measures used to eliminate them. It also discusses the useful microorganisms associated with various foods and their beneficial properties.

2.2 COMMON FOODBORNE PATHOGENS

About 250 known foodborne infections are related to microorganisms such as bacteria, viruses, protozoa, helminths, fungi, and parasites. They are individually discussed in the following sections.

2.2.1 Bacteria

2.2.1.1 Arcobacter *spp.*

Arcobacter spp. are the least studied foodborne pathogens due to the unavailability of a standard method for its isolation (Shah et al. 2011). *Arcobacter* means "arc-shaped bacterium" and are included in the family of *Campylobacteraceae*. These were previously known as "aerotolerant *Campylobacter*" (Levican and Figueras 2013; Ramees et al. 2017). In 1991, Vandamme and De Ley classified them as *Arcobacter* (Vandamme et al. 1992; Ramees et al. 2017); they are fastidious, gram-negative rods with a single polar flagellum (Ferreira et al. 2016). Due to the similarity with *Campylobacter* species, polymerase chain reaction (PCR) (particularly multiplex PCR) is usually used for the detection of *Arcobacter* spp. (Ramees et al. 2017). They have been described as a serious risk to humans by the International Commission on Microbiological Specifications for Foods (ICMSF) due to its emergence as a foodborne zoonotic pathogen in recent years. They commonly cause bacteraemia, gastroenteritis, and diarrhea in humans and mastitis and abortion in animals (Ferreira et al. 2016). Species of *Arcobacter* such as *A. butzleri*, *A. cryaerophilus*, *A. skirrowii*, and *A. cibarius* are generally considered foodborne pathogens while *A. butzleri* is more commonly associated with human infections (Amare et al. 2011; Bagalakote et al. 2014; McGrego and Wright 2015). The main source of these bacteria are animals such as cattle, sheep, and pigs (Ferreira et al. 2016). Poultry is also a source of infection (Rahimi 2014). The intestine of the animal harbors the bacteria and contaminate the meat during the processing of carcasses in slaughterhouses, thereby increasing the chance of contamination. Contaminated meat and water further facilitate the transmission to humans (Rahimi 2014). The major concern with the disease caused by *Arcobacter* is the increasing number of reports of antibiotic resistance from various parts of the world (Rahimi 2014; Ferreira et al. 2016). Multi-drug resistance has also been observed for *Arcobacter* spp. isolated from *Catla catla* in India. The association of pathogen with local fishes indicates unhygienic conditions and cross-contamination with market places (Nelapati et al. 2020). The main sources of infection for *Arcobacter* are contaminated food of animal origin such as seafood, milk, and meat.

Comparatively less information is available about the genes or factors associated with the virulence of *Arcobacter* spp. (Collado and Figueras 2011). Tabatabaei et al. (2014) have reported six genes in *Arcobacter* spp. to associated with virulence which include *cadF*, *ciaB*, *cj1349*, *mviNin*, *pldA*, and *tlyA*. In the case of piglets, *A. butzleri* has been demonstrated to impair the epithelial functioning after intestinal colonization (Bucker et al. 2009). It has also been reported to interact with red blood cells resulting in hemagglutination. The pro-inflammatory cytokines released during its interaction further the inflammatory reactions in cells (Collado and Figueras 2011). The virulence genes identified from the genome of *Arcobacter* spp. are associated with the pathogenic properties (Ramees et al. 2017; Silha et al. 2018).

Diseases caused by *Arcobacter* spp. remain unnoticed as most of the infections are asymptomatic in nature. The first reported human infection with *Arcobacter* was in 1988 and *A. cryaerophilus* was believed to be associated with the infection. The common symptoms in human infections range from diarrhea to septicemia; *Arcobacter* spp. is also an agent of traveler's diarrhea along with other organisms (McGrego and Wright 2015). *Arcobacter* associated infections have been reported mainly from chicken meat, followed by pork and beef (Shah et al. 2011) and in India its prevalence is in the range of 12%–33% (Ramees et al. 2017). *Arcobacter* spp. has also been isolated from milk samples (Shah et al. 2012) and hence is another source of infection to humans. It usually causes acute diarrhea lasting 3–15 days (Vandenberg et al. 2004); other clinical symptoms include fever, vomiting, general weakness, and abdominal pain (Vandamme et al. 1992). Special medium containing antimicrobials like cefoperazone, trimethoprim, amphotericin, novobiocin, and 5-fluorouracil is used for the isolation of *Arcobacter* spp. (Houf et al. 2001). The major preventive measures include proper cooking and good hygiene practices. Maintenance of good slaughterhouse hygiene can reduce contamination of meat at the source, thereby minimizing infection. Philips and Long (2001) have reported low temperatures reduce the growth of *Arcobacter*. Continuous monitoring of seafood is essential to eliminating incidences of foodborne infections by *Arcobacter* spp. (Rathlavath et al. 2017). As *Arcobacter* spp. have also been detected in ready-to-eat vegetables, the practice of good processing for raw vegetables is also important (Mottola et al. 2016).

2.2.1.2 Bacillus cereus

Bacillus cereus is a gram-positive, facultatively anaerobic, rod-shaped, endospore-forming, motile bacteria that is ubiquitous in soil, the natural environment, and various foods (Bottone 2010). The bacterium is known to cause food spoilage as well as foodborne infections in humans. It is transmitted mainly through foods that are stored at ambient temperature after cooking, permitting their spore germination and toxin production. Heat and temperature-resistant spores serve as the major source of transmission. Foods associated with *B. cereus* outbreaks include meat and meat products, fried and cooked rice, liquid foods, mixed food products, infant foods, starchy foods, milk and other dairy products, sterilized and processed foods, etc. (Tewari and Abdullah 2015). It is also a frequent contaminant of ready-to-eat (RTE) foods as well as fresh vegetables (Yu et al. 2020). WHO has recognized *B. cereus* as the most common foodborne bacterium in pasteurized food (Rahimi et al. 2013). These bacteria are highly stable, and some strains can remain viable even at low temperatures (psychrotolerant) making possible their transmission via cold-stored foods. The ability to form thermoduric spores and exotoxins, as well as their survival capacity at refrigeration temperatures, makes them potential contaminants of a large number of foods (Tewari and Abdullah 2015).

B. cereus causes two types of foodborne illness which are mainly characterized by diarrhea and emesis. The diarrheal symptoms usually develop 6–12 hours after ingesting contaminated food and are primarily manifested by abdominal cramps and diarrhea, subsiding within 12–14 hours. But in neonates and immunocompromised patients, the symptoms may progress into severe disease. Although the infection caused by this pathogen is usually mild, self-limiting, and rarely becomes severe, the bacteria can cause infection at a very low dose—i.e., more than 10^3 *B. cereus* cells per gram of food is not considered safe for consumption (Tewari and Abdullah 2015).

Diarrhea is caused by a series of enterotoxins produced by the organism which include hemolysin BL (Hbl), non-hemolysin enterotoxin (Nhe; tripartitie toxins), cytotoxin K (CytK), and enterotoxin FM (EntFM). The three pore-forming toxins (Hbl, Nhe, and CytK) are responsible for the onset of diarrhea. The β-barrel pore-forming toxin CytK causes serious food poisoning, skin necrosis, hemolysis, and even death. Emesis is induced by a small, heat- and acid-stable toxin known as cereulide which is pre-formed in food. It is synthesized by non-ribosomal peptide synthetases encoded by virulence *ces* gene and serves as an ionophore that disrupts cyclic phosphorylation in mitochondria. The cereulide intoxication may sometimes lead to severe clinical manifestations, such as fulminant hepatitis (acute liver failure), rhabdomylosis, acute encephalopathy, and multi-organ failure (Yu et al. 2020).

In clinical and research laboratories, the ELISA test is commonly used for routine diagnosis of *B. cereus* due to its high specificity (Zhu et al. 2016). The multiplex-PCR method allows the rapid and simultaneous detection of enterotoxigenic pathogenic strains as well as other toxin-producing bacteria from food. Early detection and quantification of cereulide toxin is critical. HEp-2 cell assay, boar sperm motility bioassay, and LC-MS analysis are also commonly used for this purpose (Yamaguchi et al. 2013) Generally, multi-locus sequence typing (MLST) and amplified fragment length polymorphism (AFLP) are considered the gold standard for the epidemiological studies of *B. cereus* (Ehling-Schulz and Messelhäusser 2013).

Treatment is for symptomatic cases only. In case of severe diarrhea or vomiting, fluid administration is required. Blood purification therapy was found effective in the quick removal of cereulide toxin from patients suspected of severe poisoning. Hygienic and safe food-handling practices can prevent the contamination of food by pathogen. Since spores of *Bacillus cereus* is heat-resistant, cooking at moderate temperature does not destroy it. Hence, rapid cooling and proper reheating (74°C) is required when reusing cooked food that has been left at room temperature. The sensitivity of toxin to low pH makes acidic foods safe from *B. cereus* poisoning (Shiota et al. 2010; Tewari and Abdullah 2015).

2.2.1.3 Brucella *spp.*

Brucellosis is a zoonotic disease caused by *Brucella* species which is a gram-negative, non-motile, and slow-growing coccobacilli. The four *Brucella* species that cause infections in different animals are: *B. abortus* in cattle, *B. canis* in dogs, *B. melitensis* in sheep and goats, and *B. suis* in swine (Bukhari 2018). Brucellosis is also known as remitting fever, undulant fever, and goat fever, Mediterranean fever, Maltese fever, Bang disease, etc. (Głowacka et al. 2018). They do not produce any toxins or spores. Since the disease is predominantly found in low-income countries, WHO has classified them as one of the leading causes of "neglected zoonotic diseases" in the world. Since 2014, brucellosis has been frequently seen in southern Brazil (Lemos et al. 2018). The studies have also confirmed the disease occurrence in two districts of northern Israel and in Jerusalem in 2014 (Megged et al. 2016). Humans become sick after the consumption of unpasteurized milk and dairy products, or undercooked meat, or through direct contact with infected animals. It is also transmitted by aerosols, blood transfusions, congenital transmission, and, rarely, from person-to-person direct contact (Hayoun et al. 2019). Brucellosis symptoms manifest as nausea, vomiting, abdominal discomfort, spontaneous bacterial peritonitis, and, rarely, causes inflammation of the ileum and colon. It is also characterized by headache, fever, myalgia, hepatomegaly, splenomegaly, miscarriage, endocarditis, meningoencephalitis, and anemia. The incubation period is quite long from three days to several weeks. *Brucella* is recognized as a potent bio-weapon and categorized as a class B bioterrorism agent due to its high virulence and contagious properties (Vassalos et al. 2009).

In endemic areas, standard agglutination is the most common test for the diagnosis of *Brucella*. It is also diagnosed by indirect ELISA and the Rose Bengal test, which is a rapid slide agglutination test. Increased levels of C-reactive protein, erythrocyte sedimentation rate, alkaline phosphatase, serum lactate dehydrogenase, and also different conditions such as anemia, leukopenia, or pancytopenia are also reported to be associated with the disease. Combination drugs, such as doxycycline with streptomycin, rifampicin, gentamicin, or sulfamethoxazole-trimethoprim, are used for the treatment of brucellosis (Meng et al. 2018). The necessary preventive measures include educating people on personal hygiene and best food-handling practices. Pasteurization of milk is very important to destroy the pathogen, but, unfortunately, this practice is still uncommon in resource-limited communities due to their cultural practices and lack of knowledge about disease transmission from raw milk (Welburn et al. 2015). Since brucellosis is a zoonotic infection, caution must be taken while handling livestock. Proper handwashing, protective clothing, and avoidance of slaughtered infected animals should be practiced on animal farms. Consumption of raw milk and meat must be avoided to control brucellosis. The best preventive method is to reduce human contact with infected animals and their raw products. The proper cooking of meat and pasteurization of milk and dairy products is strongly recommended before consumption (Vassalos et al. 2009).

2.2.1.4 Campylobacter *spp.*

Campylobacter spp. are gram-negative, non-spore-forming, spiral, and motile rod bacteria which cause campylobacteriosis. In older cultures, they form spherical or coccoid bodies (Penner 1988). *C. jejuni* and *C. coli* are the two important species related to the foodborne infections among the *Campylobacter* spp. (Fitzgerald and Nachamkin 2015). The symptoms of campylobacteriosis include abdominal cramps, fever, and diarrhea with or without blood in stools (Fitzgerald and Nachamkin 2015). The incubation period ranges from 3 to 7 days (Blaser and Engberg 2008), and, in severe cases of infection, the patient may develop post-infection complications, such as Guillain-Barré syndrome (GBS) (Fitzgerald and Nachamkin 2015). Most of the infections are caused by *C. jejuni* and about 4% by *C. coli*. People in all age groups are vulnerable to the infection, and among this, young children (<5 years) are the most vulnerable. Approximately 1% of the human population are affected each year (Denny et al. 2007). There has been an increase of 14% increase of campylobacteriosis since 2012 (Tompkins et al. 2013). A recent outbreak of *C. jejuni* was reported in Seoul, Korea due to cross-contamination from raw chicken to food handlers (Kang et al. 2019). Even though the number of cases is increasing, there are no national-level surveillance programs in developing countries where the cases are estimated at about 5%–20%. The infectious dose of *Campylobacter* spp. is comparatively low (500–800 organisms), so that the chances of infection are considerable. This dosage was calculated by a British Doctor, Robinson, by self-experiment. He swallowed pasteurized milk containing 500 organisms. Based on the results of this experiment along with additional findings, he verified the dosage of pathogen required to result in campylobacteriosis (Blaser and Engberg 2008; Epps et al. 2013). Javid (2011) has reported the emergence of multiple antibiotic resistances in *Campylobacter* spp. This is likely to be due to the overuse of antibiotics in food animal production and will have serious impact on human health (Silva et al. 2011; Otto et al. 2020).

The virulence factors associated with the pathogenesis of *Campylobacter* spp. are the motile flagella, cytolethal distending toxin (CDT), antibiotic resistance genes, several secretory proteins, etc. (Larson et al. 2008). The primary source of infection is food production animals; from them it passes into commercial food production facilities. Poultry, beef, and pork products are commonly associated with illness and among these, 50%–70% is through the consumption of contaminated poultry such as hens, ducks, turkeys, etc. (Silva et al. 2011). The contamination can also take place through unpasteurized milk, as cattle and sheep have been found to harbor the pathogen in their intestines (Besser et al. 2005; Davys et al. 2020; Kenyon et al. 2020).

Campylobacter spp. can colonize the lower gastrointestinal tract, gallbladder, and liver (Grahamm and Simmons 2005). The methods used for the identification of pathogen include PCR, random amplification of polymorphic DNA (RAPD), and pulsed-field gel electrophoresis (PFGE). The treatment of infection consists of fluid and electrolyte replacement and antibiotics in severe cases. However, in most cases, the patient will recover without specific treatment and medications (Silva et al. 2011).

To reduce or eliminate campylobacteriosis disease, proper and hygienic food processing is required. Pre-harvest interventions can reduce the concentration of pathogen in food materials. By the use of properly cooked poultry products, infection cases can be reduced to a great extent (Epps et al. 2013). As the *Campylobacter* spp. are sensitive to humidity, freezing, oxygen, etc. utilizing these will be helpful to prevent the contamination (Silva et al. 2011).

2.2.1.5 Clostridium botulinum

C. botulinum is an anaerobic, gram-positive, rod-shaped, and endospore-forming bacterium, which produce the potent neurotoxin called botulinum toxin. "Botulus" in Latin means "sausage," and the name is derived from the sausage poisoning in 18th-century Europe. *C. botulinum* is classified into four groups according to the type of illness they cause in humans or in animals. These are Group I, II, III, and IV, where the Group I and II cause human illness, and group III is associated with animal illness. Whereas, Group IV C. botulinum is not reported for any kind of illness in humans or animals so far. Botulinum toxins are classified into seven different types (A to G) on the basis of their discovery from different *C. botulinum* strains. Group I bacteria produce the toxins A, B, and F; Group II bacteria produce toxins B, E, and D. Among all these toxins, A, B, and E are the most potent neurotoxins causing human poisoning. The toxins C and D are rarely associated with human illness, whereas type F causes minimal human toxicity. Several strains other than *C. botulinum* produce botulinum toxins, such as *C. butyrricum*, *C. barati*, and *C. argentinensis* (Dhaked et al. 2010).

The toxins are produced in an inactive form, consist of heavy and light chains of molecular mass of 100 and 50 kDa, respectively, and are connected by a disulphide bond. Botulin toxin acts at the neuromuscular junction by blocking the release of acetylcholine, the principal neurotransmitter, from the presynaptic motor neurons resulting in muscle paralysis. In the body, this toxin acts at different sites: the neuromuscular junctions, autonomic ganglia, parasympathetic, and sympathetic nerve endings, which release the neurotransmitter acetylcholine. The light chain of the toxin binds with several proteins, such as synaptosomal-associated protein (SNAP) and vesicle-associated membrane protein and syntaxin, present in the nerve endings thereby inhibiting the fusion of acetylcholine-containing vesicles with the cell membrane (Sugiyama 1980; Brin 1997; Ranoux et al. 2002). On the other hand, the heavy chain (H) irreversibly interacts with high-affinity receptors in cholinergic neurons, and, through endocytosis, the toxin-receptor complex is taken up into the cell.

The symptoms of the poisoning start with weakness of the muscles followed by difficulty in breathing and speaking, and blurred and double vision. In severe cases, flaccid paralysis and respiratory muscle paralysis occurs leading to death (Lang 2003). The primary identification of botulism is through detection of the toxin in the patient's serum or stool by molecular techniques. The classical botulism of foodborne botulism is caused by the ingestion of *C. botulinum*-contaminated food—especially canned foods. The anaerobic conditions in the canned foods favor the growth of organisms and thereby increase the risk of botulism.

2.2.1.6 Cronobacter *spp.*

Cronobacter spp., previously known as *Enterobacter sakazakii*, is gram-negative, rod-shaped, motile, and facultatively anaerobic bacteria which are found widespread in the environment. They are opportunistic foodborne pathogens causing severe life-threatening infections in humans. *Cronobacter* spp. comprise of seven species, among which *C. sakazakii* is the most common pathogen associated with foodborne illness (Song et al. 2020). Although the bacteria can cause illness in all age groups, people with weak immune system, including neonates and infants, are more vulnerable to severe infections. As the bacteria are highly resistant to drying, there is risk of transmission through dried foods. Powdered infant formula (PIF) food is the major food source implicated in *Cronobacter* infection. The bacterium has been isolated from a wide range of other food sources, including dairy-based products, dried foods such as herbs, rice, fresh, frozen, RTE, fermented, and cooked food products and beverages, as well as from various food manufacturing, processing facilities, and healthcare settings. The ability to grow at a wide range of temperatures (6–45°C), efficient biofilm formation and relatively high resistance to environmental stresses such as desiccation, heat, acid stress, and osmotic stress, as well as the resistance to antibiotics and high-level disinfectants are the important factors that facilitate transmission of this bacterium (Kim et al. 2008).

C. sakazakii causes invasive systemic infection in newborns and infants characterized by local necrotizing enterocolitis, sepsis, and meningitis. In some cases, diarrhea, urinary tract infection, and septicemia have also been observed. The symptoms usually begin with fever and poor feeding, crying, and very low energy. Infants recovered from *C. sakazakii* infection often suffer from brain abscesses, developmental delays, and impaired sight and hearing. Premature or neonates with low birth weight are more susceptible to severe infection with a fatality rate of 50%–80%. Immunocompromised individuals, including cancer, HIV, and organ transplant patients are more susceptible to *C. sakazakii* infection (Feeney et al. 2014).

Clinical diagnosis of *C. sakazakii* infection is through blood sampling, testing the cerebrospinal fluid for leukocyte count, glucose and protein concentration, and other special testing with the brain (Henry and Fouladkhah 2019). Methods such as loop-mediated isothermal amplification (LAMP) assay, PCR-enzyme-linked immunosorbent assay (PCR-ELISA), real-time PCR, and immunochromatographic tests are being used for the rapid detection of *Cronobacter* spp. (Hu et al. 2016). A novel LAMP method with lateral flow assay (LFA) by probe-free label system has been proposed for the detection of *Cronobacter* spp. in PIF (Fu et al. 2018).

Centers for Disease Control and Prevention (CDC) as well as the WHO recommend breastfeeding as a leading measure for preventing the *C. sakazakii* infection in infants (i.e., infants should be exclusively breastfed for the first six months of life). Appropriate preventive measures must be taken while preparing infant formula, as it is the principal source of contamination. Using sterile liquid infant formula instead of powder, cleaning feeding bottles and working space, preparing formula fresh for each feeding, performing reconstitution at 70°C, and discarding unused formula will help reduce the risk of *C. sakazakii* contamination of PIF (Bai et al. 2019; Henry and Fouladkhah 2019).

2.2.1.7 Escherichia coli

Escherichia coli is gram-negative, non-spore-forming, rod-shaped, mesophilic, and facultatively anaerobic organism belonging to the *Enterobacteriaceae* family (Korsgaard et al. 2014). Most of the strains are harmless and only some generate a threat to humans (García et al. 2010). The pathogenic strains (pathovars) of *E. coli* are of major concern to public health, as their infectious doses are low and can transmit through various mediums, such as food and water (Croxen et al. 2013). Ruminants like cattle, sheep, and deer are the main reservoirs of the pathogen and transmission occurs through consumption of contaminated meat, milk, and dairy products. The illness can also originate from contaminated water, fruits, and vegetables (García et al. 2010). It can survive in the environment for long periods and proliferate in food materials.

Based on the pathogenic mechanism, *E. coli* is categorized into six different groups: enteropathogenic *E. coli* (EPEC), enterohemorrhagic *E. coli* (EHEC), also known as Shiga toxin-producing *E. coli* (STEC), enterotoxigenic *E. coli* (ETEC) and enteroaggregative *E. coli* (EAggEC), enteroinvasive *E. coli* (EIEC), and attaching and effacing *E. coli* (A/EEC). EHEC produces toxins that are similar to Shiga toxin (produced by *Shigella dysenteriae*). They are also characterized by the presence of virulence and antibiotic resistance genes. Every year STEC strain O157:H7 causes 63,000 illnesses and 20 deaths (García et al. 2010; Ekici and Dümen 2019). Symptoms of the infection include fever, vomiting, diarrhea, abdominal cramps, and, in some cases, it may lead to hemorrhagic colitis. The incubation period ranges from 3 to 8 days, and most of the patients recover within 10 days. Elderly people and children are categorized as high-risk groups, and, in severe cases (about 10% of patients), diseases like HUS (hemolytic uremic syndrome) may occur (Frank et al. 2011).

Various molecular methods, such PCR and ELISA, are used to detect the presence of pathogen from samples. The pathogen is susceptible to heat; therefore proper pasteurization can eliminate it from milk and milk products (D'aoust et al. 1988; Sugrue et al. 2019). Implementation of control measures along with provision of hygienic conditions can prevent the outbreak of infection efficiently.

2.2.1.8 Helicobacter pylori

Helicobacter pylori is a spiral-shaped gram-negative micro-aerophilic bacterium, colonizing the gastric mucosa of approximately half of the world's population causing severe illnesses with a high rate of morbidity and mortality (Diaconu et al. 2017). It is responsible for chronic upper gastrointestinal tract infections in humans. *H. pylori* has been recognized as a class I carcinogen by WHO and is the third leading cause of cancer death worldwide. The transmission of *H. pylori* may occur from person to person through fecal-oral (food and water), oral-oral (saliva), and gastric-oral (vomit) routes, though the exact route of transmission remains unclear. Several studies have reported the presence and survival of *H. pylori* in food and water, especially in milk and ready-to-eat foods, suggesting their role as the possible sources of infection (Quaglia and Dambrosio 2018). The prevalence of *H. pylori* infection shows large geographical variation, with a higher infection rate in developing countries due to poor socioeconomic status and overcrowded conditions (Salih Barik 2009).

The infection usually remains asymptomatic and can lead to diseases, such as chronic-active gastritis, duodenal and gastric ulcer, low-grade B-cell mucosa-associated lymphoid tissue (MALT) lymphoma of the stomach, and gastric adenocarcinoma. Among the infected individuals, about 10% develop peptic ulcer disease, 1%–3% progress to gastric cancer with a low 5-year survival rate, and 0.1% develop mucosa-associated lymphoid tissue lymphoma. The infection is usually acquired during early childhood and persists for the entire lifetime. In developing countries, middle-aged adults are more susceptible to helicobacter infection. *H. pylori* also causes extra-gastric disorders, including coronary heart disease, dermatological disorders (such as rosacea and idiopathic urticaria), autoimmune thyroid disease, thrombocytopenic purpura, and iron deficiency anemia (De Falco et al. 2015; Quaglia and Dambrosio 2018).

During pathogenesis, *H. pylori* utilizes its urease activity to neutralize hostile acidic conditions in order to enter the mucous layer. The bacterium then moves toward the host gastric epithelium cells with the aid of flagella, where specific interactions between bacterial surface proteins known as adhesins (blood group antigen-binding adhesin [BabA], sialic acid-binding adhesin [SabA] and other outer membrane proteins) and host cell receptors takes place, leading to successful colonization and persistent infection. After the successful colonization, toxins are released which causes the host tissue damage. Cytotoxin-associated gene A (Cag A) and its pathogenicity island (Cag PAI), VacA (vacuolating cytotoxin A), heat shock protein B (HspB), and the duodenal ulcer-promoting gene A (DupA) are the major toxins associated (Kao et al. 2016). The initial infection induces an inflammatory response that involves the infiltration of gastric mucosa by acute and chronic inflammatory cells, and this leads to gastritis. As the infection progresses, a series of intermediate stages of precancerous lesions develop in the following order: gastritis, atrophy, intestinal metaplasia, and eventually dysplasia. The hyperproliferation of the stomach epithelial cells leads to gastric cancer accompanied by hypochlorhydria (low-acid secretion), and atrophic gastritis (De Falco et al. 2015).

Rapid urease tests and endoscopic biopsy are the gold standards for the routine diagnosis of *H. pylori* infection. Urea breath tests and stool antigen tests are also used widely (Talebi et al. 2018). Treatment usually focuses on the eradication of *H. pylori* from the stomach. The first-line treatment for *H. pylori* infection involves a quadruple therapy containing a proton-pump inhibitor (PPI) and three antibiotics—clarithromycin, amoxicillin, and metronidazole—while the standard PPI triple therapy is restricted to areas with known low clarithromycin resistance (Gonzalez et al. 2019). Treatment is given to all patients with a positive *H. pylori* test. To confirm the eradication, the urea breath test and stool antigen test can be used, which is performed at least 4 weeks after the completion of therapy. In communities at high risk of gastric cancer, *H. pylori* "screen-and-treat" strategies are recommended (Diaconu et al. 2017).

2.2.1.9 Klebsiella pneumonia

Klebsiella pneumoniae is a major opportunistic pathogen that causes diseases, such as pneumonia, liver abscesses, septicemia, and diarrhea, in both humans and animals (Cao et al. 2014; Guo et al. 2016). In 2012, a study conducted by Translational Genomics Research Institute (TGen) and George Washington University, confirmed *K. pneumoniae* as a potential pathogen transmitted to humans via food. They collected turkey, chicken, and pork products from a retail shop in Flagstaff, AZ (US); genome sequencing from these meat sources confirmed its similarity to a patient sample (Davis and Price 2016). *K. pneumoniae*-associated extraintestinal human infections include pneumonia, septicemia, pyelonephritis, cystitis, and pyogenic liver abscess (Shon et al. 2013) with an incubation period of 1–3 weeks. Humans acquire the infection through contaminated food or via occupational transmission. Occupational transmission is mainly acquired from livestock and can spread to the community. Bovine mastitis is commonly caused by *K. pneumoniae*, which can be easily isolated from cow's milk (Osman et al. 2014). Dairy farms are mainly involved in the outbreaks of *K. pneumoniae* through contaminated milk (raw or unpasteurized) or contaminated beef products.

Commercially available poultry, beef, and pork products are usually contaminated by *K. pneumoniae*. The multidrug-resistant strains of *K. pneumoniae* have been isolated from meat that is available in the market (Kilonzo-Nthenge et al. 2013). Antimicrobial-resistant *K. pneumoniae* were isolated from fresh vegetables in the market (Falomir et al. 2013), as well as shrimp destined for international trade (Nawaz et al. 2012) and chicken farms (Wu et al. 2012). Fresh vegetables, such as carrots, tomato, sprouts, arugula, cucumber, iceberg lettuce, and herbs are also reported to act as major sources of infection when consumed in raw condition (Shahid et al. 2009). Even processed food items, such as fruit puree (Calbo et al. 2011) and beef stew (Gabida et al. 2015), can have *K. pneumoniae*. It has also been isolated from powdered infant formula, fish, street foods, and RTE food, confirming the role of food as a vehicle for the pathogen (Haryani et al. 2007; Sun et al. 2010; Puspanadan et al. 2012; Overdevest et al. 2014; Kim et al. 2015; Davis and Price 2016).

Sputum or blood samples are usually cultured in appropriate media for the identification of *K. pneumoniae* (Ergul et al. 2017; Para et al. 2018). Chest radiograph is also a useful tool for the clinical diagnosis of the pathogen. Treatment with cephalosporin (third or fourth generation) and carbapenem drugs for nosocomial infections is recommended for 14 days (Liu and Guo 2018). Raw vegetables can be washed using disinfectants, such as hydrogen peroxide, to control *K. pneumoniae* contamination (Shahid et al. 2009). It is also important to prevent cross-contamination by washing kitchen utensils and surfaces with soapy water (Perez et al. 2012). Good hygiene and food-handling practices can prevent *K. pneumoniae* from causing foodborne infections.

2.2.1.10 Listeria monocytogenes

Listeria monocytogenes is a gram-positive, microaerophilic, non-spore-forming, and rod-shaped bacteria commonly found in soil and the natural environment (Shamloo et al. 2015). The pathogen is transmitted through the ingestion of contaminated food, such as unpasteurized milk and other dairy products, raw meat, frozen vegetables, and salads, and it can also form biofilm on food-processing materials. The significant feature of *Listeria* is its ability to grow even in extreme conditions, such as in high salinity or under refrigeration and can remain viable in various food products for long periods (Camargo et al. 2017). According to CDC reports (2020), several illnesses caused by *Listeria* in the US from 2016 to 2019 have been linked with enoki mushrooms. The risk group of *Listeria* infection includes pregnant women, adults of 65 or above, and immunocompromised individuals. Despite the lower incidence of listeriosis compared to other enteric diseases, the infection is more fatal in humans. Several strains of *Listeria* isolated from food exhibited antimicrobial resistance (Radoshevich and Cossart 2018); this is a major concern for the success of treatment. Listeriosis is a major challenge to human health due to its high virulence—especially in immunocompromised children. Symptoms such as fever, headache, abdominal pain, diarrhea, vomiting, and convulsions are observed. The complications may lead to appendicitis and Meckel's diverticulitis (Li et al. 2019).

The presence of *L. monocytogenes* in meat and meat products is primarily based on the type of food product, its pH, and the number and type of inhabitant microorganisms. Poultry meat is more favorable for the growth of *L. monocytogenes* than any other meat. Several studies have also reported chicken and turkey meat to be notable sources of *L. monocytogenes* (Ayaz and Erol 2010). This infection is also transmitted from RTE foods, unpasteurized milk and dairy products, salads, uncooked meat, and meat products (Siriken et al. 2006). Sometimes *L. monocytogenes* are present in slaughtered animals symptomatically or asymptomatically; therefore, meat is most likely to be contaminated with the pathogen during or after the slaughter process. Seafood, such as fresh or frozen mussels, raw, pickled, or cold-smoked fish are also reported as sources for pathogen (Siriken et al. 2013). It is mainly transmitted to the environment through unhygienic practices in poultry flocks and carcass processing. The pathogen is rarely isolated from the poultry or chicken feces (Dhama et al. 2013).

Listeria from the contaminated samples is most effectively identified by the PFGE technique (Abdollahzadeh et al. 2016). In WHO laboratories, molecular typing methods using multilocus enzyme electrophoresis (MLEE) is employed to detect the pathogen due to its high sensitivity and ease of performance (Thomas et al. 2012). Generally, PCR-based methods are used for the epidemiological studies and also for the detection of bacteria from the poultry industries, dairy products, and other food products (Zulkifli et al. 2009). The treatment involves the use of drugs that efficiently destroy the growth of *Listeria*. Currently available treatment for human listeriosis is by the combination of drugs such as ampicillin or amoxicillin with gentamicin. The specific control measures taken are proper cleaning of refrigerators, chilling of food products at adequate temperatures, use of pasteurized milk and dairy products, and consumption of thoroughly cooked meat and meat products. Pregnant women must always be careful while consuming high-risk food such as soft cheese, smoked or raw seafood, and pate and refrigerated meat spreads. Canned or shelf-stable meat spreads and pate or smoked seafood are low-risk foods for consumption (FDA 2018, https://www.fda.gov/food/buy-store-serve-safe-food/what-you-need-know-about-preventing-listeria-infections).

2.2.1.11 Mycobacterium bovis

Mycobacterium bovis, a member of the *Mycobacterium tuberculosis* complex, is a gram-positive, acid-fast, and slow-growing aerobic rod that causes bovine tuberculosis (BTB) and many wild animals. It also causes zoonotic TB in humans. Transmission to humans occurs mainly through the consumption of unpasteurized milk or dairy products or, less frequently, through undercooked meat or direct contact with infected animals. Person-to-person direct transmission via aerosols is also possible (Lan et al. 2016). Cattle are believed to be the primary reservoirs of *M. bovis*.

Although effective control programs and pasteurization processes have significantly reduced the incidence of zoonotic TB in developed countries, it is still a burden in low-income countries. WHO has classified BTB as one of the seven neglected zoonoses. An estimated 147,000 new cases and 12,500 deaths of zoonotic tuberculosis due to *M. bovis* were reported in 2016 globally by WHO with the highest incidence in Africa, followed by Southeast Asia. However, the actual incidence of *M. bovis* TB in most cases remains underestimated, due to its similarity to TB caused by *M. tuberculosis* and lack of appropriate laboratory facilities for its isolation (Thoen et al. 2016). Poor farm biosecurity and nutrition of cattle, low socioeconomic status, illiteracy, traditional and community practices, such as consumption of raw milk, cohabitation with animals, etc., are some of the risk factors associated with transmission of *M. bovis* (Sichewo et al. 2019).

Most infection in humans presents as extra-pulmonary disease. Those affected can develop symptoms, such as cervical lymphadenopathy, affecting the tonsillar and pre-auricular nodes; intestinal lesions, chronic skin tuberculosis (lupus vulgaris), etc. The common clinical symptoms of zoonotic tuberculosis include loss of appetite, diarrhea, weight loss, intermittent fever and cough, weakness, etc. The pulmonary form of tuberculosis occurs less commonly and is characterized by fever, cough, chest pain, cavitation, and hemoptysis. It usually develops in people with reactivated infections. The incubation period is usually prolonged and may sometimes take months or years for the symptoms to develop. A few people may remain asymptomatic (Teppawar et al. 2018).

Diagnosis is made based on microscopic demonstration, isolation on selective medium, biochemical tests, and other molecular techniques. The mycobacterial culture method is considered the gold standard for the confirmation of infection. Serological tests, such as gamma interferon assay, indirect ELISA, and lymphocyte proliferation are also available. Restriction fragment length polymorphism (RFLP) and a multi-locus variable number of tandem repeats analysis (MLVA) are the molecular techniques frequently used to characterize *M. bovis* for epidemiological studies (Biffa et al. 2010; Teppawar et al. 2018). Since *M. bovis* is resistant to pyrazinamide, combinations of several other antibiotics are used for treatment. Drugs such as isoniazid®-rifampicin® or isoniazid-rifampicin-ethambutol® were found to be very effective against *M. bovis* infection (Lan et al. 2016). Implementation of biosecurity practices in herd management, intensive surveillance in endemic areas, and identification of risk pathways are some of the measures that can be adopted to prevent the transmission of zoonotic TB at the animal–human interface. Proper pasteurization of milk and continuous sanitary inspection and removal of contaminated animal products at slaughterhouses can prevent the entry of pathogen into food chain (Teppawar et al. 2018).

2.2.1.12 Plesiomonas shigelloides

Plesiomonas shigelloides are facultatively anaerobic, gram-negative, and rod-shaped bacteria (Ashkenazi 2018). It causes gastroenteritis in humans as a result of raw seafood consumption (Janda et al. 2016). *P. shigelloides* has been widely distributed in the environment mostly isolated from aquatic sources of tropical and temperate regions which serve as a natural host of bacteria (Escobar et al. 2012). There were several reports on seafood-borne outbreaks of *P. shigelloides* from Cameroon in central Africa in 2003 (Gonzalez-Rey et al. 2000; Wouafo et al. 2006). The disease is transmitted to humans through the ingestion of *P. shigelloides* contaminated food or water or via direct contact with infected animals (Wouafo et al. 2006). Molecular studies have also confirmed zoonotic disease where bacteria are found in the gut of various animals, such as cats, dogs, pigs, cows, snakes, fish, shellfish, monkeys, and vultures (Gonzalez-Rey et al. 2011). Even though 1% of fish tanks contain *P. shigelloides*, it is rarely associated with aquarium infections. The veterinarians, fish farmers, and zookeepers are placed in the occupational illness risk group. The incubation period is 1–4 days followed by acute gastroenteritis (Chen et al. 2013). Traveling to tropical countries or ingestion of raw seafood, such as shellfish, is the reason for 70% cases of *P. shigelloides*-associated gastroenteritis (Chen et al. 2013). The clinical manifestation varies and lasts for a few days with headache, fever, vomiting, abdominal pain, dehydration, and diarrhea (Escobar et al. 2012). Although rare, the symptoms associated with extraintestinal illness include bacteremia, meningitis, cellulitis, osteomyelitis, peritonitis, endophthalmitis, cholecystitis pancreatitis, and pseudoappendicitis (Xia et al. 2015). Immunodeficiency, sickle cell anemia, liver cirrhosis, and other hepatic diseases are consequences of underlying extraintestinal illnesses (Escobar et al. 2012; Chen et al. 2013).

The diagnosis of *P. shigelloides* is mainly performed by the isolation of the organism after attaining appropriate growth in its selective media. Oxidase test is primarily performed to identify *P. shigelloides* acquired through the ingestion of raw seafood or water or by traveling to tropical regions (Ashkenazi 2018). The misinterpretation of about 30% of samples by lactose fermentation tests can be verified using an oxidase test. The gastroenteritis caused by *P. shigelloides* is usually self-limited. Fluid and electrolyte replacement is the primary treatment; chemotherapeutic drugs are recommended for children with underlying illness. Drugs such as cephalosporins, amoxicillin-clavulanate, carbapenems, and fluoroquinolones are used for the treatment; the organism resists penicillin and trimethoprim-sulfamethoxazole (Maluping et al. 2005). Ceftriaxone dosage of 50 mg/kg is administered once daily for five days to treat gastroenteritis. The extraintestinal illness in children is cured using intravenous antibiotic therapy. The best method to prevent infection is by avoiding consumption of raw or undercooked seafood or contaminated water from infection-prone areas. Proper handwashing is also useful to stop the spread of this pathogen.

2.2.1.13 Pseudomonas *spp.*

Pseudomonas spp. are gram-negative, psychotropic, rod-shaped, non-sporing, and motile organisms found in soil, water, marine environments, etc. (Molina et al. 2013). One important feature of *Pseudomonas* spp. is the production of pigments, such as pyoverdine (fluorescein) by *P. aeruginosa, P. putida*, and *P. fluorescens* and pyocyanin by *P. aeruginosa* and *P. fluorescens* (Cantoni et al. 2001). They are opportunistic pathogens for animals and humans (Ridgway and Safarik 1990) and cause spoilage in foods—specially those that are stored aerobically with high water content—for example fish, meat, poultry, milk, and milk products (Raposo et al. 2017). Different groups of *Pseudomonas* spp. attacks different food products. For instance, fluorescent species like *P. fluorescens, P. aeruginosa, P. putida*, etc. attack chilled food products, whereas pectolytic strains, such as *P. viridiflava*, contaminate fresh fruits and vegetables. They are also responsible for the browning phenomenon in processed vegetables (Blackburn 2006). *Pseudomonas* spp. are predominant in milk and milk products (Schokker and Van Boekel 1999; Dogan and Boor 2003; Gunasekera et al. 2003; Raposo et al. 2017). Contamination is the result of unhygienic milk production practices and processing techniques (Richter and Vedamuthu 2015). The bacteria produce many proteolytic and lipolytic enzymes that reduce the quality as well as shelf life of milk and dairy products (Dogan and Boor 2003; Raposo et al. 2017).

P. *fragi* is associated with meat spoilage followed by *P. lundensis* and *P. fluorescens*, and that contamination comes from slaughterhouses. It causes slime and off-odor formation in raw meat and meat products during cold storage (Ercolini et al. 2009). *Pseudomonas* spp. are the most significant organisms in fish spoilage, in which *P. anguilliseptica* is most commonly associated (Toranzo et al. 2005). There have been reports that this pathogen causes "red spot disease" in pond-cultured eels in many countries (Lopez-Romalde et al. 2003).

Various media can be used to detect the presence of this bacteria from food samples; commonly used media are MacConkey agar, eosin methylene blue agar, and blood agar at suitable growth conditions. One problem associated with the isolation of pathogens is interference from non-pseudomonas which inhibit specificity and accuracy. In molecular typing methods, fluorescent oligonucleotide probe hybridization or fluorescence *in situ* hybridization (FISH) is used to identify the specific strain of pathogen from food samples (Gunasekera et al. 2003). The infections caused by *Pseudomonas* spp. are difficult to treat because of the broad antimicrobial resistance (Lister et al. 2009). However, carbapenems are usually used to treat infections despite their increasing resistance (Raposo et al. 2017). Prevention measures to reduce or eliminate the contamination include maintenance of hygiene during the processing of foods, conducting hygiene programs, and efficient use of pasteurization methods in the case of milk and dairy products (Van Tassell et al. 2012).

2.2.1.14 Salmonella *spp.*

Salmonella spp. belongs to the family of *Enterobactericeae* and it's a worldwide cause of food poisoning. They are characterized by their ability to metabolize citrate as a carbon source and lysine as a nitrogen source and production of hydrogen sulfide (Dougan and Baker 2014). Two species of *Salmonella*, *S. enterica* and *S. bongori*, cause illness in humans (Bintsis 2017). Based on the Kaufmann-White typing scheme, strains were categorized based on their surface and flagellar antigenic properties. Salmonella is divided into different serotypes, and some serovars cause serious illness, such as typhoid fever (Dekker and Frank 2015). The virulence factors of *Salmonella* spp. are encoded in pathogenicity islands (PAIs), which aids in the infection of host organisms and in which PAI 1 and PAI 2 are important for infection (Bintsis 2017).

Many wild animals and livestock harbor *Salmonella* strains in their intestines, and infections are acquired by the ingestion of contaminated food, with animals or human feces carrying the pathogen. Commonly associated foods are eggs, meat and poultry, and fruits and vegetables (Dougan and Baker 2014). The spectrum of symptoms in humans ranges from asymptomatic carriage to fatal typhoid fever. Common symptoms of infection are diarrhea, fever, and abdominal cramps happening within the incubation period of 12–72 hours. Usually the infection is self-limited, but some strains may spread systemically, which leads to various health complications (Dekker and Frank 2015). Infection risk is high with immunocompromised persons and children. In the case of typhoid fever, it is characterized by high temperature and vomiting; sometimes complications, such as neurological infection, intestinal perforation, and death occur (Hanning et al. 2009).

In laboratory diagnosis, microbial examination of fecal specimens is preferred in the case of non-typhoidal infections, whereas blood and lymph are collected for typhoidal infections. For the identification of *Salmonella* species, highly selective media like hektoen and xylose-lysine-deoxycholate (XLD) agars are used that detect hydrogen sulfide production. In serological typing, the Kauffmann-White method is widely used, which is based on the LPS O antigen, H1 and H2 flagellar antigens, and the Vi antigen (Grimont and Weill 2007). Advanced methods like mass spectrometry (MALDI-TOF MS), PFGE, and multiplex PCR are also used for the detection of the organism. To prevent infection, various steps should be taken such as thorough handwashing, hygiene maintenance in food packing, and processing and proper cooking, which can destroy the pathogen.

2.2.1.15 Shigella *spp.*

Shigella spp. belongs to the family of *Enterobacteriaceae* and possesses four serogroups: A, B, C, and D. Common organisms of this include *S. dysenteriae*, *S. flexineri*, *S. boydii*, and *S. sonnei* (Bacon and Sofos 2003). They are gram-negative, non-motile, non-spore-forming, facultatively anaerobic rods, and with a temperature requirement of 6–48°C, preferably 37°C, for growth. But some strains can tolerate lower temperatures also (Bintsis 2017). *S. dysenteriae* type 1 produces a cytotoxic Shiga toxin (Stx), causing bacillary dysentery or shigellosis ranging from watery diarrhea to dysentery. The natural reservoirs of *Shigella* are humans and large primates.

These pathogens are found in poor sanitation environments and the transmission is by person-to-person contact and the fecal-oral route. Foods like milk, chicken, shellfish, and salads are associated with *Shigella* spp. contamination (Dekker and Frank 2015). In developed countries, strains such as *S. sonnei* and *S. flexneri* are more prevalent, while

S. dysenteriae type 1 causes epidemics in African countries (Bintsis 2017). The introduction of pathogen into food materials may occur through infected food handlers during food processing and packaging. Vegetables are considered to be a major source of infection. They can also survive in chilled foods (Nygren et al. 2013).

All serogroups of *Shigella* spp. cause gastrointestinal infections with an incubation period of 12–50 hours. Patients also experience watery diarrhea, fatigue, fever, and abdominal cramps during this period. *S. dysenteriae* causes hemolytic uremic syndrome (HUS), and it is indicated by thrombocytopenia, hemolytic anemia, and acute renal failure. Another important factor for pathogenesis is the low infective dose, i.e., less than 100 cells required (Dekker and Frank 2015). Most cases are self-limited and do not cause fatality except in immunocompromised patients and in children.

In the host, they attach and penetrate into the epithelial cells of intestinal mucosa, multiply and spread intracellularly, and finally cause tissue damage. In the later stage of infection, the dead cells shed off from the mucosal surface and they appear in stools (Labbé and Carcia 2013). Conventional microbiological methods such as the use of differential and selective media and enrichment broths are used for the isolation of pathogen from food samples. Above all, PCR, immunoassays, and mass spectrometry techniques are considered to be specific and accurate in the detection. The disease can be eliminated or reduced by the use of hygienic water supplies, good personal hygiene, and proper washing of vegetables (Dekker and Frank 2015).

2.2.1.16 Staphylococcus aureus

Staphylococcus aureus is the most common pathogen in humans and animals, which colonizes the skin and mucous membranes. Production of pigmented colonies is the characteristic feature of this genus, and the name itself "aureus" means "golden" in Latin. The widespread recognition of the genus is due to the emergence of antimicrobial resistance, especially the methicillin-resistant *S. aureus* (MRSA) (Robinson et al. 2016). *S. aureus* can grow at a wide range of temperatures (7–48°C; optimum temperature 30–37°C) and also with 15% sodium chloride concentration (Kadariya et al. 2014). They can remain viable for longer durations on hands, air, and environmental surfaces (Kusumaningrum et al. 2002).

The main reservoir for the strain is the human nose and therefore nasal carriage is closely associated with Staphylococcal infections. Due to unhygienic food handling, frequent cases of staphylococcal food poisoning (SFP) are reported worldwide. According to the CDC about 2 hundred thousand cases per year are reported in the US (Scallan et al. 2011). *S. aureus* produces enterotoxins (SEs) which are water-soluble, stable polypeptides of 22–29 kDa and included in the family of pyrogenic toxins. They are able to activate T-cells and are resistant to proteolytic enzyme degradation (Fetsch and Johler 2018). The effect of SEs in SFP is initiated by intestinal inflammatory response and degranulation of mast cells (Dinges et al. 2000).

Ingestion of food containing Staphylococcal enterotoxins is the reason for SFP and food handlers are recognized as potential sources of infections (Wattinger et al. 2012; Fetsch and Johler 2018). Commonly associated foods include beef, pork, fish, oysters, and shrimp. Consumption of raw milk and dairy products are also associated with human infection, whereas *S. aureus* cause bovine mastitis in cattle (Johler et al. 2015). A low infectious dose is required for an SFP outbreak (Evenson et al. 1988). The incubation period is up to 8 hours after ingestion of contaminated food, and the main symptoms are nausea, vomiting, watery diarrhea, abdominal pain, shivering and moderate fever, and it is usually self-limiting within 24 hours. In some cases, electrolyte imbalance and morbidity occurs.

ELISA and reverse passive latex-agglutination (RPLA) are commonly used to detect the presence of enterotoxins from food samples and human samples (nasal swabs). Other tools, such as detection of SE genes by PCR, DNA microarray analysis, and whole-genome sequencing are also used in diagnostics (Fetsch and Johler 2018). SFP is associated with cross-contamination of pathogen from raw ingredients, poor hygiene, improper cleaning of equipment in food processing, and inadequate cooking time or temperature. The most effective way to irradicate the pathogen is the implementation of food safety standards like HACCP, GMPs, and GHPs. Care should also be taken in precooked foods due to the viability of *S. aureus* in the air and on environmental surfaces. Other measures, such as community food safety awareness programs (Byrd-Bredbenner et al. 2013) and food preservation under lower temperatures can prevent the growth of pathogen and, thereby, contamination (Kadariya et al. 2014).

2.2.1.17 Vibrio *spp.*

The family *Vibrionaceae* (class: Gammaproteobacteria; Order: *Vibrionales*) contains very important organisms that cause intestinal tract and extra intestinal infections in both humans and animals. They are all gram-negative, facultative anaerobes, capable of fermentation, and ubiquitous in the marine environment. The family comprises seven genera—*Vibrio, Photobacterium, Allomonas, Listonella, Enhydrobacter, Salinivibrio,* and *Enterovibrio*—of which the genus *Vibrio* has the largest number of species. Among them, ten of the species are known as human pathogens. Vibrios are important genera of the family and are free-living (autochthonous) bacterial flora in marine, estuarine, and freshwater environments, and it includes many symbiotic and pathogenic strains (Safa et al. 2010). In pathogenic species, *V. cholerae*, *V. parahaemolyticus*, and *V. vulnificus* are important in foodborne illness.

V. parahaemolyticus associated food sources are crabs, prawns, seaweed, oysters, and clams, and gastroenteritis is most commonly associated with the infection. Other symptoms include nausea, vomiting, abdominal cramps, watery diarrhea, etc. Most of the symptoms are self-limiting and, in rare cases, may lead to death (Bacon and Sofos 2003). *V. parahaemolyticus* was a leading cause of foodborne disease outbreak during the past decades (Wu et al. 2014).

V. cholerae is highly motile via a single polar flagellum and is a comma-shaped rod and non-sporing 1.4–2.6 μm long bacterium. The bacteria are oxidase-positive and reduce nitrate. They are fermentative and grow on ordinary or selective media containing bile at 37°C, and the growth is stimulated by the addition of 0.5%–1% NaCl, i.e., slightly halophilic in nature. But it is halotolerant compared to other *Vibrio* species. It is naturally found in both fresh and saltwater environments. They are capable of colonizing the surfaces of zooplankton, such as copepods with 10^4 to 10^5 cells attached to a single copepod, attaching themselves through the production of chitinase (the key component in seashells).

V. cholerae is the most prominent pathogenic species that causes watery diarrhea (cholera). Based on the differences in the composition of the major cell wall antigen (O) first described by Gardner and Venkataraman (1935), 206 serotypes have been differentiated (Shimada et al. 1993; Yamai et al. 1997). The environment is a major reservoir of this organism, and it is reported that they are associated with aquatic plants, protozoa, water birds, bivalves, animals with chitin exoskeletons, as well as abiotic substrates. They have many persistent forms including a viable but non-culturable (VBNC) state, biofilms, and a rugose survival form (Binsztein et al. 2004). The zooplankton is generally recognized as the largest environmental reservoir of the organism. Humans and animals are also carriers of *Vibrio*. Long-term carriage in humans is extremely rare, whereas short-term carriage is important in transmission of the disease. Persons with acute cholera can excrete 10^7 to 10^8 CFU/g of stool. Even after the termination of symptoms, patients who have not been treated with antibiotics may continue to excrete the pathogen for one to two weeks. The pattern of cholera transmission depends on the climate and the season of the year. Cold, acidic, and dry environments inhibit the survival of *Vibrio*, whereas warm, monsoon, alkaline, and saline conditions favor the growth. Because of the sensitivity to acid, most of the organisms die in the stomach and therefore a high infectious dose ($\sim 10^8$) of bacteria is required for the development of severe cholera symptoms; however, the infectious dose can drop to $\sim 10^4$ bacteria in persons who produce less stomach acid, including young children, the elderly, and those who take antacids in their food (Cash et al. 1974; Kitaoka et al. 2011).

Cholera is transmitted through the fecal-oral, that is, pathogens in fecal particles pass from one person to the mouth of another person. It is mainly because of the lack of adequate sanitation and poor hygiene practices. Explosive, potentially fatal, dehydrating diarrhea is the main characteristic of cholera, although most infections are mild or even asymptomatic. The incubation period ranges from several hours to five days and is dependent on the inoculum size. The onset of illness may be sudden and is characterized by watery diarrhea, or there may be premonitory symptoms such as anorexia, abdominal discomfort, and simple diarrhea. Initially, the stool is brown with fecal matter, but soon the diarrhea develops a pale grey color with an inoffensive, slightly fishy odor. Mucus in the stool confers the characteristic "rice water" appearance and is accompanied by vomiting. The most severe form of cholera is called cholera gravis where the level of diarrhea may quickly reach up to 500 to 1000 mL/hour and lead to symptoms like tachycardia, hypotension, and vascular collapse due to dehydration. Due to intensive diarrhea, skin appears to be a doughy consistency, eyes are sunken, and the hands and feet become wrinkled (as after long immersion in water, like washerwomen's hands). Symptoms like absence of tears, lethargy, unusual sleepiness or tiredness, low urine output, abdominal cramps, and nausea are also accompanied. The severe dehydration can lead to death within hours of the onset of symptoms.

V. cholerae has two virulence factors, the toxin co-regulated pilus (TCP) and the cholera toxin. TCP is encoded by the Vibrio pathogenicity island 1 (VPI1) and is a type IV pilus that allows the organism to aggregate together, which is a type of mechanism that protects individual cells from shearing forces in the small intestine (Kirn et al. 2000; Kitaoka et al. 2011). It is also essential for the *V. cholerae* to colonize in the small intestine of humans. Once the small intestine is successfully colonized by the organism, the cells begin to secrete cholera toxin (Taylor et al. 1987). The secreted toxin activates the cystic fibrosis transmembrane conductance regulator (CFTR) in the epithelial cells that line the small intestine, leading to massive fluid efflux into the lumen of the small intestine.

There are seven recorded cholera pandemics of which six have been attributed to the classical strains; the remaining one, the seventh pandemic, is attributed to the El Tor strain, which started in Indonesia (Safa et al. 2010; Kitaoka et al. 2011). Approximately 120,000 deaths occur every year because of acute cholera, mainly affecting young children between the ages of 1 and 5 years. El Tor biotype is the leading strain responsible for the annual cholera epidemics in India and Bangladesh.

Cholera can be controlled with a multi-sector approach, providing water, sanitation, hygiene (WASH) services and oral cholera vaccines (OCV). Cholera affects the poorest and most vulnerable populations around the world within communities of each affected country. Cholera control is both a matter of emergency response, in the case of outbreaks, and a matter of development when the disease is endemic in high-risk areas. Effective cholera prevention and control interventions are well known and includes both integrated and comprehensive approaches that involve activities both inside and outside of the health sector, including:

- Enhanced epidemiological and laboratory surveillance to identify endemic areas and detect, confirm, and quickly respond to outbreaks
- Universal use of safe water and basic sanitation
- Community engagement for behavioral changes and improved hygiene practices
- Quick access to treatment, such as Oral Rehydration Solution (ORS), intravenous fluids, and antibiotics for severe cases
- Protection with safe and effective OCV (Mondiale 2017)

V. cholerae do not usually cause systemic infections and antibiotics can be used as a sole treatment. A combination of oral rehydration therapy and antibiotic treatment is also used. Antibiotics are typically administered to lessen the duration of symptoms by reducing the shredding of organisms in the stool by approximately 50%, and it also reduces the severity of symptoms by reducing the volume of diarrhea. For the treatment, the antibiotics tetracycline and quinolones are commonly used, but the emergence of antibiotic resistant *V. cholerae* strains is a major risk. Therefore, administration of antibiotics should be limited to patients with severe dehydration cases (Kitaoka et al. 2011). In the severe cases, a single dose of doxycycline is given along with ORS and is usually sufficient to cure the disease.

2.2.1.18 Yersinia enterocolitica

Yersinia enterocolitica are gram-negative or gram-variable, facultatively anaerobic, and non-spore forming rods belonging to the family *Enterobacteriaceae*, causing yersiniosis. *Y. enterocolitica* also causes foodborne gastroenteritis in humans (Cary et al. 1999; Bintsis 2017). They are recognized as the third most reported zoonosis in European countries and the most commonly isolated strain in human infections (Korsgaard et al. 2014). *Y. enterocolitica* produces a heat-stable toxin, which is resistant to enzyme degradation and remains stable for long periods. These bacteria are distributed in diverse environments and are associated with foods such as milk, poultry, pork, seafood, and raw milk (Bacon and Sofos 2003). The main reservoir of *Yersinia* is pigs, from which pathogenic strains are isolated frequently (Fredriksson-Ahomaa et al. 2007). As *Y. enterocolitica* are psychotropic in nature, they can spoil chilled foods, although in low temperatures they dominate over other species. There have also been reports of *Yersinia* outbreak after the ingestion of pasteurized milk due to inadequate temperature or time of pasteurization (Longenberger et al. 2014). Foodborne infections result from the consumption of meat, especially raw or undercooked pork, and from contaminated water. The incubation period is from a few days to a week and it can be up to 4 weeks in children. The characterized symptoms are diarrhea containing bloody stools, fever, vomiting, and abdominal pain (Bacon and Sofos 2003; Bintsis 2017). However, extraintestinal infections are reported as meningitis, septicemia, myocarditis, Reiter syndrome, erythema nodosum, and glomerulonephritis. The pathogen infects all age groups of which children and immunocompromised persons are at highest risk. The unhygienic conditions in slaughterhouses, cross-contamination of carcasses and meat, and long storage of foods will allow the growth of this cold-adapted organism and increase the chances of yersiniosis outbreaks. Detection of pathogen is usually conducted by molecular methods such as conventional PCR targeting virulence genes (Fredriksson-Ahomaa et al. 2007).

As temperature is a significant factor for pathogen growth, maintaining foods in appropriate conditions is important to prevent contamination. Additionally, proper hygienic conditions, use of preservatives like sodium lactate, proper pasteurization methods for milk and dairy products, etc., can reduce the contamination (Barakat and Harris 1999; Bintsis 2017). *Yersinia* phages are reported to be effective against *Y. enterocolitica* infection and recently *Yersinia* phage, X1, is found to be able to lyse the pathogenic strains (Xue et al. 2020).

2.2.2 Viruses

2.2.2.1 Adenovirus

Adenoviruses (AdV) are large (90–100 nm), non-enveloped, and double-stranded DNA (dsDNA) viruses with icosahedral geometries, belonging to the family of *Adenoviridae*. Adenovirus has emerged as one of the leading viral causes of acute gastroenteritis (AGE), but it is less infectious than rotavirus and norovirus. Usually these viruses are the common cause of upper respiratory tract infections in children; less commonly, they can also cause gastrointestinal, ophthalmologic, genitourinary, and neurological diseases. The viral outbreak occurs mostly in closed, crowded populations. Infants, young children, and immunocompromised patients are more susceptible to HAdV-associated infectious outbreaks. The virus may also affect healthy populations, but such infections usually remain self-limited (Khanal et al. 2018).

Adenovirus-associated acute gastroenteritis occurs mostly in infants and young children. Human adenovirus types 40 and 41 have been reported to be responsible for 1%–20% of cases of diarrheal disease in children worldwide, especially among toddlers (Li et al. 2004). Several cases of adenovirus-associated gastroenteritis have been reported worldwide. In a long-term molecular study on the incidence of adenovirus infection conducted at Chiang Mai, Thailand during 2011–2017, adenovirus was detected in 7.2% of total fecal samples collected from 2312 hospitalized diarrheic cases (Kumthip et al. 2019). In another study conducted in Tehran, Iran during 2011–2015, adenovirus was detected in 4.3% (16/376) cases of hospitalized children with acute gastroenteritis (Arashkia et al. 2019).

The major route of transmission here is fecal-oral, but transmission via droplet or close contact is also possible. Transmission via water can also occur through contact with contaminated recreational fresh water or tap water. People with poor hygienic practices are also prone to adenovirus infection. Since the virus is resistant to gastric pH and bile secretions, the infected person might shed large doses of viruses (10^8–10^9 particles/g) in their stool. These viruses can remain stable for long periods in moisture-free environments and, since they are non-enveloped, they are resistant to lipid disinfectants, although they are destroyed by heat, formaldehyde, and bleach (Lion 2014; Khanal et al. 2018).

The adenoviral infection is initiated by the attachment of the virus to the host cell surface via coxsackie and adenovirus receptors (CAR) found on multiple polarized epithelial cells and membrane cofactor protein (MCP) or CD46, found on most of the host cell surfaces. The CAR-docked particles activate a secondary infection with integrin co-receptors that facilitate the virus entry via endocytosis (Stasiak and Stehle 2020). Inside the cell, following the acidification of the endosome, the viral capsid destabilizes and releases its internal minor protein (protein VI), and this, together with the toxic nature of the major capsid protein penton, eventually destroys the endosome, releasing the virions to the cytoplasm. The virus subsequently gets translocated into the nucleus via microtubules and docks at the nuclear pore where it undergoes uncoating facilitated by viral proteases that dissolve the viral capsid, releasing viral DNA. The DNA then associates with histone molecules and undergoes gene expression (Meier and Greber 2004).

HAdVs can either infect epithelial cells, cause their lysis, or remain latent in lymphoid cells. During lytic infection, the virus enters and replicates inside the host epithelial cell, inhibits its cellular macromolecule synthesis leading to cell death, and releases numerous progeny virions which result in the generation of a host inflammatory response. Following the lytic infection, HAdVs can persist in susceptible cells in a latent state for years and can eventually reactivate, reinfect, and replicate in epithelial cells, causing disease symptoms again.

The cytotoxic nature of pentons, as well as the inhibitory action of early viral proteins against tumor necrosis factor (TNF) and apoptosis, down-regulates the expression of major histocompatability (MHC) class I molecules. This prevents their recognition by cytotoxic T-cells, and hence the viral factors contribute to the pathogenicity (Crenshaw et al. 2019). Following the infection, symptoms associated with gastroenteritis usually develop after 8–10 days of incubation which include diarrhea with watery, non-bloody stool, dehydration, vomiting, and fever that lasts for 7–8 days (D'Souza 2015).

The diagnosis starts with clinical evaluation of the individual's symptoms. Generally, chest X-rays, nasal swabs, and blood or stool cultures are used to confirm HAdV infection. Since the infected person can continue the shedding of virus in their stool for about 7–14 days post-infection, feces serve as the most reliable source of isolation. The laboratory diagnostic techniques for the detection of HAdV include antigen detection, molecular detection (PCR), and viral isolation. As the virus grows poorly in cell culture, virus isolation is not a preferable method (Crenshaw et al. 2019; Ryan et al. 2019).

There is no specific treatment for HAdV infection, as most of the infections are mild. Clinical care usually focuses on alleviating the disease signs and symptoms through rehydration and fluid replacement therapy. In developing countries where children are often malnourished, supplementary nutrition is important. The antiviral agent Cidofovir® has been used in the treatment of disseminated adenovirus infections in bone marrow transplant recipients. Currently, the US Food and Drug Administration (FDA) has approved a vaccine against certain serotypes (type4 and type7) of adenovirus, but is only available for military use and not to the public. Public health measures, such as frequent handwashing, use of gloves, gowns, and goggles, disinfection with sodium hypochlorite, removal of infected feces, avoiding contaminated food or water, chlorination of swimming pools, and exclusion of infected food handlers from work are used to inhibit viral outbreaks (Ryan et al. 2019).

2.2.2.2 Enterovirus

Enteroviruses are highly infectious, emerging viral pathogens that are found everywhere in the environment. The genus *Enterovirus* belongs to the family *Picornaviridae* and comprises a highly diverse group of small, non-enveloped, icosahedral-shaped viruses with a single positive-stranded RNA genome. Based on sequence diversity, they have been divided into 15 species: enterovirus A–L and rhinovirus A–C. Human enteroviruses comprise four enterovirus species (A–D) and three rhinovirus species (A–C) (Chen et al. 2020). Based on pathogenesis, human enteroviruses are classified into four groups: Polioviruses (PVs), Coxsackie A viruses, Coxsackie B viruses, and enteric cytopathic human orphan (ECHO) viruses (Rhoades et al. 2011).

Enteroviruses are transmitted primarily by the fecal-oral route, i.e., the virus gains access to oral-mucosal epithelium and replicates in the gastrointestinal tract of susceptible individuals who are exposed to contaminated food or water. Some enteroviruses, such as rhinovirus and EV-D68, spread via respiratory secretion (Chen et al. 2020). Enteroviruses can remain in the environment for long periods and are relatively resistant to various solvents and detergents at ambient temperatures. They can survive in soil, sludge, vegetables, and freshwater sources. Poor sanitation and close living quarters are some of the risk factors for enterovirus infection (Muehlenbachs et al. 2015).

Among the human enteroviruses, the poliovirus, belonging to the species *Enterovirus C* and the best-known neurotropic enterovirus, has almost been removed from the human population by vaccination. The non-polio enteroviruses account for the majority of outbreaks in the present day (Chen et al. 2020). The association of enterovirus infection with diarrheal illness has not been truly verified. But the role of enterovirus infection in acute diarrhea has been reported in various studies. In a recent study, enteroviruses such as coxsackie virus A6 and echovirus 9 were detected in the stool samples of children with acute diarrhea in Iran. Similarly, enteroviruses (E11, CVA6, CVB2, PV3, CVB4, E18, and CVA2) were detected in the stool samples of children having diarrhea during a case study in Hebei province, China. Moreover, gastroenteritis outbreaks caused by enterovirus were also reported in Japan, Malaysia, Southern India, and Brazil during the past years (Fazelipour et al. 2019).

Once an enterovirus gains entry to the oral epithelia by contact with contaminated material, the virus infects the oropharynx and, subsequently, the gastrointestinal tract. The infection can proliferate in lymph nodes of the respiratory and gastrointestinal systems. As the clinical signs start to manifest in different organs, viremia becomes detectable and the risk of sepsis increases. In severe cases, the virus spreads to secondary sites such as the skin, mucous membranes, respiratory tract, heart, liver, pancreas, and, in some cases, enters the CNS where neurological and cardiorespiratory failure can develop (Gonzalez et al. 2019). Enteroviruses have the ability to completely shut down host translational machinery, thereby causing a cytopathic effect in the affected organs resulting in clinical disorders (Rhoades et al. 2011).

Most enterovirus infections are asymptomatic. The general symptoms include fever, headache, malaise, respiratory illness, vomiting, and diarrhea. Enterovirus infections also present with severe clinical manifestations, including hemorrhagic conjunctivitis, pharyngitis, pleurodynia, herpangina, hand-foot-and-mouth disease, paralysis, severe neonatal sepsis-like disease, hepatitis, myocarditis, pericarditis, encephalitis, aseptic meningitis, and multi-organ failure (Todd and Grieg 2015). The incubation period is usually 3–6 days. Viral shedding through feces can persist for more than 10 days, which may vary depending upon the type of enterovirus and severity of infection (Gonzalez et al. 2019).

For the diagnosis of enterovirus, specimens should be taken from feces in the early phase of the disease (shedding continues for a few weeks, following acute infection, and can be detected even well after the clinical illness has resolved), nasopharyngeal aspirates (NPA), or throat swab. The laboratory techniques for the detection of enteroviruses include virus isolation, RT-PCR, immunohistochemistry, and *in situ* hybridization techniques (Todd and Grieg 2015). There is no specific treatment for non-polio enterovirus infection. People with mild illness caused by non-polio enterovirus infection only need to treat their symptoms. Antiviral drugs such as Ribavirin®, pleconaril, and RNAi have been widely used to treat enterovirus infections (Rhoades et al. 2011).

2.2.2.3 Hepatitis A Virus

Hepatitis A virus (HAV) is one of the most recognized, common viral agents that causes foodborne illness in human. It is the causative agent of hepatitis A infection, a highly contagious liver illness. HAV belongs to the genus *Hepatovirus* of the family *Picornaviridae*. HAVs are small (27–32 nm), round, icosahedral, non-enveloped viruses (Todd and Grieg 2015). HAV exists as a single serotype, with human strains distributed into three genotypes (I, II, and III) and seven subgenotypes (IA, IB, IC, IIA, IIB, IIIA, and IIIB), respectively (Bosch et al. 2016).

Transmission of HAV is through the fecal-oral route, either by direct contact with an HAV-infected person or through HAV-contaminated food or water. Foods associated with HAV transmission include frozen, dried, or RTE foods (that do not require heating before consumption), shellfish, salads, sandwiches, vegetables, fruits, reconstituted frozen orange juice, ice cream, cheese, rice pudding, iced cake, custard, milk, bread, cookies, and other raw or undercooked foods. Contamination of food by HAV can occur in several ways, such as irrigation of fruits and vegetables with contaminated water, shellfish harvested from contaminated water, processing and preparation of food with contaminated equipment, and improper handling of food by infected individuals (Sattar et al. 2000; Hu et al. 2020).

Foodborne HAV outbreaks are becoming a serious threat to all countries with large numbers of susceptible adults. Food-associated HAV outbreaks may affect a significant number of persons and may lead to high morbidity. Recent studies have shown that the incidence of foodborne HAV outbreaks in developed countries is increasing, as the import of high-risk food prepared in HAV-endemic countries to the developed countries results in such outbreaks (Hu et al. 2020). The disease spectrum of HAV infection is wide-ranging from an asymptomatic self-limited infection to life-threatening hepatitis. Asymptomatic infection is common in children (Acheson and Fiore 2004), while adults may develop severe illness that can sometimes be fatal (Hu et al. 2020).

Hepatitis A virus survives in the acidic stomach environment after ingestion and is transported to the liver where it replicates within the hepatocytes and macrophages. It is then secreted into bile and finally excreted via feces or transferred back to the liver through an enterohepatic cycle until virus neutralization by antibodies takes place. HAV-infected people usually develop symptoms of acute hepatitis after 15–50 days of incubation by which time their serum shows elevated levels of liver enzymes aspartate/alanine aminotransferases (AST/ALTs). Before the onset of symptoms, viremia occurs, and the infected person sheds copious amounts of viruses in their feces, which declines by the onset of jaundice. The virus is also shed through saliva at lower concentrations. Following clinical hepatitis, anti-HAV IgM and, subsequently, anti-HAV IgG appear in the serum and saliva, accompanied by a marked reduction of fecal virus shedding and viremia. Once the virus has been contracted, lifelong immunity develops in the body (Shin and Jeong 2018).

HAVs are generally stable and can persist in dried feces, the environment, and mussels for long periods at room temperature. They are generally resistant to low pH, freezing, drying as well as free chlorine, and can survive for one hour at 60°C. The virus can remain stable on lettuce at 4°C for up to 12 days under various modified atmospheric conditions, as well as on frozen berries for 90 days. Hence, inactivation of the Hepatitis A virus requires heating of foods (>85°C) for 1 min or disinfecting surfaces with sodium hypochlorite (household bleach, 1:100 dilution) for 1 min (Todd and Grieg 2015).

HAV infection is generally acute, characterized by sudden onset of fever, malaise, nausea, anorexia, fatigue, and abdominal discomfort, followed by jaundice in several days. Less commonly, pruritus (associated with relapsing hepatitis), diarrhea, arthralgia, or skin rashes may develop. The infection usually remains self-limited and does not progress to chronic hepatitis. Few may show atypical clinical symptoms such as relapsing hepatitis, prolonged cholestasis, or extrahepatic manifestations. Older adults and patients with preexisting liver diseases are the high-risk groups, developing chronic infection that results in fulminant hepatitis (Todd and Grieg 2015; Shin and Jeong 2018).

HAV can be detected using several diagnostic techniques which include serological tests, antigen detection tests, and molecular methods. A serological test involves the detection of HAV-specific antibodies (IgM anti-HAV and total anti-HAV) using commercially available anti-HAV enzyme immunoassays such as ELISA, RIA, immunoblotting, etc. Immunoassays can be also used to detect HAV antigen from stool, cell culture, and environmental samples. RT-PCR, antigen capture RT-PCR, single-strand conformational polymorphism, and RFLP are some of the most rapid and sensitive molecular methods commonly used for the detection of HAV from clinical specimens, environmental samples, or food. The slow growth of HAV in cell culture and their low concentration in food samples necessitates the need for highly sensitive nucleic acid extraction methods for their detection in food. This involves viral RNA extraction using isothiocyanate followed by precipitation with polyethylene glycol (PEG) and purification using poly (dT) magnetic beads. Viruses from contaminated water are concentrated by filtration. Nucleic acid hybridization assays using labeled probes were used earlier; currently PCR is widely used for HAV detection from environmental samples (Nainan et al. 2006).

Hepatitis A can be prevented by vaccination, providing long-term immunity (as there is only a single serotype) lasting up to 20 years. Other preventive measures involve good hygiene and sanitation practices and avoidance of the consumption of shellfish from polluted waters (Todd and Grieg 2015). There is no specific antiviral therapy for hepatitis A infection, but passive protection with immunoglobulin can be provided. Supportive care, such as adequate hydration and symptomatic control of fever or vomiting with antipyretics or antiemetics, is generally performed. In severe conditions, monitoring of extrahepatic complications and renal function support via hemodialysis may be required. For controlling pruritus, administration of ursodeoxycholic acid or cholestyramine® may be considered (Shin and Jeong 2018).

2.2.2.4 Hepatitis E Virus

Hepatitis E virus (HEV) is an emerging zoonotic virus. It is small and spherical with a single-stranded positive-sense RNA genome. It is a member of the genus *Orthohepevirus* of the family *Hepeviridae*. *Orthohepevirus*, is divided into 4 species, *Orthohepevirus* A–D, of which *Orthohepevirus*A includes HEVs infecting humans. *Orthohepevirus*A contains 8 genotypes, HEV1–8. Among these, genotypes 1 and 2 are specific to humans and these viruses often cause waterborne HEV outbreaks, while 3 and 4 are zoonotic genotypes and are the causative agents of foodborne HEV infections in industrialized countries (Todd and Grieg 2015).

The major transmission mode of HEV is the fecal-oral route. The virus is generally endemic in areas with poor sanitation, where the transmission of the virus occurs mainly through contaminated drinking water, resulting in major HEV outbreaks. Unlike other common foodborne viruses, HEV3 and HEV4 strains are zoonotic, in that animals are thought to be the primary reservoirs. These viruses can infect a wide variety of hosts including humans, pigs, wild boar, deer, primates, and rabbits. Hence transmission can also occur through direct contact with these reservoir animals or consumption of contaminated meat. Sporadic cases of HEV in developed areas with better sanitation and water supply are clearly linked with this mode of transmission. High-risk foods for HEV contamination include raw or undercooked meat of infected animals and filter-feeding bivalve shellfish. HEV transmission via blood transfusion is also possible.

Typically, HEV infection is self-limiting, asymptomatic, and resolves within weeks. But among certain groups of individuals, a serious disease (fulminant hepatitis) develops, and a proportion of people with this disease may die. Immunocompromised patients, such as those with HIV infection or haematological cancer, patients with preexisting liver disease, and pregnant woman are among the high-risk groups (Harrison and DiCaprio 2018).

Following the transmission via the fecal-oral route the HEV gains entry by binding to the host cell receptor through its capsid protein. It replicates in the intestinal tract before reaching the liver. HEV also replicates in the cytoplasm of hepatocytes and is released into both blood and bile. Virions are then transported with bile from liver tissue back to the intestine. The virus is finally shed into the stool. Since HEV is not cytopathic, the liver damage induced by HEV infection may be caused by immune-mediated events which involve cytotoxic T-cells and natural killer (NK) cells. HEV remains stable in alkaline and acidic environments and remains infectious up to 60°C (Lhomme et al. 2016).

The average incubation period for HEV is about 2 weeks to 2 months. During the first phase of 1–10 days, influenza-like symptoms, abdominal pain, tenderness, nausea, vomiting, and fever develops. Some persons may also have itching (without skin lesions), skin rash, or joint pain. In the second phase of acute infection symptoms include jaundice with black urine and pale stool, anorexia, hepatomegaly (enlarged tender liver), and myalgia. This is followed by viremia, liver enzyme elevations, antibody seroconversion, and clearing of the virus from the blood, while virus may persist in stool for longer. Rarely, acute hepatitis E can be severe, and may progress to chronic infection, particularly among immunosuppressed people, such as organ transplant recipients, occasionally resulting in fulminant hepatitis. Pregnant women with Hepatitis E virus, particularly those in the second or third trimester, are at increased risk of fulminant hepatitis, fetal loss, and mortality (Todd and Grieg 2015). Infection with virus can also affect other organs resulting in extra-hepatic manifestations including aplastic anemia, acute thyroiditis, glomerulonephritis as well as neurological disorders, such as Guillain-Barre syndrome, neuralgic amyotrophy, and encephalitis (Bosch et al. 2016).

Control of animal waste, runoff, and decontaminated sewage limits the spread of HEV to coastal and surface waters (Todd and Grieg 2015). HEV 239, a recombinant vaccine produced using a genetically modified strain of *E. coli*, has been available in China since 2012 and is approved for use in people over 16 years old. No other vaccine is available for the disease (Harrison and DiCaprio 2018).

2.2.2.5 Norovirus

Norovirus (similar to the Norwalk virus, a small round-structured virus) is a non-enveloped icosahedral RNA virus with a single-stranded positive-sense RNA genome, belonging to the genus *Norovirus* of the family *Calciviridae*. Of the ten genogroups of norovirus, genogroups I (GI), GII, GIV, GVIII, and GIX can infect cattle, mice, dogs, and bats, respectively (Imai et al. 2020). Among these, strains of GII.4 have become the most common cause of major disease outbreaks (Lu et al. 2020). They are also referred to as Norwalk-like virus, calicivirus or small-round structured virus and are the leading causes of gastroenteritis, sometimes referred to as "stomach flu." More than 50% of foodborne outbreaks are caused by them (Lu et al. 2020).

Noroviruses are highly transmissible and can spread via exposure to contaminated food or water sources (fecal-oral route), person-to-person contact, aerosolized vomitus particles, and fomites. Food handlers are often suspected as the main source of foodborne outbreaks. Leafy greens (such as lettuce), fresh fruits, and shellfish are some of the foods commonly associated with norovirus outbreaks. Persons of all age groups are susceptible to norovirus infection and secondary infections are common. The low infective dose (18 viral particles) together with the lack of long-term immunity in infected people makes noroviruses the frequent cause of outbreaks, occurring mainly in closed or semi-closed settings (Todd and Grieg 2015; Ryan et al. 2019).

Norovirus infection is characterized by damage to the microvilli in the small intestine. Specific histological changes, including broadening and blunting of the villi, shortening of the microvilli, enlarged and pale mitochondria, increased cytoplasmic vacuolization, crypt cell hyperplasia, and intercellular edema are seen in norovirus infections. Decreased activity of brush border enzymes results in mild steatorrhea and transient carbohydrate malabsorption. Also the severe disruption of the epithelial barrier leads to excessive leaking back of ions and water from the sub-epithelial capillaries to the intestinal lumen, causing leak-flux diarrhea (Karst 2010; Ryan et al. 2019). The infected person may excrete noroviruses up to 30–40 days and excretion may be prolonged in the case of immunosuppressed, HIV-infected, and transplant patients (Rico et al. 2020). The infectious human norovirus particles released from feces are highly resistant to drying and heating and can survive on produce, ham, mussels, and food contact surfaces for several days but get inactivated by ultraviolet light and high-pressure processing (Todd and Grieg 2015).

The onset of illness occurs after 12–48 hours of incubation, characterized by acute onset of diarrhea, vomiting, nausea, and abdominal cramps. The diarrheal stool is watery without blood, mucus, or leukocytes. Less than 50% of patients develop low-grade fever which typically resolves within 24 hours. Other major symptoms include myalgia, headache, and chills. Usually symptoms disappear within 2 days but may become severe among high-risk groups. Infants and young children may develop more severe gastroenteritis following infection, with symptoms lasting up to six weeks. Hospitalized patients, older adults, and the immunocompromised persons are at greater risk for severe symptoms and complications, such as acute renal failure leading to hemodialysis, cardiac complications including arrhythmias, acute graft organ rejection in transplant recipients, and even death. Norovirus infections have also been involved in causing neurologic disorders including benign infantile seizures and encephalopathy (Ryan et al. 2019).

For the diagnosis of noroviruses, stool samples are usually preferred. Noroviruses can be also detected from rectal swabs and vomitus. ELISA, RT-PCR, and immunochromatographic (ICG) lateral flow assays are the commonly used diagnostic techniques (Vinje 2015). For uncomplicated gastroenteritis, the first-line therapy involves the administration of oral rehydration solutions providing essential electrolytes and sugar. If there is severe dehydration, intravenous fluid replenishment is necessary. Antiemetics and antidiarrheal agents are given to the patients. Adjuvant therapy with zinc and early intake of a calorie-rich diet may help to reduce the duration and frequency of diarrhea as well as enhances patient recovery (Ryan et al. 2019). Proper handwashing with soap and water is an effective method for reducing the transmission of the norovirus pathogen. Cleaning and disinfection of contaminated surfaces using 0.1% sodium hypochlorite and safe, hygienic practices during food preparation, processing and storage, avoiding consumption of bivalve shellfish from contaminated water, and prohibiting infected food handlers from work are some of the measures adopted to reduce the risk of norovirus transmission (Barclay et al. 2014).

2.2.2.6 Poliovirus

Poliovirus (PV) is the causal agent of paralytic poliomyelitis, an acute disease of the central nervous system (CNS) resulting in flaccid paralysis (Blondel et al. 2005). It is a human enterovirus of the genus *Enterovirus* belonging to the family *Picornaviridae*. Polioviruses are non-enveloped viruses containing a single-stranded, positive-sense RNA protected by a capsid of icosahedral symmetry (Hatib et al. 2020). They are grouped under the species *Enterovirus C* and are divided into three serotypes, poliovirus (1, 2 and 3), among which type 1 is the most virulent and common (Fernandez Garcia et al. 2018). Poliovirus is stable and can survive for several months in the environment especially, in cold or ambient temperatures, in a humid environment or when adsorbed on a solid material (Hatib et al. 2020).

Humans are the only known natural hosts of poliovirus, although non-human primates can be experimentally infected. These ubiquitous viruses are transmitted mainly by person to person, and the infected individual can shed virus in their stool in copious amounts with a load of up to 10^6 infectious units per gram of stool. Food and drinking water serve as potential vehicles for transmission where hygiene standards are low. Earlier, milk was the principal source of foodborne polio, but this route of infection had been controlled by improvements in hygiene. Transmission can also occur via droplets or contaminated surfaces (Hatib et al. 2020). In humans, poliovirus infection usually begins with oral ingestion of the virus. After oral ingestion, the virus multiplies in the alimentary mucosa, tonsils, and Peyer's patches (in the intestine). The virus then moves into the blood stream (viremia) through the putative barrier. The circulating virus then invades the CNS either through the blood-brain barrier (BBB) or via peripheral nerves and replicates within the neurons, particularly motor neurons. The motor neuron destruction due to the lytic replication of poliovirus results in paralytic poliomyelitis. Paralysis usually develops in less than 1% of those infected (Nomoto 2007).

The poliovirus gains entry into host cells by its attachment to the receptor CD155, found only on the surface of primate cells. PV binding to CD155 destabilizes the virion, induces conformational modifications, and ultimately releases RNA into the cytoplasm of the infected cell. After RNA has been released, the translation process is initiated, which produces several structural and non-structural proteins responsible for the proteolytic activities, RNA synthesis, and biochemical and structural changes that occur in the infected cell (Blondel et al. 2005).

Clinical features vary widely, ranging from mild to severe and become apparent after 7–14 days of incubation. They can be categorized into inapparent infection without symptoms, mild illness (abortive poliomyelitis), aseptic meningitis (nonparalytic poliomyelitis), and paralytic poliomyelitis. The majority (95%) of exposed patients are asymptomatic but serve as carriers. Abortive poliomyelitis (4%–8%) develops as a mild viremia characterized by gastroenteritis, influenza-like illness, and mild respiratory tract infections, which usually subside within 1 week. Around 1% of the clinical cases present as aseptic meningitis with severe muscle spasm of the neck, back, and lower limbs and recovers within 10 days. The most severe form, paralytic poliomyelitis (less than 1% of patients), presents as extreme episodes of pain in the back and lower limbs. In children, asymmetrical paralysis of limbs appears. Flaccid paralysis is the major sequelae characterized by the eventual loss of muscle reflexes. Some patients may recover completely but, if the loss of motor functions persists beyond 12 months, lifelong disability develops. Among the three forms of paralytic poliomyelitis (spinal, bulbar, and bulbospinal), bulbar poliomyelitis has the maximum fatality as the brain stem neurons are involved. Death usually occurs due to respiratory paralysis. Post-polio syndrome (PPS) can also occur, 25–30 years after the initial paralytic attack (Mehndiratta et al. 2014).

There is no cure for polio; it can only be prevented. Vaccination is an important way to prevent poliovirus infection and can establish lifelong immunity to the disease. Today poliomyelitis has been virtually eradicated from more than 99% of the world with the successful use of highly effective live oral polio vaccine (OPV) and inactivated polio vaccine (IPV). Proper treatment of drinking water, effective sewage disposal system, proper thermal processing, and good hygiene practices during production and processing of food can also help to prevent poliovirus infection (Todd and Grieg 2015).

2.2.2.7 Rotavirus

Rotaviruses (RV) are members of the genus *Rotavirus* in the *Reoviridae* family, consisting of large, non-enveloped, icosahedral wheel-shaped, viruses with segmented, double-stranded RNA genomes, surrounded by a three-layered protein capsid. They are classified into seven RV groups (A–G), among which group A causes the most human disease (Greenberg and Estes 2009).

Transmission of disease is by the fecal-oral route, through person-to-person contact, or by fomites. Less commonly, the virus gets transmitted by consuming contaminated water or food (Owino and Chu 2019). Rotaviruses are major enteric pathogens causing life-threatening gastroenteritis in infants and children worldwide, imposing a major impact on childhood morbidity and mortality. More than one-third of child deaths from diarrhea is the result of rotavirus infection (Troeger et al. 2018). The virus primarily infects mammals. Although human rotaviruses are spread mainly through the fecal-oral route, food associated infections can occur. Sandwiches, salads, conch, and deli fare are some of the foods which were found to be contaminated with human rotavirus (Todd and Grieg 2015).

Rotavirus infection can either be symptomatic or asymptomatic. Neonates rarely show symptomatic disease since they are protected by maternal antibodies. Nearly every child in the world under the age of five is at risk of infection. The immunity develops after each infection makes the subsequent occurrence less severe. Adults are less susceptible, but severe symptomatic disease may result from infections with an unusual virus strain or extremely high doses of virus (Greenberg and Estes 2009). The infected person sheds high concentrations of rotavirus in their stool (10^8–10^{11} particles per gram of feces) for about 4 days (longer for immunocompromised patients [>30 days]). The virus is stable and may persist viably in the environment for long periods if not disinfected (Owino and Chu 2019).

Poor sanitation, unsafe water, and childhood undernourishment are the major risk factors associated with diarrheal diseases (Troeger et al. 2018). Rotaviruses infect the duodenal mucosa and replicate in the mature enterocytes. The infection causes damage to mucosa that leads to malabsorption of water and electrolytes from the lumen, contributing to osmotic diarrhea, which is followed by dehydration. Rotavirus non-structural protein NSP4, a viral enterotoxin, recently recognized as a viroporin also plays an important role in the induction of diarrhea. It disrupts the tight junctions, allowing paracellular flow of water and electrolytes. It also induces Ca^{2+} ions to be released from internal stores; increases in intracellular calcium concentration trigger a cascade of events that finally results in cell lysis. Rarely, rotavirus can disseminate into extraintestinal sites, including the respiratory tract, liver, kidney, lymph nodes, and CNS (Ramig 2004; Ryan et al. 2019).

Following an incubation period of 1–3 days, the illness often begins shortly after and is characterized by the abrupt onset of fever and vomiting followed by watery diarrhea. Dehydration is the major complication that appears shortly after the onset of disease and may sometimes become severe leading to metabolic disturbances and shock that ultimately causes death. Infection is more severe in malnourished and immunocompromised persons but less severe among adults. The gastrointestinal symptoms usually get resolved in 3–7 days after the first onset of the illness (Owino and Chu 2019).

Diagnosis of rotavirus infection involves laboratory detection of fecal samples. Electron microscopy, polyacrylamide gel electrophoresis (PAGE), immunoassays, RT-PCR, virus isolation, and other advanced detection methods are used for the detection.

First-line therapy for acute rotavirus gastroenteritis involves rehydration and restoration of electrolyte balance. Children suffering from acute diarrhea should be supported with an adequate protein–calorie-rich diet. Zinc supplementation and the use of probiotics may help to decrease the diarrheal duration. No antiviral agents are currently available. Good hygiene practices, such as handwashing, safe disposal of feces, and disinfection of contaminated surfaces, are important in reducing the risk of transmission. Vaccination is one of the best approaches to prevent severe rotavirus gastroenteritis and hospitalization. Two live attenuated oral rotavirus vaccines, Rotateq® and Rotarix® are licensed globally, have been incorporated into childhood immunization schedules in 87 countries, and have demonstrated good efficacy in reducing disease by rotavirus (Ryan et al. 2019).

2.2.3 Protozoa

2.2.3.1 Cryptosporidium parvum

Cryptosporidium parvum is one of the most common parasites in soil, water, and food causing cryptosporidiosis in humans. After the rotavirus, this pathogen is the main cause of diarrheal death in children of less than five years of age (Khalil et al. 2018). According to CDC analysis, *Cryptosporidium* outbreak in the US increased by approximately 13% each year from 2009 to 2017. The transmission of disease occurs primarily through the fecal-oral route, i.e., when food of animal origin (such as milk and meat) and fresh vegetables get contaminated by *Cryptosporidium* oocysts, due to poor hygiene practices during food production and processing (Gerace et al. 2019). Travelers to and residents of developing and underdeveloped countries face greater challenges due to the lack of proper food safety measures and sanitation (Hatam-Nahavandi et al. 2015). Cryptosporidiosis generally causes self-limiting diarrhea, but the infection is serious and might be life-threatening in immunocompromised patients (Marcos and Gotuzzo 2013). Cryptosporidiosis in humans may cause symptoms such as acute gastroenteritis, persistent diarrhea, and abdominal pain followed by influenza-like illness and fever. The infection is severe in immunocompromised individuals, particularly, causing acute diarrhea and malabsorption in AIDS patients (Wang et al. 2018). Apple cider, meat, milk, seafood, and raw vegetables are identified as the major sources of infection (Vanathy et al. 2017) and the incubation period of infection ranges from 2 to 4 days.

The complex lifecycle of pathogen is initiated by the ingestion of oocysts, which are hatched inside the intestinal wall to form four infectious sporozoites. The sporozoites get attached to the epithelial cells of the intestine and undergo asexual reproduction (schizogony or merogony) to form eight merozoites (Jenkins et al. 2010). The infection is facilitated by merozoites in neighboring epithelial cells and undergoes asexual and/or sexual reproduction. Type I meront is produced by asexual reproduction while the type II meront, produced in a sexual stage, is differentiated to form micro and macrogametes and fused to form zygote. The life cycle of *Cryptosporidium* is continued by the production of four sporozoites (sporogony) from the diploid zygote. The sporozoites are protected with thick- or thin-walled oocysts and are excreted out with feces (Bouzid et al. 2013).

The diagnosis of cryptosporidiosis can be done microscopically by the detection of oocysts in a stool sample (Khurana Chaudhary2018). Due to the accurate sensitivity and specificity, enzyme-linked immunosorbent assay (ELISA) and immunochromatographic tests are more useful tools than the microscopical examination of watery or mushy stools (Hawash 2014). Commercial kits are also available for the identification of disease. The standard approach for *Cryptosporidium* diagnosis is by polymerase chain reaction (PCR) which is preferable to microscopy, ELISA, and immunochromatographic tests due to its low cost, shorter process time, and high sensitivity and specificity (Friesen et al. 2018).

US FDA approved the use of nitazoxanide for the treatment of cryptosporidiosis in children and immunocompetent patients in 2006 (Carey et al. 2004), but it is ineffective for the treatment of AIDS patients. The best methods used for the prevention of *C. parvum* contamination are the pasteurization of milk, proper sanitary disposal of excreta and sewage, appropriate wastewater treatments, and good hygiene practices.

2.2.3.2 Cyclospora cayetanensis

Cyclospora cayetanensis is a newly reported parasite that caused two nationwide food-borne outbreaks of cyclosporiasis in the United States and Canada (1996–1997) from contaminated berries. Recently, *Cyclospora* infections were also reported in the Midwest and New York in 2018 which occurred through the consumption of broccoli, cauliflower, carrots and some fast foods. Out of the 22 known *Cyclospora* species, *C. cayetanensis* is the only species identified to infect humans (Ortega and Sanchez 2010; Li et al. 2019). The primary site of infection is the small intestine and the symptoms include watery diarrhea, abdominal pain, nausea, anorexia, constipation, and fatigue (Ortega and Sanchez 2010). Cyclosporiasis has also been reported from immunocompromised patients with severe diarrhea (Li et al. 2019). The infection spreads through the ingestion of sporulated oocysts that are found in soil, through the contaminated food which includes raspberries, basil, cilantro, romaine lettuce, snow peas, etc. The incubation period of the infection ranges from 2 to 11 days (Almeria et al. 2019).

The lifecycle of *C. cayetanensis* has not been fully described and the oocyst is the only definitively identified stage. Humans are infected by the ingestion of oocyst-contaminated food and water. The oocyst excysts in the gastrointestinal tract, freeing the sporozoites, which invade the epithelial cells of jejunum and ileum of the small intestine. Inside the epithelial cells, sporozoites transform into trophozoites which subsequently form two types of meronts (schizonts) asexually. The type I schizonts are with 8–12 small merozoites and type II schizonts have 4 long merozoites. Type II merozoites produced from type II schizont develops in to microgametes and macrogametes. The gametes fuse to form a zygote, which later develops into an unsporulated oocyst and is excreted with feces into the soil, where sporulation occurs (Almeria et al. 2019).

Generally, young children, foreigners, and immunocompromised individuals are at higher risk of infection. Sometimes the disease is followed by autoimmune disorders such as Guillain–Barre syndrome (GBS) and reactive arthritis syndrome (Reiter's syndrome) (Richardson et al. 1998). The diagnosis is usually detection of oocysts by conventional bright field microscopy, phase contrast microscopy, and epifluorescence microscopy. Several molecular methods have also been developed for the identification of cyclosporiasis such as real-time PCR (qPCR) (Varma et al. 2003), multiplex PCR (Shapiro et al. 2019; Temesgen et al. 2019), and MLST (Guo et al. 2016).

Treatment of the infection includes the use of trimethoprim-sulfamethoxazole which has proven more effective than ciprofloxacin, which was used in the past (Goldberg and Bishara 2012). Heating and pasteurization of food and milk products before consumption and proper food-handling practices during food processing are necessary to prevent the illness.

2.2.3.3 Entamoeba histolytica

Entamoeba histolytica is an invasive parasite causing amoebiasis or amoebic dysentery. The rate of infection is highest in India, Africa, and Central and North America, especially in Mexico (Prakash and Bhimji 2017). The parasite resides primarily in the intestinal epithelium (Cornick and Chadee 2017) and the disease is asymptomatic in most cases (Watnabe and Petri 2015). Due to the attachment of trophozoites in the intestinal mucosa, patients show symptoms such as severe bloody diarrhea (amoebic colitis), liver abscess, abdominal pain, fever, flatulence, vomiting, and weight loss. The incubation period ranges from 2 to 4 weeks.

Intestinal amoebiasis is a common type of asymptomatic *E. histolytica* infection, while amoebic colitis develops specific symptoms which include mild diarrhea to severe dysentery, abdominal pain, anorexia, weight loss, and chronic fatigue (Haque et al. 2003). Improper treatment or diagnosis may lead to serious complications such as fatal amoebic enterocolitis, toxic enlargement of colon and fistulizing perianal ulcerations (Gardiner et al. 2015).

E. histolytica is also associated with another prevailing infection, amoebic liver abscess (Prakash and Bhimji 2017). The mortality rate from liver abscess is 89% and from necrotizing colitis is 40% (Ortiz-Castillo et al. 2012). In a report Wuerz et al. (2012) have suggested that around 50%–80% of patients develop symptoms such as fever and constant pain on the right upper quadrant and in almost 50% of patients, symptoms such as severe diarrhea, abdominal pain, and weight loss are experienced. Pulmonary amoebiasis is the second most common infection caused by the spread of amoebic liver abscess through the lymphatic system (Ackers and Mirelman 2006). It affects the right lower or middle lobe of the lungs and shows symptoms such as fever, hemoptysis, right upper quadrant pain, and pain in the right shoulder.

The transmission of the parasite is via the fecal-oral route by the ingestion of matured cysts. In the small intestine, trophozoites (12–60 μm in diameter) are formed by excystation, which migrate to the large intestine. The trophozoites multiply by binary fission and produce new cysts which cause infection, and both stages are passed in the feces (Haque et al. 2003). The cysts form can exist longer in the external environment (at least 8 days) compared to the trophozoite stage that is readily destroyed when excreted from the body. Trophozoites invade the colonic epithelium and migrate through the blood to different sites—in particular in the liver, resulting in tissue damage, ulceration, and, ultimately, liver abscess (Wuerz et al. 2012; Prakash and Bhimji 2017).

Different diagnostic methods of *E. histolytica* include microscopical and serological examination, stool antigen detection, molecular techniques like PCR, and colonoscopy. The best way to diagnose *E. histolytica* infection is by the combination of serology and PCR or stool antigen detection (Cheepsattayakorn and Cheepsattayakorn 2014). PCR is a quick and accurate method used to differentiate *E. histolytica* from *E. despars*, which is a non-invasive parasite. ELISA and radioimmunoassay or immunofluorescence methods are available tests for stool antigen detection; the indirect hemagglutination (IHA) test is more sensitive to detect serum antigen (Guerrant et al. 2011). While in colonoscopy, direct visualization of the colon has been employed for the detection of amoebiasis with non-specific gastrointestinal symptoms (Horiki et al. 2015).

Metronidazole, a drug that is active against the trophozoite stage of the parasite, is used for the treatment of amoebic dysentery and liver abscess, whereas Paromomycin and Diloxanide furoate are the drugs recommended for the treatment of asymptomatic intestinal amoebiasis. Proper sanitation facilities and good hygiene practices, thorough handwashing, and proper washing of fruits and vegetables before cooking are important preventive measures of amoebiasis.

2.2.3.4 Giardia lamblia

The human intestinal parasite *Giardia lamblia* causes giardiasis and was first discovered by Antonie Van Leeuwenhoek in 1681. According to a WHO report, *G. lamblia* is the third most common cause of diarrhea and it affects children more than adults (Lanata et al. 2013). The disease occurs predominantly in developing countries, and about 2%–3% in developed countries (Feng and Xiao 2011). The pathogen is commonly transmitted through the fecal-oral route by the ingestion of contaminated food and water infected with giardia cyst (Hooshyar et al. 2017). The incubation period of the infection ranges from 4 to 25 days. The disease is asymptomatic in some cases, while symptomatic giardiasis produces symptoms such as acute watery diarrhea, nausea, epigastric pain, abdominal cramps, fatigue, anorexia, nausea, and weight loss (Feng and Xiao 2011; Ryan and Caccio 2013).

The life cycle of *G. lamblia* consists of two stages—vegetative trophozoite and infectious cysts forms. There are two intermediate stages—encyzoites and excyzoites (Ankarklev et al. 2010). The infectious cyst is ingested by humans through contaminated food and water, and it undergoes the excystation stage triggered by the acidic environment of the stomach. The excyzoites undergo divisions to release four trophozoites into the small intestine (Bernander et al. 2001). Later, the trophozoites undergo encystation and develop into mature cyst forms (Barash et al. 2017). Cysts are released with the fecal matter and can survive for long periods outside the host body.

The diagnosis of *Giardia* can be made by microscopic examination of the cyst in stool sample. The sensitivity of the method varies according to the use of direct or concentration methods, number of fecal samples, and trained professionals (Gutiérrez-Cisneros et al. 2011; Soares and Tasca 2016). Immunoassays such as ELISA or immunofluorescence techniques are used for the detection of antibodies in fecal samples (Heyworth 2014); the String test (Entero test) detects the presence of trophozoites in fluid obtained from duodeno-jejunal intubation; while counter immunoelectrophoresis (CIE), ELISA, and rapid antigen detection tests (RDTs) are used to detect *Giardia* fecal antigens (Heyworth 2014). Several immunoassay kits are available for the detection of *Giardia* as well as *Cryptosporidium* and *Entamoeba* species antigens in fecal samples (Soares and Tasca 2016). According to studies, the specificity and sensitivity of molecular methods, such as nested PCR, in the identification of giardiasis is 82.9% and 89.9%, respectively (Hijjawi et al. 2018).

Chemotherapeutic drugs used for the treatment of infection are metronidazole, benzimidazoles (BIs) derivatives, quinacrine, furazolidone, paromomycin, and nitazoxanide (Lalle 2010). Preventive measures of giardiasis are similar to that of amoebiasis.

2.2.3.5 Sarcocystis

Sarcocystis species belong to a group of coccidian parasites and are the causative agent of the zoonotic infection called sarcocystosis. The two main species of *Sarcocystis* which cause human infections are *S. hominis* and *S. suihominis*. Sarcocystosis is transmitted by the ingestion of beef contaminated with matured cysts of *S. hominis* and raw pork contaminated with cysts of *S. suihominis* (Djurković-Djaković et al. 2013; Fayer et al. 2015). The disease is predominantly found in areas near livestock farming in tropical and subtropical countries (Djurković-Djaković et al. 2013). People get infected by muscular sarcocystosis through the ingestion of sporocysts-containing food items, such as beef and pork (Fayer et al. 2015). Several cases of acute muscular sarcocystosis were reported in Malaysia in 2011–2012 (AbuBakar et al. 2013). David et al. (2013) observed that the disease has appeared to be asymptomatic or symptomatic with nausea, vomiting, severe enteritis, stomach pains, and diarrhea within 3–6 hours after the consumption of infected beef, and continued for 48 hours. Mild stomach pain and diarrhea were also shown after the ingestion of infected pork which prevailed for 2 to 3 weeks until sporocysts were shed in feces.

The life cycle of *Sarcocystis* is completed in two hosts, namely definitive host and intermediate host. The definitive host for both *S. hominis* and *S. suihominis* is man, while the intermediate hosts are cattle and pigs. The infection of intermediate hosts occurs through the ingestion of sporocyst-contaminated food. Sporozoites released by the rupture of sporocysts in the intestine enter into the endothelial cells of the blood vessel and undergo asexual reproduction or schizogony, producing first-generation merozoites. The first generation merozoites produced from schizogony invade small capillaries and blood vessels and again reproduce asexually, producing second-generation merozoites. The second-generation merozoites invade muscle cells and develop into sarcocysts containing bradyzoites, which are the infective stage for the definitive host. Humans get infected when they eat undercooked meat containing sarcocysts. The bradyzoites released by the rupture of sarcocysts in the small intestine invade the intestinal epithelium and differentiate into microgametes and macrogametes. The oocysts formed by the fusion of gametes sporulate and the sporulated oocysts are excreted from the intestine through the feces.

Sarcocystosis is diagnosed by the identification of cysts using microscopic examination of stool (Fayer et al. 2015). *S. suihominis* in pork is detected by direct examination of a macroscopic cyst (Djurković-Djaković et al. 2013). Dot-ELISA is preferred for the serodiagnosis of *Sarcocystis* in cattle due to its high sensitivity than double immunodiffusion and counter immunoelectrophoresis (CIE) (Singh et al. 2004). Molecular methods, such as PCR, are used for species-specific identification of cysts in infected pigs and cattle (Gonzalez et al. 2006).

The best way to control the life cycle of *Sarcocystis* is the prevention of consumption of uncooked meat by the definitive host and fecal contamination of the feed and water for livestock (Bhatia et al. 2010). Anticoccidials, such as amprolium and salinomycin, are used as chemotherapeutic drugs. To some extent, albendazole is efficacious in the treatment of muscular sarcocystosis. Adjuvant treatment with corticosteroids and nonsteroidal anti-inflammatory medications may help to ameliorate symptoms.

2.2.3.6 Toxoplasma gondii

Toxoplasmosis is the infection caused by the coccidian parasite *Toxoplasma gondii*. The infection results from the ingestion of sporulated oocysts in contaminated food and water or by the intake of raw or undercooked meat or pork-carrying latent bradyzoite cysts. It is also transmitted from mother to fetus (congenital toxoplasmosis) or during organ transplantation via infected allograft (Dunay et al. 2018). In Brazil, an outbreak of toxoplasmosis through the consumption of fruits contaminated with *T. gondii* was reported in 2013 (Morais et al. 2016). Toxoplasmosis has been reported in the environment in about 350 wild host species of mammals and birds (Lindsay and Dubey 2020). The environment gets contaminated by the excretion of oocysts by felids, mostly stray or domestic cats. Infection is asymptomatic in most cases, and the incubation period is 5–23 days. In acute infections, lymphadenopathy and lymphocytosis persist for longer periods. Congenital toxoplasmosis may lead to abortion or stillbirth, chorioretinitis, and brain damage. Toxoplasmosis in immunocompromised individuals results in cerebritis (commonly in AIDS patients), chorioretinitis, pneumonia, myocarditis, and, sometimes, it leads to death.

The sexual phase of the life cycle is initiated by the ingestion of oocysts or tissues that are infected with bradyzoite cysts by members of the feline family. Upon completion of the sexual phase in the cat intestine, oocysts are released in the cat's feces. The oocyst sporulates in the environment and the sporulated oocyst, which is tolerant to external conditions for longer periods, actively transmits the infection. The asexual phase of the life cycle takes place in an intermediate host (any warm-blooded animal). The oocysts are ingested by the intermediate host and, in the intestine, sporozoites are liberated by the breaking of cysts. The sporozoites invade the intestinal epithelium and differentiate into tachyzoite. Then tachyzoite invade muscle and neural tissue and develop into bradyzoites in tissue cysts. The infection is transferred to a definitive host (cats) either by the consumption of an intermediate host or directly by the ingestion of sporulated oocysts. Tissue cysts may develop in domestic animals like pigs and sheep after ingestion of sporulated cysts from the environment. Humans get infected by consuming undercooked meat of animals containing tissue cysts, ingestion of food contaminated with sporulated cysts, and during blood transfusion and organ transplantation. Transplacental transfer from mother to fetus is also reported. In humans, cysts exist as bradyzoites inside the skeletal muscles, myocardium, brain, and eyes.

Serology is preferred for the diagnosis in immunocompetent patients. In the case of congenital toxoplasmosis, molecular methods like PCR are performed on DNA taken from the amniotic fluid of pregnant women. Stained biopsy specimens also help to detect tissue cysts. Recently, diagnosis has also been made by the detection of sporozoite-specific antigen from the sample (Hill et al. 2011). Combinations of inhibitory drugs, such as pyrimethamine and trimethoprim (dihydrofolate reductase) sulfadiazine, sulfamethoxazole, and sulfadoxine (dihyopteroate synthetase), which inhibit the folic acid synthesis, are currently used for effective treatment (Dunay et al. 2018). Cooking the meat properly at an appropriately high temperature is necessary to kill the infectious pathogen. It is found that the pathogen survives at –10°C for 106 days and at 35°C and 40°C for 32 days and 9 days, respectively. But when vegetables and meat are cooked at 55–60°C for 1–2 min, the pathogen is destroyed (Dubey 2010). It is very important to maintain personal hygiene, with proper handwashing, particularly after contact with raw meat, soil, or cats. Also, fruits and vegetables should be washed carefully before use.

2.2.4 Helminths

2.2.4.1 Trematodes

2.2.4.1.1 Clonorchis sinensis *Clonorchis sinensis* known as Chinese liver fluke is commonly found in the regions of South Korea, China, northern Vietnam, and far-eastern Russia (Qian et al. 2016). It is categorized as group I biocarcinogen by the International Agency for Research on Cancer (IARC), subpart of WHO (Bouvard et al. 2009). In humans, infection appears through the ingestion of encysted metacercariae (late infective larval stage) from fresh water fishes. Infection rate is higher in those who consume raw or uncooked meat or fish or those who regularly eat at restaurants. The risk group of infection includes females, people of 40–60 years of age, fishermen, farmers, and travelers, who increase the transmission risk to other countries (Qian et al. 2016). The incubation period varies based on the number of worms present inside the body, and symptoms include jaundice, indigestion, biliary inflammation, bile duct obstruction, liver cirrhosis, cholangiocarcinoma (CCA), and hepatic carcinoma (Hong and Fang 2012). Certain non-specific symptoms appear as fatigue, nausea, jaundice, and hepatosplenomegaly (Qian et al. 2013). The acute phase prevails for at least a month with symptoms such as fever, diarrhea, pain in the upper abdomen, pain and enlargement of the liver, anorexia, and leucocytosis; eosinophilia and jaundice are also seen (Diemert 2017).

The life cycle of *C. sinensis* consists of three hosts: first intermediate host such as freshwater snails; second intermediate host, like freshwater fishes or shrimp; and the definitive host viz, humans or cats, dogs, or other carnivorous mammals. In the definitive host and in water, it is an egg, while in freshwater snails, four stages are seen which are miracidium, sporocyst, redia, and cercaria. Metacercaria form is present in the intermediate host and as adult worms in definitive hosts (Petney et al. 2013). The infection is transmitted to the definitive host via the consumption of metacercariae from raw or undercooked fish. In the duodenum, metacercariae undergo excystation, and later, in bile ducts, they develop into adult worms (Hong and Fang 2012). They exist in bile ducts as adult worms for an extended period of time and cause clonorchiasis in humans.

The reproductive cycle of *C. sinensis* has both sexual and asexual stages (Qian et al. 2016). The adult worms produce embryonated eggs which are excreted via fecal matter by the definitive host. In freshwater snails (intermediate host), eggs are divided to form miracidia, and undergo different stages of asexual development to form sporocysts, rediae, and cercariae which are finally released into the water. After some time, cercariae come in contact with and penetrate the flesh of freshwater fishes (second intermediate host) and develop into encysted metacercariae in the muscles and subcutaneous tissues of the host. When humans or other definitive hosts consume half-cooked, salted, pickled, or dried infected fish, excysted metacercariae in the duodenum invade the intrahepatic bile duct and develop into infectious adult flukes (Kim et al. 2011).

Diagnosis is commonly done by stool examination to detect the presence of embryonated eggs in the sample (Hong and Fang 2012), using techniques such as direct fecal smear, the Kato-Katz (KK) method, and the formalin-ether concentration technique (FECT). KK method is reported to be more reliable for the diagnosis of clonorchiasis even though FECT is more sensitive to diagnose mild infections (Qian et al. 2013). Serological methods are also used for the diagnosis of clonorchiasis. Specific antibody detection is preferred to the detection of an antigen due to the low presence of antigen. An egg yolk immunoglobulin (IgY)-based immunomagnetic bead enzyme-linked immunosorbent assay system (IgY-IMB-ELISA) is a sensitive and specific method to identify circulating antigen in human clonorchiasis (Nie et al. 2014). Various PCR techniques such as conventional PCR, real-time PCR (qPCR), multiplex PCR, restriction fragment length polymorphism (RFLP)-PCR, and fluorescence resonance energy transfer (FRET)-PCR can also be used for diagnosis. LAMP is another sensitive and rapid method useful in the diagnosis of *C. sinensis*.

Clonorchiasis can be treated effectively with a praziquantel (PZQ) drug (Choi et al. 2004). Recent studies have asserted that tribendimidine is more effective than praziquantel for the treatment of *C. sinensis* infection (Xu et al. 2014). It is important to control infection in the reservoir host by feeding pets with thoroughly cooked food and disposing of pet feces properly. Construction of toilets and pigsties at appropriate distances away from fresh water bodies and avoiding the use of raw freshwater fish/shrimp are some other ways to reduce infection by *C. sinensis* (Tang et al. 2016). Clean water for fish aquiculture and an anti-clonorchiasis fish vaccine are recommended preventive measures.

2.2.4.1.2 Fasciola hepatica

Fascioliasis is a trematode infection caused by the liver fluke, *Fasciola hepatica* and giant liver fluke, *F. gigantica*. Human fascioliasis is categorized as a food or plant trematode zoonosis under neglected tropical disease (NTD) (Mas-Coma et al. 2014). The infection leads to cholangitis and biliary obstruction (Chatterjee 1969). It appears as asymptomatic or symptomatic with gastrointestinal symptoms, such as cough, bronchial asthma, chronic cholecystitis, and liver abscesses followed by biliary colic, epigastric pain, and jaundice (Arora and Arora 2010). Acute infection-associated symptoms in children are upper right quadrant pain, abdominal pain, fever, and anemia. Humans get infected by drinking water infected with viable metacercariae of trematodes (Chatterjee 1969) or by eating raw vegetables, such as water cress (*Nasturtium officinale*) with encysted metacercaria on its leaves (Mas-Coma et al. 2014). The incubation period is 4–6 weeks.

In the life cycle of *F. hepatica*, cattle, sheep, buffalo, horses, camels, and humans can act as definitive hosts; the developed ovulated eggs are passed out in stool into fresh-water bodies (Fentie et al. 2013). The egg hatches in water into a ciliated miracidium larva, which enters the snail *Lymnaea*, the intermediate host. The part of the life cycle in the snail is completed in 30–60 days and, during this period, miracidium passes through three different developmental stages—namely, sporocyst, redia, and cercaria. The mature cercaria comes out of the snail body and gets encysted as metacercaria on aquatic vegetation and other surfaces. Humans and other definitive hosts are infected while drinking contaminated water or by ingesting infected grasses or leafy vegetables. Metacercariae undergo excystation in the duodenum, penetrate the intestinal wall, and ends up at the peritoneal cavity. The immature flukes then migrate through the liver parenchyma into bile ducts, where they mature into adult flukes. After 3–4 months of infection period, eggs are produced, which are excreted in feces, and the lifecycle is continued (Sah et al. 2017).

The most common diagnostic method involves the detection of eggs in stool samples by light-microscopy (Mas-Coma et al. 2014). However, antibody detection is preferred as *Fasciola* eggs are produced only after 3–4 months of infection. Antigen tests, serology, and ELISA tests (Santana et al. 2013) are some of the improved diagnostic approaches for the detection of pathogen. In serology, immunoblot assay is used to identify IgG antibody which is complementary to FhSAP2 (recombinant antigen) derived from *F. hepatica*. Other types of testing include the use of radiological and abdominal imaging using ultrasonography, computed axial tomography (CAT scan), and specific molecular techniques (Prichard et al. 2012).

Triclabendazole® (TCBZ) is a recommended drug, effective against different stages of trematode (Mas-Coma et al. 2014). Bithionol administered orally can also be used for the successful treatment of fascioliasis (Diemert 2017). It is best to avoid the consumption of raw aquatic vegetation infected with metacercariae of *F. hepatica*. The control of animal access to commercial watercress is another way to prevent the transmission of disease.

2.2.4.1.3 Fasciolopsis buski

Fasciolopsis buski, a giant intestinal fluke that causes fasciolopsiasis is predominantly found in China, East Asian countries, and in the southwest Pacific regions. The trematode is transmitted to humans through the ingestion of aquatic vegetation like bamboo shoots, watercress, and chestnuts infected with metacercariae. After ingestion, metacercariae excyst in the duodenum, get attached to the intestinal mucosa, and develop into adult worms within 3 months (Chai et al. 2005). The fluke attachment to the intestinal mucosa results in conditions such as local inflammation and ulcerations followed by hemorrhage (Diemert 2017). The symptoms associated with infections are diarrhea, constipation, epigastric pain, nausea, and vomiting. Intestinal obstruction, leucocytosis, eosinophilia, and edema due to malabsorption may also occur. The majority of cases are asymptomatic in nature.

In the lifecycle of *F. buski*, the metacercariae in the duodenum and jejunum get attached to the mucosal epithelium and develop into adult flukes. The adult worm is a flat, fleshy, ovate trematode with 2–7.5 cm long by 0.8–2 cm wide. About 25,000 immature eggs are produced and excreted in the feces and are deposited in freshwater sources. Eggs are ellipsoid, yellow to brown in color, with an operculum at one end. In fresh water, the miracidium larva hatches out from the embryonated egg, and it invades the intermediate host snail (genus *Segmentina* or *Hippeutis*). Then the miracidia develops into different stages, such as sporocysts, rediae, and cercariae after 4–7 weeks of penetration in the intermediate host. The cercariae undergo encystation to form metacercariae on aquatic plants. Humans get infected after ingestion of metacercariae from aquatic vegetation, such as chestnuts and watercress (Bogitsh et al. 2018).

Fecal examination is commonly used for the identification of *F. buski* eggs (O'Dempsey 2010). It rarely involves the detection of adult flukes in the stool or vomitus. Direct smear examination, Stoll's dilution, formalin ethyl acetate centrifugation, and Kato-Katz technique (KKT) are other diagnostic approaches (Sripa et al. 2010).

Praziquantel is the drug administered for effective treatment, as it inhibits the lifecycle of the fluke (WHO 2009). It is necessary to use thoroughly cooked vegetables instead of fresh vegetables to feed pigs or for human intake to reduce the transmission of metacercariae, which is sensitive to heat. Also, prevention of food contamination and proper sanitation facilities helps to inhibit the fluke lifecycle.

2.2.4.1.4 Opisthorchis viverrini *Opisthorchis viverrini* (Southeast Asian liver fluke) and *Opisthorchis felineus* (cat liver fluke), considered deadly, are parasitic trematodes found in tropical regions. *O. felineus* usually infect cats and dogs in Central and Eastern Europe and Siberia (Diemert 2017). *O. viverrini* is prevalent in northeast Thailand and Laos. It is transmitted from fermented uncooked fish and cyprinid fish dishes (Pla som) that are infected with metacercariae (Onsurathum et al. 2016). Infection is usually asymptomatic and in symptomatic cases, indications like right upper quadrant abdominal pain, flatulence, fatigue, urticaria, and mild hepatomegaly are seen (Mairiang et al. 2003). In severe cases, fluke will be attached to bile ducts of the liver, extrahepatic ducts, and the gall bladder causing epithelial hyperplasia, biliary obstruction leading to cholangitis, obstructive jaundice, and cholecystitis. The most prevalent liver cancer in Thailand is known to be Cholangiocarcinoma (CCA), which is the fifth most common cause of death in males and eighth in females (Bundhamcharoen et al. 2011). CCA is considered to have association with *O. verrini* and, rarely, *O. felineus*.

Both *O. viverrini* and *O. felineus* have similar life cycle as that of *C. sinensis*. The life cycle of *O. viverrini* begins when eggs are excreted in the feces of the definitive hosts; humans, cats, and, rarely, dogs (Petney et al. 2013). These are then ingested by freshwater snails (first intermediate host) where eggs undergo asexual reproduction to form cercariae. Three known taxa of snails that act as intermediate hosts are *Bithynia funiculata*, *B. siamensis*, and *B. goniomphalos* (Kiatsopit et al. 2013). The cercariae further develop into metacercariae in the second intermediate host, cyprinid fish. Humans become sick when they consume uncooked infected fish. In countries where opisthorchiasis is common, eating raw or fermented fish is a common practice (Grundy-Warr et al. 2012). In humans, the metacercariae mature to form adult worms which invade the biliary duct and, later attach to the epithelium to cause infection. About 50–200 eggs/feces/day are produced for many years by the adult worms (Khuntikeo et al. 2018).

Direct examination of the fecal sample to detect the presence of *O. viverrini* eggs is one of the traditional methods used for the diagnosis. But low infection intensities, irregular egg excretion, and cholangiocarcinoma are some of the challenges faced by this method. Modified formalin ether concentration technique, modified thick Kato smear, and Stoll's dilution egg count techniques are successfully used for diagnosis. The LAMP technique is used for the identification of *Opisthorchis*, which offers higher sensitivity (Phung et al. 2014), though it is easy to distinguish *O. viverrini* from *O. felineus* using species-specific PCR. The intradermal test, immunoelectrophoresis, indirect hemagglutination assay, indirect fluorescent antibody test, and indirect ELISA are some of the serological tests used for the detection of liver fluke (Johansen et al. 2010). Indirect ELISA, an antibody-antigen detection assay is performed using antigens such as somatic extracts of adult worms (Hong and Fang 2012) and excretory-secretory antigens. The use of recombinant antigens for the diagnosis of eggs and worms has helped to reduce the cross reactivity of proteins in parasites and for improved performance (Li et al. 2012).

Praziquantel has been used to treat the disease. The best method for the prevention is to stop the consumption of raw or undercooked fish; measures should be taken to control the transmission of liver fluke from human feces to snail by improving public sanitation systems. It is important to promote public awareness to the risk groups about food safety management, which is the primary method to control opisthorchiasis infection. But it is difficult in areas where the consumption of raw fish is related to traditional practices (Grundy-Warr et al. 2012).

2.2.4.1.5 Paragonimus westermani *Paragonimus westermani*, lung fluke, causes foodborne parasitic infection paragonimiasis in humans and mammals. It is one of the common human parasites in many parts of Asia, Africa, and South America (Procop 2009). Infection occurs in humans through the consumption of raw or insufficiently cooked crustaceans and the raw meat of wild boar or deer with encysted metacercariae (Yoshida et al. 2019). At the early stages, infection appears to be asymptomatic. In severe infections, it typically produces symptoms similar to tuberculosis and causes fever, fatigue, generalized myalgia, and abdominal pain. Pleuropulmonary paragonimiasis induces sub-acute to chronic inflammatory disease of the lungs with symptoms such as chronic coughing, thoracic pain, blood-stained sputum, dyspnea and hemoptysis. Cerebral hemorrhage, edema or meningitis is common in ectopic paragonimiasis, and abdominal paragonimiasis, as the name indicates, causes abdominal pain, diarrhea, and ulcers (Singh et al. 2012).

The life cycle of *Paragonimus* begins with the release of unembryonated eggs in the feces or sputum of humans. The egg is hatched within two weeks to form miracidia which is ingested by the first intermediate host that belongs to the freshwater molluscan species of the genera *Semisulcospira*, *Oncomelania*, *Brotia*, and *Thiara*. The miracidium undergoes different developmental stages to form sporocyst, rediae, and, by the end of 9–13 weeks, it develops into cercariae. The cercariae are further formed into metacercariae in the second intermediate host, crustaceans like crab or crayfish. Later metacercariae get encysted in the liver, gills, intestine, skeleton muscles, and in the heart of the crab. The definitive host, mammals including humans and species of *Canidae* and *Felidae* family are infected via the consumption of raw or undercooked crab or cray fish with meta-cercariae. The excysted metacercariae in the small intestine pass through the intestinal walls, abdominal cavity, and diaphragm and finally get attached to the thoracic cavity. The adult worms are produced in the lung parenchyma which further penetrates into the organs and tissues like brain and striated muscles. Generally, in the absence of specific treatment the parasite may survive longer and infection is persisted in the final host from 1 to 20 years (Singh et al. 2012).

Diagnosis is mainly based on the microscopic identification of dark golden-brown eggs in stool or sputum sample (Diemert 2017). Paragonimiasis is easily detected by sero-diagnosis rather than microscopy. Complement fixation test is a standard test for the diagnosis. EIA and immunoblot (IB) test are used to identify antibody titre in the sample. In the case of pulmonary paragonimiasis, chest X-ray is taken to detect patchy infiltrate with nodular cystic shadows. Praziquantel is the drug administered for the treatment (Biltricide® oral tablet). Triclabendazole® is an alternative drug used for infected adults and children (CDC 2019, https://www.cdc.gov/parasites/paragonimus/health_profes sionals/index.html#tx). Consumption of properly boiled crab or crayfish dishes is safe. Thorough cooking and proper hygienic practices are necessary to stop the spread of the infection.

2.2.4.2 Cestodes

2.2.4.2.1 Diphyllobothrium spp. Diphyllobothriasis is an intestinal tapeworm infection caused by the cestode *Diphyllobothrium* in mammals, including humans, canids, felids, bears, and mustelids. *D. latum*, known as fish tapeworm or broad tapeworm, is one of the longest tapeworms at 4–5m long and 10–20mm wide (Lee et al. 2015). Diphyllobothriasis is transmitted in humans by the ingestion of infectious larvae of pathogen in the under-cooked freshwater fish (Ito et al. 2014). Recently the disease has shown rapid re-emer-gence due to the changing food habits of people particularly the regular consumption of raw fish. The disease is mostly asymptomatic in humans and fishermen and middle-aged men are included in the higher risk group. The symptoms, mild in nature, include fatigue, constipation, abdominal pain, and watery diarrhea. Some other symptoms are pernicious anemia, headache, and allergy (Scholz et al. 2009).

The life cycle of the tapeworm is initiated when unembryonated eggs are passed through the stool of humans and are hatched in water as coracidia, a free-swimming larval stage. This is ingested by first intermediate host (crustaceans). Then the coracidia develops into the second larval stage called procercoid. Later these infected freshwater crustaceans are ingested by the second intermediate host (freshwater fishes) and the pro-cercoid stage undergoes further development inside the host tissues into the third larval stage called plerocercoid. When small fishes are ingested by their predators, plerocer-coid invades the muscle tissues of these predators (large fishes). Finally, the definite host (humans) acquire infection by the consumption of infected raw fishes. The plerocercoid infective larvae further develop into adult tapeworms in the small intestine of humans. The adult worms are more than 10m long and are attached to the intestinal mucosa. Mature tapeworms release unembryonated eggs in feces to continue the life cycle (Lee et al. 2015).

The diagnosis of *D. latum* is mainly by the microscopic identification of eggs or pro-glottids with characteristic features from the stool sample (Scholz et al. 2009). Molecular methods used are far more reliable tools for the identification of *D. latum* infection, which includes species level identification with the use of PCR and RFLP. The endoscopy, abdominal magnetic resonance imaging (MRI), and colonoscopy are some of the expensive diagnostic approaches (An et al. 2017).

Praziquantel is the drug of choice for diphyllobothriasis (Scholz et al. 2009). Niclosamide is an alternative drug administered with 2 gm dosage for adults or 1 gm for children above six years of age. The drug is highly effective and rarely causes side effects such as abdominal cramps, diarrhea, nausea, and vomiting. Humans are the primary hosts of *D. latum* so that proper sanitation and hygienic habits are necessary to prevent the spread of infection. Proper cooking or refrigeration at −10°C for 24 to 48 hours of fish is helpful to prevent diphyllobothriasis infection in humans.

2.2.4.2.2 Echinococcus *spp.* Echinococcosis is a cyst-forming zoonotic infection caused by cestode of the genus *Echinococcus* in its larval stage (Otero-Abad et al. 2013). Echinococcosis has been categorized as one of the 17 neglected diseases targeted for control or elimination by 2050 by WHO. Echinococcosis is reported worldwide (Cucher et al. 2016) even though huge efforts have been taken to reduce the spread of infection, in the past twenty years. The two common clinical forms of this cestode infection are alveolar echinococcosis (AE) caused by *E. multilocularis* and cystic echinococcosis (CE) caused by *E. granulosus* (Grosso et al. 2012). During the period of 2014–2016, the incidence of AE and CE increased in Kyrgyzstan (Paternoster et al. 2020). CE patients with liver cysts show symptoms such as discomfort of the upper abdomen, loss of appetite, and bile duct compression followed by jaundice. In AE, symptoms appear as weight loss, abdominal pain, general malaise, and signs of hepatic failure. The infection is transmitted into humans via the fecal-oral route by the ingestion of *Echinococcus* egg-contaminated food or water (Mehta et al. 2016).

In the life cycle of *Echinococcus*, the adult worms are found in the intestines of definitive hosts (canids and felids). The proglottids of adult worms mature and release eggs that are passed into the feces. The infection in humans (intermediate hosts) is acquired by the consumption of grasses, berries, herbs, or water contaminated with *Echinococcus* eggs. These eggs hatch into oncospheres in the intestine. The oncospheres penetrate the intestinal wall and migrate to various organs, especially to the liver and lungs through the blood. In these organs, the oncospheres develop into thick-walled hydatid cysts or metacestodes (Mehta et al. 2016). Several protoscoleces develop inside the metacestodes and are ingested by the definitive host. The cyst wall is digested in the host intestine liberating the protoscolices. The protoscoleces then attach to the intestinal wall and develop into adults (Wen et al. 2019).

The cost-effective and portable UltraSound (US) imaging techniques are widely used for the identification of liver lesions of CE and AE, and it is extremely useful for early diagnosis in endemic areas. X-rays are used for diagnosing lung cysts (Tamarozzi et al. 2018). In the serological methods the detection of specific antibodies is conducted using antigens present in the hydatid fluid (HF) of *Echinococcus* (Kern et al. 2017). For the diagnosis of asymptomatic AE in European countries, mass screening, regular health checkups, and systematic follow-up of associated diseases with ultra sound examination, are performed (Piarroux et al. 2011). The CT scan and fluorodeoxyglucose-positron emission tomography (FDG-PET) are also involved in the early detection of AE (Chauchet et al. 2014). In the case of CE, MRI is the preferred diagnostic tool over CT scanning (Stojkovic et al. 2016).

Currently used treatment strategies for echinococcosis are surgery for the cyst removal, PAIR (Puncture, Aspiration, Injection of protoscolicidal agent, and Reaspiration) technique; chemotherapy using antihelminthic drugs like mebendazole, albendazole, and praziquantel; and watch and wait method for inactive cysts (Velasco-Tirado et al. 2018). The cystic echinococcosis can be relieved by safe animal slaughtering conditions, dosing dogs with praziquantel, and preventing animals from consuming meat infected with cysts (Cvejic et al. 2016). AE is prevented by avoiding contact with wild animals, like foxes or stray dogs, and proper handwashing with soap before and after contact with dogs or cats.

2.2.4.2.3 Taenia solium *and* Taenia saginata   Taeniasis and cysticercosis in humans are caused by *Taenia solium* (pork tapeworm) and *Taenia saginata* (beef tapeworm). WHO has listed both worms as the most important foodborne parasites of humans that negatively affect public health and have socioeconomic and trade impact. Humans get infected with taeniasis via the consumption of raw or undercooked meat and act as the definitive hosts for three different *Taenia* species (*Taenia solium*, *Taenia saginata* and *Taenia asiatica*) (Van et al. 2014). Cysticercosis infects humans through the ingestion of raw or insufficiently cooked meat. *T. solium* eggs and larvae act as the infectious agents in humans (Okello et al. 2017). Studies have found a recent history of cysticercosis in the United Republic of Tanzania with about 17,853 cases in 2012 (Trevisan et al. 2017). Generally, in humans, taeniasis appears as mild and shows symptoms such as abdominal distension, abdominal pain, digestive disorders, and anal pruritis (Gonzales et al. 2016). Rarely, in humans, taeniasis may lead to severe health consequences as intestinal taeniasis including cholecystitis (Hakeem et al. 2012), appendicitis (Kulkarni et al. 2014) , and intestinal obstruction (Li et al. 2015). Neurocysticercosis (NCC) is a condition caused by *T. solium* in which one or more cysts are present within the CNS, especially in the brain; the symptoms shown by NCC are seizures, paralysis, dementia, chronic headache, blindness, or death (Carabin et al. 2011). The consumption of raw vegetables, raw or undercooked meat and organs of wild or domesticated animals, use of night soil (excrement) as fertilizers, and use of wastewater to irrigate crops are the factors that result in the high risk of transmission of taeniasis and cysticercosis (Carrique-Mas and Bryant 2013).

The eggs or proglottids that are excreted with stool can survive for long periods in the external environment. The released eggs contaminate water and vegetation which are then ingested by cattle (*T. saginata*) and pigs (*T. solium*). The eggs are hatched as oncospheres in the animal intestine where they penetrate the intestinal wall and striated muscles. Then they develop into cysticerci. The infection occurs in humans via the intake of raw or undercooked beef or pork with cysticerci. They attach to the human intestine and form matured adult worms. The adult worms may live for 25 years or more according to the host lifespan (Silva et al. 2011). *T. saginata* adults usually produce 1000 to 2000 proglottids, while *T. solium* adults can produce an average of 1,000 proglottids. Taeniasis is caused by the ingestion of cysticerci, while cysticercosis is caused by the direct ingestion of *T. solium* eggs passed into stool due to poor hygiene practices (Garcia 2018).

Traditional methods used for the diagnosis of taeniasis include direct microscopic examination of eggs in feces, with or without using formal ether concentration technique. The drawback of this method is the reduced sensitivity in diagnosis due to intermittent egg shedding. DNA-based tools are used for species-specific diagnosis and molecular methods such as rapid nested PCR are also useful for the diagnosis (Okello et al. 2017).

Antihelminithic drugs such as niclosamide, praziquantel, and tribendimidine are commonly used in the treatment of taeniasis. Prevention strategies taken are proper disposal of sewage, good hygiene practices while handling meat, and thorough cooking of beef and pork. The use of well-constructed latrines, management of sewerage sludge and wastewater, and good animal husbandry practices are some other ways to ensure proper management of disease transmission (Okello et al. 2017).

2.2.4.3 Nematodes

2.2.4.3.1 Anisakis *spp.* The nematode of *Anisakis* spp. causes a zoonotic disease known as anisakiasis (Nieuwenhuizen and Lopata 2013). According to WHO, approximately 56 million individuals are infected with the nematode *Anisakis* spp. through the consumption of contaminated fish products. The humans acquire infection through the uptake of raw or undercooked marine fishes and crustaceans with the third larval stage of *Anisakis* spp. (Ivanovic et al. 2015). This infection can be mild or severe with gastric, intestinal, and ectopic anisakiasis (Baird et al. 2014). The acute anisakiasis is characterized by severe abdominal pain, nausea, and vomiting. Sometimes symptoms that mimic peptic ulcer, appendicitis, or peritonitis with severe allergic reactions that range from urticaria to anaphylactic shock are also accompanied (Villazanakretzer et al. 2016). The disease is usually misdiagnosed or poorly diagnosed due to the lack of typical characteristic symptoms (Carrascosa et al. 2015).

The cetaceans (marine mammals) act as the definitive host where the parasites develop into the adult stage. The life cycle of the parasite is initiated with the release of eggs along with the feces from the definitive host into water bodies. The eggs are then hatched to form the second larval stage of the parasite (Baird et al. 2014; Buchmann and Mehrdana 2016). The larvae are ingested by the intermediate hosts, such as crustaceans and molluscs and are developed into third larval stage (Baird et al. 2014). These are taken by fishes or squids (paratenic host); encysted in the intestines and visceral organs and remains at the third larval stage of development. The lifecycle is completed when these fishes or squids are ingested by humans or marine mammals (whales, seals, and dolphins) in which larvae further develop to fourth larval stage and matured to the adult stage. The nematode releases eggs with the host feces into the marine environment to continue its life cycle (Pozio 2013).

The diagnosis of anisakiasis is initially based on the patient's food habits. The most prevalent method in diagnosis is the demonstration of *Anisakis* spp. worms obtained through gastroscopy or surgery. Endoscopy, radiography, or surgeries are general methods used in diagnosis. The preservation of vomited larvae helps to differentiate three types of *Anisakis* larvae by identifying the structure of the digestive tract (Diemert 2017).

The best method to treat anisakiasis is by the removal of the worm from the body using endoscopy or surgery. Albendazole, 400 mg twice daily for 6–21 days, has been reported as a successful drug to cure anisakiasis (Moore et al. 2002). The US FDA recommends cooking fish at a temperature of 63–74 °C before consumption (Beldsoe and Oria 2001). The larval invasion into muscles can be prevented by the immediate evisceration of fish (Chen et al. 2018) and proper cooking of fish at higher or lower temperature is helpful to kill the infectious larval form (Chen et al. 2014).

2.2.4.3.2 Ascaris lumbricoides Ascariasis is a common nematode infection caused by *Ascaris lumbricoides* which is one of the largest nematode colonizing in human intestines (Jourdan et al. 2018). Transmission occurs in humans via the consumption of raw vegetables and other food sources or water contaminated with *Ascaris* eggs. *A. suum* which is genetically similar to *A. lumbricoides* causes round worm intestinal infection in both humans and pigs (Shao et al. 2014). The infection caused by *A. lumbricoides* is usually asymptomatic and occur predominantly in developing and tropical countries especially in rural populations due to poor sanitation facilities (Wright et al. 2018). The clinical manifestation of symptomatic ascariasis includes growth retardation, pneumonitis, bowel obstruction, hepatobiliary and pancreatic injury. While intestinal ascariasis produces atypical abdominal symptoms such as pain, discomfort, dyspepsia, distention, nausea, vomiting, diarrhea, and anorexia (Dold and Holland 2011). Symptoms of larval ascariasis occur within 4–16 days after the infection.

In the life cycle, humans are the definitive host and get infected with *A. lumbricoides* embryonated eggs, through the fecal-oral route. The eggs are hatched in the small intestine to form larvae within 4 days after ingestion of contaminated food. The larvae get penetrated to the intestinal walls, enter the portal venous system and lymphatic channels, and via lymphatic system, hematogenously released to liver and finally lungs. It is matured further in the alveoli of the lungs within 10–14 days and through the tracheobronchial tree it is released into hypopharynx, and is swallowed again. Later it is returned back to the small intestine and develops into adult worms. The adult worms exist in the lumen (jejunum/ileum) and female worms produce fertilized eggs which are passed with the feces. A female adult of *A. lumbricoides* worm can produce approximately 200,000 eggs per day that can remain viable in suitable conditions for up to 10 years in the soil (De Lima Corvinoy and Bhimji 2018).

Intestinal ascariasis is diagnosed by identifying eggs or adult worms in the stool sample. During respiratory and early intestinal infections, instead of stool, sputum, or gastric washings are used for larval identification (Claus et al. 2018). The antihelminthic drugs, such as oral albendazole (400 mg) as a single dose or mebendazole, are drugs of choice for ascariasis treatment (Shah et al. 2018). Mass drug administration (MDA) conducted along with improved sanitation facilities, good hygienic practices, and health education all contribute to better treatment and elimination of the nematode (Strunz et al. 2014). WASH (Water, Sanitation, and Hygiene) programs have been implemented in schools and have proven more effective than prevention by medication alone (Freeman et al. 2013).

2.2.4.3.3 Trichuris trichuria *Trichuris trichuria* is a common human intestinal nematode (whipworm) with worldwide distribution. WHO and the CDC have listed trichuriasis in the NTD category (Viswanath et al. 2019). The disease is caused by ingestion of embryonated eggs present in fresh vegetables or from the environment and results in whipworm colonization in the human intestine (Guerrant et al. 2011). School children are more prone to the infection due to the increased chance of exposure to the worms, lack of immunity, and poor sanitation and hygiene practices. Mild infections are usually asymptomatic, although severe infection can present with abdominal pain, rectal prolapse, anemia, anorexia, bloody diarrhea, and inflammatory bowel disease (IBD) (Perez et al. 2012). Trichuriasis dysentery syndrome (TDS) is prevalent in children with a combination of mucoid diarrhea and occasional rectal bleeding (Ok et al. 2009).

Infection is acquired in humans through the fecal-oral route via the consumption of infected eggs. The ingested eggs hatch in the small intestine as larvae. As adult worms in the large intestine, their anterior portions are attached to colonic mucosa. This causes cell destruction and results in the activation of eosinophils, lymphocytes, and plasma cells as immune defense mechanisms. This is indicated by rectal bleeding and abdominal pain. The adult worm usually lives in the terminal ileum and cecum, while in some cases they invade the entire colon and rectum. The worm may remain alive in the body for 1–4 years without treatment (Viswanath et al. 2019).

Diagnosis is done through the identification of eggs or worms in feces combined with peripheral eosinophil count (5%–20 %). Colonoscopy is rarely used in diagnosis particularly when the number of worms or eggs is low in stool (Perez-Luna et al. 2013). The most widely used stool sample examination techniques with high sensitivity include Kato-Katz and filtration/concentration (Guerrant et al. 2011). PCR assay is also used due to its high sensitivity and specificity for detecting trichuriasis (Pilotte et al. 2016).

Albendazole (400 mg) is used for effective treatment and mebendazole, levamisole, ivermectin, and pyrantel are also used as alternative drugs for the treatment. However, various studies have shown inefficiency of mebendazole and albendazole for the treatment of infections with *T. trichiura* (Levecke et al. 2014b). Preventive measures include proper sanitation and safety measures of food and water. The use of urea and solar heat to inactivate embryonated eggs in the soil has also been proposed (Sharad et al. 2012). It is also important to educate people about hygienic practices, such as handwashing, thorough washing of vegetables, and proper cooking of food to reduce contamination with embryonated eggs.

2.2.4.3.4 Trichinella spiralis   Trichinellosis or trichinosis is a foodborne nematode infection from *Trichinella spiralis*. The infection occurs in humans by eating raw or insufficiently cooked pork contaminated with larvae of *Trichinella* species (Tang et al. 2015). The main source of human infection is pork and its products (Liu 2017). In 2017, California reported a few cases of infection associated with the consumption of raw bear meat (Heaton et al. 2018). The incubation period of trichinellosis is 7–21 days and the infection may be asymptomatic in most cases due to the low number of parasites. If symptomatic, mild non-bloody diarrhea, nausea, vomiting, and abdominal discomfort appears in the initial phase. In severe cases, symptoms range from fever, abdominal pain, diarrhea, nausea, vomiting, and myalgia to serious myocarditis and encephalitis.

Trichinella larvae are encysted in the muscle tissue of domestic or wild animals such as pigs, bear, moose, etc. The consumption of meat with contaminated larvae by humans results the infection. The infective muscle larvae (ML) are passed into the stomach and then to intestinal columnar epithelial cells of the host. The larvae molt four times to attain sexual maturity. Approximately 1500 newborn larvae (NBL) are released by fertilized females. They penetrate the intestine and are released to skeletal muscle via the blood and lymphatic system. Then NBL are developed into infective ML (Tang et al. 2015). The life span of adult worms in the intestinal mucosa is 4–6 weeks in humans and 10–20 days in mice and rats (Campbell 2012). While the ML may live from months to years in the muscle fibers.

According to the World Organization for Animal Health (OIE), to maintain food safety or for disease surveillance regular examination of *Trichinella* larvae in muscle tissues is essential. Several immunological assays have been developed for the detection of trichenellosis in both domestic and wild animals. A common approach for the detection of *Trichinella* infection is ELISA and is based on excretory or secretory antigens (ES) from muscle larvae (Gottstein et al. 2009). However, false negative results in animals at the early stage of infection is a drawback of the ML-ES ELISA technique (Yang et al. 2016b). ELISA performed along with ES antigens of adult worms of *Trichinella* species is helpful for the early and specific serodiagnosis of the infection (Sun et al. 2015b). The LAMP assay was also used for the detection of *T. spiralis* larvae, especially in mouse muscle samples, to directly detect the larvae during meat inspection (Li et al. 2012).

Anthelmintic drugs such as albendazole and mebendazole, are primarily used for the treatment of trichinellosis. Health education is provided to eliminate the pathogen to an extent (Troiano et al. 2017). Heating at 77°C until the disappearance of pink color in meat and freezing at –15°C for 20–30 days, or –23°C for 10–20 days, and –29°C for 6–12 days are also effective in killing the larva (Troiano and Nante 2019). Proper cooking of meat, good dietary habits, ensuring meat safety, and improved techniques for diagnosis and treatment all contribute to prevent trichinellosis.

2.3 MICROBIAL TOXINS IN FOOD

2.3.1 Ciguatera

Ciguatera is a natural toxin (ciguatoxin) that causes intoxication in humans after consumption of fish containing toxins from poisonous dinoflagellate of *Gambierdiscus* spp. These microorganisms are found in algae that are ingested by crustacea and little fishes and serve as food to predator fishes (Zlateva et al. 2017). The consumption of bigger fishes with a high level of toxin can result in food poisoning. Ciguatera fish poisoning (CFP) in humans is indicated by the presence of gastrointestinal, neurological, and cardiac symptoms and may lead to death (Chan 2015). CFP outbreaks have been reported from many countries like Germany, Vietnam, and India. In Germany, CFP outbreaks were associated with *Gambierdiscus* spp. imported from Vietnam (Roeder et al. 2010), while India reported nearly 200 cases of CFP in 2015–2016 due to the consumption of red snapper fish (Rajeish et al. 2016). Gastrointestinal symptoms such as nausea, vomiting, abdominal pain, and diarrhea usually begin within 6–12 hours of food poisoning and are cured within 1–4 days. The neurological symptoms can be dizziness, headache, myalgia, arthralgia, generalized itching, and paraesthesia in hands, feet, or mouth. When the toxin affects the cardiac system, immediate medical care is required due to the hypotension and bradycardia conditions (Friedman et al. 2017).

The challenges faced during the detection of ciguatoxin are due to its colorless, odorless, and tasteless nature. Currently, due to the absence of reliable biomarkers, CFP diagnosis is mainly based on the clinical manifestations and patient's history of eating coral reef fish. Extraction and purification techniques are also used for toxin identification. The mouse bioassay is the most commonly used method for detection. Biomolecular assay methods, such as cytotoxicity, receptor binding, and immunoassay are also used for diagnostic purposes. Recently, ELISA-based methods have been developed and the liquid chromatography-mass spectrometry (LC-MS) techniques are helpful to confirm the presence of ciguatera toxin (Farrell et al. 2016).

No specific drugs are available for CFP treatment. The treatment is based upon supportive care and the patient's symptoms. After several clinical trials, intravenous (IV) administration of mannitol was found to be an effective method for CFP treatment (Friedman et al. 2017). Only limited control measures are available due to the stability and difficulty in toxin identification. The food poisoning is not caused due to the improper handling of fish and cannot be prevented by improving storage or cooking methods. It is necessary to educate people about the risk of eating coral reef fish, especially barracuda, grouper, and snapper fishes and to avoid consumption of fish from CFP-prone regions.

2.3.2 Mushroom Toxins

Mushrooms, the visible fruiting body of fungi are rich in nutrients and used as a food source. Even though edible mushrooms are available, approximately 100 known species of mushrooms produce toxins that are poisonous to humans and new toxic species still continue to be identified (Lima et al. 2012). Mushroom poisoning, also known as mycetism or mycetismus, refers to the consumption of misidentified mushrooms which leads to intoxication in humans (Schenk-Jaeger et al. 2012). There have been many reports regarding the ingestion of wild, toxic mushrooms (Zhou et al. 2012), hundreds of which

TABLE 2.1   Classifications of Mushrooms Based on Their Toxin Production

Mushrooms	Toxins	Clinical Manifestations
Chlorophyllum molybdites, Omphalotus illudens	Gastrointestinal toxin (GI toxin)	Nausea, vomiting, abdominal pain, diarrhea
Galerina, Amanita and *Lepiota* species	Hepatotoxin	Severe hepatic failure, kidney failure, coagulopathy, pancreatitis
Russula subnigricans, Tricholoma equestre	Mycotoxin	Myalgia, muscle weakness, myocarditis, respiratory failure, kidney failure
Amanita proxima, A. smithiana	Nephrotoxin	Acute renal failure with mild hepatitis
Morchella sp.	Neurotoxin	Ataxia, paresthesia, tremor, visual disorders, confusion

Adapted and modified from Graeme (2014).

have been fatal each year (Yardan et al. 2010). The mushroom named *Amanita phalloides* is most associated with fatality (Levine et al. 2011). The incubation period ranges from a day to several weeks. The toxicity of mushrooms commonly produces gastrointestinal distress which is the first reported symptom in mushroom toxin poisoning. In clinical manifestation, four stages are observed: (i) latency or quiescent phase; (ii) gastrointestinal phase; (iii) clinical remission (despite of clinical improvement, progressive organ damage happens in due course); and (iv) fulminant hepatitis or multiorgan failure (Santi et al. 2012). Treatment is given for coagulation disorders, hypoglycaemia, and fluid and electrolyte imbalance (Trabulus and Altiparmak 2011).

Mushroom poisoning is a major public health problem in many countries due to the difficulty of differentiating edible and nonedible (poisonous) mushrooms. Some of the factors considered for the identification of poisonous mushrooms are (a) they turn green or purple when they are cut; (b) they taste bitter and sting or burn the tongue; (c) they have a bad odor (Ukwuru et al. 2018). Since there are many poisonous mushrooms, proper awareness should be given to both people and healthcare workers. Children are at higher risk to mushroom poisoning. *Amanita phalloides* (deathcap), *Cortinarius* sp. (web caps), *Calerina marginata* (autumn skullcap), *Podostroma cornu-damae* (poison fire coral), and *Lepiota brunneoincarnata* (deadly dapperling) are some of the most dangerous mushrooms seen in wild forests. Mushrooms are classified based on their toxin production (Table 2.1).

2.3.3 Mycotoxins

Mycotoxins are the low molecular weight, secondary metabolites mainly produced by the fungi like *Aspergillus*, *Penicillium*, and *Fusarium* that are poisonous to both humans and animals. Mycotoxicosis is a toxic condition in humans and animals. According to United Nations Food and Agriculture Organization (FAO) and WHO, about 25% of the world's crops such as nuts, cereals, and rice are contaminated by mycotoxins each year (Pandya and Arade 2016). The consumption of mycotoxin-contaminated food negatively affects human and animal health and the development of symptoms in toxin exposed persons are based on the type of toxin, extent of exposure to the toxin (dosage and time period), physiological and nutritional properties, and synergistic effects due to prior exposure to

TABLE 2.2 Important Food-Associated Mycotoxins

Mycotoxins	Food Items Involved	Fungal Species
Aflatoxin	Wheat, rice, maize, barley, oats, peanuts, ground nuts, walnuts, sorghum, spices, almond, figs, and cottonseed	*Aspergillus flavus*, *Aspergillus parasiticus*
Aflatoxin M1	Milk and other dairy products	Metabolism of Aflatoxin B1
Deoxynivalenol	Cereals and cereal products	*Fusarium* species
Fumonisins	Rice, wheat, barely, maize and maize products, sorghum, millets, oats, asparagus	*Fusarium proliferatum*, *Fusarium verticilloides*
Ochratoxin A	Cereals, pulses, dried fruits, grapes, wine, cheese, spices, cocoa, coffee	*Aspergillus ochraceus*, *Penicillium verrucosum*
Patulin	Apples, apple juice, grapes, peaches, pears, olives	*Penicillium expansum*
Trichothecenes	Cereals, feeds, legumes, fruits and vegetables	*Fusarium garminearum*
Zearalenone	Wheat, maize, barley, cereals and cereal products, corn, rye, sorghum	*Fusarium culmorum*, *Fusarium graminearum*

Adapted and modified from Alshannaq and Yu (2017).

any other chemicals (Gajecka et al. 2013). There are approximately 400 known mycotoxins, but still aflatoxin (AFs), ochratoxins (OT), fumonisins (FBs), zearalenone (ZEA), deoxynivalenol (DON), and trichothecenes are the ones that mostly affect public health (Table 2.2) (Ates et al. 2013). Various conditions, such as carcinogenic, hemorrhagic, mutagenic, immunotoxic, hepatotoxic, and neurotoxic have been reported for the mycotoxins (Milićević et al. 2010). It is difficult to prevent mycotoxin contamination in food due to high heat stability and physical and chemical treatments (Marin et al. 2013). The prevention strategies include improved harvesting practices, proper storage of food, and maintenance of appropriate conditions during transportation, processing, and marketing of food. Thereby the challenges in food spoilage caused by mycotoxins can be reduced drastically (Khazaeli et al. 2014).

2.3.3.1 Aflatoxin

Aflatoxin is produced mainly by the fungi, *Aspergillus flavus* and *Aspergillus parasiticus*. There are different types of AFs, and, among them, AFB1 is the predominant mycotoxin that causes food spoilage. AFB1 is categorized as Group I carcinogen by International Agency for Research on Cancer (IARC) and is generally considered a hepatocarcinogen in mammals (Muhammad et al. 2017). Aflatoxin spoilage takes place in a variety of food items, such as cereals (maize, rice, barley, oats, and sorghum), peanuts, ground nuts, pistachios, almonds, walnuts, and cottonseeds (De Boevre et al. 2012). Aflatoxin contamination in maize has been reported from Central Tanzania in 2016 with a 30% mortality rate out of 67 cases reported (Kamala et al. 2018). This toxin primarily affects liver function, whereas AFM1 is the aflatoxin found in milk which is categorized in Group II B carcinogen (Ostry et al. 2017). Acute aflatoxicosis is characterized by abdominal pain, vomiting, edema, and eventually death (Sabran et al. 2013). The effects of aflatoxin on hepatic cells or tissue injury and other minute abnormalities are based on the patient's extended exposure to aflatoxin-contaminated food (Williams et al. 2011; Gholami-Ahangaran et al. 2016).

2.3.3.2 Ochratoxin

Ochratoxin is mainly produced by *Aspergillus ochraceus* and *Penicillium verrucosum*. Ochratoxin A (OTA) is the most common ochratoxin found in cereals, dried fruits, grapes, coffee, etc. (Liuzzi et al. 2017). The kidney is the primary target of ochratoxin action, and previous studies have reported OTA as a renal carcinogen (Russo et al. 2016). OTA has been listed under Group2B carcinogen by IARC due to its carcinogenic nature in humans, and it is immunosuppressive, teratogenic, and nephrotoxic in nature (Ladeira et al. 2017). Balkan endemic nephropathy (BEN) is a kidney disease caused by OTA that, in severe cases, leads to renal failure in humans. BEN is also associated with upper urothelial tract cancer (Roupret et al. 2015).

2.3.3.3 Fumonisins

Fumonisins are mainly synthesized by *Fusarium verticillioides* and *Fusarium proliferatum*. Fumonisin B1 (FB1) is the most prevalent fumonisin (Lerda 2017). Due to their high toxicity, fumonisins are categorized in Group2B by IARC (Ostry et al. 2017). The liver and kidney are the main sites affected by fumonisins (Mazzoni et al. 2011). Toxicity in several organs of animals, such as the liver, lungs, nervous system, cardiovascular system, and kidney, has also been reported (Bertero et al. 2018). Maize and maize-based products are often contaminated with fumonisins. Other commonly exposed grains are rice, wheat, barley, maize, rye, oat, millet, and sorghum (Cendoya et al. 2018).

2.3.3.4 Zearalenone

Fusarium graminearum and *Fusarium culmorum* are the primary producers of estrogenic mycotoxin Zearalenone (ZEA) found in human foods and animal feeds. ZEA has been associated with scabby grain toxicosis in Japan, China, Australia, and the USA and is listed under Group 3 carcinogen by IARC. The toxin is indicated by the presence of symptoms such as nausea, vomiting, and diarrhea (Liao et al. 2009). ZEA mainly contaminates corn, wheat, rice, barley, sorghum, and rye. Reproductive disorders and fertility disorders are also caused by the zearalenone due to its higher affinity to estrogen receptors, hence the estrogenic dysfunction (Tralamazza et al. 2016).

2.3.3.5 Deoxynivalenol

Deoxynivalenol (DON), also known as vomitoxin, is synthesized by *Fusarium* species and is a major trichothecene contaminant in wheat and wheat products (Mishra and Das 2013). The presence of DON was reported in 81% of cattle feed from 81 countries by BIOMIN World Mycotoxin Survey, and it was recognized as the most common mycotoxin found worldwide (Cinar and Onbasi 2019). The consumption of DON-contaminated food results in nausea, diarrhea, gastrointestinal tract lesions, and reduced weight gain. DON toxicity can cause symptoms such as vomiting and reduced feed intake by animals, and, in severe cases, hematopoietic damage and immune dysfunction is also seen (Maresca 2013; Wu et al. 2015a; Wu et al. 2015b).

2.3.3.6 Patulin

Penicillium, Aspergillus, and *Byssochkamys* species of fungi found on fruits and vegetables are involved in the synthesis of patulin mycotoxin (PAT). Among this, *P. expansum* is the major producer of PAT and contaminates apples and apple products, and other fruits such as pears, peach, oranges and grapes (Yang et al. 2014). It was initially identified as a potential antibiotic, but later symptoms, such as convulsions, agitation, ulceration, edema, bowel inflammation, vomiting, and brain, liver, and kidney damages in humans was observed (de Melo et al. 2012). PAT level can be controlled in fruits and vegetables by proper fruit washing and removal of decayed portions from fruits (Forouzan and Madadlou 2014).

2.3.3.7 Trichothecenes

Trichothecenes are mycotoxins produced predominantly by *Fusarium graminearum*. There are four types of tricothecenes of which types A and E are commonly found in cereals (Villafana et al. 2019). Toxicity in humans and animals is caused by the consumption of type A toxin (Nathanail et al. 2015). A decrease in plasma glucose level, blood cell, and leukocyte count, as well as weight loss, alimentary toxic aleukia (ATA), and certain other pathological changes in the liver and stomach due to the toxicity are induced by trichothecenes in humans and animals (Adhikari et al. 2017).

2.3.3.8 Ergot

Ergot alkaloids are toxic compounds produced in the sclerotia of the fungi of the *Claviceps* species, which are pathogens of wheat, sorghum, and other grasses. In humans and animals, ergotism is caused by the ingestion of cereals infected with ergot sclerotia (Belser-Ehrlich et al. 2013). The ergotism is historically known as St. Anthony's Fire and is caused by the consumption of bread produced from contaminated flour. The two main clinical conditions of ergotism are (1) Gangrene, which affects blood supply to the extremities, and (2) convulsions, which damage the CNS. Ergot alkaloids have pharmaceutical properties also (Ashiq 2015; Jeswal and Kumar 2015). The best method to reduce the disease is by proper cleaning of grains before use.

2.3.4 Shellfish Toxins

Among the thousands of known marine microalgae, around 300 are harmful while more than 100 are involved in toxin production (Visciano et al. 2016). As discussed in ciguatera, toxins produced by marine algae cause human illness during the ingestion of contaminated seafood. This intoxication can lead to complications and even death in fish, birds, and marine mammals. The human illness is acquired by the consumption of contaminated shellfish and can result in a wide range of symptoms (Richter and Fidler 2015; Turner and Goya 2015), mainly affecting the nervous and intestinal systems. Severe damage and death to marine organisms and fish are caused by some of the dinoflagellates and diatom species which include *Noctiluca scintillans* and *Skeletonema costatum*. In humans, health problems are due to the toxin production by diatoms which belong to genera of *Alexandrium, Gymnodinium, Dinophysis*, and *Pseudo-nitzchia* (Berti and Milandri 2014; Bruce et al. 2015). The new analytical methods like LC-MS are used to determine the toxicity of shellfish and help to maintain national and international efforts to control the toxin level. Human toxic syndromes caused from exposure to marine biotoxins can be of different types (Table 2.3) (Wei-Dong et al. 2009).

2.3.5 Scombroid Poisoning

Scombroid poisoning or histamine fish poisoning is one of the most common foodborne illnesses in the United States (Nordt and Pomeranz et al. 2016). The histidine content is higher in saltwater fish with darker meat than lighter meat fish (Tortorella et al. 2014).

TABLE 2.3   Classification of Shellfish Toxic Syndromes in Humans, Toxins Produced, and Seafood Intoxicated

Toxic Syndromes	Intoxicated Seafoods	Major Toxins Produced
Amnesic shellfish poisoning (ASP)	Oysters, clams, mussels, cockles, whelks	Domoic acid (DA) and related compounds
Azaspiracid shellfish poisoning (AZP)	Oysters and mussels	Azaspiracids (AZAs) and related compounds
Diarrhetic shellfish poisoning (DSP)	Mussels, gastropods, clams, scallops, oysters, geoduck	Dinophysis toxins (DTXs), Okadaic acid, Pectenotoxins (PTXs), Yessotoxins (YTXs)
Neurotoxic shellfish poisoning (NSP)	Mussels, whelks, oysters, cockles	Brevetoxins (PbTxs)
Paralytic shellfish poisoning (PSP)	Clams, crabs, fish, lobster, oysters, mussels	Gonyautoxin (GTX), Neosaxitoxin (NEO), Saxitoxins (STXs)

Adapted and modified from Cetinkaya and Mus (2012).

In 2016, scombroid fish poisoning occurred in an elementary school in Korea with an attack rate of 5.4% via the consumption of yellowtail fish (Kang et al. 2018). The histidine in fish is converted to histamine with the help of enzyme histidine decarboxylase and due to inadequate storage and refrigeration conditions. The increased level of histamine can cause allergic reactions like bronchial spasm, flushed skin, dyspnea, hypotension, and urticaria in humans. Supportive care is the treatment for milder cases with symptoms such as rashes, flushing, nausea, cramps, and diarrhea. Antihistamines and steroids are used for therapy, and in, severe or incurable conditions, epinephrine can be administered (Hungerford et al. 2010). To prevent scombroid fish poisoning, histamine production should be controlled by proper handling and storage practices of fish in which histidine is naturally present The storage temperature of fish should be maintained <4°C in order to avoid poisoning (Ferris et al. 2018). Due to the improper handling of fish, histidine can easily be converted to histamine. Providing consumers with proper education regarding scombroid poisoning is essential to prevent the risk to public health.

2.4 BENEFICIAL MICROORGANISMS IN FOOD

Food is also the source of diverse types of microorganisms with health-promoting and beneficial effects.

2.4.1 Acetic Acid Bacteria

Acetic acid bacteria (AAB) are a group of bacteria that oxidizes ethanol to acetic acid. AAB are obligatory aerobic, catalase-positive, oxidase-negative, non-spore forming, and rod-shaped bacteria that can occur as single, pairs or in short chains (Malimas et al. 2017). AAB are mesophilic, with optimum temperature for growth between 25–30°C. Growth is mitigated at higher temperatures (Saichana et al. 2015). The optimum pH for growth is 5.0–6.5, and growth is possible even at low pH values (Wang et al. 2015). They

TABLE 2.4 Role of Acetic Acid Bacteria (AAB) in Fermentation of Alcoholic Beverages

Acetic Acid Bacteria	Fermented Alcoholic Products
Acetobacter lambici and *Gluconobacter cerevisiae*	Lambic beer
A. aceti, A. pasteurianus, Komagataeibacter xylinus, K. hansenii, Gluconobacter oxydans, Gluconacetobacter saccharivorans	Kombucha
Acetobacter aceti, A. pasteurianus, Gluconacetobacter liquefaciens, Gluconobacter oxydans, Komagataeibacter hansenii, K. europaeus, K. xylinus	Vinegar
Acetobacter aceti, A. cerevisiae, A. fabarum, A. ghanensis, A. lovaniensis, A. sicerae, several other species of *Gluconacetobacter* and *Gluconobacter*	Water kefir

Adapted and modified from Lynch et al. (2019).

are generally isolated from fruits and vegetables rich in carbohydrates which have the acidic environment that favor the growth of bacteria.

AAB are responsible for the production of a variety of fermented foods and beverages, such as vinegar, kombucha, lambic beers, kefir, cocoa, etc. (Pothakos et al. 2016) (Table 2.4). *Acetobacter, Gluconacetobacter, Gluconobacter,* and *Komagataeibacter* species are mainly involved in vinegar production due to the high efficiency to produce acetic acid from ethanol and high resistance of acetic acid in fermentation medium (Andrés-Barrao et al. 2013). AAB species exhibit a higher capacity to oxidise alcohols, aldehydes, or sugars in aerobic conditions. On the other hand, AAB are also related to the spoilage of alcoholic beverages, such as beer, wine, and cider, and fruit juices (Taban and Saichana 2017). The spoilage of AAB during production and fermentation of alcoholic beverages, results in undesirable odor and sour taste (Kregiel et al. 2018).

2.4.2 Lactic Acid Bacteria

Lactic acid bacteria (LAB) are industrially useful. They are non-motile, non-spore forming clusters of rod- or cocci-shaped fastidious bacteria. LAB is acid-tolerant as it produces huge amounts of lactic acid in the medium. Species of *Cornybacterium, Lactobacillus, Lactococcus, Clostridium,* and *Streptococcus* are major members of the LAB (Marco et al. 2017). They are common inhabitants of the gastrointestinal tract of different animals and are also present in dairy products, food products, and soil. LAB is classified into two types: homofermentative and heterofermentative based on their difference in end products. In homofermentative LAB, lactic acid is the only end product produced from glucose fermentation. In the heterofermentative type, equimolar amounts of CO_2, alcohol, and lactic acid are produced (Rattanachaikunsopon and Phumkhachorn 2010).

2.4.2.1 Role of LAB in Fermented Food

Lactic acid bacteria have an important role in the preservation of milk, fermentation of vegetables, cheese, yogurt, kefir, and buttermilk (Rakhmanova et al. 2018). LAB used in fermentation includes *Lactobacillus, Leuconostoc, Lactococcus, Enterococcus, Streptococcus, Pediococcus* spp., etc. The improved nutritional value, acceptability, quality, and durability have made lactic acid fermentation of vegetables as preferred methods for preservation of finished and semi-finished products (Kingston et al. 2010). *Pediococcus pentasaceous, Lactobacillus plantarum, Lactobacillus fermentum, Lactobacillus brevis,*

Enterobacter faecalis, *Lactococcus lactis*, *Enterococcus faecium*, and *Lactobacillus acidophilus* are some of the LAB species used in the fermentation of vegetables (Tamang et al. 2009a), showing strong acidification and coagulation properties. *L. plantarum* isolated from inziangsang, a fermented dry leafy vegetable product, also exhibited antimicrobial activity toward *Pseudomonas aeruginosa* and *Staphylococcus aureus*. Also, *L. plantarum* from Kanji/Kanjika, an ayurvedic medicinal plant has been reported to be a source of Vitamin B_{12} (Madhu et al. 2010). *L. plantarum*, *L. corniformis*, *L. brevis*, *L. fermentum*, *L. delbrueckii*, *Lactococcus lactis*, and *Enterococcus durans* are involved in bamboo shoot fermentation, and they also exhibit functional probiotic properties (Tamang et al. 2009b). Many of the LAB are Generally Recognized as Safe (GRAS) for consumption, thus promoting its use as a probiotic.

LAB are also found in milk and milk products (Delavenne et al. 2012). The frequent spoilage of milk by pathogenic microorganisms promotes the use of LAB for milk preservation. LAB are involved in the conversion of lactose into lactic acid and in the production of bacteriocin to inhibit the growth of pathogens in milk. Based on temperature of growth, LAB are of two types: mesophilic with a temperature range of 20–30°C and thermophiles with optimum temperature of 30–45°C. *Streptococcus cremoris*, *S. thermophilus*, *S. lactis*, *Lactobacillus helveticus*, *L. alimentarius*, *L. hilgardii*, *L. pseudoplantarum*, *L. acidophilus*, *L. bulgaricus*, *L. kefir*, *L. brevis*, *Leuconostoc mesenteroides*, *Lactococcus lactis* ssp. *cremoris*, *Enterococcus faecium*, *L. cremoris*, *L. bifermentans*, and *L. casei* ssp. *casei* are LAB species predominantly found in dairy products (Sarwar et al. 2018).

2.4.3 Yeast

Yeasts are widely used for the fermentation of food and beverages (Figure 2.1). *Saccharomyces cerevisiae*, known as "baker's yeast," is widely used in bakery, winemaking, and brewing industries and confectionary processes worldwide (Romano and Capece 2010; Akbari et al. 2012). They are involved in the fermentation of alcoholic beverages like wine, beer, and distilled liquor. The secondary metabolites produced in low amount by yeast during wine fermentation provide flavor and other sensory properties to wine (Cortes and Blanco 2011; De Benedictis et al. 2011; Navarrete-Bolanos 2012). The high nutritional value of yeast also marks its importance in single-cell protein (SCP) production. *Saccharomyces*, *Pichia*, *Debaryomyces*, *Candida*, and *Kluyveromyces* species of yeast are involved in the fermentation of table olives (Arroyo-López et al. 2008; Tofalo et al. 2013). The yeast species commonly found in dairy products are *Candida lusitaniae*, *Candida krusei*, *Kluveromyces lactis*, *Kluyveromyces marxianus*, *Debaryomyces hansenii*, *Yarrowia lipolytica*, *S. cerevisiae*, *Galactomyces geotrichum*, *Pichia*, etc (Erten et al. 2014). Natural fermentation methods are used in cereal-based fermented foods using mixed cultures of yeast, bacteria, and fungi (Gotcheva et al. 2000; Hammes et al. 2005; Settani et al. 2011). Undesirable growth of yeasts on meat and processed meat products may lead to spoilage.

2.4.4 Health Benefits of Probiotics, Prebiotics, and Synbiotics

The term probiotic originates from the Latin word "pro" and a Greek word "biotikos" which means "for life." It is defined as live non-pathogenic microorganisms that are truly beneficial to the host. *Lactobacillus rhamnosus*, *Lactobacillus reuteri*, *Bifidobacteria*

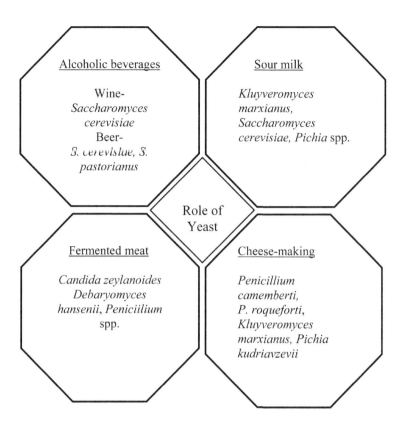

FIGURE 2.1   Role of yeast along with other organisms in food fermentation Adapted and modified from Tofalo et al. (2019).

spp., *Lactobacillus casei, Lactobacillus acidophilus*, etc. are some examples of probiotic microorganisms (Pandey et al. 2015). They are safe, cost effective, and inhibit microbial pathogens and induce the release of gastrointestinal hormones when consumed (Bongaerts and Severijnen 2016). They also play a main role in the regulation of brain behavior through the bidirectional signaling of neurons and immunomodulation in the host (Kristensen et al. 2016). Probiotics also play an important role in reducing weight gain and obesity and improving human health via the regulation of gut microflora (Kobyliak et al. 2016) (Table 2.5). They have the ability to prevent pathogenicity, diabetes, inflammatory reactions, allergies, cancer, and angiogenic responses that regulate the activities of the brain and CNS. Probiotics such as *Saccharomyces boulardii, Bifidobacteria* spp., *Lactobacillus reuter*, and some other *Lactobacillus* spp. are effective in treatment of diarrheal disease (Isolauri 2004). Some of the mechanisms of action of probiotics to provide health benefits on humans are:

(a) Destroy pathogens with the production of inhibitory substances like H_2O_2, bacteriocin, organic acids, etc.
(b) Competitively block the site of adhesion for pathogenic bacteria
(c) Limit nutrient availability to pathogens
(d) Degrade toxins as well as block their receptors and inhibit toxin production of pathogens
(e) Immunomodulation (Pandey et al. 2015)

TABLE 2.5   Commercially Available Probiotic Products

Products	Probiotics Strains Used
Align®	*Bifidobacterium infantis* 35,264
Activia® yogurt	*Bifidobacterium animalis* DN173 010
BioGaia® probiotic chewable tablets or drops	*Lactobacillus reuteri* ATCC 55,730
Culturelle®, Dannon Danimals®	*Lactobacillus rhamnosus* GG
DanActive® fermented milk	*Lactobacillus casei* DN-114001
EcoVag®	*Lactobacilli rhamnosus* PBO1, *Lactobacilli gasseri* EB01
Fem-dophilus®	*Lactobacillus reuteri* RC-14, *Lactobacillus rhamnosus* GR-1
EvoraPlus®	*Streptococcus uberis* KJ2
Florastor®	*Saccharomyces cerevisiae,* S. boulardii
LC1®	*Lactobacillus johnsonii* Lj-1
ProBiora3®	*Streptococcus oralis* KJ3
Sold as ingredient	*Lactobacillus acidophilus* NCFM, *Lactobacillus acidophilus* LA5, *Bifidobacterium lactis* Bb-12, *Lactobacillus rhamnosus* 271, *Lactobacillus rhamnosus* R0011, *Lactobacillus plantarum* OM, *Bifidobacterium lactis* HN019, *Lactobacillus fermentum* VRI003, *Lactobacillus acidophilus* R0052
Sustenex®, digestive advantage and sold as ingredient	*Bacillus coagulans* BC30
Yakult®	*Lactobacillus casei* strain Shirota, *Bifidobacterium breve* strain Yakult

Adapted and modified from Kerry et al. (2018).

Prebiotics provide a different way of promoting beneficial microflora in the gastro-intestinal tract. Here, the complex carbohydrates, such as inulin and some fructo-oligosaccharides, are used for the growth of a healthy microbial population (Gibson et al. 2017). Examples of commonly used prebiotics are insulin (with bifidogenic factor), synthetically produced oligofructose, and fructo-oligosaccharides (FOS) from sucrose, oligosaccharides with galactose and xylulose content (Hutkins et al. 2016). Vegetables, fruits, and grains are the natural sources of prebiotics which when consumed provide several health benefits, such as prevention of diarrhea, intestinal inflammation, and symptoms associated with it, and also help to prevent colon cancer (Pena 2007). They are also used to lower the risk of cardiovascular diseases, increase the bioavailability and uptake of minerals, and enhance weight loss (Pokusaeva et al. 2011).

Synbiotics is a combination of probiotics and prebiotics. They promote the survival of intestinal microorganisms with live microbial dietary supplements (Tufarelli et al. 2016). The synergistic activity of probiotics and prebiotics are effective in improving human health. The number of commercially available synbiotics has increased due to the public awareness of their health benefits, disease control, and therapy (Table 2.6).

TABLE 2.6 Commonly Known Combinations of Synbiotics

Probiotics	Prebiotics	References
Bifidobacterium bifidum, B. lactis, Klebsiella, Bacteroides fragilis, Lactobacillus acidophilus, Peptostreptococcaceae	Fructo-oligosaccharides	Bartosch et al. 2005; Hoseinifar et al. 2017
Bifidobacterium animalis, B. lactics, Lactobacillus acidophilus, L. casei, L. paracasei	Inulin	Dixit et al. 2016
Bifidobacterium adolescentis IVS-1, Bifidobacterium bifidum, B. longum, B. catenulatum	Galactooligosaccharide	Macfarlane et al. 2008; Krumbeck et al. 2016
Bifidobacterium sp., Lactococcus latics	Arabinoxylan and Arabinoxylan oligosaccharides (AXOS)	Le Barz et al. 2015
Bifidobacteria, Lactobacillus rhamnosus, Bacteroides fragilis group	Isomalto-oligosaccharides	Patel and Goyal 2012
Lactobacillus plantarum, L. casei, Bifidobacterium adolescentis	Xylo-oligosaccharides	Aachary and Prapulla 2011
Lactobacillus acidophilus, L. rhamnosus, L. bulgaricus, Bifidobacteria lactis	Lactulose	Oliveira et al. 2011
Lactobacillus and Bifidobacterium genus, Zymomonas mobilis	Lactosucrose	Han et al. 2009
Lactobacillus casei subsp. casei JCM1134™	Dextran	Ogawa et al. 2005; 2006

2.5 CONCLUSION

This chapter describes different pathogenic microorganisms associated with foodborne diseases, their disease mechanisms, treatment, and preventive measures. A detailed representation of bacteria, viruses, protozoa, and helminths (which includes trematodes, cestodes, and nematodes) with relevance to food are discussed. Moreover, toxins in food and beneficial organisms related to various food and food products are also analyzed. With the increasing incidences of foodborne outbreaks, understanding on causative agents is critical in the food safety and quality sector. Food safety must be of prime importance in every step of the food chain, from production to consumption. About 90% of foodborne hazards can be prevented by taking appropriate control measures and implementing food standards at each sector. Therefore, the acquisition of information regarding the epidemiology of harmful pathogens is helpful to ensure the safety, nutritional quality, and acceptability of foods.

BIBLIOGRAPHY

Aachary, A. A. and Prapulla, S. G. 2011. Xylooligosaccharides (XOS) as an emerging prebiotic: Microbial synthesis, utilization, structural characterization, bioactive properties, and applications. *Comprehensive Reviews in Food Science and Food Safety* 10(1):2–16.

Abdollahzadeh, E., Ojagh, S. M., Hosseini, H., Ghaemi, E. A., Irajian, G. and Heidarlo, M. N. 2016. Antimicrobial resistance of *Listeria monocytogenes* isolated from seafood and humans in Iran. *Microbial Pathogenesis* 100:70–4.

AbuBakar, S., Teoh, B. T., Sam, S. S., et al. 2013. Outbreak of human infection with *Sarcocystis nesbitti*, Malaysia, 2012. *Emerging Infectious Diseases* 19(12):1989–91.

Acheson, D. and Fiore, A. E. 2004. Hepatitis A transmitted by food. *Clinical Infectious Diseases* 38(5):705–15.

Ackers, J. P. and Mirelman, D. 2006. Progress in research on *Entamoeba histolytica* pathogenesis. *Current Opinion in Microbiology* 9(4):367–73.

Adhikari, M., Negi, B., Kaushik, N., Adhikari, A., Al-Khedhairy, A. A., Kaushik, N. K. and Choi, E. H. 2017. T-2 mycotoxin: Toxicological effects and decontamination strategies. *Oncotarget* 8(20):33933–52.

Akbari, H., Karimi, K., Lundin, M. and Taherzadeh, M. J. 2012. Optimization of baker's yeast drying in industrial continuous fluidized bed dryer. *Food and Bioproducts Processing* 90(1):52–7.

Almeria, S., Cinar, H. N. and Dubey, J. P. 2019. *Cyclospora cayetanensis* and cyclosporiasis: An update. *Microorganisms* 7(9):317.

Alshannaq, A. and Yu, J. H. 2017. Occurrence, toxicity, and analysis of major mycotoxins in food. *International Journal of Environmental Research and Public Health* 14(6):632.

Amare, L. B., Saleha, A. A., Zunita, Z., Jalila, A. and Hassan, L. 2011. Prevalence of *Arcobacter* spp. on chicken meat at retail markets and in farm chickens in Selangor, Malaysia. *Food Control* 22(5):732–6.

Andrés-Barrao, C., Benagli, C., Chappuis, M., Pérez, R. O., Tonolla, M. and Barja, F. 2013. Rapid identification of acetic acid bacteria using MALDI-TOF mass spectrometry fingerprinting. *Systematic and Applied Microbiology* 36(2):75–81.

Ankarklev, J., Jerlström-Hultqvist, J., Ringqvist, E., Troell, K. and Svärd, S. G. 2010. Behind the smile: Cell biology and disease mechanisms of *Giardia* species. *Nature Reviews in Microbiology* 8(6):413–22.

Arashkia, A., Bahrami, F., Farsi, M., et al. 2019. Molecular analysis of human adenoviruses in hospitalized children <5 years old with acute gastroenteritis in Tehran, Iran. *Journal of Medical Virology* 91(11):1930–6.

Arora, D. R. and Arora, B. B. 2010. *Textbook of medical parasitology*. CBS Publishers, India, p. 271.

Arroyo-López, F. N., Querol, A., Bautista-Gallego, J. and Garrido-Fernández, A. 2008. Role of yeasts in table olive production. *International Journal of Food Microbiology* 128(2):189–96.

Ashiq, S. 2015. Natural occurrence of mycotoxins in food and feed: Pakistan perspective. *Comprehensive Reviews in Food Science and Food Safety* 14(2):159–75.

Ashkenazi, S. 2018. *Plesiomonas shigelloides. Principles and practice of pediatric infectious diseases*. Elsevier, UK, pp. 832–3.

Ates, E., Mittendorf, K., Stroka, J. and Senyuva, H. 2013. Determination of *Fusarium* mycotoxins in wheat, maize and animal feed using on-line clean-up with high resolution mass spectrometry. *Food Additives and Contaminants: Part A* 30(1):156–65.

Ayaz, N. D. and Erol, I. 2010. Relation between serotype distribution and antibiotic resistance profiles of *Listeria monocytogenes* isolated from ground turkey. *Journal of Food Protection* 73(5):967–72.

Bacon, R. T. and Sofos, J. N. 2003. Characteristics of biological hazards in foods. *Food Safety Handbook* 10:157–95.

Bagalakote, P. S., Rathore, R. S., Ramees, T. P., et al. 2014. Molecular characterization of *Arcobacter* isolates using randomly amplified polymorphic DNA-polymerase chain reaction (RAPD-PCR). *Asian Journal of Animal and Veterinary Advances* 9(9):543–55.

Bai, Y., Yu, H., Guo, D., Fei, S. and Shi, C. 2019. Survival and environmental stress resistance of *Cronobacter sakazakii* exposed to vacuum or air packaging and stored at different temperatures. *Frontiers in Microbiology* 10:303.

Baird, F. J., Gasser, R. B., Jabbar, A. and Lopata, A. L. 2014. Foodborne anisakiasis and allergy. *Molecular and Cellular Probes* 28(4):167–74.

Barakat, R. K. and Harris, L. J. 1999. Growth of *Listeria monocytogenes* and *Yersinia enterocolitica* on cooked modified-atmosphere-packaged poultry in the presence and absence of a naturally occurring microbiota. *Applied and Environmental Microbiology* 65(1):342–5.

Barash, N. R., Nosala, C., Pham, J. K., McInally, S. G., Gourguechon, S., McCarthy-Sinclair, B. and Dawson, S. C. 2017. Giardia colonizes and encysts in high-density foci in the murine small intestine. *mSphere* 2(3):e00343–16.

Barclay, L., Park, G. W., Vega, E., Hall, A., Parashar, U., Vinjé, J. and Lopman, B. 2014. Infection control for Norovirus. *Clinical Microbiology and Infection* 20(8):731–40.

Bartosch, S., Woodmansey, E. J., Paterson, J. C., McMurdo, M. E. and Macfarlane, G. T. 2005. Microbiological effects of consuming a synbiotic containing *Bifidobacterium bifidum*, *Bifidobacterium lactis*, and oligofructose in elderly persons, determined by real-time polymerase chain reaction and counting of viable bacteria. *Clinical Infectious Diseases* 40(1):28–37.

Beldsoe, G. E. and Oria, M. P. 2001. Potential hazards in cold smoked fish: Parasites. *Journal of Food Science* 66:S1100–3.

Belser-Ehrlich, S., Harper, A., Hussey, J. and Hallock, R. 2013. Human and cattle ergotism since 1900: Symptoms, outbreaks, and regulations. *Toxicology and Industrial Health* 29(4):307–16.

Bernander, R., Palm, J. D. and Svärd, S. G. 2001. Genome ploidy in different stages of the *Giardia lamblia* life cycle. *Cellular Microbiology* 3(1):55–62.

Bertero, A., Moretti, A., Spicer, L. J. and Caloni, F. 2018. Fusarium molds and mycotoxins: Potential species-specific effects. *Toxins* 10(6):244.

Berti, M. and Milandri, A. 2014. Le biotossine marine. *Igiene degli alimenti. Aspetti Igienico-Sanitari Degli Alimenti di Origine Animale. Cap* 7:163–96.

Besser, T. E., LeJeune, J. T., Rice, D. H., et al. 2005. Increasing prevalence of *Campylobacter jejuni* in feedlot cattle through the feeding period. *Applied and Environmental Microbiology* 71(10):5752–8.

Bhatia, B. B., Pathak, K. M. L. and Juyal, P. D. 2010. Textbook of veterinary parasitology. *Journal of Veterinary Parasitology* 24(2):211–21.

Biffa, D., Skjerve, E., Oloya, J., et al. 2010. Molecular characterization of *Mycobacterium bovis* isolates from Ethiopian cattle. *BMC Veterinary Research* 6(1):28.

Binsztein, N., Costagliola, M. C., Pichel, M., et al. 2004. Viable but nonculturable *Vibrio cholerae* O1 in the aquatic environment of Argentina. *Applied and Environmental Microbiology* 70(12):7481–6.

Bintsis, T. 2017. Foodborne pathogens. *AIMS Microbiology* 3(3):529.

Blackburn, C. D. W. 2006. *Food spoilage microorganisms*. Ed. C. de W. Blackburn. Woodhead Publishing, London.

Blaser, M. J. and Engberg, J. 2008. Clinical aspects of *Campylobacter jejuni* and *Campylobacter coli* infections. In: *Campylobacter*, 3rd ed. (pp. 99–121). American Society of Microbiology, 99–121.

Blondel, B., Colbere-Garapin, F., Couderc, T., Wirotius, A. and Guivel-Benhassine, F. 2005. Poliovirus, pathogenesis of poliomyelitis, and apoptosis. In: *Role of apoptosis in infection* (pp. 25–56). Springer, Berlin.

Bogitsh, B. J., Carter, C. E. and Oeltmann, T. N. 2018. *Human parasitology*. Academic Press, New York.

Bongaerts, G. P. and Severijnen, R. S. 2016. A reassessment of the propatria study and its implications for probiotic therapy. *Nature Biotechnology* 34(1):55–63.

Bosch, A., Pintó, R. M. and Guix, S. 2016. Foodborne viruses. *Current Opinion in Food Science* 8:110–9.

Bottone, E. J. 2010. *Bacillus cereus*, a volatile human pathogen. *Clinical Microbiology Reviews* 23(2):382–98.

Bouvard, V., Baan, R., Straif, K., et al. 2009. A review of human carcinogens – Part B: Biological agents. *The Lancet Oncology* 10(4):321.

Bouzid, M., Hunter, P. R., Chalmers, R. M. and Tyler, K. M. 2013. *Cryptosporidium* pathogenicity and virulence. *Clinical Microbiology Reviews* 26(1):115–34.

Brin, M. F. 1997. Botulinum toxin: Chemistry, pharmacology, toxicity, and immunology. *Muscle and Nerve: Official Journal of the American Association of Electrodiagnostic Medicine* 20(S6):146–68.

Bruce, K. L., Leterme, S. C., Ellis, A. V. and Lenehan, C. E. 2015. Approaches for the detection of harmful algal blooms using oligonucleotide interactions. *Analytical and Bioanalytical Chemistry* 407(1):95–116.

Buchmann, K. and Mehrdana, F. 2016. Effects of anisakid nematodes *Anisakis* simplex (sl), *Pseudoterranova decipiens* (sl) and *Contracaecum osculatum* (sl) on fish and consumer health. *Food and Waterborne Parasitology* 4:13–22.

Bucker, R., Troeger, H., Kleer, J., Fromm, M. and Schulzke, J. D. 2009. *Arcobacter butzleri* induces barrier dysfunction in intestinal HT-29/B6 cells. *The Journal of Infectious Diseases* 200(5):756–64.

Bukhari, E. E. 2018. Pediatric brucellosis: An update review for the new millennium. *Saudi Medical Journal* 39(4):336–41.

Bundhamcharoen, K., Odton, P., Phulkerd, S. and Tangcharoensathien, V. 2011. Burden of disease in Thailand: Changes in health gap between 1999 and 2004. *BMC Public Health* 11(1):53.

Byrd-Bredbenner, C., Berning, J., Martin-Biggers, J. and Quick, V. 2013. Food safety in home kitchens: A synthesis of the literature. *International Journal of Environmental Research and Public Health* 10(9):4060–85.

Calbo, E., Freixas, N., Xercavins, M., et al. 2011. Foodborne nosocomial outbreak of SHV1 and CTX-M-15–producing *Klebsiella pneumoniae*: Epidemiology and control. *Clinical Infectious Diseases* 52(6):743–9.

Camargo, A. C., Woodward, J. J., Call, D. R. and Nero, L. A. 2017. *Listeria monocytogenes* in food-processing facilities, food contamination, and human listeriosis: The Brazilian scenario. *Foodborne Pathogens and Disease* 14(11):623–36.

Campbell, W. 2012. *Trichinella and Trichinosis*. Springer Science and Business Media.

Cantoni, C., Stella, S., Ripamonti, B. and Marchese, R. 2001. Anomalous colouration of mozzarella cheese. *Industrie Alimentari* 40(399):33–5.

Cao, X., Xu, X., Zhang, Z., Shen, H., Chen, J. and Zhang, K. 2014. Molecular characterization of clinical multidrug-resistant *Klebsiella pneumoniae* isolates. *Annals of Clinical Microbiology and Antimicrobials* 13(1):1–5.

Carabin, H., Ndimubanzi, P. C., Budke, C. M., et al. 2011. Clinical manifestations associated with neurocysticercosis: A systematic review. *PLoS Neglected Tropical Diseases* 5(5):e1152.

Carey, C. M., Lee, H. and Trevors, J. T. 2004. Biology, persistence and detection of *Cryptosporidium parvum* and *Cryptosporidium hominis* oocyst. *Water Research* 38(4):818–62.

Carrascosa, M. F., Mones, J. C., Salcines-Caviedes, J. R. and Román, J. G. 2015. A man with unsuspected marine eosinophilic gastritis. *The Lancet Infectious Diseases* 15(2):248.

Carrique-Mas, J. J. and Bryant, J. E. 2013. A review of foodborne bacterial and parasitic zoonoses in Vietnam. *EcoHealth* 10(4):465–89.

Cary, J. W., Linz, J. E. and Bhatnagar, D., eds. 1999. *Microbial Foodborne Diseases: Mechanisms of Pathogenesis and Toxin Synthesis*. CRC Press.

Cash, R. A., Music, S. I., Libonati, J. P., Snyder, M. J., Wenzel, R. P. and Hornick, R. B. 1974. Response of man to infection with *Vibrio cholerae*. I. Clinical, serologic, and bacteriologic responses to a known inoculum. *Journal of Infectious Diseases* 129(1):45–52.

Cendoya, E., Chiotta, M. L., Zachetti, V., Chulze, S. N. and Ramirez, M. L. 2018. Fumonisins and fumonisin-producing *Fusarium* occurrence in wheat and wheat by products: A review. *Journal of Cereal Science* 80:158–66.

Cetinkaya, F. and Mus, T. E. 2012. Shellfish poisoning and toxins. *Journal of Biological and Environmental Sciences* 6(17):115–9.

Chai, J. Y., Murrell, K. D. and Lymbery, A. J. 2005. Fish-borne parasitic zoonoses: Status and issues. *International Journal for Parasitology* 35(11–12):1233–54.

Chan, T. Y. 2015. Ciguatera fish poisoning in East Asia and Southeast Asia. *Marine Drugs* 13(6):3466–78.

Chandran, A., Williams, K., Mendum, T., et al. 2019. Development of a diagnostic compatible BCG vaccine against Bovine tuberculosis. *Scientific Reports* 9(1):1–11.

Chatterjee, K. D. 1969. *Parasitology (protozoology and helminthology) in relation to clinical medicine*. 7th Edition.

Chauchet, A., Grenouillet, F., Knapp, J., et al. 2014. Increased incidence and characteristics of alveolar echinococcosis in patients with immunosuppression-associated conditions. *Clinical Infectious Diseases* 59(8):1095–104.

Cheepsattayakorn, A. and Cheepsattayakorn, R. 2014. Parasitic pneumonia and lung involvement. *BioMed Research International* 2014:874021.

Chen, B. S., Lee, H. C., Lee, K. M., Gong, Y. N. and Shih, S. R. 2020. Enterovirus and encephalitis. *Frontiers in Microbiology* 11:261.

Chen, H. X., Zhang, L. P., Gibson, D. I., et al. 2018. Detection of ascaridoid nematode parasites in the important marine food-fish *Conger myriaster* (Brevoort) (Anguilliformes: Congridae) from the Zhoushan Fishery, China. *Parasites and Vectors* 11(1):274.

Chen, H. Y., Cheng, Y. S., Grabner, D. S., Chang, S. H. and Shih, H. H. 2014. Effect of different temperatures on the expression of the newly characterized heat shock protein 90 (Hsp90) in L3 of *Anisakis* spp. isolated from *Scomber australasicus*. *Veterinary Parasitology* 205(3–4):540–50.

Chen, X., Chen, Y., Yang, Q., et al. 2013. *Plesiomonas shigelloides* infection in Southeast China. *PLoS ONE* 8(11):e77877.

Choi, B. I., Han, J. K., Hong, S. T. and Lee, K. H. 2004. Clonorchiasis and cholangiocarcinoma: Etiologic relationship and imaging diagnosis. *Clinical Microbiology Reviews* 17(3):540–52.

Cinar, A. and Onbaşi, E. 2019. Mycotoxins: The hidden danger in foods. In: *Mycotoxins and food safety*. IntechOpen, 89001.

Claus, P. E., Ceuppens, A. S., Cool, M. and Alliet, G. 2018. *Ascaris lumbricoides*: Challenges in diagnosis, treatment and prevention strategies in a European refugee camp. *Acta Clinica Belgica* 73(6):431–4.

Collado, L. and Figueras, M. J. 2011. Taxonomy, epidemiology, and clinical relevance of the genus *Arcobacter*. *Clinical Microbiology Reviews* 24(1):174–92.

Coordinating Office of the National Survey on the Important Human Parasitic 2005. A national survey on current status of the important parasitic diseases in human population. *Chinese Journal of Parasitology and Parasitic Diseases* 23(5):332.

Corbel, M. J. 2006. *Brucellosis in humans and animals*. World Health Organization.

Cornick, S. and Chadee, K. 2017. *Entamoeba histolytica*: Host parasite interactions at the colonic epithelium: Host parasite interactions at the colonic epithelium. *Tissue Barriers* 5(1):e1283386.

Cortes, S. and Blanco, P. 2011. Yeast strain effect on the concentration of major volatile compounds and sensory profile of wines from *Vitis vinifera* var. Treixadura. *World Journal of Microbiology and Biotechnology* 27(4):925–32.

Crenshaw, B. J., Jones, L. B., Bell, C. R., Kumar, S. and Matthews, Q. L. 2019. Perspective on adenoviruses: Epidemiology, pathogenicity, and gene therapy. *Biomedicines* 7(3):61.

Croxen, M. A., Law, R. J., Scholz, R., Keeney, K. M., Wlodarska, M. and Finlay, B. B. 2013. Recent advances in understanding enteric pathogenic *Escherichia coli*. *Clinical Microbiology Reviews* 26(4):822–80.

Cucher, M. A., Macchiaroli, N., Baldi, G., et al. 2016. *Cystic echinococcosis* in South America: Systematic review of species and genotypes of *Echinococcus granulosus* sensu lato in humans and natural domestic hosts. *Tropical Medicine and International Health* 21(2):166–75.

Cvejic, D., Schneider, C., Fourie, J., de Vos, C., Bonneau, S., Bernachon, N. and Hellmann, K. 2016. Efficacy of a single dose of milbemycin oxime/praziquantel combination tablets, Milpro®, against adult *Echinococcus multilocularis* in dogs and both adult and immature *E. multilocularis* in young cats. *Parasitology Research* 115(3):1195–202.

D'aoust, J. Y., Park, C. E., Szabo, R. A., Todd, E. C. D., Emmons, D. B. and McKellar, R. C. 1988. Thermal inactivation of *Campylobacter* species, *Yersinia enterocolitica*, and hemorrhagic *Escherichia coli* 0157: H7 in fluid milk. *Journal of Dairy Science* 71(12):3230–6.

D'Souza, D. H. 2015. Update on foodborne viruses: Types, concentration and sampling methods. In: *Advances in microbial food safety* (pp. 102–16). Woodhead Publishing.

Davis, G. S. and Price, L. B. 2016. Recent research examining links among *Klebsiella pneumoniae* from food, food animals, and human extraintestinal infections. *Current Environmental Health Reports* 3(2):128–35.

Davys, G., Marshall, J. C., Fayaz, A., Weir, R. P. and Benschop, J. 2020. Campylobacteriosis associated with the consumption of unpasteurised milk: Findings from a sentinel surveillance site. *Epidemiology and Infection* 148:e16.

De Benedictis, M., Bleve, G., Grieco, F., Tristezza, M., Tufariello, M. and Grieco, F. 2011. An optimized procedure for the enological selection of non-*Saccharomyces* starter cultures. *Antonie van Leeuwenhoek* 99(2):189–200.

De Boevre, M., Di Mavungu, J. D., Landschoot, S., et al. 2012. Natural occurrence of mycotoxins and their masked forms in food and feed products. *World Mycotoxin Journal* 5(3):207–19.

De Falco, M., Lucariello, A., Iaquinto, S., Esposito, V., Guerra, G. and De Luca, A. 2015. Molecular mechanisms of *Helicobacter pylori* pathogenesis. *Journal of Cellular Physiology* 230(8):1702–7.

de Lima Corvinoy, D. F. and Bhimji, S. S. 2018. *Ascariasis*. Treasure Island.

de Melo, F. T., de Oliveira, I. M., Greggio, S. et al. 2012. DNA damage in organs of mice treated acutely with patulin, a known mycotoxin. *Food and Chemical Toxicology* 50(10):3548–55.

Dekker, J. P. and Frank, K. M. 2015. Salmonella, Shigella, and Yersinia. *Clinics in Laboratory Medicine* 35(2):225–46.

Delavenne, E., Mounier, J., Déniel, F., Barbier, G. and Le Blay, G. 2012. Biodiversity of antifungal lactic acid bacteria isolated from raw milk samples from cow, ewe and goat over one-year period. *International Journal of Food Microbiology* 155(3):185–90.

Denny, J., Boelaert, F., Borck, B., Heuer, O. E., Ammon, A. and Makela, P. 2007. Zoonotic infections in Europe: Trends and figures - A summary of the EFSA-ECDC annual report. *Weekly Releases (1997–2007)* 12(51):3336.

Dhaked, R. K., Singh, M. K., Singh, P. and Gupta, P. 2010. Botulinum toxin: Bioweapon and magic drug. *The Indian Journal of Medical Research* 132(5):489–503.

Dhama, K., Verma, A. K., Rajagunalan, S., Kumar, A., Tiwari, R., Chakraborty, S. and Kumar, R. 2013. *Listeria monocytogenes* infection in poultry and its public health importance with special reference to foodborne zoonoses. *Pakistan Journal of Biological Sciences* 16(7):301–8.

Diaconu, S., Predescu, A., Moldoveanu, A., Pop, C. S. and Fierbințeanu-Braticevici, C. 2017. *Helicobacter pylori* infection: Old and new. *Journal of Medicine and Life* 10(2):112–7.

Diemert, D. J. 2017. Cestode and trematode infections. In: *Infectious diseases* (pp. 1032–7). Elsevier.

Dinges, M. M., Orwin, P. M. and Schlievert, P. M. 2000. Exotoxins of *Staphylococcus aureus*. *Clinical Microbiology Reviews* 13(1):16–34.

Dixit, Y., Wagle, A. and Vakil, B. 2016. Patents in the field of probiotics, prebiotics, Synbiotics: A review. *Journal of Food: Microbiology, Safety and Hygiene* 1(2):111.

Djurković-Djaković, O., Bobić, B., Nikolić, A., Klun, I. and Dupouy-Camet, J. 2013. Pork as a source of human parasitic infection. *Clinical Microbiology and Infection* 19(7):586–94.

Dogan, B. and Boor, K. J. 2003. Genetic diversity and spoilage potentials among *Pseudomonas* spp. isolated from fluid milk products and dairy processing plants. *Applied and Environmental Microbiology* 69(1):130–8.

Dold, C. and Holland, C. V. 2011. Ascaris and ascariasis. *Microbes and Infection / Institut Pasteur* 13(7):632–7.

Dougan, G. and Baker, S. 2014. *Salmonella enterica* serovar Typhi and the pathogenesis of typhoid fever. *Annual Review of Microbiology* 68: 317–36.

Dubey, J. P. 2010. Toxoplasmosis of animals and humans. *Parasites and Vectors* 3: 112.

Dunay, I. R., Gajurel, K., Dhakal, R., Liesenfeld, O. and Montoya, J. G. 2018. Treatment of toxoplasmosis: Historical perspective, animal models, and current clinical practice. *Clinical Microbiology Reviews* 31(4):e00057-17.

Ehling-Schulz, M. and Messelhäusser, U. 2013. *Bacillus* "next generation" diagnostics: Moving from detection toward subtyping and risk-related strain profiling. *Frontiers in Microbiology* 4: 32.

Ekici, G. and Dümen, E. 2019. *Escherichia coli* and food safety. In: *The universe of Escherichia coli*. IntechOpen 82375.

Epps, S. V., Harvey, R. B., Hume, M. E., Phillips, T. D., Anderson, R. C. and Nisbet, D. J. 2013. Foodborne *Campylobacter*: Infections, metabolism, pathogenesis and reservoirs. *International Journal of Environmental Research and Public Health* 10(12):6292–304.

Ercolini, D., Russo, F., Nasi, A., Ferranti, P. and Villani, F. 2009. Mesophilic and psychrotrophic bacteria from meat and their spoilage potential in vitro and in beef. *Applied and Environmental Microbiology* 75(7):1990–2001.

Ergul, A. B., Cetin, S., Altintop, Y. A. et al. 2017. Evaluation of microorganisms causing ventilator-associated pneumonia in a pediatric intensive care unit. *The Eurasian Journal of Medicine* 49(2):87–91.

Erten, H., Ağirman, B., Gündüz, C. P. B., Çarşanba, E., Sert, S., Bircan, S. and Tangüler, H. 2014. Importance of yeasts and lactic acid bacteria in food processing. In: *Food processing: Strategies for quality assessment* (pp. 351–78). Springer.

Escobar, J. C., Bhavnani, D., Trueba, G., Ponce, K., Cevallos, W. and Eisenberg, J. 2012. *Plesiomonas shigelloides* infection, Ecuador, 2004–2008. *Emerging Infectious Diseases* 18(2):322–4.

Evenson, M. L., Hinds, M. W., Bernstein, R. S. and Bergdoll, M. S. 1988. Estimation of human dose of staphylococcal enterotoxin A from a large outbreak of staphylococcal food poisoning involving chocolate milk. *International Journal of Food Microbiology* 7(4):311–6.

Falomir, M. P., Rico, H. and Gozalbo, D. 2013. *Enterobacter* and *Klebsiella* species isolated from fresh vegetables marketed in Valencia (Spain) and their clinically relevant resistances to chemotherapeutic agents. *Foodborne Pathogens and Disease* 10(12):1002–7.

Farrell, H., Zammit, A., Manning, J., et al. 2016. Clinical diagnosis and chemical confirmation of ciguatera fish poisoning in New South Wales, Australia. *Communicable Diseases Intelligence Quarterly Report* 40(1):E1.

Fayer, R., Esposito, D. H. and Dubey, J. P. 2015. Human infections with *Sarcocystis* species. *Clinical Microbiology Reviews* 28(2):295–311.

Fazelipour, M., Makvandi, M., Samarbafzadeh, A., et al. 2019. Detection of enteroviruses in children with acute diarrhea. *Archives of Clinical Infectious Diseases* 14(4):e83916.

Feeney, A., Kropp, K. A., O'Connor, R. and Sleator, R. D. 2014. *Cronobacter sakazakii*: Stress survival and virulence potential in an opportunistic foodborne pathogen. *Gut Microbes* 5(6):711–8.

Feng, Y. and Xiao, L. 2011. Zoonotic potential and molecular epidemiology of *Giardia* species and giardiasis. *Clinical Microbiology Reviews* 24(1):110–40.

Fentie, T., Erqou, S., Gedefaw, M. and Desta, A. 2013. Epidemiology of human fascioliasis and intestinal parasitosis among school children in Lake Tana Basin, northwest Ethiopia. *Transactions of the Royal Society of Tropical Medicine and Hygiene* 107(8):480–6.

Fernandez-Garcia, M. D., Majumdar, M., Kebe, O., et al. 2018. Emergence of vaccine-derived polioviruses during Ebola virus disease outbreak, Guinea, 2014–2015. *Emerging Infectious Diseases* 24(1):65–74.

Ferreira, S., Queiroz, J. A., Oleastro, M. and Domingues, F. C. 2016. Insights in the pathogenesis and resistance of *Arcobacter*: A review. *Critical Reviews in Microbiology* 42(3):364–83.

Ferris, H. A., Ryan, F. M., Byrne, K., Fleming, E., O'Sullivan, H. and Hamilton, D. 2018. Scombrotoxic fish poisoning secondary to Tuna ingestion. *Irish Medical Journal* 111(6):773.

Fetsch, A. and Johler, S. 2018. *Staphylococcus aureus* as a foodborne pathogen. *Current Clinical Microbiology Reports* 5(2):88–96.

Fitzgerald, C. and Nachamkin, I. 2015. Campylobacter and Arcobacter. In: *Manual of clinical microbiology*, 11th ed. (pp. 998–1012). American Society of Microbiology.

Forouzan, S. and Madadlou, A. 2014. Incidence of patulin in apple juices produced in West Azerbayjan Province, Iran. *Journal of Agricultural Science and Technology* 16: 1613–22.

Frank, C., Faber, M. S., Askar, M., et al. 2011. Large and ongoing outbreak of haemolytic uraemic syndrome, Germany, May 2011. *Eurosurveillance* 16(21):19878.

Fredriksson-Ahomaa, M., Stolle, A. and Stephan, R. 2007. Prevalence of pathogenic *Yersinia enterocolitica* in pigs slaughtered at a Swiss abattoir. *International Journal of Food Microbiology* 119(3):207–12.

Freeman, M. C., Clasen, T., Brooker, S. J., Akoko, D. O. and Rheingans, R. 2013. The impact of a school-based hygiene, water quality and sanitation intervention on soil-transmitted helminth reinfection: A cluster-randomized trial. *The American Journal of Tropical Medicine and Hygiene* 89(5):875–83.

Friedman, M. A., Fernandez, M., Backer, L. C., et al. 2017. An updated review of ciguatera fish poisoning: Clinical, epidemiological, environmental, and public health management. *Marine Drugs* 15(3):72.

Friesen, J., Fuhrmann, J., Kietzmann, H., Tannich, E., Müller, M. and Ignatius, R. 2018. Evaluation of the Roche LightMix Gastro parasites multiplex PCR assay detecting *Giardia duodenalis*, *Entamoeba histolytica*, cryptosporidia, *Dientamoeba fragilis*, and *Blastocystis hominis*. *Clinical Microbiology and Infection* 24(12):1333–7.

Fu, S., Jiang, Y., Jiang, X., Zhao, Y., Chen, S., Yang, X. and Man, C. 2018. Probe-free label system for rapid detection of *Cronobacter* genus in powdered infant formula. *AMB Express* 8(1):155.

Gabida, M., Gombe, N. T., Chemhuru, M., Takundwa, L., Bangure, D. and Tshimanga, M. 2015. Foodborne illness among factory workers, Gweru, Zimbabwe, 2012: A retrospective cohort study. *BMC Research Notes* 8(1):493.

Gajęcka, M., Stopa, E., Tarasiuk, M., Zielonka, Ł. and Gajęcki, M. 2013. The expression of type-1 and type-2 nitric oxide synthase in selected tissues of the gastrointestinal tract during mixed mycotoxicosis. *Toxins* 5(11):2281–92.

García, A., Fox, J. G. and Besser, T. E. 2010. Zoonotic enterohemorrhagic *Escherichia coli*: A one health perspective. *ILAR Journal* 51(3):221–32.

Garcia, H. H. 2018. Neurocysticercosis. *Neurologic Clinics* 36(4):851–64.

Gardiner, B. J., Simpson, I. and Woolley, I. J. 2015. Caught in the act, a case of fulminant amoebic colitis. *JMM Case Reports* 2(4):e000081.

Gardner, A. D. and Venkatraman, K. V. 1935. The antigens of the cholera group of vibrios. *Epidemiology and Infection* 35(2):262–82.

Gerace, E., Presti, V. D. M. L. and Biondo, C. 2019. Cryptosporidium infection: Epidemiology, pathogenesis, and differential diagnosis. *European Journal of Microbiology and Immunology* 9(4):119–23.

Gholami-Ahangaran, M., Rangsaz, N. and Azizi, S. 2016. Evaluation of turmeric (*Curcuma longa*) effect on biochemical and pathological parameters of liver and kidney in chicken aflatoxicosis. *Pharmaceutical Biology* 54(5):780–7.

Gibson, G. R., Hutkins, R. W., Sanders, M. E., et al. 2017. Expert consensus document: The International Scientific Association for Probiotics and Prebiotics (ISAPP) consensus statement on the definition and scope of prebiotics. *Nature Reviews: Gatroenterology and Hepatology* 14(8):491–502.

Glowacka, P., Żakowska, D., Naylor, K., Niemcewicz, M. and Bielawska-Drozd, A. 2018. Brucella–Virulence factors, pathogenesis and treatment. *Polish Journal of Microbiology* 67(2):151–61.

Goldberg, E. and Bishara, J. 2012. Contemporary unconventional clinical use of co-trimoxazole. *Clinical Microbiology and Infection* 18(1):8–17.

Gonzales, I., Rivera, J. T., Garcia, H. H. and Cysticercosis Working Group in Peru 2016. Pathogenesis of *Taenia solium* taeniasis and cysticercosis. *Parasite Immunology* 38(3):136–46.

González, A., Salillas, S., Velázquez-Campoy, A., Espinosa Angarica, V., Fillat, M. F., Sancho, J. and Lanas, Á 2019. Identifying potential novel drugs against *Helicobacter pylori* by targeting the essential response regulator HsrA. *Scientific Reports* 9(1):1–13.

Gonzalez, G., Carr, M. J., Kobayashi, M., Hanaoka, N. and Fujimoto, T. 2019. Enterovirus-associated hand-foot and mouth disease and neurological complications in Japan and the rest of the world. *International Journal of Molecular Sciences* 20(20):5201.

González, L. M., Villalobos, N., Montero, E., et al. 2006. Differential molecular identification of *Taeniid* spp. and *Sarcocystis* spp. cysts isolated from infected pigs and cattle. *Veterinary Parasitology* 142(1–2):95–101.

González-Rey, C., Siitonen, A., Pavlova, A., Ciznar, I., Svenson, S. B. and Krovacek, K. 2011. Molecular evidence of *Plesiomonas shigelloides* as a possible zoonotic agent. *Folia Microbiologica* 56(2):178.

González-Rey, C., Svenson, S. B., Bravo, L., Rosinsky, J., Ciznar, I. and Krovacek, K. 2000. Specific detection of *Plesiomonas shigelloides* isolated from aquatic environments, animals and human diarrhoeal cases by PCR based on 23S rRNA gene. *FEMS Immunology and Medical Microbiology* 29(2):107–13.

Gotcheva, V., Pandiella, S. S., Angelov, A., Roshkova, Z. G. and Webb, C. 2000. Microflora identification of the Bulgarian cereal-based fermented beverage boza. *Process Biochemistry* 36(1–2):127–30.

Gottstein, B., Pozio, E. and Nöckler, K. 2009. Epidemiology, diagnosis, treatment, and control of trichinellosis. *Clinical Microbiology Reviews* 22(1):127–45.

Graeme, K. A. 2014. Mycetism: A review of the recent literature. *Journal of Medical Toxicology* 10(2):173–89.

Graham, C. and Simmons, N. L. 2005. Functional organization of the bovine rumen epithelium. *American Journal of Physiology-Regulatory, Integrative and Comparative Physiology* 288(1):R173–81.

Greenberg, H. B. and Estes, M. K. 2009. Rotaviruses: From pathogenesis to vaccination. *Gastroenterology* 136(6):1939–51.

Grimont, P. A. and Weill, F. X. 2007. Antigenic formulae of the *Salmonella* serovars. *WHO Collaborating Centre for Reference and Research on Salmonella* 9, 1–166.

Grosso, G., Gruttadauria, S., Biondi, A., Marventano, S. and Mistretta, A. 2012. Worldwide epidemiology of liver hydatidosis including the Mediterranean area. *World Journal of Gastroenterology* 18(13):1425.

Grundy-Warr, C., Andrews, R. H., Sithithaworn, P., Petney, T. N., Sripa, B., Laithavewat, L. and Ziegler, A. D. 2012. Raw attitudes, wetland cultures, life-cycles: Socio-cultural dynamics relating to *Opisthorchis viverrini* in the Mekong Basin. *Parasitology International* 61(1):65–70.

Guerrant, R. L., Walker, D. H. and Weller, P. F. 2011. *Tropical infectious diseases: Principles, pathogens and practice*. Elsevier Health Sciences.

Gunasekera, T. S., Dorsch, M. R., Slade, M. B. and Veal, D. A. 2003. Specific detection of *Pseudomonas* spp. in milk by fluorescence in situ hybridization using ribosomal RNA directed probes. *Journal of Applied Microbiology* 94(5):936–45.

Guo, Y., Roellig, D. M., Li, N., et al. 2016. Multilocus sequence typing tool for *Cyclospora cayetanensis*. *Emerging Infectious Diseases* 22(8):1464–7.

Guo, Y., Zhou, H., Qin, L., et al. 2016. Frequency, antimicrobial resistance and genetic diversity of *Klebsiella pneumoniae* in food samples. *PLoS ONE* 11(4):e0153561.

Gutiérrez-Cisneros, M. J., Martínez-Ruiz, R., Subirats, M., Merino, F. J., Millán, R. and Fuentes, I. 2011. Assessment of two commercially available immunochromatographic assays for a rapid diagnosis of *Giardia duodenalis* and *Cryptosporidium* spp. in human fecal specimens. *Enfermedades Infecciosas y Microbiologia Clinica* 29(3):201–3.

Guzmán-Hernández, R. L., Contreras-Rodríguez, A., Ávila-Calderón, E. D. and Morales-García, M. R. 2016. Brucelosis: Zoonosis de importancia en México. *Revista Chilena de Infectología* 33(6): 656–62.

Hakeem, S. Y., Rashid, A., Khuroo, S. and Bali, R. S. 2012. *Taenia saginata*: A rare cause of gall bladder perforation. *Case Reports in Surgery* 2012: 572484.

Hammes, W. P., Brandt, M. J., Francis, K. L., Rosenheim, J., Seitter, M. F. and Vogelmann, S. A. 2005. Microbial ecology of cereal fermentations. *Trends in Food Science and Technology* 16(1–3):4–11.

Han, W. C., Byun, S. H., Kim, M. H., et al. 2009. Production of lactosucrose from sucrose and lactose by a levansucrase from *Zymomonas mobilis*. *Journal of Microbiology and Biotechnology* 19(10):1153–60.

Hanning, I. B., Nutt, J. D. and Ricke, S. C. 2009. Salmonellosis outbreaks in the United States due to fresh produce: Sources and potential intervention measures. *Foodborne Pathogens and Disease* 6(6):635–48.

Haque, R., Huston, C. D., Hughes, M., Houpt, E. and Petri Jr, W. A. 2003. Amebiasis. *New England Journal of Medicine* 348(16):1565–73.

Harrison, L. C. and DiCaprio, E. 2018. Hepatitis E virus: An emerging foodborne pathogen. *Frontiers in Sustainable Food Systems* 2: 1–14.

Haryani, Y., Noorzaleha, A. S., Fatimah, A. B., et al. 2007. Incidence of *Klebsiella pneumonia* in street foods sold in Malaysia and their characterization by antibiotic resistance, plasmid profiling, and RAPD–PCR analysis. *Food Control* 18(7):847–53.

Hatam-Nahavandi, K., Mohebali, M., Mahvi, A. H., et al. 2015. Evaluation of *Cryptosporidium* oocyst and *Giardia* cyst removal efficiency from urban and slaughterhouse wastewater treatment plants and assessment of cyst viability in wastewater effluent samples from Tehran, Iran. *Journal of Water Reuse and Desalination* 5(3):372–90.

Hatib, A., Hassou, N., Benchekroun, M. N., et al. 2020. The waterborne and foodborne viral diseases related to reemerging of poliovirus. In: *Emerging and reemerging viral pathogens* (pp. 999–1015). Academic Press.

Hawash, Y. 2014. Evaluation of an immunoassay-based algorithm for screening and identification of *Giardia* and *Cryptosporidium* antigens in human faecal specimens from Saudi Arabia. *Journal of Parasitology Research* 2014: 213745.

Hayoun, M. A., Smith, M. E. and Shorman, M. 2019. *Brucellosis*. StarPearls Publishing.

Heaton, D., Huang, S., Shiau, R., et al. 2018. Trichinellosis outbreak linked to consumption of privately raised raw boar meat-California, 2017. *Morbidity and Mortality Weekly Report* 67(8):247.

Henry, M. and Fouladkhah, A. 2019. Outbreak history, biofilm formation, and preventive measures for control of *Cronobacter sakazakii* in infant formula and infant care settings. *Microorganisms* 7(3):77.

Heyworth, M. F. 2014. Diagnostic testing for *Giardia* infections. *Transactions of the Royal Society of Tropical Medicine and Hygiene* 108(3):123–5.

Hijjawi, N., Yang, R., Hatmal, M. M., et al. 2018. Comparison of ELISA, nested PCR and sequencing and a novel qPCR for detection of *Giardia* isolates from Jordan. *Experimental Parasitology* 185: 23–8.

Hill, D., Coss, C., Dubey, J. P., et al. 2011. Identification of a sporozoite-specific antigen from *Toxoplasma gondii*. *The Journal of Parasitology* 97(2):328–37.

Hiroi, S., Kawahata, T. and Furubayashi, K. 2020. First isolation of human adenovirus type 85 by molecular analysis of adenoviruses in cases of urethritis. *Journal of Medical Microbiology* 69(2):265–9.

Hoeflinger, J. L. and Miller, M. J. 2017. *Cronobacter sakazakii* ATCC 29544 autoaggregation requires FliC flagellation, not motility. *Frontiers in Microbiology* 8: 301.

Hong, S. T. and Fang, Y. 2012. *Clonorchis sinensis* and clonorchiasis, an update. *Parasitology International* 61(1):17–24.

Hooshyar, H., Ghafarinasab, S., Arbabi, M., Delavari, M. and Rasti, S. 2017. Genetic variation of *Giardia lamblia* isolates from food-handlers in Kashan, Central Iran. *Iranian Journal of Parasitology* 12(1):83–9.

Horiki, N., Furukawa, K., Kitade, T., et al. 2015. Endoscopic findings and lesion distribution in *Amebic colitis*. *Journal of Infection and Chemotherapy* 21(6):444–8.

Hoseinifar, S. H., Ahmadi, A., Raeisi, M., Hoseini, S. M., Khalili, M. and Behnampour, N. 2017. Comparative study on immunomodulatory and growth enhancing effects of three prebiotics (galactooligosaccharide, fructooligosaccharide and inulin) in common carp (*Cyprinus carpio*). *Aquaculture Research* 48(7):3298–307.

Houf, K., Devriese, L. A., Van Hoof, J., Vandamme, P. and Vandamme, P. 2001. Susceptibility of *Arcobacter butzleri*, *Arcobacter cryaerophilus*, and *Arcobacter skirrowii* to antimicrobial agents used in selective media. *Journal of Clinical Microbiology* 39(4):1654–6.

Hu, S., Yu, Y., Li, R., Wu, X., Xiao, X. and Wu, H. 2016. Rapid detection of *Cronobacter sakazakii* by real-time PCR based on the cgcA gene and TaqMan probe with internal amplification control. *Canadian Journal of Microbiology* 62(3):191–200.

Hu, X., Collier, M. G. and Xu, F. 2020. Hepatitis A outbreaks in developed countries: Detection, control, and prevention. *Foodborne Pathogens and Disease* 17(3):166–71.

Hungerford, J. M. 2010. Scombroid poisoning: A review. *Toxicon* 56(2):231–43.

Hutkins, R. W., Krumbeck, J. A., Bindels, L. B., et al. 2016. Prebiotics: Why definitions matter. *Current Opinion in Biotechnology* 37: 1–7.

Imai, K., Hagi, A., Inoue, Y., Amarasiri, M. and Sano, D. 2020. Virucidal efficacy of Olanexidine gluconate as a hand antiseptic against human Norovirus. *Food and Environmental Virology* 12(2):180–90.

Isolauri, E. 2004. Dietary modification of atopic disease: Use of probiotics in the prevention of atopic dermatitis. *Current Allergy and Asthma Reports* 4(4):270–5.

Ito, A. and Budke, C. M. 2014. Culinary delights and travel? A review of zoonotic cestodiases and metacestodiases. *Travel Medicine and Infectious Disease* 12(6):582–91.

Ivanovic, J., Baltic, M. Z., Boskovic, M., et al. 2015. Anisakis infection and allergy in humans. *Procedia Food Science* 5: 101–4.

Janda, J. M., Abbott, S. L. and McIver, C. J. 2016. *Plesiomonas shigelloides* revisited. *Clinical Microbiology Reviews* 29(2):349–74.

Javid, M. H. 2011. *Campylobacter* infections. In: *Medscape reference drugs and diseases and procedures* (pp. 1–7). Emedicine Medscape.

Jenkins, M. B., Eaglesham, B. S., Anthony, L. C., Kachlany, S. C., Bowman, D. D. and Ghiorse, W. C. 2010. Significance of wall structure, macromolecular composition, and surface polymers to the survival and transport of *Cryptosporidium parvum* oocysts. *Applied and Environmental Microbiology* 76(6):1926–34.

Jeswal, P. and Kumar, D. 2015. Mycobiota and natural incidence of aflatoxins, ochratoxin A, and citrinin in Indian spices confirmed by LC-MS/MS. *International Journal of Microbiology* 2015: 242486.

Johansen, M. V., Sithithaworn, P., Bergquist, R. and Utzinger, J. 2010. Towards improved diagnosis of zoonotic trematode infections in Southeast Asia. In: *Advances in parasitology* (Vol. 73, pp. 171–95). Academic Press.

Johler, S., Weder, D., Bridy, C., Huguenin, M. C., Robert, L., Hummerjohann, J. and Stephan, R. 2015. Outbreak of staphylococcal food poisoning among children and staff at a Swiss boarding school due to soft cheese made from raw milk. *Journal of Dairy Science* 98(5):2944–8.

John, D. T. and Petri, W. A. 2013. *Markell and voge's medical parasitology: Arabic Bilingual Edition*. Elsevier Health Sciences.

Jourdan, P. M., Lamberton, P. H., Fenwick, A. and Addiss, D. G. 2018. Soil-transmitted helminth infections. *The Lancet* 391(10117):252–65.

Kadariya, J., Smith, T. C. and Thapaliya, D. 2014. *Staphylococcus aureus* and staphylococcal food-borne disease: An ongoing challenge in public health. *BioMed Research International* 2014:827965.

Kamala, A., Shirima, C., Jani, B., et al. 2018. Outbreak of an acute aflatoxicosis in Tanzania during 2016. *World Mycotoxin Journal* 11(3):311–20.

Kang, C. R., Bang, J. H. and Cho, S. I. 2019. *Campylobacter jejuni* foodborne infection associated with cross-contamination: Outbreak in Seoul in 2017. *Infection and Chemotherapy* 51(1):21–7.

Kang, C. R., Kim, Y. Y., Lee, J. I., Joo, H. D., Jung, S. W. and Cho, S. I. 2018. An outbreak of scombroid fish poisoning associated with consumption of yellowtail fish in Seoul, Korea. *Journal of Korean Medical Science* 33(37):e235.

Kao, C. Y., Sheu, B. S. and Wu, J. J. 2016. *Helicobacter pylori* infection: An overview of bacterial virulence factors and pathogenesis. *Biomedical Journal* 39(1):14–23.

Karst, S. M. 2010. Pathogenesis of noroviruses, emerging RNA viruses. *Viruses* 2(3):748–81.

Kenyon, J., Inns, T., Aird, H., Swift, C., Astbury, J., Forester, E. and Decraene, V. 2020. Campylobacter outbreak associated with raw drinking milk, North West England, 2016. *Epidemiology and Infection* 148: e13.

Kern, P., da Silva, A. M., Akhan, O., Müllhaupt, B., Vizcaychipi, K. A., Budke, C. and Vuitton, D. A. 2017. The echinococcoses: Diagnosis, clinical management and burden of disease. In: *Advances in parasitology* (Vol. 96, pp. 259–369). Academic Press.

Kerry, R. G., Pradhan, P., Samal, D., et al. 2018. Probiotics: The ultimate nutritional supplement. In *Microbial biotechnology* (pp. 141–52). Springer.

Khalil, I. A., Troeger, C., Rao, P. C., et al. 2018. Morbidity, mortality, and long-term consequences associated with diarrhoea from Cryptosporidium infection in children younger than 5 years: A meta-analyses study. *The Lancet Global Health* 6(7):e758–68.

Khanal, S., Ghimire, P. and Dhamoon, A. S. 2018. The repertoire of adenovirus in human disease: The innocuous to the deadly. *Biomedicines* 6(1):30.

Khazaeli, P., Najafi, M. L., Bahaabadi, G. A., Shakeri, F. and Naghibzadeh, A. 2014. Evaluation of aflatoxin contamination in raw and roasted nuts in consumed Kerman and effect of roasting, packaging and storage conditions. *Life Science Journal* 10: 578–83.

Khuntikeo, N., Titapun, A., Loilome, W., et al. 2018. Current perspectives on opisthorchiasis control and cholangiocarcinoma detection in Southeast Asia. *Frontiers in Medicine* 5: 117.

Khurana, S. and Chaudhary, P. 2018. Laboratory diagnosis of cryptosporidiosis. *Tropical Parasitology* 8(1):2–7.

Kiatsopit, N., Sithithaworn, P., Saijuntha, W., Petney, T. N. and Andrews, R. H. 2013. *Opisthorchis viverrini*: Implications of the systematics of first intermediate hosts, Bithynia snail species in Thailand and Lao PDR. *Infection, Genetics and Evolution* 14: 313–9.

Kilonzo-Nthenge, A., Rotich, E. and Nahashon, S. N. 2013. Evaluation of drug-resistant Enterobacteriaceae in retail poultry and beef. *Poultry Science* 92(4):1098–107.

Kim, H. S., Chon, J. W., Kim, Y. J., Kim, D. H., Kim, M. S. and Seo, K. H. 2015. Prevalence and characterization of extended-spectrum-β-lactamase-producing *Escherichia coli* and *Klebsiella pneumoniae* in ready-to-eat vegetables. *International Journal of Food Microbiology* 207: 83–6.

Kim, J. B., Jeong, H. R., Park, Y. B., Kim, J. M. and Oh, D. H. 2010. Food poisoning associated with emetic-type of *Bacillus cereus* in Korea. *Foodborne Pathogens and Disease* 7(5):555–63.

Kim, J. H., Choi, M. H., Bae, Y. M., Oh, J. K., Lim, M. K. and Hong, S. T. 2011. Correlation between discharged worms and fecal egg counts in human clonorchiasis. *PLoS Neglected Tropical Diseases* 5(10):e1339.

Kim, K. P. and Loessner, M. J. 2008. *Enterobacter sakazakii* invasion in human intestinal Caco-2 cells requires the host cell cytoskeleton and is enhanced by disruption of tight junction. *Infection and Immunity* 76(2):562–70.

Kingston, J. J., Radhika, M., Roshini, P. T., Raksha, M. A., Murali, H. S. and Batra, H. V. 2010. Molecular characterization of lactic acid bacteria recovered from natural fermentation of beet root and carrot Kanji. *Indian Journal of Microbiology* 50(3):292–8.

Kirk, M. D., Pires, S. M., Black, R. E., et al. 2015. World Health Organization estimates of the global and regional disease burden of 22 foodborne bacterial, protozoal, and viral diseases, 2010: A data synthesis. *PLoS Medicine* 12(12):e1001921.

Kirn, T. J., Lafferty, M. J., Sandoe, C. M. and Taylor, R. K. 2000. Delineation of pilin domains required for bacterial association into microcolonies and intestinal colonization by *Vibrio cholerae*. *Molecular Microbiology* 35(4):896–910.

Kitaoka, M., Miyata, S. T., Unterweger, D. and Pukatzki, S. 2011. Antibiotic resistance mechanisms of *Vibrio cholerae*. *Journal of Medical Microbiology* 60(4):397–407.

Kobyliak, N., Conte, C., Cammarota, G., et al. 2016. Probiotics in prevention and treatment of obesity: A critical view. *Nutrition and Metabolism* 13(1):14.

Korsgaard, H., Helwigh, B., Sørensen, A. I. V., Skiby, J. E. and Høg, B. B. 2014. EFSA (European Food Safety Authority) and ECDC (European Centre for Disease Prevention and Control), 2014. The European Union summary report on trends and sources of zoonoses, zoonotic agents and food-borne outbreaks in 2012.

Kregiel, D., James, S. A., Rygala, A., Berlowska, J., Antolak, H. and Pawlikowska, E. 2018. Consortia formed by yeasts and acetic acid bacteria *Asaia* spp. in soft drinks. *Antonie van Leeuwenhoek* 111(3):373–83.

Kristensen, N. B., Bryrup, T., Allin, K. H., Nielsen, T., Hansen, T. H. and Pedersen, O. 2016. Alterations in fecal microbiota composition by probiotic supplementation in healthy adults: A systematic review of randomized controlled trials. *Genome Medicine* 8(1):52.

Krumbeck, J. A., Maldonado-Gomez, M. X., Ramer-Tait, A. E. and Hutkins, R. W. 2016. Prebiotics and synbiotics: Dietary strategies for improving gut health. *Current Opinion in Gastroenterology* 32(2):110–9.

Kulkarni, A. S., Joshi, A. R., Shere, S. K. and Bindu, R. S. 2014. Appendicular taeniasis presenting as acute appendicitis a report of two cases with review of literature. *International Journal of Health Science and Research* 4(4):194–7.

Kumthip, K., Khamrin, P., Ushijima, H. and Maneekarn, N. 2019. Enteric and non-enteric adenoviruses associated with acute gastroenteritis in pediatric patients in Thailand, 2011 to 2017. *PLoS ONE* 14(8):e0220263.

Kusumaningrum, H. D., Van Putten, M. M., Rombouts, F. M. and Beumer, R. R. 2002. Effects of antibacterial dishwashing liquid on foodborne pathogens and competitive microorganisms in kitchen sponges. *Journal of Food Protection* 65(1):61–5.

Labbé, R. G. and García, S. eds. 2013. *Guide to Foodborne Pathogens*. Wiley Blackwell.

Ladeira, C., Frazzoli, C. and Orisakwe, O. E. 2017. Engaging one health for non-communicable diseases in Africa: Perspective for mycotoxins. *Frontiers in Public Health* 5: 266.

Lalle, M. 2010. Giardiasis in the post genomic era: Treatment, drug resistance and novel therapeutic perspectives. *Infectious Disorders-Drug Targets* 10(4):283–94.

Lan, Z., Bastos, M. and Menzies, D. 2016. Treatment of human disease due to *Mycobacterium bovis*: A systematic review. *European Respiratory Journal* 48(5):1500–3.

Lanata, C. F., Fischer-Walker, C. L., Olascoaga, A. C., Torres, C. X., Aryee, M. J., Black, R. E. and Child Health Epidemiology Reference Group of the World Health Organization and UNICEF 2013. Global causes of diarrheal disease mortality in children <5 years of age: A systematic review. *PLoS ONE* 8(9):e72788.

Lang, A. M. 2003. Focused review: Botulinum toxin type A therapy in chronic pain disorders. *Archives of Physical Medicine and Rehabilitation* 84(3):S69–73.

Larson, C. L., Christensen, J. E., Pacheco, S. A., Minnich, S. A. and Konkel, M. E. 2008. *Campylobacter jejuni* secretes proteins via the flagellar type III secretion system that contribute to host cell invasion and gastroenteritis. In *Campylobacter*, 3rd ed. (pp. 315–332). American Society of Microbiology.

Le Barz, M., Anhê, F. F., Varin, T. V., et al. 2015. Probiotics as complementary treatment for metabolic disorders. *Diabetes and Metabolism Journal* 39(4):291–303.

Lee, S. H., Park, H. and Yu, S. T. 2015. *Diphyllobothrium latum* infection in a child with recurrent abdominal pain. *Korean Journal of Pediatrics* 58(11):451–3.

Lemos, T. S., Cequinel, J. C., Costa, T. P., Navarro, A. B., Sprada, A., Shibata, F. K., Gondolfo, R. and Tuon, F. F., 2018. Outbreak of human brucellosis in Southern Brazil and historical review of data from 2009 to 2018. *PLoS Neglected Tropical Diseases* 12(9):e0006770.

Lerda, D. 2017. Fumonisins in foods from Cordoba (Argentina). *Toxicology Open Access* 3(2):125.

Levecke, B., Montresor, A., Albonico, M., et al. 2014. Assessment of anthelmintic efficacy of mebendazole in school children in six countries where soil-transmitted helminths are endemic. *PLoS Neglected Tropical Diseases* 8(10):e3204.

Levican, A., Collado, L. and Figueras, M. J. 2013. *Arcobacter cloacae* sp. nov. and *Arcobacter suis* sp. nov., two new species isolated from food and sewage. *Systematic and Applied Microbiology* 36(1):22–7.

Levine, M., Ruha, A. M., Graeme, K., Brooks, D. E., Canning, J. and Curry, S. C. 2011. Toxicology in the ICU: Part 3: Natural toxins. *Chest* 140(5):1357–70.

Lhomme, S., Marion, O., Abravanel, F., Chapuy-Regaud, S., Kamar, N. and Izopet, J. 2016. Hepatitis E pathogenesis. *Viruses* 8(8):212.

Li, J., Shi, K., Sun, F., et al. 2019. Identification of human pathogenic *Enterocytozoon bieneusi, Cyclospora cayetanensis*, and *Cryptosporidium parvum* on the surfaces of vegetables and fruits in Henan, China. *International Journal of Food Microbiology* 307: 108292.

Li, L., Shimizu, H., Doan, L. T. P., et al. 2004. Characterizations of adenovirus type 41 isolates from children with acute gastroenteritis in Japan, Vietnam, and Korea. *Journal of Clinical Microbiology* 42(9):4032–9.

Li, M. H., Li, Y. J., Hu, B., et al. 2019. Clinical characteristics and next generation sequencing of three cases of *Listeria monocytogenes* meningitis with complications. *Zhonghua Er Ke za Zhi = Chinese Journal of Pediatrics* 57(8):603–7.

Li, P., Xu, L., Xiang, J., et al. 2015. Taeniasis related frequent intestinal obstruction: Case report and mini-review. *Journal of Gastroenterology and Hepatology Research* 4(1):1455–8.

Li, X., Liu, W., Wang, J., et al. 2012. Rapid detection of *Trichinella spiralis* larvae in muscles by loop-mediated isothermal amplification. *International Journal for Parasitology* 42(13–14):1119–26.

Li, Y., Hu, X., Liu, X., et al. 2012. Serological diagnosis of clonorchiasis: Using a recombinant propeptide of cathepsin L proteinase from *Clonorchis sinensis* as a candidate antigen. *Parasitology Research* 110(6):2197–203.

Liao, C. D., Chiueh, L. C. and Shih, D. Y. C. 2009. Determination of zearalenone in cereals by high-performance liquid chromatography and liquid chromatography-electrospray tandem mass spectrometry. *Journal of Food and Drug Analysis* 17(1).

Lima, A. D., Fortes, R. C., Novaes, M. G. and Percário, S. 2012. Poisonous mushrooms; a review of the most common intoxications. *Nutrición Hospitalaria* 27(2):402–8.

Lindsay, D. S. and Dubey, J. P. 2020. Toxoplasmosis in wild and domestic animals. In *Toxoplasma gondii* (pp. 293–320). Academic Press.

Lion, T. 2014. Adenovirus infections in immunocompetent and immunocompromised patients. *Clinical Microbiology Reviews* 27(3):441–62.

Lister, P. D., Wolter, D. J. and Hanson, N. D. 2009. Antibacterial-resistant *Pseudomonas aeruginosa*: Clinical impact and complex regulation of chromosomally encoded resistance mechanisms. *Clinical Microbiology Reviews* 22(4):582–610.

Liu, C. and Guo, J. 2018. Characteristics of ventilator-associated pneumonia due to hypervirulent *Klebsiella pneumoniae* genotype in genetic background for the elderly in two tertiary hospitals in China. *Antimicrobial Resistance and Infection Control* 7(1):95.

Liu, D. ed. 2017. *Laboratory models for foodborne infections.* CRC Press.

Liuzzi, V. C., Fanelli, F., Tristezza, M., et al. 2017. Transcriptional analysis of *Acinetobacter* sp. neg1 capable of degrading ochratoxin A. *Frontiers in Microbiology* 7: 2162.

Longenberger, A. H., Gronostaj, M. P., Yee, G. Y., et al. 2014. *Yersinia enterocolitica* infections associated with improperly pasteurized milk products: Southwest Pennsylvania, March–August, 2011. *Epidemiology and Infection* 142(8):1640–50.

López-Romalde, S., Magariños, B., Núñez, S., Toranzo, A. E. and Romalde, J. L. 2003. Phenotypic and genetic characterization of *Pseudomonas anguilliseptica* strains isolated from fish. *Journal of Aquatic Animal Health* 15(1):39–47.

Lu, Y., Ma, M., Wang, H., et al. 2020. An outbreak of norovirus-related acute gastroenteritis associated with delivery food in Guangzhou, southern China. *BMC Public Health* 20(1):1–7.

Lynch, K. M., Zannini, E., Wilkinson, S., Daenen, L. and Arendt, E. K. 2019. Physiology of acetic acid bacteria and their role in vinegar and fermented beverages. *Comprehensive Reviews in Food Science and Food Safety* 18(3):587–625.

Macfarlane, G. T., Steed, H. and Macfarlane, S. 2008. Bacterial metabolism and health related effects of galacto-oligosaccharides and other prebiotics. *Journal of Applied Microbiology* 104(2):305–44.

Madhu, A. N., Giribhattanavar, P., Narayan, M. S. and Prapulla, S. G. 2010. Probiotic lactic acid bacterium from kanjika as a potential source of vitamin B 12: Evidence from LC-MS, immunological and microbiological techniques. *Biotechnology Letters* 32(4):503–6.

Mairiang, E. and Mairiang, P. 2003. Clinical manifestation of opisthorchiasis and treatment. *Acta Tropica* 88(3):221–7.

Malimas, T., Vu, H. T. L., Muramatsu, Y., Yukphan, P. and Tanasupawat, S. 2017. Systematics of acetic acid bacteria. In: *Acetic acid bacteria* (pp. 3–43). CRC Press.

Maluping, R. P., Lavilla-Pitogo, C. R., DePaola, A., Janda, J. M., Krovacek, K. and Greko, C. 2005. Antimicrobial susceptibility of *Aeromonas* spp., *Vibrio* spp. and *Plesiomonas shigelloides* isolated in the Philippines and Thailand. *International Journal of Antimicrobial Agents* 25(4):348–50.

Marco, M. L., Heeney, D., Binda, S., et al. 2017. Health benefits of fermented foods: Microbiota and beyond. *Current Opinion in Biotechnology* 44: 94–102.

Marcos, L. A. and Gotuzzo, E. 2013. Intestinal protozoan infections in the immunocompromised host. *Current Opinion in Infectious Diseases* 26(4):295–301.

Maresca, M. 2013. From the gut to the brain: Journey and pathophysiological effects of the food-associated trichothecene mycotoxin deoxynivalenol. *Toxins* 5(4):784–820.

Marin, S., Ramos, A. J., Cano-Sancho, G. and Sanchis, V. 2013. Mycotoxins: Occurrence, toxicology, and exposure assessment. *Food and Chemical Toxicology* 60: 218–37.

Mas-Coma, S., Valero, M. A. and Bargues, M. D. 2014. Fascioliasis. In *Digenetic trematodes* (pp. 77–114). Springer.

Mazzoni, E., Scandolara, A., Giorni, P., Pietri, A. and Battilani, P. 2011. Field control of *Fusarium* ear rot, *Ostrinia nubilalis* (Hübner), and fumonisins in maize kernels. *Pest Management Science* 67(4):458–65.

McGregor, A. C. and Wright, S. G. 2015. Gastrointestinal symptoms in travellers. *Clinical Medicine* 15(1):93.

Megged, O., Chazan, B., Ganem, A., Ayoub, A., Yanovskay, A., Sakran, W., Miron, D., Dror-Cohen, A., Kennes, Y., Berdenstein, S. and Glikman, D. 2016. Brucellosis outbreak in children and adults in two areas in Israel. *The American Journal of Tropical Medicine and Hygiene* 95(1):31–4.

Mehndiratta, M. M., Mehndiratta, P. and Pande, R. 2014. Poliomyelitis: Historical facts, epidemiology, and current challenges in eradication. *The Neurohospitalist* 4(4):223–9.

Mehta, P., Prakash, M. and Khandelwal, N. 2016. Radiological manifestations of hydatid disease and its complications. *Tropical Parasitology* 6(2):103.

Meier, O. and Greber, U. F. 2004. Adenovirus endocytosis. *The Journal of Gene Medicine: A* 6(S1):S152–63.

Meng, F., Pan, X. and Tong, W. 2018. Rifampicin versus streptomycin for brucellosis treatment in humans: A meta-analysis of randomized controlled trials. *PLoS ONE* 13(2):e0191993.

Milićević, D. R., Škrinjar, M. and Baltić, T. 2010. Real and perceived risks for mycotoxin contamination in foods and feeds: Challenges for food safety control. *Toxins* 2(4):572–92.

Mishra, H. N. and Das, C. 2003. A review on biological control and metabolism of aflatoxin. *Critical Reviews in Food Science and Nutrition* 43(3):245–64.

Molina, G., Pimentel, M. R. and Pastore, G. M. 2013. *Pseudomonas*: A promising biocatalyst for the bioconversion of terpenes. *Applied Microbiology and Biotechnology* 97(5):1851–64.

Mondiale de la Santé Organisation, and World Health Organization 2017. Cholera vaccines: WHO position paper–August 2017–Vaccins Anticholériques: Note de Synthèse de l'OMS–Août 2017. *Weekly Epidemiological Record Relevé Epidémiologique Hebdomadaire* 92(34):477–98.

Moore, D. A., Girdwood, R. W. A. and Chiodini, P. L. 2002. Treatment of anisakiasis with albendazole. *The Lancet* 360(9326):54.

Morais, R. D. A. P. B., Freire, A. B. C., Barbosa, D. R. L., et al. 2016. *Acute Toxoplasmosis Outbreak in the Municipality of Ponta de Pedras, Marajó Archipelago, Pará State, Brazil/ Surto de Toxoplasmose Aguda no Município de Ponta de Pedras, Arquipélago do Marajó, Estado do Pará, Brasil: Características Clínicas, Laboratoriais e Epidemiológicas.*

Mottola, A., Bonerba, E., Bozzo, G. et al. 2016. Occurrence of emerging food-borne pathogenic *Arcobacter* spp. isolated from pre-cut (ready-to-eat) vegetables. *International Journal of Food Microbiology* 236: 33–7.

Muehlenbachs, A., Bhatnagar, J. and Zaki, S. R. 2015. Tissue tropism, pathology and pathogenesis of enterovirus infection. *The Journal of Pathology* 235(2):217–28.

Muhammad, I., Sun, X., Wang, H., et al. 2017. Curcumin successfully inhibited the computationally identified CYP2A6 enzyme-mediated bioactivation of aflatoxin b1 in arbor acres broiler. *Frontiers in Pharmacology* 8: 143.

Nainan, O. V., Xia, G., Vaughan, G. and Margolis, H. S. 2006. Diagnosis of hepatitis A virus infection: A molecular approach. *Clinical Microbiology Reviews* 19(1):63–79.

Nathanail, A. V., Varga, E., Meng-Reiterer, J., et al. 2015. Metabolism of the Fusarium mycotoxins T-2 toxin and HT-2 toxin in wheat. *Journal of Agricultural and Food Chemistry* 63(35):7862–72.

Navarrete-Bolaños, J. L. 2012. Improving traditional fermented beverages: How to evolve from spontaneous to directed fermentation. *Engineering in Life Sciences* 12(4):410–8.

Nawaz, M., Khan, S. A., Tran, Q., Sung, K., Khan, A. A., Adamu, I. and Steele, R. S. 2012. Isolation and characterization of multidrug-resistant *Klebsiella* spp. isolated from shrimp imported from Thailand. *International Journal of Food Microbiology* 155(3):179–84.

Nelapati, S., Tumati, S. R., Thirtham, M. R., Ramani Pushpa, R. N., Kamisetty, A. K. and Ch, B. K. 2020. Occurrence, virulence gene and antimicrobial susceptibility profiles of *Arcobacter* sp. isolated from catla (*Catla catla*) in India. *Letters in Applied Microbiology* 70(5):365–71.

Nie, G., Wang, T., Lu, S., Liu, W., Li, Y. and Lei, J. 2014. Detection of *Clonorchis sinensis* circulating antigen in sera from Chinese patients by immunomagnetic bead ELISA based on IgY. *PLoS ONE* 9(12):e113208.

Nieuwenhuizen, N. E. and Lopata, A. L. 2013. Anisakis–a food-borne parasite that triggers allergic host defences. *International Journal for Parasitology* 43(12–13):1047–57.

Nigam, P. K. and Nigam, A. 2010. Botulinum toxin. *Indian Journal of Dermatology* 55(1):8–14.

Nomoto, A. 2007. Molecular aspects of poliovirus pathogenesis. *Proceedings of the Japan Academy, Series B* 83(8):266–75.

Nordt, S. P. and Pomeranz, D. 2016. Scombroid poisoning from tilapia. *The American Journal of Emergency Medicine* 34(2):339.

Nygren, B. L., Schilling, K. A., Blanton, E. M., Silk, B. J., Cole, D. J. and Mintz, E. D. 2013. Foodborne outbreaks of shigellosis in the USA, 1998–2008. *Epidemiology and Infection* 141(2):233–41.

O'Dempsey, T. 2010. Helminthic infections. In: *Antibiotic and chemotherapy* (pp. 842–59). WB Saunders.

Ogawa, T., Asai, Y., Tamai, R., Makimura, Y., Sakamoto, H., Hashikawa, S. and Yasuda, K. 2006. Natural killer cell activities of synbiotic *Lactobacillus casei* ssp. *casei* in conjunction with dextran. *Clinical and Experimental Immunology* 143(1):103–9.

Ogawa, T., Asai, Y., Yasuda, K. and Sakamoto, H. 2005. Oral immunoadjuvant activity of a new synbiotic *Lactobacillus casei* subsp. *casei* in conjunction with dextran in BALB/c mice. *Nutrition Research* 25(3):295–304.

Ok, K. S., Kim, Y. S., Song, J. H., et al. 2009. *Trichuris trichiura* infection diagnosed by colonoscopy: Case reports and review of literature. *The Korean Journal of Parasitology* 47(3):275–80.

Okello, A. L. and Thomas, L. F. 2017. Human taeniasis: Current insights into prevention and management strategies in endemic countries. *Risk Management and Healthcare Policy* 10: 107.

Oliveira, R. P. D. S., Florence, A. C. R., Perego, P., De Oliveira, M. N. and Converti, A. 2011. Use of lactulose as prebiotic and its influence on the growth, acidification profile and viable counts of different probiotics in fermented skim milk. *International Journal of Food Microbiology* 145(1):22–7.

Onsurathum, S., Pinlaor, P., Haonon, O., et al. 2016. Effects of fermentation time and low temperature during the production process of Thai pickled fish (pla-som) on the viability and infectivity of *Opisthorchis viverrini* metacercariae. *International Journal of Food Microbiology* 218: 1–5.

Ortega, Y. R. and Sanchez, R. 2010. Update on *Cyclospora cayetanensis*, a food-borne and waterborne parasite. *Clinical Microbiology Reviews* 23(1):218–34.

Ortiz-Castillo, F., Salinas-Aragón, L. E., Sánchez-Aguilar, M., et al. 2012. Amoebic toxic colitis: Analysis of factors related to mortality. *Pathogens and Global Health* 106(4):245–8.

Osman, K. M., Hassan, H. M., Orabi, A. and Abdelhafez, A. S. 2014. Phenotypic, antimicrobial susceptibility profile and virulence factors of *Klebsiella pneumoniae* isolated from buffalo and cow mastitic milk. *Pathogens and Global Health* 108(4):191–9.

Ostry, V., Malir, F., Toman, J. and Grosse, Y. 2017. Mycotoxins as human carcinogens - The IARC monographs classification. *Mycotoxin Research* 33(1):65–73.

Otero-Abad, B. and Torgerson, P. R. 2013. A systematic review of the epidemiology of echinococcosis in domestic and wild animals. *PLoS Neglected Tropical Diseases* 7(6):e2249.

Otto, S. J., Levett, P. N., Reid-Smith, R. J., et al. 2020. Antimicrobial resistance of human *Campylobacter* species infections in Saskatchewan, Canada (1999–2006):A historical provincial collection of all reported cases. *Foodborne Pathogens and Disease* 17(3):178–86.

Overdevest, I. T. M. A., Heck, M., Van Der Zwaluw, K., et al. 2014. Extended spectrum β-lactamase producing *Klebsiella* spp. in chicken meat and humans: A comparison of typing methods. *Clinical Microbiology and Infection* 20(3):251–5.

Owino, C. O. and Chu, J. J. H. 2019. Recent advances on the role of host factors during non-poliovirus enteroviral infections. *Journal of Biomedical Science* 26(1):47.

Pandey, K. R., Naik, S. R. and Vakil, B. V. 2015. Probiotics, prebiotics and synbiotics - A review. *Journal of Food Science and Technology* 52(12):7577–87.

Pandya, J. P. and Arade, P. C. 2016. Mycotoxin: A devil of human, animal and crop health. *Advances in Life Science* 5: 3937–41.

Para, R. A., Fomda, B. A., Jan, R. A., Shah, S. and Koul, P. A. 2018. Microbial etiology in hospitalized North Indian adults with community-acquired pneumonia. *Lung India: Official Organ of Indian Chest Society* 35(2):108.

Patel, S. and Goyal, A. 2012. The current trends and future perspectives of prebiotics research: A review. *3 Biotech* 2(2):115–25.

Paternoster, G., Boo, G., Wang, C., et al. 2020. Epidemic cystic and alveolar echinococcosis in Kyrgyzstan: An analysis of national surveillance data. *The Lancet Global Health* 8(4):e603–11.

Peña, A. S. 2007. Intestinal flora, probiotics, prebiotics, synbiotics and novel foods. *Revista Española de Enfermedades Digestivas* 99(11):653.

Penner, J. L. 1988. The genus *Campylobacter*: A decade of progress. *Clinical Microbiology Reviews* 1(2):157–72.

Perez, K. L., Lucia, L. M., Cisneros-Zevallos, L., Castillo, A. and Taylor, T. M. 2012. Efficacy of antimicrobials for the disinfection of pathogen contaminated green bell pepper and of consumer cleaning methods for the decontamination of knives. *International Journal of Food Microbiology* 156(1):76–82.

Petney, T. N., Andrews, R. H., Saijuntha, W., Wenz-Mücke, A. and Sithithaworn, P. 2013. The zoonotic, fish-borne liver flukes *Clonorchis sinensis*, *Opisthorchis felineus* and *Opisthorchis viverrini*. *International Journal for Parasitology* 43(12–13):1031–46.

Phillips, C. A. and Long, C. 2001. The survival of *Arcobacter butzleri* on chicken. *International Journal of Medical Microbiology* 291(S31):93.

Phung, L. T., Loukas, A., Brindley, P. J., Sripa, B. and Laha, T. 2014. Retrotransposon OV-RTE-1 from the carcinogenic liver fluke *Opisthorchis viverrini*: Potential target for DNA-based diagnosis. *Infection, Genetics and Evolution* 21: 443–51.

Piarroux, M., Piarroux, R., Giorgi, R., Knapp, J., Bardonnet, K., Sudre, B., Watelet, J., Dumortier, J., Gérard, A., Beytout, J. and Abergel, A. 2011. Clinical features and evolution of alveolar echinococcosis in France from 1982 to 2007: Results of a survey in 387 patients. *Journal of Hepatology* 55(5):1025–33.

Pilotte, N., Papaiakovou, M., Grant, J. R., Bierwert, L. A., Llewellyn, S., McCarthy, J. S. and Williams, S. A. 2016. Improved PCR-based detection of soil transmitted helminth infections using a next-generation sequencing approach to assay design. *PLoS Neglected Tropical Diseases* 10(3):e0004578.

Pokusaeva, K., Fitzgerald, G. F. and van Sinderen, D. 2011. Carbohydrate metabolism in bifidobacteria. *Genes and Nutrition* 6(3):285–306.

Pothakos, V., Illeghems, K., Laureys, D., Spitaels, F., Vandamme, P. and De Vuyst, L. 2016. Acetic acid bacteria in fermented food and beverage ecosystems. In *Acetic acid bacteria* (pp. 73–99). Springer.

Pozio, E. 2013. Integrating animal health surveillance and food safety: The example of Anisakis. *Revue Scientifique et Technique (International Office of Epizootics)* 32(2):487–96.

Prakash, V. and Bhimji, S. S. 2017. *Abscess, amebic liver.* StarPearls Publishing.

Prichard, R. K., Basáñez, M. G., Boatin, B. A., et al. 2012. A research agenda for helminth diseases of humans: Intervention for control and elimination. *PLoS Neglected Tropical Diseases* 6(4):e1549.

Procop, G. W. 2009. North American paragonimiasis (caused by *Paragonimus kellicotti*) in the context of global paragonimiasis. *Clinical Microbiology Reviews* 22(3):415–46.

Puspanadan, S., Afsah-Hejri, L., Loo, Y. Y., et al. 2012. Detection of *Klebsiella pneumoniae* in raw vegetables using most probable number-polymerase chain reaction (MPN-PCR). *International Food Research Journal* 19(4):1757.

Qian, M. B., Chen, Y. D. and Yan, F. 2013. Time to tackle clonorchiasis in China. *Infectious Diseases of Poverty* 2(1):4.

Qian, M. B., Utzinger, J., Keiser, J. and Zhou, X. N. 2016. Clonorchiasis. *The Lancet* 387(10020):800–10.

Quaglia, N. C. and Dambrosio, A. 2018. *Helicobacter pylori*: A foodborne pathogen? *World Journal of Gastroenterology* 24(31):3472.

Radoshevich, L. and Cossart, P. 2018. *Listeria monocytogenes*: Towards a complete picture of its physiology and pathogenesis. *Nature Reviews in Microbiology* 16(1):32–46.

Rahimi, E. 2014. Prevalence and antimicrobial resistance of *Arcobacter* species isolated from poultry meat in Iran. *British Poultry Science* 55(2):174–80.

Rahimi, E., Abdos, F., Momtaz, H., Torki Baghbadorani, Z. and Jalali, M. 2013. *Bacillus cereus* in infant foods: Prevalence study and distribution of enterotoxigenic virulence factors in Isfahan Province, Iran. *The Scientific World Journal* 2013.

Rajeish, M., Shekar, M., Madhushree, H. N. and Venugopal, M. N. 2016. Presumptive case of ciguatera fish poisoning in Mangalore, India. *Current Science* 111(9):1543–7.

Rakhmanova, A., Khan, Z. A. and Shah, K. 2018. A mini review fermentation and preservation: Role of lactic acid bacteria. *MOJ Food Process Technology* 6(5):414–7.

Ramees, T. P., Dhama, K., Karthik, K., et al. 2017. *Arcobacter*: An emerging foodborne zoonotic pathogen, its public health concerns and advances in diagnosis and control – A comprehensive review. *Veterinary Quarterly* 37(1):136–61.

Ramig, R. F. 2004. Pathogenesis of intestinal and systemic rotavirus infection. *Journal of Virology* 78(19):10213–20.

Ranoux, D., Gury, C., Fondarai, J., Mas, J. L. and Zuber, M. 2002. Therapy with botulinum toxin. *Journal of Neurology, Neurosurgery and Psychiatry* 72(459):62.

Raposo, A., Pérez, E., de Faria, C. T., Ferrús, M. A. and Carrascosa, C. 2017. Food spoilage by *Pseudomonas* spp. - An overview. *Foodborne Pathogens and Antibiotic Resistance*: 41–58.

Rathlavath, S., Kohli, V., Singh, A. S., Lekshmi, M., Tripathi, G., Kumar, S. and Nayak, B. B. 2017. Virulence genotypes and antimicrobial susceptibility patterns of *Arcobacter butzleri* isolated from seafood and its environment. *International Journal of Food Microbiology* 263: 32–7.

Rattanachaikunsopon, P. and Phumkhachorn, P. 2010. Lactic acid bacteria: Their antimicrobial compounds and their uses in food production. *Annals of Biological Research* 1(4):218–28.

Rhoades, R. E., Tabor-Godwin, J. M., Tsueng, G. and Feuer, R. 2011. Enterovirus infections of the central nervous system. *Virology* 411(2):288–305.

Richardson Jr, R. F., Remler, B. F., Katirji, B. and Hatem Murad, M. 1998. Guillain–Barré syndrome after Cyclospora infection. *Muscle and Nerve: Official Journal of the American Association of Electrodiagnostic Medicine* 21(5), 669–71.

Richter, I. and Fidler, A. E. 2015. Detection of marine microalgal biotoxins using bioassays based on functional expression of tunicate xenobiotic receptors in yeast. *Toxicon* 95: 13–22.

Richter, R. L. and Vedamuthu, E. R. 2015. Milk and milk products. In: *Compendium of methods for the microbiological examination of foods*. F. P. Downes and K. Ito, eds. (pp. 483–95). American Public Health Association.

Rico, E., Pérez, C., Belver, A., et al. 2020. Norovirus detection in environmental samples in Norovirus outbreaks in closed and semi-closed settings. *Journal of Hospital Infection* 105(1):3–9.

Ridgway, H. F., Safarik, J., Phipps, D., Carl, P. and Clark, D. 1990. Identification and catabolic activity of well-derived gasoline-degrading bacteria from a contaminated aquifer. *Applied and Environmental Microbiology* 56(11):3565–75.

Robinson, T. P., Bu, D. P., Carrique-Mas, J., et al. 2016. Antibiotic resistance is the quintessential one health issue. *Transactions of the Royal Society of Tropical Medicine and Hygiene* 110(7):377–80.

Roeder, K., Erler, K., Kibler, S., et al. 2010. Characteristic profiles of ciguatera toxins in different strains of *Gambierdiscus* spp. *Toxicon* 56(5):731–8.

Romano, P. and Capece, A. 2010. *Saccharomyces cerevisiae* as Bakers' yeast. In: *Encyclopedia of biotechnology in agriculture and food* (pp. 1–4). CRC Press.

Rouprêt, M., Babjuk, M., Compérat, E., et al. 2015. European association of urology guidelines on upper urinary tract urothelial cell carcinoma: 2015 update. *European Urology* 68(5):868–79.

Russo, P., Capozzi, V., Spano, G., Corbo, M. R., Sinigaglia, M. and Bevilacqua, A. 2016. Metabolites of microbial origin with an impact on health: Ochratoxin A and biogenic amines. *Frontiers in Microbiology* 7: 482.

Ryan, E. T., Hill, D. R., Solomon, T., Endy, T. P. and Aronson, N. 2019. *Hunter's tropical medicine and emerging infectious diseases*. Elsevier Health Sciences.

Ryan, U. and Cacciò, S. M. 2013. Zoonotic potential of *Giardia*. *International Journal for Parasitology* 43(12–13):943–56.

Sabran, M. R., Jamaluddin, R., Ahmad, Z. and Ahmad, Z. 2013. A mini review on aflatoxin exposure in Malaysia: Past, present and future. *Frontiers in Microbiology* 4: 334.

Safa, A., Nair, G. B. and Kong, R. Y. 2010. Evolution of new variants of *Vibrio cholerae* O1. *Trends in Microbiology* 18(1):46–54.

Sah, R., Khadka, S., Khadka, M., et al. 2017. Human fascioliasis by *Fasciola hepatica*: The first case report in Nepal. *BMC Research Notes* 10(1):1–4.

Saichana, N., Matsushita, K., Adachi, O., Frébort, I. and Frebortova, J. 2015. Acetic acid bacteria: A group of bacteria with versatile biotechnological applications. *Biotechnology Advances* 33(6):1260–71.

Salih, B. A. 2009. *Helicobacter pylori* infection in developing countries: The burden for how long? *Saudi Journal of Gastroenterology* 15(3):201.

Santana, B. G., Dalton, J. P., Camargo, F. V., Parkinson, M. and Ndao, M. 2013. The diagnosis of human fascioliasis by enzyme-linked immunosorbent assay (ELISA) using recombinant cathepsin L protease. *PLoS Neglected Tropical Diseases* 7(9):e2414.

Santi, L., Maggioli, C., Mastroroberto, M., Tufoni, M., Napoli, L. and Caraceni, P. 2012. Acute liver failure caused by *Amanita phalloides* poisoning. *International Journal of Hepatology* 2012.

Sarwar, A., Aziz, T., Din, J., Khalid, A., Rahman, T. and Daudzai, Z. 2018. Pros of lactic acid bacteria in microbiology: A review. *Biomed Letters* 4. 59–66.

Sattar, S. A., Tetro, J., Bidawid, S. and Farber, J. 2000. Foodborne pread of hepatitis A: Recent studies on virus survival, transfer and inactivation. *Canadian Journal of Infectious Diseases* 11(3).

Scallan, E., Hoekstra, R. M., Angulo, F. J., et al. 2011. Foodborne illness acquired in the United States - Major pathogens. *Emerging Infectious Diseases* 17(1):7.

Schenk-Jaeger, K. M., Rauber-Lüthy, C., Bodmer, M., Kupferschmidt, H., Kullak-Ublick, G. A. and Ceschi, A. 2012. Mushroom poisoning: A study on circumstances of exposure and patterns of toxicity. *European Journal of Internal Medicine* 23(4):e85–91.

Schokker, E. P. and van Boekel, M. A. 1999. Kinetics of thermal inactivation of the extracellular proteinase from *Pseudomonas fluorescens* 22F: Influence of pH, calcium, and protein. *Journal of Agricultural and Food Chemistry* 47(4):1681–6.

Scholz, T., Garcia, H. H., Kuchta, R. and Wicht, B. 2009. Update on the human broad tapeworm (genus *Diphyllobothrium*), including clinical relevance. *Clinical Microbiology Reviews* 22(1):146–60.

Settanni, L., Tanguler, H., Moschetti, G., Reale, S., Gargano, V. and Erten, H. 2011. Evolution of fermenting microbiota in tarhana produced under controlled technological conditions. *Food Microbiology* 28(7):1367–73.

Shah, A. H., Saleha, A. A., Murugaiyah, M., Zunita, Z. and Memon, A. A. 2012. Prevalence and distribution of *Arcobacter* spp. in raw milk and retail raw beef. *Journal of Food Protection* 75(8):1474–8.

Shah, A. H., Saleha, A. A., Zunita, Z. and Murugaiyah, M. 2011. *Arcobacter* – An emerging threat to animals and animal origin food products? *Trends in Food Science and Technology* 22(5):225–36.

Shah, J. and Shahidullah, A. 2018. *Ascaris lumbricoides*: A startling discovery during screening colonoscopy. *Case Reports in Gastroenterology* 12(2):224–9.

Shahid, M., Malik, A., Adil, M., Jahan, N. and Malik, R. 2009. Comparison of beta-lactamase genes in clinical and food bacterial isolates in India. *The Journal of Infection in Developing Countries* 3(8):593–8.

Shamloo, E., Jalali, M., Mirlohi, M., Madani, G., Metcalf, D. and Merasi, M. R. 2015. Prevalence of *Listeria species* in raw milk and traditional dairy products in Isfahan, Iran. *International Journal of Environmental Health Engineering* 4(1):1.

Shao, C. C., Xu, M. J., Alasaad, S., Song, H. Q., Peng, L., Tao, J. P. and Zhu, X. Q. 2014. Comparative analysis of microRNA profiles between adult *Ascaris lumbricoides* and *Ascaris suum*. *BMC Veterinary Research* 10(1):1–6.

Shapiro, K., Kim, M., Rajal, V. B., Arrowood, M. J., Packham, A., Aguilar, B. and Wuertz, S. 2019. Simultaneous detection of four protozoan parasites on leafy greens using a novel multiplex PCR assay. *Food Microbiology* 84, 103252.

Sharad, S. M., Ligia Maria, C. E. and Ricardo, I. 2012. A novel strategy for environmental control of soil transmitted helminthes. *Health Environment Journal* 3(3):28.

Shimada, T., Baker, C. S., Morton, O. and Hawk, J. L. 1993. Outbreak of *Vibrio cholerae* non-O1 in India and Bangladesh. *Lancet* 341(8856):1347.

Shin, E. C. and Jeong, S. H. 2018. Natural history, clinical manifestations, and pathogenesis of hepatitis A. *Cold Spring Harbor Perspectives in Medicine* a031708.

Shiota, M., Saitou, K., Mizumoto, H., et al. 2010. Rapid detoxification of cereulide in *Bacillus cereus* food poisoning. *Pediatrics* 125(4):e951–5.

Shon, A. S., Bajwa, R. P. and Russo, T. A. 2013. Hypervirulent (hypermucoviscous) *Klebsiella pneumoniae*: A new and dangerous breed. *Virulence* 4(2):107–18.

Sichewo, P. R., Etter, E. M. C. and Michel, A. L. 2019. Prevalence of *Mycobacterium bovis* infection in traditionally managed cattle at the wildlife-livestock interface in South Africa in the absence of control measures. *Veterinary Research Communications* 43(3):155–64.

Šilha, D., Vacková, B. and Šilhová, L. 2018. Occurrence of virulence–associated genes in *Arcobacter butzleri* and *Arcobacter cryaerophilus* isolates within the Czech Republic. *Folia Microbiology* 64(1):25–31.

Silva, J., Leite, D., Fernandes, M., Mena, C., Gibbs, P. A. and Teixeira, P. 2011. *Campylobacter* spp. as a foodborne pathogen: A review. *Frontiers in Microbiology* 2: 200.

Singh, B. B., Sharma, J. K., Juyal, P. D. and Gill, J. P. S. 2004. Seroprevalence of *Sarcocystis* sp. of cattle in Punjab. *Journal of Veterinary Parasitology* 18(1):75–6.

Singh, T. S., Sugiyama, H. and Rangsiruji, A. 2012. Paragonimus and paragonimiasis in India. *The Indian Journal of Medical Research* 136(2):192.

Siriken, B., Ayaz, N. D. and Erol, I. 2013. Prevalence and serotype distribution of *Listeria monocytogenes* in salted anchovy, raw anchovy, and raw mussel using IMS-based cultivation technique and PCR. *Journal of Aquatic Food Product Technology* 22(1):77–82.

Siriken, B., Pamuk, Ş., Ozakin, C., Gedikoglu, S. and Eyigör, M. 2006. A note on the incidences of *Salmonella* spp., *Listeria* spp. and *Escherichia coli* O157: H7 serotypes in Turkish sausage (Soudjouck). *Meat Science* 72(1):177–81.

Soares, R. and Tasca, T. 2016. Giardiasis: An update review on sensitivity and specificity of methods for laboratorial diagnosis. *Journal of Microbiological Methods* 129: 98–102.

Song, J. R., Fu, Y. W., Li, P., Du, T., Du, X. J. and Wang, S. 2020. Protective effect of recombinant proteins of *Cronobacter sakazakii* during pregnancy on the offspring. *Frontiers in Cellular and Infection Microbiology* 10: 1–15.

Sripa, B., Kaewkes, S., Intapan, P. M., Maleewong, W. and Brindley, P. J. 2010. Foodborne trematodiases in Southeast Asia: Epidemiology, pathology, clinical manifestation and control. In: *Advances in parasitology* (Vol. 72, pp. 305–50). Academic Press.

Stasiak, A. C. and Stehle, T. 2020. Human adenovirus binding to host cell receptors: A structural view. *Medical Microbiology and Immunology* 209(3):325–33.

Stojkovic, M., Junghanss, T., Veeser, M., Weber, T. F. and Sauer, P. 2016. Endoscopic treatment of biliary stenosis in patients with alveolar echinococcosis–report of 7 consecutive patients with serial ERC approach. *PLoS Neglected Tropical Diseases* 10(2):e0004278.

Strunz, E. C., Addiss, D. G., Stocks, M. E., Ogden, S., Utzinger, J. and Freeman, M. C. 2014. Water, sanitation, hygiene, and soil-transmitted helminth infection: A systematic review and meta-analysis. *PLoS Medicine* 11(3):e1001620.

Sugiyama, H. 1980. *Clostridium botulinum* neurotoxin. *Microbiological Reviews* 44(3):419.

Sugrue, I., Tobin, C., Ross, R. P., Stanton, C. and Hill, C. 2019. Foodborne pathogens and zoonotic diseases. In: *Raw milk* (pp. 259–72). Academic Press.

Sun, F., Wu, D., Qiu, Z., Jin, M., Li, J. and Li, J. 2010. Development of real-time PCR systems based on SYBR Green for the specific detection and quantification of *Klebsiella pneumoniae* in infant formula. *Food Control* 21(4):487–91.

Sun, G. G., Wang, Z. Q., Liu, C. Y., et al. 2015. Early serodiagnosis of trichinellosis by ELISA using excretory–secretory antigens of *Trichinella spiralis* adult worms. *Parasites and Vectors* 8(1):484.

Taban, B. M. and Saichana, N. 2017. Physiology and biochemistry of acetic acid bacteria. In *Acetic Acid Bacteria* (pp. 71–91). CRC Press.

Tabatabaei, M., Aski, H. S., Shayegh, H. and Khoshbakht, R. 2014. Occurrence of six virulence-associated genes in *Arcobacter* species isolated from various sources in Shiraz, Southern Iran. *Microbial Pathogenesis* 66: 1–4.

Talebi Bezmin Abadi, A. 2018. Diagnosis of *Helicobacter pylori* using invasive and non-invasive approaches. *Journal of Pathogens 2018*, 9064952, 1-13.

Tamang, J. P., Chettri, R. and Sharma, R. M. 2009. Indigenous knowledge of Northeast women on production of ethnic fermented soybean foods. *Indian Journal of Traditional Knowledge* 8(1):122–6.

Tamang, J. P., Tamang, B., Schillinger, U., Guigas, C. and Holzapfel, W. H. 2009. Functional properties of lactic acid bacteria isolated from ethnic fermented vegetables of the Himalayas. *International Journal of Food Microbiology* 135(1):28–33.

Tamarozzi, F., Akhan, O., Cretu, C. M., et al. 2018. Prevalence of abdominal cystic echinococcosis in rural Bulgaria, Romania, and Turkey: A cross-sectional, ultrasound-based, population study from the HERACLES project. *The Lancet Infectious Diseases* 18(7):769–78.

Tang, B., Liu, M., Wang, L., et al. 2015. Characterisation of a high-frequency gene encoding a strongly antigenic cystatin-like protein from *Trichinella spiralis* at its early invasion stage. *Parasites and Vectors* 8(1):78.

Tang, Z. L., Huang, Y. and Yu, X. B. 2016. Current status and perspectives of *Clonorchis sinensis* and clonorchiasis: Epidemiology, pathogenesis, omics, prevention and control. *Infectious Diseases of Poverty* 5(1):71.

Taylor, R. K., Miller, V. L., Furlong, D. B. and Mekalanos, J. J. 1987. Use of phoA gene fusions to identify a pilus colonization factor coordinately regulated with cholera toxin. *Proceedings of the National Academy of Sciences of the United States of America* 84(9):2833–7.

Temesgen, T. T., Robertson, L. J. and Tysnes, K. R. 2019. A novel multiplex real-time PCR for the detection of *Echinococcus multilocularis*, *Toxoplasma gondii*, and *Cyclospora cayetanensis* on berries. *Food Research International* 125: 108636.

Teppawar, R. N., Chaudhari, S. P., Moon, S. L., Shinde, S. V., Khan, W. A. and Patil, A. R. 2018. Zoonotic tuberculosis: A concern and strategies to combat. In: *Basic biology and applications of actinobacteria*. IntechOpen, DOI: 10.5772/intechopen.76802.

Tewari, A. and Abdullah, S. 2015. *Bacillus cereus* food poisoning: International and Indian perspective. *Journal of Food Science and Technology* 52(5):2500–11.

Thoen, C. O., Kaplan, B., Thoen, T. C., Gilsdorf, M. J. and Shere, J. A. 2016. Zoonotic tuberculosis. A comprehensive one HEALTH approach. *Medicina (Buenos Aires)* 76(3).

Thomas, D. J. I., Strachan, N., Goodburn, K., Rotariu, O. and Hutchison, M. L. 2012. A review of the published literature and current production and processing practices in smoked fish processing plants with emphasis on contamination by *Listeria monocytogenes*. Final FSA Report.

Todd, E. C. and Grieg, J. D. 2015. Viruses of foodborne origin: A review. *Virus Adaptation and Treatment* 7: 25–45.

Tofalo, R., Fusco, V., Böhnlein, C., et al. 2019. The life and times of yeasts in traditional food fermentations. *Critical Reviews in Food Science and Nutrition* 60(18):1–30.

Tofalo, R., Perpetuini, G., Schirone, M., Suzzi, G. and Corsetti, A. 2013. Yeast biota associated to naturally fermented table olives from different Italian cultivars. *International Journal of Food Microbiology* 161(3):203–8.

Tompkins, B. J., Wirsing, E., Devlin, V., Kamhi, L., Temple, B., Weening, K., Cavallo, S., Allen, L., Brinig, P., Goode, B. and Fitzgerald, C. 2013. Multistate outbreak of *Campylobacter jejuni* infections associated with undercooked chicken livers—northeastern United States, 2012. *Morbidity and Mortality Weekly Report* 62(44):874.

Toranzo, A. E., Magariños, B. and Romalde, J. L. 2005. A review of the main bacterial fish diseases in mariculture systems. *Aquaculture* 246(1–4):37–61.

Tortorella, V., Masciari, P., Pezzi, M., et al. 2014. Histamine poisoning from ingestion of fish or scombroid syndrome. *Case Reports in Emergency Medicine 2014*.

Trabulus, S. and Altiparmak, M. R. 2011. Clinical features and outcome of patients with amatoxin-containing mushroom poisoning. *Clinical Toxicology* 49(4):303–10.

Tralamazza, S. M., Bemvenuti, R. H., Zorzete, P., de Souza Garcia, F. and Corrêa, B. 2016. Fungal diversity and natural occurrence of deoxynivalenol and zearalenone in freshly harvested wheat grains from Brazil. *Food Chemistry* 196: 445–50.

Traylor, J. and Mathew, D. 2020. Histamine (scombroid toxicity, mahi-mahi flush) toxicity. In: *StatPearls [Internet]*. StatPearls Publishing, 1–14.

Trevisan, C., Devleesschauwer, B., Schmidt, V., Winkler, A. S., Harrison, W. and Johansen, M. V. 2017. The societal cost of *Taenia solium* cysticercosis in Tanzania. *Acta Tropica* 165: 141–54.

Troeger, C., Khalil, I. A., Rao, P. C., et al. 2018. Rotavirus vaccination and the global burden of rotavirus diarrhea among children younger than 5 years. *JAMA Pediatrics* 172(10):958–65.

Troiano, G., Mercone, A., Bagnoli, A. and Nante, N. 2017. International travelers' sociodemographic, health, and travel characteristics: An Italian study. *Annals of Global Health* 83(2):380–5.

Troiano, G. and Nante, N. 2019. Human trichinellosis in Italy: An epidemiological review since 1989. *Journal of Preventive Medicine and Hygiene* 60(2):E71.

Tufarelli, V. and Laudadio, V. 2016. An overview on the functional food concept: Prospectives and applied researches in probiotics, prebiotics and synbiotics. *Journal of Experimental Biology and Agricultural Sciences* 4(3):273–8.

Turner, A. D. and Goya, A. B. 2015. Occurrence and profiles of lipophilic toxins in shellfish harvested from Argentina. *Toxicon* 102: 32–42.

Ukwuru, M. U., Muritala, A. and Eze, L. U. 2018. Edible and non-edible wild mushrooms: Nutrition, toxicity and strategies for recognition. *Journal of Clinical Nutrition and Metabolites* 2: 1–20.

Van De, N., Le, T. H., Lien, P. T. H. and Eom, K. S. 2014. Current status of taeniasis and cysticercosis in Vietnam. *The Korean Journal of Parasitology* 52(2):125–9.

Vanathy, K., Parija, S. C., Mandal, J., Hamide, A. and Krishnamurthy, S. 2017. Cryptosporidiosis: A mini review. *Tropical Parasitology* 7(2):72–80.

Vandamme, P., Pugina, P., Benzi, G., et al. 1992. Outbreak of recurrent abdominal cramps associated with *Arcobacter butzleri* in an Italian school. *Journal of Clinical Microbiology* 30(9):2335–7.

Vandenberg, O., Dediste, A., Houf, K., et al. 2004. *Arcobacter* species in humans. *Emerging Infectious Diseases* 10(10):1863–7.

Van Tassell, J. A., Martin, N. H., Murphy, S. C., Wiedmann, M., Boor, K. J. and Ivy, R. A. 2012. Evaluation of various selective media for the detection of *Pseudomonas* species in pasteurized milk. *Journal of Dairy Science* 95(3):1568–74.

Varma, M., Hester, J. D., Schaefer, F. W., Ware, M. W. and Lindquist, H. A. 2003. Detection of *Cyclospora cayetanensis* using a quantitative real-time PCR assay. *Journal of Microbiological Methods* 53(1):27–36.

Vassalos, C. M., Economou, V., Vassalou, E. and Papadopoulou, C. 2009. Brucellosis in humans: Why is it so elusive? *Reviews in Medical Microbiology* 20(4):63–73.

Velasco-Tirado, V., Alonso-Sardón, M., Lopez-Bernus, A., et al. 2018. Medical treatment of cystic echinococcosis: Systematic review and meta-analysis. *BMC Infectious Diseases* 18(1):306.

Villafana, R. T., Ramdass, A. C. and Rampersad, S. N. 2019. Selection of *Fusarium trichothecene* toxin genes for molecular detection depends on TRI gene cluster organization and gene function. *Toxins* 11(1):36.

Villazanakretzer, D. L., Napolitano, P. G., Cummings, K. F. and Magann, E. F. 2016. Fish parasites: A growing concern during pregnancy. *Obstetrical and Gynecological Survey* 71(4):253–9.

Vinjé, J. 2015. Advances in laboratory methods for detection and typing of Norovirus. *Journal of Clinical Microbiology* 53(2):373–81.

Visciano, P., Schirone, M., Berti, M., Milandri, A., Tofalo, R. and Suzzi, G. 2016. Marine biotoxins: Occurrence, toxicity, regulatory limits and reference methods. *Frontiers in Microbiology* 7: 1051.

Viswanath, A. and Williams, M. 2019. *Trichuris trichiura* (whipworm, roundworm). In *StatPearls [Internet]*. StatPearls Publishing, 1–16.

Wang, B., Shao, Y. and Chen, F. 2015. Overview on mechanisms of acetic acid resistance in acetic acid bacteria. *World Journal of Microbiology and Biotechnology* 31(2):255–63.

Wang, Z. D., Liu, Q., Liu, H. H., Li, S., Zhang, L., Zhao, Y. K. and Zhu, X. Q. 2018. Prevalence of *Cryptosporidium*, *Microsporidia* and *Isospora* infection in HIV-infected people: A global systematic review and meta-analysis. *Parasites and Vectors* 11(1):28.

Watanabe, K. and Petri Jr, W. A. 2015. Molecular biology research to benefit patients with *Entamoeba histolytica* infection. *Molecular Microbiology* 98(2):208–17.

Wattinger, L., Stephan, R., Layer, F. and Johler, S. 2012. Comparison of *Staphylococcus aureus* isolates associated with food intoxication with isolates from human nasal carriers and human infections. *European Journal of Clinical Microbiology and Infectious Diseases* 31(4):455–64.

Wei-Dong, Y., Min-Yi, W., Jie-Sheng, L., Xi-Chun, P. and Hong-Ye, L. 2009. Reporter gene assay for detection of shellfish toxins. *Biomedical and Environmental Sciences* 22(5):419–22.

Welburn, S. C., Beange, I., Ducrotoy, M. J. and Okello, A. L. 2015. The neglected zoonoses - The case for integrated control and advocacy. *Clinical Microbiology and Infection* 21(5):433–43.

Wen, H., Vuitton, L., Tuxun, T., Li, J., Vuitton, D. A., Zhang, W. and McManus, D. P. 2019. Echinococcosis: Advances in the 21st century. *Clinical Microbiology Reviews* 32(2):e00075–18.

WHO 2009. WHO model formulary. In: *World Health Organization* (p. 644). WHO Press, Geneva.

Williams, J. G., Deschl, U. and Williams, G. M. 2011. DNA damage in fetal liver cells of turkey and chicken eggs dosed with aflatoxin B 1. *Archives of Toxicology* 85(9):1167–72.

Wouafo, M., Pouillot, R., Kwetche, P. F., Tejiokem, M. C., Kamgno, J. and Fonkoua, M. C. 2006. An acute foodborne outbreak due to *Plesiomonas shigelloides* in Yaounde, Cameroon. *Foodborne Pathogens and Disease* 3(2):209–11.

Wright, J. E., Werkman, M., Dunn, J. C. and Anderson, R. M. 2018. Current epidemiological evidence for predisposition to high or low intensity human helminth infection: A systematic review. *Parasites and Vectors* 11(1):65.

Wu, H., Liu, B. G., Liu, J. H., Pan, Y. S., Yuan, L. and Hu, G. Z. 2012. Phenotypic and molecular characterization of CTX-M-14 extended-spectrum beta-lactamase and plasmid-mediated ACT-like AmpC beta-lactamase produced by *Klebsiella pneumoniae* isolates from chickens in Henan Province, China. *Genetics and Molecular Research* 11(3):3357–64.

Wu, L., Liao, P., He, L., et al. 2015b. Dietary L-arginine supplementation protects weanling pigs from deoxynivalenol-induced toxicity. *Toxins* 7(4):1341–54.

Wu, L., Liao, P., He, L., Ren, W., Yin, J., Duan, J. and Li, T. 2015a. Growth performance, serum biochemical profile, jejunal morphology, and the expression of nutrients transporter genes in deoxynivalenol (DON)-challenged growing pigs. *BMC Veterinary Research* 11(1):144.

Wu, Y., Wen, J., Ma, Y., Ma, X. and Chen, Y. 2014. Epidemiology of foodborne disease outbreaks caused by *Vibrio parahaemolyticus*, China, 2003–2008. *Food Control* 46: 197–202.

Wuerz, T., Kane, J. B., Boggild, A. K., et al. 2012. A review of amoebic liver abscess for clinicians in a nonendemic setting. *Canadian Journal of Gastroenterology = Journal Canadien de Gastroenterologie* 26(10):729–33.

Xia, F. Q., Liu, P. N. and Zhou, Y. H. 2015. Meningoencephalitis caused by *Plesiomonas shigelloides* in a Chinese neonate: Case report and literature review. *Italian Journal of Pediatrics* 41(1):3.

Xu, L. L., Jiang, B., Duan, J. H., et al. 2014. Efficacy and safety of praziquantel, tribendimidine and mebendazole in patients with co-infection of *Clonorchis sinensis* and other helminths. *PLoS Neglected Tropical Diseases* 8(8):e3046.

Xue, Y., Zhai, S., Wang, Z., et al. 2020. The Yersinia phage x1 administered orally efficiently protects a murine chronic enteritis model against *Yersinia enterocolitica* infection. *Frontiers in Microbiology* 11: 351.

Yamaguchi, M., Kawai, T., Kitagawa, M. and Kumeda, Y. 2013. A new method for rapid and quantitative detection of the *Bacillus cereus* emetic toxin cereulide in food products by liquid chromatography-tandem mass spectrometry analysis. *Food Microbiology* 34(1):29–37.

Yamai, S., Okitsu, T., Shimada, T. and Katsube, Y. 1997. Distribution of serogroups of *Vibrio cholerae* non-O1 non-O139 with specific reference to their ability to produce cholera toxin, and addition of novel serogroups. *Kansenshogaku Zasshi. The Journal of the Japanese Association for Infectious Diseases* 71(10):1037–45.

Yan, Q. and Fanning, S. 2015. Strategies for the identification and tracking of *Cronobacter* species: An opportunistic pathogen of concern to neonatal health. *Frontiers in Pediatrics* 3: 38.

Yang, J., Li, J., Jiang, Y., et al. 2014. Natural occurrence, analysis, and prevention of mycotoxins in fruits and their processed products. *Critical Reviews in Food Science and Nutrition* 54(1):64–83.

Yang, Y., Cai, Y. N., Tong, M. W., et al. 2016. Serological tools for detection of *Trichinella* infection in animals and humans. *One Health* 2. 25–30.

Yardan, T., Baydin, A., Eden, A. O., Akdemir, H. U., Aygun, D., Acar, E. and Arslan, B. 2010. Wild mushroom poisonings in the Middle Black Sea region in Turkey: Analyses of 6 years. *Human and Experimental Toxicology* 29(9):767–71.

Yoshida, A., Doanh, P. N. and Maruyama, H. 2019. Paragonimus and paragonimiasis in Asia: An update. *Acta Tropica* 199: 105074.

Yu, S., Yu, P., Wang, J., et al. 2020. A study on prevalence and characterization of *Bacillus cereus* in ready-to-eat foods in China. *Frontiers in Microbiology* 10: 3043.

Zhou, Z. Y., Shi, G. Q., Fontaine, R., et al. 2012. Evidence for the natural toxins from the mushroom *Trogia venenata* as a cause of sudden unexpected death in Yunnan Province, China. *Angewandte Chemie* 124(10):2418–20.

Zhu, L., He, J., Cao, X., Huang, K., Luo, Y. and Xu, W. 2016. Development of a double-antibody sandwich ELISA for rapid detection of *Bacillus cereus* in food. *Scientific Reports* 6(1):1–10.

Zlateva, S., Marinov, P., Yovcheva, M., Bonchev, G., Ivanov, D. and Georgiev, K. 2017. Ciguatera poisoning: Pacific disease, foodborne poisoning from fish in warm seas and oceans. *Journal of IMAB–Annual Proceeding (Scientific Papers)* 23(1):1474–9.

Zulkifli, Y., Alitheen, N. B., Son, R., Raha, A. R., Samuel, L., Yeap, S. K. and Nishibuchi, M. 2009. Random amplified polymorphic DNA-PCR and ERIC PCR analysis on *Vibrio parahaemolyticus* isolated from cockles in Padang, Indonesia. *International Food Research Journal* 16(2):141–50.

CHAPTER 3

Molecular Epidemiology of Foodborne Diseases

Afshan Shafi, Umar Farooq, Kashif Akram,
Muhammad Zaki Khan, Zafar Hayat, and Khizar Hayat

CONTENTS

3.1 INTRODUCTION

Epidemiology is a branch of medicine responsible for the study of cause, distribution, and determinants of disease that helps in preventing and controlling the disease. Many studies regarding epidemiology were initiated in the 1940s that influence the prevention of many human diseases; for example, in the 1940s many community trials were conducted on fluoride supplementation for the prevention of dental caries (Arnold et al. 1953). Later in 1947, the Framingham Heart Study was initiated to identify determinants and risk factors of heart diseases (Oppenheimer 2005). And in 1954, polio vaccine trials were initiated that led to the prevention and eradication of poliomyelitis (Francis et al. 1955).

Traditional epidemiological studies are less effective for a deeper understanding of diseases and their spread (Susser 1998), while molecular epidemiology (ME) can give a

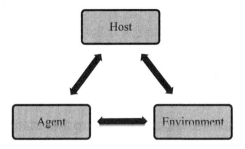

FIGURE 3.1 Epidemiological triad for foodborne disease.

deeper understanding of the disease and its spread. Molecular epidemiology was developed by merging epidemiology with molecular biology. This merging led to the development of laboratory techniques which could be applied on large sample sizes, allowing epidemiologists to conduct epidemiological studies at the molecular level (Haghdoost 2008). It is said that the term "molecular epidemiology" was introduced in 1979 by Lower (Vineis 2007), although some studies show that Kilbourne used the term in 1973 in describing influenza subtype distribution around the world (Kilbourne 1973). Subsequently, a dramatic rise in molecular epidemiological studies was observed in various research areas.

Any disease which is caused by ingesting food is called a "foodborne disease." Today, it is the most common public health problem, as millions of people are affected worldwide, and the most difficult for public health organizations to combat. Majowicz et al. (2016) stated that "public health specialists working in infectious foodborne infection, obesity, food insecurity, food allergy, and dietary contaminants should actively consider how their seemingly targeted public health actions may produce unintended positive or negative population health impacts." Therefore, among all public health-related problems, the issue of foodborne disease needs to be focused on.

The basic concept of molecular epidemiology is that there may be harmful microorganism or chemicals present in food that can cause disease. The three points of the triad in Figure 3.1 are equally important and must be addressed.

3.1.1 Prevalence of Foodborne Diseases

Contaminated food causes illness in 1/10 of the people in the world. Africa is the most affected region, Southeast Asia is the second-highest region, while Europe has the least number of patients affected by foodborne illness, globally (WHO 2015). According to WHO's report (2016), about 420,000 people die due to foodborne illness every year. The death rate is highest in children under age 5 comprising about 30% of total deaths. Almost all the regions of the world are affected by foodborne illness but the number of cases is highest in low-income regions (WHO 2015). Newman et al. (2015) stated that food preparation and handling is performed by people having low socioeconomic status. Those persons received specific hygienic instructions while working in food or other related industries.

The Food Safety News (2013) reported that there are two main factors–less access to the healthcare system and poor nutrition—responsible for (bacterial and viral) infections and foodborne illness. It is clear from literature that *Campylobacter* and *Salmonella* incidence are high in those groups of low socioeconomic status (Newman et al. 2015).

An important foodborne pathogen, *Campylobacter*, is a serious concern in New Zealand, where they plan to reduce the per capita incidences of foodborne campylobacteriosis to 10% by 2020 (Lopez et al. 2016). In 2013, in Israel it was reported that, among all foodborne illness, *Campylobacter* was responsible for 8000 cases (Times of Israel 2016). A study was conducted in Australia which stated that most of the foodborne diseases occurred as gastroenteritis, but the effect of non-gastrointestinal illnesses and sequelae were substantial. Kirk et al. (2014) stated that the number of cases of campylobacteriosis and salmonellosis increased from 2000 to 2010.

The number of cases of foodborne illness in South Korea was 336,138 during 2008 to 2012; among these, about 2.3% of the infected were hospitalized, about 14.4% were treated by outpatient visits, and about 83.3% were not seen by any provider (Park et al. 2015). Almost 30,840 gastroenteritis patients were hospitalized due to the consumption of contaminated food. Each year in Australia, about 4.1 million people are affected by the consumption of contaminated food (Krik 2014). In the United States (US) about 76 million people are affected by foodborne illness each year, with 3,000 deaths; Malaysia has lower numbers than the US, United Kingdom, and Australia; however, this may be due to under-reporting of cases (Soon et al. 2011). Soon et al. (2011) also stated that in Malaysia unsanitary food handling procedures were responsible for foodborne diseases that caused more than 50% of food poisoning events.

The first reported foodborne disease in the United Arab Emirates (UAE) was investigated in 2011; the results found that 63 cases were reported in the first nine months, but the number of confirmed cases dropped dramatically to 518 by 2013 (Khaleej Times 2014). A study in Catalonia, Spain, October 2004–October 2005 found 181 reported cases. This same report asserted that Norovirus outbreaks are under-reported compared to bacterial outbreaks (Martinez et al. 2008). There was an outbreak of listeriosis from January 2015 to February 2016 in Italy, which was associated with contaminated pork product (Marini et al. 2016). In Canada, about 1.6 million people are affected by foodborne illness each year (Thomas et al. 2013).

According to the Food Advisory Consumer Service (2009), successful reporting systems and epidemiological surveillance in developed countries have raised awareness among people regarding the dangers of foodborne diseases and have also taught them preventive measures. These investigations illustrate the importance for countries to report all suspected infections, in line with the international food security agenda, to inform all countries' planning and budgeting (Crush and Frayne 2011). The number of deaths in 6 regions of the world during 2015 are shown in Figure 3.2.

3.1.2 Important Foodborne Diseases

Different foodborne diseases have different etiologies. Kirk et al. (2015) stated that children are most affected by foodborne diseases. There is a great need of food safety interventions for avoiding foodborne diseases, especially in middle- and low-income countries. There are two groups of foodborne diseases on the basis of the causative agent (biological or chemical). Some important foodborne diseases are:

1. Bacterial foodborne diseases
2. Fungal foodborne diseases
3. Viral foodborne diseases
4. Parasitic foodborne diseases

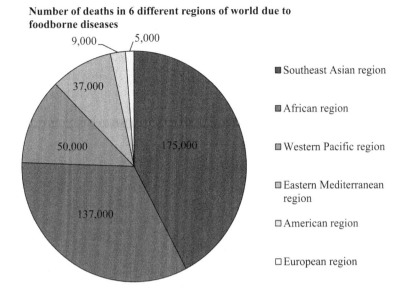

Number of deaths in 6 different regions of world due to foodborne diseases

9,000 5,000

37,000

50,000   175,000

137,000

- Southeast Asian region
- African region
- Western Pacific region
- Eastern Mediterranean region
- American region
- European region

FIGURE 3.2   Foodborne diseases—2015 global burden (WHO 2015).

3.2 BACTERIAL FOODBORNE DISEASES

Gastrointestinal diseases are caused by many bacterial strains. These bacteria cause contamination which ultimately affects human health. Some foodborne diseases which are caused by bacteria are salmonellosis (paratyphoid and typhoid), tuberculosis, and some infections such as caused by *Escherichia coli* and *Staphylococcus* spp. The disease develops (pathogenesis) by direct ingestion, then these harmful bacteria produce toxins that affect humans and animals. *Clostridium botulinum* is a bacterium that produces *Botulinum* toxin that is responsible for botulism. Among all of the foodborne diseases, millions of people affected by a group of bacterial foodborne diseases each year (Fleckenstein et al. 2010). Food safety is achieved by identifying the possible pathogenic microbe that causes foodborne disease. Narsaiah et al. (2012) stated that "the traditional methods, such as culturing, colony counting, immunoassay, and chromatography, for food contaminant detection, are time-consuming and boring" and proposed the need for new techniques, like biosensors, for help in screening.

3.2.1 Gastroenteritis

Based on the causative microbes there are different types of the gastroenteritis.

3.2.1.1 *Gastroenteritis Induced by* Aeromonas hydrophila

Aeromonas-induced gastroenteritis is caused by bacterium *Aeromonas hydrophila* a gram-negative, motile (moving), non-spore forming, and facultatively anaerobic bacteria. In various concentrations, this *Aeromonas* spp. are present in the aquatic environment, drinking water, sewage, and foods like seafood, chicken, raw milk, vegetables, ground beef, veal, lamb, and pork (WHO 2008a; Janda and Abbott 2010). Abdominal pain, fever, watery stools, and vomiting are the symptoms of this infection. In many cases, cholecystitis and bronchopneumonia are observed (Mossel et al. 1999). Four groups are affected by *Aeromonas* infection: (1) bloodborne dyscrasias, (2) wound problems and infections of the connective tissues, (3) gastrointestinal tract syndromes, and (4) metabolic diseases (Janda and Abbott 2010). Chronic dysentery can be treated by either of 2 different therapies–antimicrobial treatment and rehydration therapy (Farrar et al. 2014). Piperacillin-tazobactam, cefepime, ceftazidime, and imipenem are all known as β–lactams and strains of *A. hydrophila* shows resistance against them (Janda and Abbott 2010). Fluoroquinolone is an antimicrobial drug that is effective against *A. hydrophila* (Farrar et al. 2014).

3.2.1.2 *Gastroenteritis Induced by* Bacillus cereus

Bacillus cereus is usually defined as gram-positive, mostly mesophilic, heat-resistant, rod-shaped, facultatively anaerobic, spore-producing, and omnipresent in the environment. Its natural environment comprises marine and fresh water, decomposed organic matter, and foods, like spices, boiled or fried rice, milk, dairy products, dried food, fomites, sauces, and vegetables dishes; and invertebrates' intestinal tracts are the reservoirs of *B. cereus* (WHO 2008a; Bottone 2010). Two toxins, emetic and diarrheal, are associated with the pathogenicity of the *B. cereus*. Diarrheal syndrome which is characterized by acute diarrhea, belly pain, and nausea is caused by a diarrheal toxin. Emetic syndrome, described as severe nausea, vomiting, abdominal cramps, and (rarely) diarrhea, is due to emetic toxin (Ehling-Schulz et al. 2004). Due to β-lactamase formation, *B. cereus* strains are usually resistant to erythromycin, penicillin, tetracycline, and carbapenem. In *B. cereus* disease, the recommended treatment is empiric antibiotic therapy when awaiting the antibiotic susceptibility. Many methods are adopted for empirical treatment of patients like broad-spectrum cephalosporins and vancomycin when treated against *B. cereus* infection (Bottone 2010).

3.2.2 Botulism

Botulism is the disease caused by the anaerobic spore-forming, gram-positive, motile, rod-shaped, and protein neurotoxin-producing bacterium, *Clostridium botulinum*. Seven kinds (A to G) of toxins have been discovered; botulism is associated with type F. *C. botulinum* is responsible for four syndromes: infant botulism (infects the intestine, forms colonies, and produces toxin), wound botulism, foodborne botulism, and adult intestinal toxemia botulism (bacteria colonize the intestine and produce toxins) (Sobel 2005). Foodborne infection is the result of the consumption of contaminated food, like fish, vegetables, condiments, meat and meat products, and fish products (infant botulism is caused by honey consumption) (Fratamico et al. 2005). By intentional and accidental exposure of the botulinum toxins botulism occurs. In foodborne botulism, the patient suffers from symptoms like heart failure, paralysis and respiratory illness, difficulty in speaking and swallowing, constipation, dry mouth, dizziness, visual disturbance, headache, muscle weakness, abdominal pain, fatigue, and vomiting. Toxins are very injurious to health even in small doses, resulting in the binding of the neuromuscular pathway, blockage of acetylcholine transmission, and neuromuscular and flaccid paralysis. An intensive care regimen of regular monitoring and mechanical ventilation is necessary and required for the botulism patient. Paralysis from botulism is prolonged, lasting from weeks to months, and careful protection is important during the time of debilitation and administration of antitoxin therapy. Antitoxin can arrest spread and reduce the time of paralysis, and it depends on mechanical ventilation treatment (Sobel 2005).

3.2.3 Brucellosis

Undulant/Mediterranean fever, also known as Malta fever, is caused by *Brucella* spp. and known as brucellosis (Gul and Erdem 2015). Brucellosis in humans is caused by *Brucella suis*, *B. abortus*, and *B. melitensis* (Hossain et al. 2014). Important features of *Brucella* spp. are that they are gram-negative, short, oval, non-spore-forming, aerobic, and non-moving rods that usually grow in a pH of 6.6–7.7 and 37°C and are heat-labile (Hui et al. 2001). *Brucella* spp. grow on cattle, sheep, goats, camels, dogs, buffalo, host species, milk and milk products, and the offal that is derived from those animals (WHO 2006b). Continuous, intermittent, or irregular fever, sweating, headache, chills, arthralgias, constipation, weight reduction, anorexia, and generalized ache are the clinical symptoms of the brucellosis. These symptoms can persist for weeks up to months. In 20%–60% of cases, osteoarticular problems are the most frequent. Other problems due to brucellosis, like cardiovascular and neurological conditions, insomnia, and depression can also occur (Hui et al. 2001).

3.2.4 Campylobacteriosis

Gastrointestinal infections in human are caused by two dominant species *Campylobacter coli* and *C. jejuni*. Contaminated water, cattle, sheep, pigs, birds, livestock, and wild and domestic animals are carriers of *Campylobacter* which is a gram-negative, enteric, and non-spore-forming bacterium. It transmits to humans by consumption of raw milk and meat (especially chicken and turkey). Development of campylobacteriosis starts 2–5 days after exposure. The main symptoms of the campylobacteriosis are watery and (rarely) bloody diarrhea, nausea, severe abdominal pain, and fever. Campylobacteriosis is a very common foodborne illness in industrialized countries and it is the main reason for child and traveler's diarrhea; other diseases caused by this bacterium are cholecystitis, meningitis, pancreatitis, reactive arthritis, erythema nodosum, and endocarditis and are present in 2%–10% of patients. Frequently misdiagnosed as appendicitis, campylobacteriosis is commonly asymptomatic, and requires no treatment with antibiotics–except in children and infants (WHO 2008a; Williams et al. 2015).

3.2.5 Cholera

Vibrio cholera is a gram-negative bacterium that usually presents in the environment and can ferment sucrose. There are four subgroups of *cholera* spp. –O139, non-O139, non-O1, and O1 (Chowdhury et al. 2016). *V. cholerae* O139 and O1 can cause severe dehydration as the result of profuse watery diarrhea. Collapse and death are prevented by fluid and salt replacement. The main two subgroups, non-O1 and non-O139, present with cholera-similar diarrhea. Bacterial infections of cholera are commonly due to the consumption of seafood, fruits, and vegetables, cooked rice, and ice that are contaminated with the *V. cholera*. It is transmitted through the fecal-oral route, as the infection spreads when an infected person comes into contact with a healthy person (Finkelstein 1996). Alternating doses of antibiotics with fluid replacement is an effective treatment regimen, reducing the time and amount of diarrhea and the time of hospitalization.

3.2.6 Necrotic Enteritis

Clostridium perfringens bacterium is a gram-positive, non-motile, anaerobic, and spore-producing bacteria that is the cause of necrotic enteritis (Wells and Wilkins 1996). Although it can be found in many types of environments, it is most common in the intestines of ill and healthy people (Lacey et al. 2016). Transmission is through the fecal-oral pathway. Housing structures, insects, water, and infected food are the sources of transmission of this bacterium (Lee et al. 2011). Many problems, like abdominal cramps, frequent vomiting, diarrhea, and fever, and many kinds of diseases, like kidney and liver infection, gastrointestinal syndromes, dermatitis, and gas gangrene are results of the toxins formed by the bacterium *C. perfringens* (WHO, 2008a; Lacey et al. 2016). Another syndrome associated with these toxins is morbidity. The symptoms of food poisoning are caused by the consumption of the *C. perfringens* infected diet (Lee et al. 2011; Omernik and Płusa 2015). Due to the present lack of treatment like antimicrobial treatment, radical surgery and hyperbaric treatment, the rate of spread of the microbe increased that ultimately leads to death. Epsilon-toxin, formed by the B and D strains of bacterium *C. perfringens* is the third strongest clostridial toxin (Omernik and Płusa 2015).

3.2.7 Infection Caused by *Enterobacter sakazakii*

Enterobacter sakazakii is a gram-negative, moving, rod-shaped and nonsporulating pathogenic toxin-producing bacteria that is currently grouped into *Cronobacter*. It can cause foodborne diseases, usually in children but also adults with a weak immune system. The presence of *E. sakazakii* in dried foods, like powdered infant formula, poses a high risk for infants. Contaminated food is considered the primary route of infection. *Cronobacter* cannot exist in pasteurized powdered milk, but it has been reported in some other foods, like sausages, meat, teas, fish, herbs, spices, fruit, vegetables, milk, legume products, cereal, rice, and bread (Hunter et al. 2008; FDA, 2012). Although only powdered infant formula is related to cases of disease, this bacterium can grow in many other foods. The US Food and Drug Administration (FDA) reported an important warning, in 2012, related to the occurrence of the *E. sakazakii* in the child diet (FDA 2012). *E. sakazakii* can survive for 2 years in dried milk formula. Hydrocephalus, seizures, development delay, brain abscesses, body temperature changes, irritability, jaundice, grunting respirations, poor feeding, are the main symptoms of infection by this bacterium (Hunter et al. 2008; FDA 2012). Antibiotics like gentamicin and ampicillin are effective against *E. sakazakii* disease (Hunter et al. 2008).

3.2.8 Infection Caused by *Enterococcus faecalis*

Enterococcus faecalis bacterium is the catalase-negative, round-shape, facultatively anaerobic, gram-negative, ovoid bacterium. A few strains of *E. faecalis* produce toxin known as cytolysin, this toxin is associated with hemolytic and bactericidal function. The main reason of the foodborne disease from this bacterium is not well known. Raw milk, pasteurized milk, meat pie, pudding, sausage, evaporated milk, and cheese are the main food sources of this bacterium. After the consumption of contaminated food, different symptoms occur such as chills, fever, nausea, diarrhea, abdominal cramps. This infection presents symptoms similar to staphylococcal intoxication. There are many drugs like ampicillin, ureidopenicillin, vancomycin, and gentamicin that can be consumed singly or in combination to combat enterococcal infection (FDA 2012).

3.2.9 Infection Caused by *Escherichia coli*

The family of the *Escherichia coli* is Enterobacteriaceae. The bacterium belongs to the prokaryotic family. Other important characteristics of *E. coli* are that they are gram-negative, non-spore forming, and have capsules or microcapsules (Fratamico et al. 2005; Riemann and Cliver 2006). Human and animal feces are the main source of this bacterium and fecal material is responsible for the spread of this bacterium in the environment. Infected food, including meat and contaminated water, is the major source of the spread of this bacterium (Riemann and Cliver 2006). Suitable temperature for the growth of the *E. coli* is 37°C. There are 6 different strains of *E. coli*: (1) enterohemorrhagic *E. coli* (EHEC) related to traveler's diarrhea, (2) one which is responsible for hemolytic uremic syndrome and hemorrhagic colitis known as enterohemorrhagic *E. coli* (EHEC), (3) enteroaggregative *E. coli* (EAEC) which is associated with diarrhea in children, (4) enteropathogenic *E. coli* (EPEC) the main cause of watery diarrhea in young children and infants, (5) diffusely adherent *E. coli* (DAEC) causes diarrhea in mammals, and (6) enteroinvasive *E. coli* (EIEC) which has similar biochemistry to *Shigella* (Fratamico et al. 2005; Riemann and Cliver 2006).

There are many symptoms of the EHEC like stomach pain, vomiting, and watery diarrhea that can change to bloody diarrhea; and this bacterium can also affect the central nervous system (CNS). Diseases, like the hemolytic uremic disease, can result in death, the main reason of kidney infection and failure; young children and elderly persons are at high risk of thrombocytopenia and hemolytic anemia. Thrombotic thrombocytopenic purpura and erythema nodosum are also associated with this bacterium (Fratamico et al. 2005). Abdominal cramps and watery diarrhea symptoms occur due to the toxins produced by ETEC. Myalgia, vomiting, bloating, headache, tremor, and anorexia are the major symptoms of this bacterial infection. Poor people, infants, and young children are the most affected by this disease. Due to poor nutrition and severe dehydration, the death rate is usually high in children who are affected with ETEC. In adults, this diarrhea, known as traveler's diarrhea, usually lasts 1–5 days. For the treatment of this traveler's diarrhea antibiotics are not important.

3.2.10 Listeriosis

This *Listeria monocytogenes* bacterium is the gram-positive bacterium, does not form spores, and it is a facultatively anaerobic type of bacterium. This bacterium spreads many diseases like influenza. The major symptoms are high fever, sometimes gastrointestinal symptoms, and headache. The worldwide mortality rate for listeriosis is approximately 25% (Noordhout et al. 2014). The reservoirs of this bacterium are spoiled vegetables, water, sewage, soil, plant material, domestic animals, and other affected people and animals. Due to poor hygienic practices during the processing and cooking of food, it can be found in the cooked diet. It can also penetrate and be found in foods like raw, cooked, and smoked fish and seafood, unpasteurized milk, cooked sausages, raw and cooked poultry, ice cream, and cheeses (Allen et al. 2016; Leong et al. 2016). *L. monocytogenes* are responsible for many diseases, like meningoencephalitis and septicemia in newborn babies. Listeriosis can also pass through the feto–placental barrier, resulting in spontaneous abortion or stillbirth during pregnancy. People with weak immune systems, pregnant women, newborns, fetuses, and the elderly are at high risk of *L. monocytogenes* infection. The mortality rate is usually 20%–30%; when untreated, it can go up to 70% (WHO 2008a; Leong et al. 2016). The existence of *L. monocytogenes* in food processing and diet may affect resistance to antimicrobials, like ampicillin, penicillin, and trimethoprim-sulfamethoxazole. The use of the detergent, salt, and temperature are not effective methods for inactivating *L. monocytogenes*. Food preservation seems to help control this organism and avoid foodborne infection (Allen et al. 2016).

3.2.11 Infection Caused by *Mycobacterium bovis*

Mycobacterium bovis is a straight or slightly curved, aerobic, rod-shaped, gram-positive bacterium. An outer cover or layer that is called *Mycobacterium tuberculosis* is not present in the *M. bovis* (FDA 2012; Epstein 2015). A cell that is usually classified as the acid-fast bacteria is, not as the gram-negative or gram-positive bacterium, is also present in mycobacterium. Human and animal tuberculosis are usually spread due to both *M. bovis* and *M. tuberculosis*. It is also the cause of foodborne illness like tuberculosis in humans. This disease usually spreads through the ingestion of contaminated food or the inhalation of infected droplets. Other causing agents of this disease are raw milk, unpasteurized milk, cheese, and other food and food products. A significant path of infection is contaminated milk. Intake of unpasteurized milk increases the risk of transmission of tuberculosis to humans. Raw or uncooked meat from certain infected deer and other animals also increase the chance of spreading of this tuberculosis. In some cases, the symptoms of tuberculosis appear for months or years, but, in general, this foodborne tuberculosis is asymptomatic. General symptoms of this disease are weight loss, fever, loss of appetite, night sweats, and fatigue.

The affected part of the body plays a vital role in the manifestation of symptoms like weight loss and fatigue: pulmonary tuberculosis manifests as severe cough, blood stained sputum, chest pain, fever, malaise, night sweats, wasting, dyspnea, and hemoptysis; if the gastrointestinal tract is affected then symptoms are abdominal swelling, diarrhea, and abdominal cramps. Human infections are usually asymptomatic. If the *M. bovis* remains untreated in humans, it can cause death. Infants, children, the elderly, and people who have weak immune systems are at the highest risk of infection. This bacterium is resistant to current antimicrobial medicines. Now the recommended therapy is the Anti-tumor necrosis factor (TNF) that is usually started after standard medical therapies like pyrazinamide, isoniazid, ethambutol, and rifampicin (FDA 2012; Ali et al. 2013).

3.2.12 Q Fever

The causing agent of the Q fever is an important obligate intracellular named *Coxiella burnetii* that is a gram-negative bacterial pathogen. The most common way of spreading this bacterial infection is breathing the aerosolized bacteria. The main reason of the spreading of this bacterium by the oral route is the ingestion of unpasteurized milk or other dairy products. Ticks are another main source or reservoir for *C. burnetii*; a tick can spread the bacteria directly via a bite or indirectly by infected feces. There are two main stages of transmission of this disease: the first stage is the acute stage which is usually less severe; the other stage is the chronic stage which is very complicated and can result in high mortality rates. The main symptoms of acute Q fever are chills, fatigue, headache, and major respiratory problems. The most common symptoms of the Q fever are muscle aches, headache, nausea, vomiting, very heavy sweating, dry cough, high fever (105°F/40.6°C), abdominal or chest cramps, and diarrhea. The more serious is the chronic Q fever and the symptoms are hepatitis, myocarditis, and pneumonia (Ryan et al. 2011).

3.2.13 Salmonellosis

Salmonellosis is the most common and the most encountered foodborne illness and is the main foodborne illness-related cause of death in the world. Due to the salmonellosis infection, almost 93.8 million people get sick and 155,000 deaths occur worldwide annually (Majowicz et al. 2010). There are 2500 identified strains of *salmonella,* and the causative agent is non-typhoid *salmonella* serotypes. The two most important serotypes of this bacterium are the *Salmonella enterica* and the *S. typhimurium* which are usually transmitted from animals to humans (Switt et al. 2009; WHO 2013). The bacterium is widely distributed in the wild and among domestic animals. Many foods are responsible for salmonellosis in humans, including poultry and poultry products (like eggs), raw meat, green vegetables, and contaminated water (Switt et al. 2009; Reddy et al. 2016; Rey Matias et al. 2016). Care is necessary to avoid cross-contamination from animal to human (WHO 2013). The many symptoms of salmonellosis include headache, fever, abdominal cramps, vomiting, nausea, and diarrhea. Major symptoms appear within 6–72 hours and last up to 7 days (Switt et al. 2009; Majowicz et al. 2010).

3.2.14 Shigellosis

It has been observed that children of 2–5 years are at higher risk of shigellosis. Studies reveal that, among all patients of shigellosis, 55.6% are children of 2–5 years old. About 1 million people around the world die due to shigellosis annually. Different species of *Shigella* are responsible for causing shigellosis which includes *Shigella dysenteriae, S. sonnei, S. boydii,* and *S. flexneri.* Shigellosis has many symptoms like abdominal pain, fever, tenesmus, dysentery with mucoid bloody stools, and watery diarrhea (Niyogi 2005; Talebreza et al. 2015). *Shigella* spp. is responsible for causing inflammation, mucosal ulceration, bleeding, and erythema nodosum or hemolytic uremic syndrome in 2%–3% of cases. There are various methods of transmission of the diseases, but mainly it is transmitted through water or foods containing fecal matter (it is also transmitted through person-to-person contact). Sources of food contamination include fertilization via wastewater or sewage water. Shigellosis could be caused by consuming uncooked foods, vegetables, salads, contaminated water, and raw milk. Children under 5 years of age are highly affected by this and it is the most common form of diarrhea in infants. Developing countries are highly affected by shigellosis; mortality and morbidity rate is also high in children less than 5 years old. Severe symptoms are observed in elderly individuals that lead to death due to malnutrition-induced shigellosis. Travelers have a greater chance of getting infected by shigellosis. Shigellosis is mostly treated by antimicrobial agents (Niyogi 2005; WHO 2008a).

3.2.15 Intoxication Caused by *Staphylococcus aureus*

Staphylococcus aureus is a non-spore-forming, gram-positive facultatively anaerobic, non-motile bacterium. There are many cases reported all around the world due to *S. aureus*-related food poisoning. Many food products are responsible for *Staphylococcus aureus* intoxication including cheese, ice-cream, chicken, ham, salad, eggs, and cream-filled products. The symptoms can be observed 2–6 hours after contaminated food ingestion. Symptoms include prostration, vomiting, severe nausea, and cramps, and in severe cases, diarrhea or acute gastroenteritis. Antimicrobial resistance is high in many microbes due to complications in the food chain, when these resistant microbes are transferred to the body with food then it can develop disease. Many microbes are resistant to antimicrobials for example *S. aureus* were isolated from some food-producing animals like chicken, pigs, and cattle, and it was observed that they are methicillin-resistant. Oxacillin and penicillin resistance are also observed in many strains (Hennekinne et al. 2012; Johler et al. 2015). They were resistant to antimicrobial agents, and, due to this, treatment failure and mortality rate was high. Anti-staphylococcal antibiotics were used for the treatment (Kaye et al. 2008).

3.2.16 Typhoid and Paratyphoid Fever

Typhoid fever and paratyphoid fever are enteric fevers that are caused by *Salmonella typhi* and *S. paratyphi*. Symptoms include abdominal pain, vomiting, headache, high fever, and diarrhea; in some cases, rashes and constipation which last for months have also been observed. Humans are the reservoir of these organisms and serve as vehicles as well as cases. Infection can be caused by contaminated food or water and contact with the stools of an infected person. Many foods like dairy products, prepared food, vegetables, salads, and shellfish carry this bacterium. The spread of the disease in a community is greatly affected by chronic carriers and treatment is a difficult task. Enteric fever can be treated with antibiotic therapy. Resistance against some antibiotics like tetracycline, trimethoprim-sulfamethoxazole, streptomycin, ampicillin, and chloramphenicol is observed in *S. typhi*. High mortality and morbidity rate has been observed due to multiple drug resistance. Vaccination is another effective method of controlling typhoid fever; children up to 15 years old are vaccinated for prevention (WHO, 2008a; Wain et al. 2015).

3.2.17 Gastroenteritis Induced by *Vibrio parahaemolyticus*

Vibrio parahaemolyticus and *V. cholera* are biochemically similar to each other (Kim et al. 1999). Many foods are responsible for *V. parahaemolyticus*–associated gastroenteritis such as fish (raw or minimally processed) and fishery products; it can also be caused by cooked foods that are cross-contaminated with raw fish. Symptoms of this disease are vomiting, headache, fever, nausea, abdominal pain, and profuse watery diarrhea, and these can be observed 4–96 hours after consuming contaminated food (Liu et al. 2015). Septicemia, a blood-poisoning disease, is also caused by *V. parahaemolyticus* after entering the bloodstream via the intestine, but occurrences of this disease are very rare (Alouf et al. 2015). There is no need of antimicrobial treatment, as this is responsible for increasing resistance in microbes. In case of fever with bloody or febrile diarrhea, empiric antimicrobial therapy is advised; symptoms can persist for more than one week in persons having weak immunity (Zollner-Schwetz and Krause 2015).

3.2.18 Infection Caused by *Vibrio vulnificus*

Vibrio vulnificus is a halophilic (salt-loving), non-spore-forming rod, virulent, gram-negative, motile bacterium found in oysters (seafood). *V. vulnificus*-related infections are mostly mild, and they are observed rarely, but they could be fatal in severe cases. Bloody and profuse diarrheas are common symptoms of this infection. Septicemia and wound infections are the two main clinical manifestations and their origin is traumatized epithelial surfaces or the gastrointestinal tract. The risk of *V. vulnificus*-induced septicemia is high in those persons that have immunosuppression, hemochromatosis, alcoholic liver disease, and chronic liver disease. The fatality rate is high at 40%–60%, and it goes up to 90% in patients that have a hypotensive disease. Primarily *V. vulnificus* infections are treated with surgical interventions, but this seems to be ineffective in those patients who have wound infections (WHO 2008a; Tsao et al. 2013).

3.2.19 Yersiniosis

Yersinia enterocolitica and *Y. pseudotuberculosis* are major causes of yersiniosis but *Y. pseudotuberculosis* cases are less frequent compared to *Yersinia enterocolitica*. *Y. enterocolitica* is a gram-negative, facultatively anaerobic, non-spore-forming bacterium that belongs to the family *Enterobacteriaceae*. Many food products like pork products (tongue, tonsils, gut), uncured and cured meat, and milk and milk products are responsible for yersiniosis. Symptoms include vomiting, abdominal pain, mild fever, and diarrhea. These may be chronic in several cases; for example in immunocompromised individuals, this results in reactive arthritis, eye complaints, erythema nodosum, septicemia, uveitis, cholangitis, glomerulonephritis, splenic abscesses, myocarditis, spondylitis, pneumonia, and lymphadenitis. The chances of misdiagnosis with appendicitis are high due to similar symptoms. Yersinia infections also similar symptoms as terminal ileitis, tumoral lesions, and Crohn's disease. Healthy people with healthy immunity can overcome this disease easily. Young children and persons that have weak immunity are at high risk of death, and mortality and morbidity rates are high in this group (WHO 2008a; Galindo et al. 2011). Transmission through person-to-person contact is rare. No clear treatment for *Yersinia* spp. is observed through literature. Mainly, antimicrobial treatment is used for treating this disease, which appears effective against this infection, and patients should receive therapy for 3 weeks with septicemia (Tauxe 2015).

3.3 FUNGAL FOODBORNE INFECTIONS/DISEASES

Fungi are organisms that play vital roles for maintaining the ecosystems of the Earth, and they are widely distributed in the world (Buckley 2008). Fungi are also the main pillar of many food and pharmaceuticals as they are important for making many products including beverages. *Saccharomyces cerevisiae* is a common yeast used in the brewing industry. There are many cheese varieties, soybean products (including soy sauce and fermented soybeans) in which use of fungi is common. Fungi could be consumed as food like mushrooms, truffles, and corn smuts. There are, however, some fungi that are dangerous for human beings; out of about 1.5 million fungi present on this planet, about 300 are considered harmful for human consumption as they cause serious illness which may lead to death (Hawksworth 2001). Secondary metabolites are synthesized from fungi that could be otherwise harmful in many cases. There is fungi toxin, mycotoxin, that is a contaminant of crops like wheat, rice, barley, maize, etc. There are many health hazards related to the ingestion of mycotoxins like neural tube defects, acute poisoning, liver disease, and cancer, but mycotoxin-related disease burden in still unknown (Marroquin-Cardona et al. 2014).

Fungi are responsible for gastrointestinal diseases as well. Many food contaminations are caused by fungi which ultimately affect health of an individual. Fungi produce toxins which are known as mycotoxins, and many fungal foodborne diseases are caused by the ingestion of these mycotoxins. Long-term health effects of these toxins are observed. A good example is liver cancer which is caused by long-term ingesting of aflatoxins from fungal-contaminated food (Anantaphruti 2001; Dorny et al. 2009).

According to WHO estimates, about 600 million people are affected by foodborne diseases each year (WHO 2015). There two main paths to fungal food poisoning–direct consumption of fungi (eating mushrooms), or indirect, due to their byproducts (mycotoxins). There are some fungi which are responsible for spoilage and contamination of food and those are known as pathogens which include *Fusarium*, *Alternaria*, *Mucormycetes*, *Aspergillus*, and *Candida* (Tomsikova 2002; Brenier-Pinchart et al. 2006; Pitt and Hocking 2009).

Persons that have weak immunity are at high risk of pathogen infection. Invasive fungal infections (IFIs) are responsible for causing illness and an increase in mortality rate and impose an extra burden on public health (Vallabhaneni et al. 2015a). The frequency and risk that are associated with foodborne IFIs are still unknown.

3.3.1 Mold Infections

Acute myelogenous leukemia was reported in a person who has diabetes and some other symptoms like vomiting, fever, headache, and nausea. Later on, it was discovered that the main cause of the symptoms was a fungus named *Mucor circinelloides* which was previously transmitted to that man via ingestion of Greek yogurt (Lazar et al. 2014). This case was reported during a recall process of Greek yogurt. In late August 2013 it was announced by a manufacturer of Greek yogurt that there was a quality issue in their yogurt as it was contaminated with *M. circinelloides*. In early September the manufacturer recalled the entire contaminated yogurt line (FDA 2013). Although almost 300 persons presented the same symptoms, it was later found that there was no proof relating the illness to the yogurt (Cadotte 2013). *M. circinelloides* infection is very rare in people who have weak immune systems; this is also not the basic cause of gastrointestinal problems. There are some studies that suggest that *M. circinelloides* are able to produce secondary metabolites, named as toxins, which may cause illness (Lee et al. 2014).

Many cases of mucormycotic have been reported that relate to the consumption of contaminated food and beverages and the main reason of disseminated and gastrointestinal infections. In Australia, a similar case was reported 3 hours after the patient had eaten a meal that had been unrefrigerated for a day. Symptoms included fever, abdominal pain, and diarrhea (Aboltins et al. 2006). *Mucor indicus* was found in the blood and feces sample of the infected person; this person was treated with amphotericin B lipid preparations, and they recovered after 6 weeks. As the symptoms emerged rapidly, it was suspected that this illness was caused by toxins. In South Africa, a similar type of case was reported where the patient was facing symptoms like dysfunction of organs, septic shock, and abdominal pain, later it was discovered that *Rhizopus* species were the main cause (Martinello et al. 2012). The patient had consumed home-brewed beer, but other factors were also responsible for this: he was misusing the beer and was already infected with bowel syndrome; in addition to these, he was suffering from *Typhimurium enteritis* predisposed and *Salmonella enterica* serotype infections.

Another similar infection was reported where the patient was suffering with abdominal pain and pneumonia for 3 weeks (Sutherland and Jones 1960). After hospitalization, the patient died within 3 days. The postmortem report of that person indicated that he died from pneumonia; later, though, upon histopathology of stomach cells, it was discovered the person was suffering mucormycosis and also had a gastric ulcer. It was supposed that mucormycosis was the result of the over-consumption of fermented corn beverages.

Hong Kong has a hematological oncology unit that contains documentation of possible foodborne mold infection outbreaks (Cheng et al. 2009). Samples from 5 different patients suffering from abdominal pain were taken and cultured; their histopathology proved that the main cause of those infections was ingestion of *Rhizopus microsporus*-contaminated mucormycotic food. An additional 7 patients were also diagnosed (5 asymptomatic and 2 with abdominal symptoms); their stool was tested and found to have substantial amounts of *R. microsporus*. Fungal isolates were taken from two different sources – ready-to-eat prepackaged contaminated food and allopurinol tablets which are made from cornstarch – and it was found that these were identical to the isolates obtained from patients by DNA sequencing (Cheng et al. 2009).

Many other foodborne infections were also identified and diagnosed as nonmucormycete mold infection. For example, a case of *Fusarium moniliforme* infection was reported in a man who was suffering from lymphoblastic leukemia, and his diet was completely composed of cereal-based products (Karam et al. 2005). Gastric cancer was reported in a person from French Guiana after consuming dried salted fish and it was found that *Monascus ruber* was the main reason behind this (Iriart et al. 2010). In France, 3 cases were reported of pulmonary illness and the root cause of that infection was *Aspergillus fumigatus* (Vermorel-Faure et al. 1993).

3.3.2 Yeast Infections

There are few cases of yeast infection found in literature. Compared to mold infections, the infections which are associated with yeast are only found in dairy products. It was observed that a pregnant woman who consumed organic dairy products above the prescribed limit then developed a *Candida kefyr* infection and transmitted it to her premature twin babies through the bloodstream (Pineda et al. 2012). It was unclear about pasteurization of those dairy products. Literature suggested that *C. kefyr* was possibly found in raw fermented dairy products as well as in cow's milk that is ultra-pasteurized (Gadaga et al. 2001; Pineda et al. 2012).

Some dairy products are natural sources of *Candida catenulata*, and this is responsible for human infections (Pineda et al. 2012). A case of gastric cancer was reported, and it was believed that *C. catenulata* fungemia was responsible (Radosavljevic et al. 1999). In year 2001–2002, four patients were hospitalized in Spain, due to a disease caused by *Blastoschizomyces capitatus* which is now known as *Saprochaete capitata*. Among the 4 patients, 2 had oropharyngeal colonization and the other 2 had a fatal disseminated infection. It was discovered that the source was vacuum flasks that contained milk because molecular genotyping techniques indicated that both isolates (from patient and from flask) were identical (Gurgui et al. 2011).

3.3.3 Dietary Supplements—A Cause of Fungal Infection

Apart from *Saccharomyces* infections, literature shows that there are some other fungi which are sources of IFI. A study revealed that a patient with transplanted bone marrow started taking oral naturopathic supplements after 5 months of transplant, and, about 2 months later, he developed hepatic mucormycosis infection (Oliver et al. 1996). The patient was consuming 10 different supplements, and, out of the 10 supplements, 4 contained fungi including *Rhizopus*, *Aspergillus*, and *Mucor*. A strain of *Mucor indicus* was obtained from the supplement and checked against strain obtained from the patient's liver through DNA sequencing; it was found that both were genetically identical. Another report indicated that a 10-year-old girl suffering from acute lymphatic leukemia started chemotherapy as a treatment and later it was found that she developed multiple liver abscesses and appendicitis after 4 months of chemotherapy (Bellete et al. 2006). After histopathological examination of the liver abscesses and appendix, mucormycetes were revealed and identified by DNA sequencing as *Absidia corymbifera* which is now reclassified as *Lichtheimia corymbifera*. It was found that she used probiotics and that were infected with *A. corymbifera* with 98% similarity to clinical strain (Bellete et al. 2006). A fatal case of gastrointestinal mucormycosis was observed in a premature neonate who was suffering necrotizing enterocolitis and was on a special diet. The neonate's cecum sample and dietary supplements were analyzed by DNA sequencing and the presence of *Rhizopus oryzae* was found (Vallabhaneni et al. 2015b).

3.4 VIRAL FOODBORNE DISEASES

Gastrointestinal diseases can be caused by the ingestion of viruses. Many viruses are identified as contaminants that can cause foodborne diseases. Norwalk virus infection (Aliabadi et al. 2015), Hepatitis E infection, and Hepatitis A infection are caused by viruses present in food (Sridhar et al. 2015). Viral foodborne diseases are caused by direct ingestion of viruses. Viruses are also responsible for foodborne diseases. Clinical features, causes, classification, risk reduction, and prevention are also discussed in this section.

3.4.1 Hepatitis A

Hepatitis A virus (HAV) belongs to the family *Picornaviridae* and it is a non-enveloped small virus of 28 nm diameter and has a single-strand RNA. HAV is transmitted through person-to-person contact and spread through the fecal-oral route. The incubation period for HAV is 25–28 days. In the latter part of the incubation period, the virus is shed in the feces. HAV enters through the intestine and then migrates toward the liver via the bloodstream; the period during which the virus is detectable in the blood is called the viremic stage. There are various symptoms of this disease like malaise, vomiting, loss of appetite, nausea, fever, fatigue, abdominal pain, and headache; older individuals also suffer acute liver dysfunction. Many foods, such as raw fruits and vegetables, frozen strawberries, shellfish (mussels and clams), bakery products, and water that are contaminated with fecal matter are responsible for the spread of HAV as is swimming in contaminated water (in pools or lakes). Although individuals older than 50 are at high risk, compared to children and young ones, the fatality rate is very low at 0.3%. Effective vaccines are available for HAV infection prevention, and are preferred in some developing countries due to the low cost. If one is affected with HAV than Immune-serum globulin is used for curing this disease and is best method of preventing infection within 14 days; this method could also be used in travelers as a pre-exposure prophylaxis (Koopmans et al. 2002; WHO 2008a).

3.4.2 Hepatitis E

This virus, the main cause of acute hepatitis (epidemic and endemic) in humans, is transmitted via the fecal-oral route. Improper food handling, consumption of contaminated foods, and animal-to-human contact are a few of the transmission pathways. Infection may be caused by consuming and handling infected meat and meat products, pig products, contaminated drinking water, and the use of raw or uncooked meat. There is much evidence that hepatitis E could be transmitted from wild animals such as pork (wild boar), deer, and game birds (Yugo and Meng 2013). Virus ultimately reaches the liver after entering the intestine through the bloodstream. Persons that have weak immunity and pregnant women are at greater risk of this infection than the rest of the population, with higher rates of mortality in these individuals (0.5%, 4%, and 20%, respectively). There are various symptoms of this disease, including hepatomegaly, jaundice, anorexia, myalgia, malaise, sometimes fever, abdominal pain, vomiting, and nausea (FDA 2012; Yugo and Meng 2013). There is currently no treatment as with other viruses, however, a vaccine is available (FDA 2012); China was the first country to introduce the vaccine for hepatitis E virus (Yugo and Meng 2013).

3.4.3 Gastroenteritis Induced by Norovirus

Acute gastroenteritis is associated with noroviruses, and they are responsible for foodborne outbreaks. They possess different physical and chemical characteristics. These viruses have complex epidemiology which is based on different factors like environment, population immunity, seasonality and virus evolution. Norovirus is transmitted through person-to-person contact, via the fecal-oral route, and causes norovirus-induced gastroenteritis (WHO 2008a; Ramani et al. 2014). Diarrhea and vomiting are common in this infection and may lead to dehydration, which can be fatal in children and elderly people. Care is mandatory to avoid this infection. This infection can not be cured by the use of antibiotics (FDA 2012). Vaccines are available against norovirus (WHO 2008a,b).

3.4.4 Poliomyelitis

Poliovirus belongs to the family *Picornaviridae*, is a small round-shaped virus with a single strand of RNA, and is responsible for poliomyelitis, which is an infectious disease. Primarily, this virus infects the intestinal tract and then is transmitted through the blood toward the brain (CNS) and regional lymph nodes. In this way, the virus increases in the blood (viremia), causing symptoms like malaise and fever. The main route of invasion of the virus in humans is the fecal-oral route. Contaminated food and drinking water are the main cause of virus spread. Poliovirus infection is asymptomatic in those persons having good immunity. Children and young adults are at greater risk of being infected by this virus. Although there is no cure for Polio infection, vaccines are available that are effective against poliovirus infection (WHO 2005).

3.4.5 Gastroenteritis Induced by Rotavirus

Children are at high risk of rotavirus infection that causes diarrhea leading to dehydration; almost half a million children under 5 die annually due to this disease. The virus is transmitted through the fecal-oral route. Many foods like raw vegetables and fruits and salad are the source of rotavirus-induced gastroenteritis. The symptoms are many, such as fever, vomiting, and watery diarrhea that starts within 2 days after infection, leads to dehydration and hypovolemic shocks, which is followed by death. Electrolyte therapy and fluid consumption are recommended Rotavirus is responsible for activating secretomotor neurons which stimulates the secretion of fluids and solutes (FDA 2012). Vaccines are available and are the best way to protect children against infection of this virus (Smulders et al. 2013).

3.5 PARASITIC FOODBORNE DISEASES

Parasites are living organisms that can cause food contamination and are responsible for foodborne diseases, mainly gastrointestinal diseases (Dorny et al. 2009). Dorny et al. (2009) noted that foodborne diseases that are caused by parasites were under-recognized, although common; hence there is a need of greater awareness. Torgerson et al. (2015) stated that "food is a source of transmission of parasites to humans" and "the burden of foodborne diseases that results in morbidity and mortality among populations is high." Some diseases that are caused by parasites are liver fluke infestation named as opisthorchiasis, sparganosis, cysticercosis, and paragonimiasis. Ito and Budke (2014) stated that "rural areas are at high risk of parasitic foodborne diseases, and this is also common in developing countries." There is a need of awareness for preventing parasitic foodborne diseases. In this section, we discuss the clinical features, causes, prevention and risk reduction, and classification of foodborne diseases (Table 3.1).

3.5.1 Amebiasis

Entamoeba histolytica is a parasite which causes amebiasis dysentery. This parasite is found in contaminated water, vegetables, and fruits. Ingestion of contaminated food and water with cysts is the main source of transmission of *E. histolytica*. The infection may

TABLE 3.1 Major Foodborne Diseases and Causing Organisms*

Organism	Common Name of Illness	Signs and Symptoms	Food Sources
Bacteria			
Aeromonas hydrophila	*Aeromonas*-induced gastroenteritis	Abdominal cramps, watery diarrhea, nausea, vomiting	Chicken, seafood, meat, raw milk, contaminated water, contaminated fruits and vegetables, uncooked meat of lamb, goat, and pig
Bacillus cereus	*B. cereus* food poisoning	Abdominal cramps, watery diarrhea, nausea	Meats, stews, gravies, vanilla sauce
Brucella spp.	Brucellosis	Profuse sweating and joint and muscle pain	Raw milk and soft cheeses made with unpasteurized goat or cow milk
Campylobacter jejuni	Campylobacteriosis	Diarrhea, cramps, fever, and vomiting; diarrhea may be bloody	Raw and undercooked poultry, unpasteurized milk, contaminated water
Clostridium botulinum	Botulism	Vomiting, diarrhea, blurred vision, double vision, difficulty swallowing, muscle weakness; can result in respiratory failure and death	Improperly canned foods, especially home-canned vegetables, fermented fish, baked potatoes in aluminum foil
Clostridium perfringens	Perfringens food poisoning	Intense abdominal cramps, watery diarrhea	Meats, poultry, gravy, dried or precooked foods, time and/or temperature-abused foods
Escherichia coli	Food poisoning	Watery or bloody diarrhea, abdominal cramps, with or without fever	Varied: Water or food contaminated with human or animal feces
Listeria monocytogenes	Listeriosis	Fever, muscle aches, and nausea or diarrhea; pregnant women may have mild flu-like illness, and infection can lead to premature delivery or stillbirth; the elderly or immunocompromised patients may develop bacteremia or meningitis	Unpasteurized milk, soft cheeses made with unpasteurized milk, ready-to-eat meats
Mycobacterium bovis	Bovine tuberculosis	Tuberculosis-like symptoms	Raw milk and soft cheeses made with unpasteurized cow milk

(Continued)

TABLE 3.1 (CONTINUED) Major Foodborne Diseases and Causing Organisms*

Organism	Common Name of Illness	Signs and Symptoms	Food Sources
Salmonella spp.	Salmonellosis	Diarrhea, fever, abdominal cramps, vomiting	Eggs, poultry, meat, unpasteurized milk or juice, cheese, contaminated raw fruits and vegetables, minimally processed poultry meat products
Shigella spp.	Shigellosis, bacillary dysentery	Abdominal cramps, fever, and diarrhea; stools may contain blood and mucus	Raw produce, contaminated drinking water, uncooked foods, and cooked foods that are not reheated after contact with an infected food handler
Staphylococcus aureus	Staphylococcal food poisoning	Sudden onset of severe nausea and vomiting; abdominal cramps; diarrhea and fever may be present	Unrefrigerated or improperly refrigerated meats, potato and egg salads, cream pastries
Vibrio spp.	Vibriosis	Diarrhea, vomiting, abdominal pain	Raw or undercooked seafood
Yersinia enterocolitica	Yersiniosis	Diarrhea, vomiting, abdominal pain	Raw or undercooked pork, unpasteurized milk or contaminated water
Fungi			
Mucor indicus	Mucormycotic infection	Fever, abdominal pain and diarrhea	Unrefrigerated food, contaminated food and beverages
Salmonella enterica	*Salmonella enterica* serotype infections	Abdominal pain and pneumonia	Fermented corn beverages and beer
Fusarium moniliforme	*Fusarium moniliforme* infection	Abdominal cramps, diarrhea and vomiting	Contaminated cereal-based products
Viruses			
Astrovirus	Variously called stomach flu, viral gastroenteritis	Diarrhea, followed by nausea, vomiting, fever, malaise, and abdominal pain	Fecal-oral transmission and handler-contaminated food
Hepatitis A	Jaundice	Lethargy, loss of appetite, nausea, vomiting, fever, jaundice	Seafood and handler-contaminated food

(*Continued*)

TABLE 3.1 (CONTINUED) Major Foodborne Diseases and Causing Organisms*

Organism	Common Name of Illness	Signs and Symptoms	Food Sources
Noroviruses	Variously called viral gastroenteritis, winter diarrhea, acute nonbacterial gastroenteritis, food poisoning, and food infection	Nausea, vomiting, abdominal cramping, diarrhea, fever, headache; diarrhea is more prevalent in adults, vomiting more common in children	Raw produce, contaminated drinking water, uncooked foods and cooked foods that are not reheated after contact with infected food handler; shellfish from contaminated waters
Rotaviruses	Variously called stomach flu, viral gastroenteritis	Diarrhea in infants and children	Fecal-oral transmission, and handler-contaminated food
Sapovirus	Variously called stomach flu, viral gastroenteritis	Nausea, diarrhea, vomiting, abdominal cramp, headache, myalgia and fever	Fecal-oral transmission, and handler-contaminated food
Parasites			
Cryptosporidium spp.	Cryptosporidiosis	Diarrhea (usually watery), stomach cramps, upset stomach, slight fever	Uncooked food or food contaminated by an ill food handler after cooking, contaminated drinking water
Cyclospora cayetanensis	Cyclosporiasis	Diarrhea, loss of appetite, weight loss, stomach cramps/pain, bloating, increased gas, nausea, fatigue, vomiting, body aches, headache, fever, and other flu-like symptoms	Uncooked food or food contaminated by an ill food handler after cooking, contaminated drinking water
Giardia intestinalis	Giardiasis	Diarrhea, flatulence, stomach or abdominal cramps, nausea, dehydration	Contaminated meat
Toxoplasma gondii	Toxoplasmosis	Flu-like symptoms	Undercooked meat such as pork, lamb, and venison
Trichinella spp.	Trichinellosis	Diarrhea, fever, profuse sweating, weakness, muscular pain, swelling around eyes	Contaminated meat, especially pork and wild game

*Adley and Dillon (2011), CDC (2015), FDA (2015), Kaper et al. (2004), Ryan et al. (2011), Vallabhaneni et al. (2015a), and Girgui et al. (2011)

be asymptomatic and may cause high fever, bloody diarrhea, vomiting, and stomach pain. Dysentery of amebic fulminant is often fatal (55%–88%). This may cause other problems like colonic ulcer, liver abscess, perforation of the colon, and chronic carriage (WHO, 2008a). In children and adults, metronidazole has shown effective results in the management of amebic dysentery (Löfmark et al. 2010).

3.5.2 Anisakiasis

Anisakiasis is caused by *Anisakis* spp., and it is a human parasitic infection. Clinical signs and symptoms of anisakiasis are similar to appendicitis. Cases of anisakiasis infection are reported all over the world, but Japan has the highest prevalence of infections in humans. Undercooked or raw fish, including infected sashimi and sushi, are the main sources of anisakiasis infection. The motile larvae make holes in the walls of the intestine and cause nausea, epigastric pain, acute ulceration, vomiting, and hematemesis. Consumption of *Anisakis* spp.-infected food may also cause allergic reactions, like broncoconstriction, rhinitis, cough, anaphylaxis, urticarial, and gastrointestinal responses. Serological and molecular approaches are warranted for better diagnosis of anisakiasis disease, since diagnosis is often confused with bacterial gastroenteritis (WHO 2008a; Baird et al. 2014). The US FDA recommends that raw fish be stored at –20°C or below for 7 days and alternatively could be stored at –35°C or below for 15 hours (FDA 2012). There is no treatment for this infection and in some patients the symptoms resolve after only symptomatic treatment. Albendazole alone may be used in anisakiasis treatment (Moore et al. 2002).

3.5.3 Ascariasis

Ascaris lumbricoides is the causing agent of ascariasis. Ascariasis is an intestinal infection. It is asymptomatic and arises after eating food infected with *Ascaris* spp. The nematode lays eggs which penetrate the intestinal walls. The larvae then migrate to the lungs via blood. They break into alveoli and move with the tubes of bronchi and trachea of the pharynx. Intestinal worms may cause malnutrition. Signs of infection include fever, colic, vomiting, gastrointestinal discomfort, and rare cases of neurological disorders and pulmonary symptoms (Cross 1996). Levamisole, albendazole, pyrantel, and mebendazole are recommended by WHO to cure the infection (WHO 2006a).

3.5.4 Clonorchiasis

Clonorchis sinensis is the causative agent of clonorchiasis. *C. sinesis* is a carcinogenic human liver fluke that is transmitted by the consumption of infected raw freshwater fish. As is, WHO control programs against clonorchiasis infection. Though some of the patients are asymptomatic, in some cases, eosinophilia may occur. Other symptoms include abdominal pain, anorexia, loss of appetite, indigestion, irregular bowel movement, malaise, and fever. Epigastric discomfort, abdominal fullness, diarrhea, anemia, and weight loss are observed in severe conditions. Clonorchiasis is endemic to Malaysia, Russia, Japan, China, Taiwan, and Korea. Tribendimidine and Praziquantel found effective results both in vivo and in vitro for the treatment of clonorchiasis (WHO 2008a; Hong and Fanq 2012).

3.5.5 Cryptosporidiosis

There are 23 different species of *Cryptosporidium*, but *Cryptosporidium parvum* is the only one that causes human infection (Xiao et al. 2000). *C. parvum* is a zoonotic (transmits from animal to humans) protozoan parasite which causes cryptosporidiosis disease. The symptoms of this infectious disease include nausea, abdominal pain, diarrhea, vomiting, and influenza. This infection may cause severe infection in children under 5 years, AIDS patients, immunocompromised, and immunocompetent patients. The infection may spread by infected patients or consumption of contaminated food and water. The oocyst of *Cryptosporidium* spp. show non-significant results against chlorination, but cooking methods at high temperature or washing solutions, like Alconox, are effective methods and show significant results in eliminating the protozoan oocyst (Xiao et al. 2000; Hong et al. 2014).

3.5.6 Cyclosporiasis

Cyclosporiasis is an infectious disease caused by *Cyclospora cayetanensis* which is an intracellular coccidian parasite that causes intestinal illness. Consumption of *C. cayetanensis*-contaminated fresh food like basil, raspberries, and lettuce may cause cyclosporiasis in humans. Signs and symptoms emerge after 1 week of exposure including cramping, weight loss, body aches, increased gas, headache, nausea, watery diarrhea, vomiting, fever, fatigue, and other flu-like symptoms. In some cases, infection may be asymptomatic. This infection is mostly found in people who live in tropical and subtropical regions. Patients that have weak immunity and other diseases, like cancer or HIV/AIDS, young, and elder persons are also at high risk. Trimethoprim-sulfamethoxazole is an antibiotic which is used in the treatment of cyclosporiasis (FDA 2012; CDC 2015b,c).

3.5.7 Diphyllobothriasis

Diphyllobothriasis is an infectious disease caused by *Diphyllobothrium latum* also known as tapeworm. *Diphyllobothrium* are parasites of the human intestine and other fish-eating mammals and birds. It is the longest (about 10m) tapeworm in humans. Infection may spread by the infected person or through consumption of undercooked fresh fish (ceviche, sushi, tartare, and sashimi) and meat. Symptoms of diphyllobothriasis include altered appetite, abdominal pain, and diarrhea. A great amount of vitamin B12 is absorbed by the tapeworm in the human intestine, which may cause vitamin B12 deficiency in humans leading to anemia. All raw fish (contaminated with worms) consumers are at risk of infectious disease. Niclosamide and Praziquantel are used to treat the tapeworm infectious disease and also used in the treatment of diphyllobothriasis (FDA 2012).

3.5.8 Fascioliasis

Fascioliasis is mainly caused by the trematodes *Fasciola gigantica* and *Fasciola hepatica*, of which *F. gigantica* is the larger. While *F. hepatica* is mostly found in temperate zones, tropical regions of Asia and Africa are big reservoirs for *F. gigantica*. Symptoms of this disease include dizziness, abdominal tenderness, up to 60% leucocytosis with eosinophilia, bronchial asthma, epigastric pain, cough, fever, obstructive jaundice, urti-caria, bronchial asthma, and may cause immunosuppression (Carrada-Bravo 2003; WHO 2008a,b; Valero et al. 2009). Fascioliasis diagnosis is crucial and based on egg classification found in stools or bile. Fascioliasis can be cured with intramuscular emetine praziquantel, oral bithionol, and hydrochloride, as recommended by WHO (Carrada-Bravo 2003; WHO 2008b).

3.5.9 Giardiasis

Giardia lamblia (protozoan parasite) is the causing agent of giardiasis. Giardiasis is an infection of the intestine characterized by weight loss, abdominal pain, nausea, fatigue, nausea, and chronic and relapsing diarrhea and anorexia. This infection may be severe in AIDS patients, immunocompromised people, and patients who are deficient in immu-noglobulin. Severe dehydration may cause death, mainly in malnourished children and infants. Antibiotic therapy including benzimidazole compounds, nitroimidazole deriva-tives (tinidazole or metronidazole), or acridine dyes are the standard treatment for giar-diasis (WHO 2008a; CFSPH 2012).

3.5.10 Nanophyetiasis

Nanophyetiasis is a parasitic disease of the intestine caused by *Nanophyetus salmincola*. *N. salmincola* is a small trematode parasite in the flatworm phylum. This disease is spread due to consuming raw and contaminated fish (especially salmon), hence the mon-icker "fish flu." Symptoms of this disease including gas/bloating, unexplained peripheral blood (eosinophilia), nausea, diarrhea, vomiting, and abdominal discomfort. Signs and symptoms may remain for several months without medication or treatment. Worms can be killed by proper treatment and medication. Anthelminthic drugs can cease the pro-duction of eggs, thereby halting the infection (Harrell and Deardorff 1990; FDA 2012).

3.5.11 Opisthorchiasis

Opisthorchis felineus and O. *viverrini* are the causative agents of opisthorchiasis. Severe signs and symptoms occur 2–4 weeks after eating undercooked or raw fish. Symptoms of this disease include arthralgia, urticaria, high-grade fever, abdominal pain, lymphadenop-athy, and dizziness. These symptoms are similar to acute viral hepatitis. Opisthorchiasis is associated with gallstones, cholecystitis, cholangiocarcinoma, liver abscess, and cholan-gitis. Laboratory findings include increased liver enzymes and eosinophilia. Albendazole, mebendazole, and praziquantel are drugs of choice to treat and eradicate the infection (Mairiang and Mairiang 2003; WHO 2008a).

3.5.12 Paragonimiasis

Paragonimus trematode is the causative agent of paragonimiasis. It is a parasitic food-borne disease caused by the consumption of raw or undercooked foods, such as crusta-ceans or crayfish (CDC 2010). Severe infection of this disease may cause eosinophilia with abdominal pain, fever, generalized myalgia, and fatigue (WHO 2008a). Paragonimiasis mostly infects the lungs, but worms can also invade other organs, including the brain and skin. Praziquantel is used in the treatment of paragonimiasis (CDC 2010).

3.5.13 Taeniasis and Cysticercosis

There are two main parasitic infections caused by *Taenia solium*-taeniasis (adult tapeworm in the intestine) and cysticercosis which is caused by larval cysts. Heart-related and CNS complications may occur if the larval parasite localizes in the eye, causing cysticercosis. Cysts are asymptomatic in most patients but visual activity is impaired due to ophthalmic cysticercosis (infection of the eye). Symptoms of human cysticercosis depend on the parasite's location. Consumption of undercooked or raw beef or pork can cause the infection. Niclosamide or Praziquantel drugs are used to treat taeniasis. There is currently no treatment for neurocysticercosis (CNS infection). The WHO recommends surgical treatments with antiepileptic, albendazole, corticosteroids, and praziquantel drugs (Gonzales et al. 2016; WHO 2016).

3.5.14 Congenital Toxoplasmosis and Toxoplasmosis

Toxoplasmosis is an infectious disease caused by *Toxoplasma gondii*. Infection may be transmitted through 3 different modes: eating undercooked meat, exposure to cat feces, and trans-placental infection during pregnancy (congenital toxoplasmosis) (Martin 2001; WHO 2008a). Immunocompromised individuals and congenitally infected fetuses are at a high risk of infection. Intrauterine infections may cause brain damage, chorioretinitis, and abortion or stillbirth (Martin 2001; Saadatnia and Golkar 2012). In immunocompromised individuals, myocarditis, chorioretinitis, cerebritis, pneumonia, and death have been reported. Patients suffering from AIDS and those who have undergone organ transplants are at high risk of cerebral toxoplasmosis that can lead to death (WHO 2008a; Saadatnia and Golkar 2012). Pyrimethamine-sulfadiazine and spiramycin drugs are used fetus during pregnancy. Sulfadiazine and spiramycin are suggested for those patients with symptoms of meningoencephalitis, pneumonitis, and myocarditis (Rajapakse et al. 2013).

3.5.15 Trichinellosis

Trichinella spp. is the causing agent of trichinellosis. Trichinellosis is a parasitic roundworm disease and is also known as trichiniasis or trichinosis. Ingestion of undercooked or raw meat contaminated with larvae may cause this infectious disease. Symptoms of this disease depend on the number of ingested larvae. Ingested larvae grow in the intestine and then penetrate into blood vessels. They spread into the body via blood and invade skeletal muscles. Primary symptoms are fever, vomiting, diarrhea, and nausea. Neurological and cardiac complications can occur when parasites enter into tissues. Severe complications, like pneumonitis, myocarditis, myocardial failure, and meningoencephalitis can result in death. In the early stage of infection, albendazole is recommended. Steroids with albendazole have been used during the severe infection (WHO 2008a; CDC 2015a).

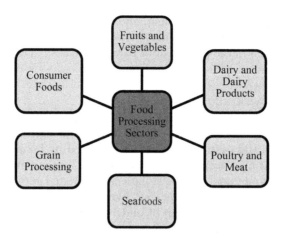

FIGURE 3.3   Food sectors that demand food pathogen detection and control.

3.6 SECTORS THAT DEMAND FOOD PATHOGEN DETECTION AND CONTROL

Health and food safety are directly impacted by pathogens; that is why their detection is critical for safe food and good health. There are many food sectors that demand detection of food pathogens, including dairy, grains, fermentation, fisheries, alcoholic and non-alcoholic beverages, fruits and vegetables, health-related foods and supplements, convenience and packaged foods, as shown in Figure 3.3 (Pankhurst et al. 2011; Garcia et al. 2013; Das et al. 2014). Food pathogens are found everywhere–in the natural environment, at work, in homes–but they are present in large concentrations in contaminated foods. They grow on different media, but water is necessary for their growth. Many deaths are reported due to illness caused by foodborne pathogens. In the above-mentioned food sectors, the chances of food pathogen occurrence are high. Different natural disasters are responsible for elevating pathogen levels, such as floods and cyclones.

3.7 METHODS FOR DETECTION OF FOOD PATHOGENS AND FOODBORNE ILLNESS

There are different methods that are used for the detection of foodborne pathogens and their toxins. Different methods are also implied for their classification. ISO (International Organization for Standardization) gives criteria for performing food safety tests, and there are six measurement parameters which should be fulfilled: accuracy, cost, speed, ease of use, approved methods, and type. According to the user's needs, these six standards are useful. There are two methods named traditional and advanced (on the basis of molecular techniques) that are used for the detection of foodborne pathogens and their toxins. All these techniques are explained below (De Boer and Beumer 1999) (Figure 3.4 and 3.5).

3.7.1 Conventional Methods

Many traditional and conventional techniques are being used for detecting food pathogen which includes PCR (polymerase chain reaction), culturing and plating (Zarain

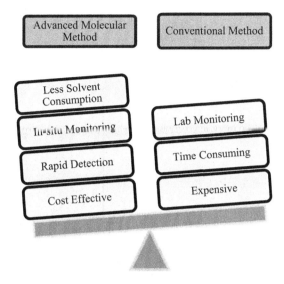

FIGURE 3.4 Comparative analysis between advanced molecular and conventional methods.

et al. 2012), and immunological processes. In traditional methods, microbes are grown on media, followed by colony counting, and after that, selected microbes (colonies) are grown on special media that allows growth of specific microbes. Antigen antibody reaction-based technique is known as an immunological technique. In the PCR technique, DNA isolation is carried out with the help of molecular techniques, and, in this way, the analysis of the desired DNA molecule is possible (Velusamy et al. 2010).

3.7.1.1 Non-Culturing Method

Gram-staining is a non-culturing method which can be used for detecting foodborne pathogenic bacteria in contaminated food. It is an important method as it gives rapid information about bacteria (gram-positive or gram-negative) allowing for quick antimicrobial treatment. Detection of microbes is related with nitrite production which is a unique feature of Enterobacteriaceae. Negative aspects of this technique are also observed, such as low concentrations of microbes can not be measured using this method.

3.7.1.2 Culture-Based Method

Culturing and plating are two traditional methods of detecting food pathogenic microbes that provide additional accuracy. These techniques are used for growing and isolating microbes on specific media. Literature has revealed that *L. monocytogenes*, a foodborne pathogen, could be cultured by this method (De Boer and Beumer 1999; Artault et al. 2001; Stephan et al. 2003) as well as other microbes like *Yersinia enterocolitica* (Weagant 2008), *C. jejuni* (Sanders et al. 2007), *S. aureus*, *Salmonella* sp., and *E. coli*. There are several drawbacks to this method: it is the most time-consuming, as it requires 2–4 days for initial results; 6–10 additional days are required for authentication; and each type of microbe grows on specific media with specific conditions. Growth rate and time is also different for different microbes, for example, 5–10 days are required for growth of *Campylobacter*, and 14–16 days are required for constructive effect (Brooks et al. 2004). Other factors like endotoxin production and bacterial count and spore formation are also influenced.

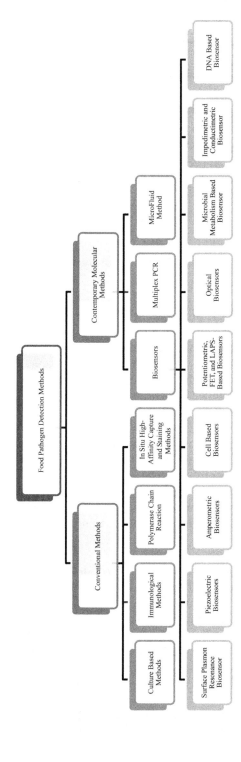

FIGURE 3.5 Different methods (conventional and contemporary molecular) for detection of foodborne pathogens.

3.7.1.3 Polymerase Chain Reaction (PCR)

PCR is widely used all over the world for its high accuracy. This method can be used for making thousands of copies of small amounts of DNA, and it can also be easily amplified. There are many modified PCR techniques which are being used for the identification of endotoxins produced by microbes; they include real-time PCR and reverse transcriptase PCR (RT-PCR) (Mafu et al. 2009). At present, multiplex PCR (Josefson et al. 2011) and nested PCR (Zarain et al. 2012) are in hot debate. There are many other techniques which can be combined with PCR and are most effective, including LightCycler® real-time PCR (LC-PCR), fluorescence in situ hybridization (FISH), most probable number PCR (MPN-PCR), surface acoustic wave (SAW) sensor, PCR-enzyme-linked immunosorbent assay (PCR-ELISA), and sandwich hybridization assay (Beumer and Brinkman 1989). In contrast to older methods, PCR is much faster and shorter. There are many other benefits of using this technique: it is fully automated so there is no chance of human error or additional human aids to perform an experiment; this method could be adopted with several different methods to enhance its performance and it also gives verifiable outcomes; many microbes can be identified by using multiplex PCR.

3.7.1.4 Rapid or Alternative Detection Techniques

There are many rapid techniques for the identification of microbes which are also termed alternative detection techniques. These processes include NA-based (nucleic acid-based) assays, immunoassays, and bacterial cultures. There are two major standards for alternative detection techniques for microbes–detection limit and enhancement time. For example, in food, we can determine *L. monocytogenes*. At the AFNOR NF certification website, detection limit of ISO 11290 is documented. The process takes 48 hours and is responsible for a verified detection edge of 104 cells, examined in food.

3.7.1.5 Immunological Methods

Different forms of microbes like their toxins (endo-, exo-, and entero-), spores, viruses, and bacterial cells can be detected by immunological methods. Various microbes like *Staphylococcus*, *E. coli*, *L. monocytogenes*, and *Salmonella* can also be detected by these methods (Velusamy et al. 2010). Serological typing is an immunological technique that is widely used for detection of various strains of bacteria (Das et al. 2014). Bacterial cells can be detected by latex agglutination, while reverse passive agglutination is used for detection of soluble toxins. Foodborne pathogens can be detected by the latest immunological method of detection known as biosensor (Velusamy et al. 2010). Many pathogens and their toxins can be determined–especially bacterial foodborne pathogens–using this method. Antigen/antibody (Ag/Ab) are those detection methods that are most practical and widely acknowledged in the literature for the detection of foodborne pathogens. Since the immunological method is based on the interaction between antigen and antibody, it is also responsible for detecting metabolites (Musher et al. 2007). These immunological methods depend on similarities and require compulsory interactions, like protein-DNA interfaces, protein-protein interfaces, and protein-carbohydrates interfaces (Lazcka et al. 2007). This method makes it easy to isolate food pathogens by protein examination because antibodies have the ability to isolate food pathogens from a matrix; and the method is effective on either large or small samples. Antibody preparation methods and their properties are responsible for their nomenclature; they may be monoclonal or polyclonal depending on preparation and properties. Antibodies that have identical immune cells and are cloned from their parent cells are termed monoclonal antibodies (moAbs or mAbs). There is a monovalent affinity in monoclonal antibodies that binds with the same epitope. Monoclonal antibody-based sensors are required where we need high specificity (Byrne et al. 2009; Velusamy et al. 2010). B-cells of immunized animal antibodies are called polyclonal antibodies (pAbs). These polyclonal antibodies are a combination of antigen and immunoglobulin particles, which have different epitopes. Animals like sheep, goats, and rabbits are major sources of pAbs. In a single cell, there may be many epitopes that define the character of pAbs. There is a need for high specificity like that provided by recombinant antibody-based sensors (Byrne et al. 2009). Different microbes can be detected by pAbs like *Campylobacter* sp. and *L. monocytogenes*. Recombinant antibodies are better compared to conservative antibodies. There are many benefits of recombinant antibodies such as animals are not utilized in their production, required production time is short, and antibodies obtained through this method are of high quality. The use of recombinant antibodies is responsible for a reduction in price and time for label-free biosensors development. Microfluidic protein analysis is a sensitive method for the detection of foodborne pathogens. This method is responsible for counting and discriminating a limited number of cells; it can accurately distinguish between strains of interest. There are many homologs of exterior base proteins that any pathogenic microbial culture contains; this may indicate the presence of different kinds of cells by a specific antibody. So it is recommended that a simultaneously expressed antigen, which is species-specific, is expected (Byrne et al. 2009). The great benefits of using immunological methods are that they are computable and less expensive.

3.7.1.6 Immunoassays by Molecular Imprinted Polymers, Molecular Probes, and Aptamers

Antibodies are commonly used for food safety immunoassays. There are some disadvantages of using antibodies: they have weak metal and chemical consistency, they are very expensive, and animals are needed for their manufacture. Many techniques are widely used in place of antibody treatment, including protein-based molecular imprinted polymers (MIPs), enzyme-substrate interaction, aptamers, small-molecule probes (Ahmad et al. 2011), and antimicrobial peptides (AMPs). Enzyme-substrate reactions are preferred over Ab-Ag reactions because it can be renewed with no possible loss. Proteins can be targeted by using enzymes. Another technique, MIPs, is more reliable and has high predictability with low cost compared to antibodies. Other methods of pathogen detection that are based on proteins are critical for shielding the basic position and direction of protein organization, which is best for high sensitivity and specificity. For instance, hsp60 is designed for the determination of *Listeria* adhesion protein (LAP) in *L. monocytogenes*, and works by means of a receptor of LAP. The hsp60 gives more sensitive and compatible results compared to monoclonal antibodies. This protein could be synthesized by *E. coli* when it combines with complementary DNA (cDNA), which makes it commercially viable. Aptamers can also be used in place of antibodies. Aptamers are NA particles that can be produced in labs and combine easily with proteins, cells, and molecular particles. Aptamers have great similarities and specificity allowing an enhanced edge of recognition designed for bio-sensing functions. Epitomes are unreachable to antibodies; aptamers are responsible for combining epitomes and antibodies. Aptamers and actual matrices have similar features, making them an excellent choice to replace antibodies; and because they can be modified at immobilization, with no change in similarities, they can be used over and over again.

3.7.1.7 In Situ High-Affinity Capture and Staining Method

This is an easy and exclusive method for the detection of pathogens with affinity capture markers or peripherals which are used for the prediction of pathogens present on surfaces. The solid phase surface (SPS) method involves working with proteins placed into food fortification bags soon after stomaching under desired pathogen attack. During evaluation, the main challenge was to collect appropriate SPS. The main differences were (1) colored dye would permit exact visualization of the target pathogen, (2) an elevated flctionalization ability of the exterior surface on the way to permit perceptive recognition, and (3) enough constancy of the functionalized shell for the whole period of the incubation phase. Two types of SPSs are used: glass made SPS and plastic made SPS. Plastic made SPS are more economical compared to glass slides that are used for weak reliability at once functionalized and employed under the mandatory incubation state, this also has an elevated outside layer and it is appropriate for low-polluted surface uniqueness.

3.7.2 Food Pathogen Detection through Advanced Molecular Techniques

Contaminated food lacks nutrition and sanitation. Some chemicals and pathogens are responsible for food contamination. All over the world, diseases transmitted through contaminated food cause serious trouble for humans and may be responsible for millions of deaths. Approximately 250 plus diseases are recognized in which the foodstuff acts as a powerful commutable arbitrator. We have to develop a good detection method for better monitoring. The detection method must be fast, easy, inexpensive, and commercially viable. Different types of advanced molecular methods, such as LC-PCR hybridization, different forms of immunosensors, and many others, are available for the perfect monitoring of contaminated foodstuff. The IUPAC differentiate the different forms of biosensors for food pathogen detection like the receptor-retransducer procedure. It is the biological means of detection method which is used to distinguish the quantifiable and half-quantifiable information of related pathogen. By applying the appropriate signal transduction method, the sensors act as a biological component. When we compare the biosensor with the convectional detection method, the major distinction is the separate response of objective and translated elements. These biosensors are used to detect the ecological pathogen or pathogen in foodstuff. These kinds of sensors provide real-time detection of pathogen and are fast and cost-effective. That's why it is important to improve detection techniques–for better results. Different types of microorganism detection methodologies like SPR, optical, whole cell, electrochemical biosensors, and others are used for monitoring. That's the basic reason they are used for the detection of food pathogen and some other biochemical applications.

3.7.2.1 Microfluidics for Pathogen Detection

This pathogen detection method targets the protein, DNA, and cells. To detect the food-borne pathogen and their toxins, the microfluidic biosensors play a fundamental role. In this biosensor, the transducer and biochemical radiants are used for the section of analyte. A number of biosensors are used for the detection of pathogen in foodstuff. Different types of transducers are used for the detection of signals that indicate the presence of specific antibodies. To free the foodstuff from pathogen, sterilizations techniques follow.

3.7.2.2 Multiplex PCR Assay

Food pathogens cause gastroenteritis all over the world, and one of the main causes of increased mortality rates. More than 1.5–2.5 million deaths are due to transmittable gastroenteritis, which infects children and the elderly the most because their immunity is weak. For analysis of different types of bacterial species like *Campylobacter* spp., *Vibrio* spp., *E. coli*, and different strains of *Salmonella* that cause foodborne disease, multiplex molecular technology known as multiplex PCR has been developed. It can be used to examine more than 30 species of entero-pathogenic bacteria. To know the global study position of *Vibrio cholera* we use this test for better analysis. For detection of group B microbial product and water-contaminating pathogens we use multiplex PCR.

3.7.2.3 Biosensors for Food Pathogen Detection

Rules and consequences have been developed to ensure food security and protection. Several airborne pathogens and contaminants makes the food unhygienic. Different types of endotoxin and exotoxin are commonly known airborne pathogens, but some other strains of these toxins, like *Clostridium*, *Vibrio cholerae*, and *Anthrax* are also common examples of contaminations. Children, pregnant women, old people, and those who have weak immune systems are most susceptible to these diseases. Biosensor is a powerful technique for the detection of pathogens. Biosensor is a logical device that is used to translate the genetic response into electrical signals. It consists of two parts, a bioreceptor and a transducer. The bioreceptor detects the response of objective analyte, while the transducer converts these responses into electrical signals. Biosensors are capable of quick detection of the pathogen and are powerful enough to detect even single-cell contaminants like liposomes (Chen and Durst 2006). According to recent data, it is clear that different types of label-free biosensors are used for the early detection of airborne contaminants. Some biosensors work on the basis of genetic identification, so there is the need to enhance the utilization of SPR, potentio matrix and ampometrix biosensors, and some electrochemical base biosensors for copying their fluorescent response. Different types of enzyme and antibodies are used for this purpose. Different types of techniques are used for microbial detection such as mass spectrometry, Raman and infrared spectroscopy, MALDI-TOF, and auto-fluorescentable procedure (Estes et al. 2003). A number of biosensor devices are used to determine the hereditary properties of microbes and their physiological assortment.

3.7.2.3.1 Surface Plasmon Resonance Biosensors Multiple forms of biosensors are used but the ocular illumination of metal is the most useful one because it is used for the detection of foodborne pathogen. The basic principle of SPR depends on the observational data. First we have to immobilize the antibodies on gold-contigu and capture them after assorting them with the food pathogenic bacteria. And after attracting with an electron at a specific wavelength a particular type of resonance is generated. Then analysis of the quality of wavelength indicates the presence of pathogen. The SPR biosensor is used to detect the immunological efforts of the pathogen (Rasooly and Herold 2006; Bhunia 2007). SPR is used to detect the different kind of food pathogenic bacteria like *E. coli*, *Salmonella* and *L. monocytogenes*.

3.7.2.3.2 Optical Biosensors Optical biosensors are the most acceptable form of biosensor for the detection of contaminant due to their high selective process. The most commercially used biosensor by researchers is the fiber optic. According to Bhunia (2007) and Kumar et al. (2016), its working principle depends on the fluorescent labeling of toxin. When the pathogen is attached at the outer side of the optic-fiber then the laser beam is passing through it with the wavelength of 635 nm and it can produce illumine (Bhunia 2007) which can be detected with the help of illumine detector. For quick recognition of foodborne pathogen and its toxin a number of optical biosensors are developed (Baeumner et al. 2003). In this form of biosensor, the illumination plays a vital role to attach with antibodies and detect the particular bacteria or pathogen. Some commonly use optical biosensors are ELISA and PCR.

3.7.2.3.3 Piezoelectric Biosensors Piezoelectric biosensor is the combination of various suspected detectors, in which a semi-precious stone is used to influence the base of the expected resonance frequency. Lithium neonate quartz can act as a basic raw substance for piezoelectric biosensors. The distinct antibodies are attaching the outer surface of this biosensor and cover the whole surface. The foodborne pathogens are attached with the antibodies and a large number of quartz crystals is accumulated which can be analyze by the quartz system microbalance (QCM) in real-time. This is an easy or trouble-free method for the detection of foodborne pathogen like *S. typhimurium.*

3.7.2.3.4 Cell-Based Biosensors For detection of foodborne pathogen the cell-based biosensors act as a consistent device. On this form of biosensor an electrical cell is used to develop a modification in neighboring cells. In this whole process the cell membrane acts as a capacitor while the fluid present in the cell acts as a resistant component. This electrical disturbance in the surrounding cells shows that the whole functionality of the cell is altering like change in cell concentration or cell intensification. Each cell is designed to detect the harmful bacteria in foodstuff. For understanding the cell morphological process and development process of tissue culture the cell-based biosensors are commonly used. It is an easy approach that is used to recognize the various enzyme and cofactors that is produced by the cell. It is also used to gather the information for the recombination methodology of DNA.

3.7.2.3.5 Amperometric Biosensors Amperometric biosensors are used to observe an electrochemically lively complex. In this process the electrodes are used and various forms of analytes are condensed on this electrode. The thin film tools are composed of gold, platinum, and carbon residues. For developing the diverse form of biosensors different types of ink are used. For developing the Amperometric biosensors the screen-printed electrochemical mixture is used because this is cost effective and commonly found in large quantities. These are non-reusable forms of biosensor and due to this they are less toxic. It is an enzymatic catalytic process that can be used to translate the electrochemical non-reactive compound into a reactive one. HRP and alkaline phosphate are the most commonly used enzyme in this process. Amperometric biosensors are also used to detect the enzyme-linked compounds which is why in some cases it is known as an immunosensor. A major side effect of this sensor is that it can produce false signals in cells. This can be overcome by using a typical membrane to analyze the analytes on the basis of charge to mass ratio by using the biosensor. Different type of DNA-based biosensors and antibodies-based biosensors are used for this purpose. Amperometric biosensors are also used to detect pathogens in airborne foodstuff.

3.7.2.3.6 Microbial Metabolism-Based Biosensor Amperometric biosensors are used to carry out the series of biochemical reactions in the bacterial cell during its cellular processing. Different type oxygen electrodes are used for the in situ screening of bacterial pathogen. These electrode are used to develop the clerk which basically use to measure the oxygen neutralization in bacterial cell like *S. typhimurium*. The quantity of cathode rays is indicating the complexity of the reaction. High quantity of cathode rays is generate during the microbial metabolism in bacterial cell. By using this process different type of enzyme can also be identified with the appropriate use of different type of transducers. GUS (glucuronosohydrolase) is a catalytic enzyme. It is used to develop the different isoforms in water. The *E. coli* recognition with GUS is based on the spectrometric identification and it is lengthy and conventional method. It is a time taking process and this method is developing through the electro oxidative components. In this process the carbon attach electrode are used for the reduction of *Moraxella* spp. by p-nitro phenol (PNP) and it is used for the production of hydro-quinine. Different types of coli-forms are observed in water and the d-GAL can indicate the quality results. All that phenomena is used for the enzymatic breakdown and conversation of galactose and glucose molecule. Microbial metabolism based biosensors are high through put sensors which provide the quality results.

3.7.2.3.7 DNA Based Biosensors This kind of biosensors are directly used for the airborne detection of foodstuff pathogen. The probe (nucleotide sequence) is present at the exterior side of the transducer that is used for the immobilization of the bacteria. The opposite sequence of nucleotide attach with the probe by the process of hybridization. The degree of hybridization is used for the presence and absence of opposite nucleotide sequence in foodstuff. Recent studies indicated the different type of electrochemical based b transducers combined with the DNA based sensors for the high quality results. In this approach we can determine the bacterial toxin by identifying the specific gene coding probes, these probes are specific for the detection of foodborne pathogen. This can act as fundamental step of evolution for encoded sequences. This process is use for the toxin in detection inside the foodstuff. *E. coli*, *Mycobacterium tuberculosis*, *Cryptosporidium* and *Giardia* are some bacterial species that is use for the production of novel geno-sensors. The phenomena of DNA amplification is established by utilizing the various type of magnetic primers. Different type of terminologies like PCR, 16S rRNA microarray technologies are used for the automatic detection of different strains of bacteria like *E. coli*, *Pseudomonas aeruginosa* and *S. aureus*. Different types of oligonucleotide sequences are also used for the analysis. And phenomena indicate the development of different type of hybridization techniques. By the action of the phosphate molecule different type of biotin labeled AP and amino phosphates are released in this way. This method is also used to prevent the amplification by PCR and it is a reliable and simple method for the detection of foodborne pathogen. Now different type of RNA sense sensors are developed on the DNA RNA based hybridization method, these biosensors show quick response and give better consequences (Baeumner et al. 2003).

3.7.2.3.8 Impedimetric and Conductimetric Biosensors Impedimeric and conductimeric biosensors are used for the detection of electrochemical signaling. For carrying out the procedures, the impedimeric capacitance and conductance can be altered. Biosensors are an appropriate method for checking the metabolic activity of organism. To differentiate the development of microbes primary impedance quantities are used. The first bacteria that was distinguished between other food pathogens using the impedance method was *Salmonella*. In 1996, *Salmonella* in foodstuff was discovered by using the different types of screening techniques. These screening techniques are based on the acknowledgement of impedance procedure. Some other forms of bacteria, like *Entrobacteriaceae*, *L. monocytogenes*, and *Listeria* spp. are also detected by using the impedance-based microbial method. This method is widely use to differentiate between microorganisms in the experiment. The protective coating of thermal oxides is done by means of silicon. It can be used for the detection of *E. coli* strain and is composed of gold electrodes. Different types of polyclonal antibodies can be used to recognize microbial immobilization. In short, our data indicate that this particular type of biosensor is effective for the detection of pathogens in foodstuff.

3.7.2.3.9 Potentiometry, FET, and LAPS   For the detection of pathogens and toxins in food, the field effective transistors (FET), light addressable potentiometric sensors (LAPS), and potentiometry biosensors are used. The principle is based on the ion resolution method. Fluctuation in pH and change in the concentration of ion can indicate the presence of pathogen in food. The optimum range is 10^{-5}–10^{-2} mol/L. They are accessible and cost-effective methods. The major disadvantage is that its selectivity is dependent on the different types of ecological samples. For enlightening the genetic procedures, the ion-sensitive FET is used. These biosensors show weak identification or low recognition, and these sensors can act as industrial biosensors (Lazcka et al. 2007). For analyzing the pathogen in foodstuff the ISFET is used in combination with potentiometry. At the commercial level, different types of immunological techniques are used for the detection of pathogen. Different type of illumines are spread on the nitrocellulose plate for the detection of pathogen on foodstuff.

3.7.2.3.10 Electrochemical Immunosensors   The quick quantification of food pathogenic bacteria, such as *Salmonella* spp., *E. coli*, and others is performed with the help of electrochemical immunosensors. By immobilization of the different strain of *E. coli*, *Salmonella* and *Campylobacter* antibodies we are able to develop the immunological biosensors. Different type of immunological test is performed for the pathogen detection but the most commonly used is sandwich immunological test. This test is performing in the presence of three antibodies that linked with each other and form the nanocrystal like structure. These techniques are used for the airborne detection of pathogen.

3.8 CONCLUSION

Molecular epidemiology is basically the merging of biological molecular tools and epidemiological studies. Similar to the statistical tools, the molecular tools play a key role in the diagnosis and treatment of pandemics of foodborne diseases. There are different foodborne diseases having different etiologies including bacterial, fungal, viral, and parasitic. These diseases often result in serious illness with prevalence of 1/10 people and 420,000 deaths (among which 30% are children under 5 years, and 14.5% are hospital cases) annually worldwide. Significantly, the molecular epidemiology through various biological molecular tools such as immunoassay, in-situ high-affinity capture and staining method, biosensors, optical biosensors, PCR assay, potentiometry etc. play a key role in the diagnosis and treatment of foodborne diseases as well as in the adoption of preventive measures.

BIBLIOGRAPHY

Aboltins, C. A., Pratt, W. A. and Solano, T. R. 2006. Fungemia secondary to gastrointestinal *Mucor indicus* infection. *Clinical Infectious Diseases: An Official Publication of the Infectious Diseases Society of America* 42(1):154–5.

Adley, C. C. and Dillon, C. 2011. Listeriosis, salmonellosis and verocytoxigenic *E. coli*: Significance and contamination in processed meats. In: Kerry, J. P. and Kerry, J. F. (eds.), *Processed meats: Improving safety, nutrition and quality*. Woodhead Publishing Ltd, Oxford, 72–108.

Ahmad, F., Seyrig, G., Tourlousse, D. M., Stedtfeld, R. D., Tiedje, J. M. and Hashsham, S. A. 2011. A CCD-based fluorescence imaging system for real-time loop-mediated isothermal amplification-based rapid and sensitive detection of waterborne pathogens on microchips. *Biomedical Microdevices* 13(5):929–37.

Ali, T., Kaitha, S., Mahmood, S., Ftesi, A., Stone, J. and Bronze, M. S. 2013. Clinical use of anti-TNF therapy and increased risk of infections. *Drug, Healthcare and Patient Safety* 5:79–99.

Aliabadi, N., Lopman, B. A., Parashar, U. D. and Hall, A. J. 2015. Progress toward Norovirus vaccines: Considerations for further development and implementation in potential target populations. *Expert Review of Vaccines* 14(9):1241–53.

Allen, K. J., Wałecka-Zacharska, E., Chen, J. C., et al. 2016. *Listeria monocytogenes* an examination of food chain factors potentially contributing to antimicrobial resistance. *Food Microbiology* 54:178–89.

Alouf, J. E., Popoff, M. R. and Ladant, D. 2015. *The comprehensive sourcebook of bacterial protein toxins*. Academic Press, Burlington, MA.

Anantaphruti, M. T. 2001. Parasitic contaminants in food. *Southeast Asian Journal of Tropical Medicine and Public Health* 32(S2):218–28.

Arnold, F. A., Dean, H. T., Jay, P. and Knutson, J. W. 1953. Effects of fluoridated water supplies on dental caries incidence: Results of the seventh year of study at Grand Rapids and Muskegon, Mich. *Public Health Reports* 68:141–8.

Artault, S., Blind, J. L., Delaval, J., Dureuil, Y. and Gaillard, N. 2001. Detecting *Listeria monocytogenes* in food. *International Journal of Food Hygiene* 12:23.

Baeumner, A. J., Cohen, R. N., Miksic, V. and Min, J. 2003. RNA biosensor for the rapid detection of viable *Escherichia coli* in drinking water. *Biosensor and Bioelectronics* 18(4):405–13.

Baird, F. J., Gasser, R. B., Jabbar, A. and Lopata, A. L. 2014. Foodborne anisakiasis and allergy. *Molecular and Cellular Probes* 28(4):167–74.

Baumann, A. and Sadkowska-Todys, M. 2007. Foodborne infections and intoxications in Poland in 2005. *Przeglad Epidemiologyczny* 61(2):257.

Begum, A., Bari, S., Chowdhury, F. R., Ahmed, N. and Sayeed, K. 2015. Pattern of antimicrobial sensitivity and resistance against *Salmonella* species in a tertiary hospital in Dhaka. *Journal of Enam Medical College* 5(2):88–92.

Bellete, B., Raberin, H., Berger, C., et al. 2006. Molecular confirmation of an absidiomycosis following treatment with a probiotic supplement in a child with leukemia. *Journal of Mycology Medicine* 16(2):72–6.

Beumer, R. R. and Brinkman, E. 1989. Detection of *Listeria* spp. with a monoclonal antibody-base enzyme-linked immunosorbent assay (ELISA). *Food Microbiology* 6(3):171–7.

Bezirtzoglou, E. and Stavropoulou, E. 2011. Immunology and probiotic impact of newborn and young children intestinal microflora. *Anaerobe* 17:370–3.

Bhunia, A. K. 2007. Biosensor and bio-based methods for the separation and detection of food borne pathogen. *Advances in Food and Nutrition Research* 54:1–44.

Bottone, E. J. 2010. *Bacillus cereus*, a volatile human pathogen. *Clinical Microbial Review* 23(2):382–98.

Brenier-Pinchart, M. P., Faure, O., Garban, F., et al. 2006. Ten-year surveillance of fungal contamination of food within a protected haematological unit. *Mycoses* 49(5):421–5.

Brooks, B. W., Devenish, J., Lutze-Wallace, C. L., Milnes, D., Robertson, R. H. and Berlie-Surujballi, G. 2004. Evaluation of a monoclonal antibody-based enzyme-linked immunosorbent assay for detection of *Campylobacter fetus* in bovine preputial washing and vaginal mucus samples. *Veterinary Microbiology* 103(1–2):77–84.

Buckley, M. 2008. *The Fungal Kingdom: Diverse and Essential Roles in Earth's Ecosystem*. American Academy of Microbiology, Washington, DC.

Busl, K. M. and Bleck, T. P. 2012. Treatment of neuroterrorism. *Expert Review of Neurotherapeutics* 9(1):139–57.

Byrne, B., Stack, E., Gilmartin, N. and O'Kennedy, R. 2009. Antibody-based sensors: Principles, problems and potential for detection of pathogens and associated toxins. *Sensors* 9(6):4407–45.

Cadotte, J. 2013. FDA gets nearly 300 reports of illness from chobani yogurt. Twin Falls, Idaho: Times-News-Southern Idaho Local News.

Campos, A. K. C., Cardonha, A. M. S., Pinheiro, L. B. G., Ferrira, N. R., Azevedo, P. R. M. and Stamford, T. L. M. 2009. Assessment of personal hygiene and practice of food handlers in municipal public schools of Natal, Brazil. *Food Control* 20(9):807–9.

Carrada-Bravo, T. 2003. Fascioliasis: Diagnosis, epidemiology and treatment. *Revista de Gastroenterologia de Mexico* 68(2):135–42.

CDC 2005. *Bioterrorism Agents/Diseases by Categories*. Centers for Disease Control and Prevention. http://www.bt.cdc.gov/agent/agentlist-category.asp.

CDC 2010. *Human paragonimiasis after eating raw or undercooked crayfish: July 2006–September 2010*. Centers for Disease Control and Prevention, Missouri. https://www.cdc.gov/mmwr/preview/mmwrhtml/mm5948a1.htm.

CDC 2015a. *CDC diseases and conditions home page*. Centers for Disease Control and Prevention, Atlanta, GA. http://www.cdc.gov/DiseasesConditions/.

CDC 2015b. *Cyclosporiasis*. Centers for Disease Control and Prevention. http://www.cdc.gov/parasites/cyclosporiasis/resources/pdf/cyclosporiasis_general-public_061214.pdf.

CDC 2015c. *Trichinellosis surveillance—United States 2008–12*. Centers for Disease Control and Prevention. http://www.cdc.gov/mmwr/preview/mmwrhtml/ss6401a1.htm.

CFSPH 2012. *Giardiasis*. Center for Food Security and Public Health. http://www.cfsph.iastate.edu/Factsheets/pdfs/giardiasis.pdf.

Chen, C. S. and Durst, R. A. 2006. Simultaneous detection of *Escherichia coli* O157: H7, *Salmonella* spp. and *Listeria monocytogenes* with an array-based immunosorbent assay using universal protein G-liposomal nanovesicles. *Talanta* 69(1):232–8.

Cheng, V. C., Chan, J. F., Ngan, A. H., et al. Outbreak of intestinal infection due to *Rhizopus microsporus*. *Journal of Clinical Microbiology* 47(9):2834–43.

Chowdhury, G., Sangeeta, J., Sanjay, B., et al. 2016. Extraintestinal infections caused by non-toxigenic *Vibrio cholerae* non-O1/non-O139. *Frontiers in Microbiology* 7:1–5.

Clark, W. F., Macnab, J. J. and Sontrop, J. M. 2008. The Walkerton health study 2002–2008. http://www.food-label-compliance.com/Sites/5/Downloads/Clark-Wm-Walkerton-Health-Study-2008.

Correia, A. M., Gonçalves, G. and Saraiva, M. M. 2004. Foodborne outbreaks in northern Portugal, 2002. *Eurosurveillance* 9(3):18–20.

Cross, J. H. 1996. Enteric nematodes of humans. In: Baron, S. (ed.), *Medical microbiology*. University of Texas Medical Branch at Galveston, Galveston, TX.

Crush, S. J. and Frayne, B. G. 2011. Urban food insecurity and the new international food security agenda. *Development Southern Africa* 28(4):528–32.

Das, A. P., Kumar, P. S. and Swain, S. 2014. Recent advances in biosensor based pyrogen detection. *Biosensors and Bioelectronics* 51:62–75.

De Boer, E. and Beumer, R. R. 1999. Methodology for detection and typing of foodborne microorganisms. *International Journal of Food Microbiology* 50(1–2):119–30.

Dhama, K., Rajagunalan, S., Chakraborty, S., et al. 2013. Foodborne pathogens of animal origin: Diagnosis, prevention, control and their zoonotic significance: A review. *Pakistan Journal of Biological Sciences* 16(20):1076–85.

Dorny, P., Praet, N., Deckers, N. and Gabriel, S. 2009. Emerging foodborne parasites. *Veterinary Parasitology* 163(3):196–206.

Eguale, T., Gebreyes, W. A., Asrat, D., Alemayehu, H., Gunn, J. S. and Engidawork, E. 2015. Non-typhoidal *Salmonella* serotypes, antimicrobial resistance and co-infection with parasites among patients with diarrhea and other gastrointestinal complaints in Addis Ababa, Ethiopia. *BMC Infectious Diseases* 15:497.

Ehling-Schulz, M., Fricker, M. and Scherer, S. 2004. Identification of emetic toxin producing *Bacillus cereus* strains by a novel molecular assay. *FEMS Microbiology Letters* 232(2):189–95.

Environmental Health National Norms and Standards 2013. *Government Gazette.* South Africa. http://www.nicd.ac.za/assets/files/ environmental health national_ Norms_ and_standards/.

Epstein, E. 2015. *Disposal and management of solid waste: pathogens and diseases.* CRC Press, Boca Raton, FL.

Estes, C., Duncan, A., Wade, B., Lloyd, C., Ellis, J. W. and Powers, L. 2003. Reagentless detection of microorganisms by intrinsic fluorescence. *Biosensor and Bioelectronics* 18(5–6):511–9.

Evans, A. S. and Brachman, P. S. 1998. *Bacterial infections of humans: Epidemiology and control.* Springer, New York.

Farrar, J., Hotez, P., Junghanss, T., Kang, G., Lalloo, D. and White, N. J. 2014. *Manson's tropical diseases*, 23rd ed. Elsevier Saunders, Philadelphia, PA.

Fasoro, A. A., Faeji, C. O., Oni, O. I. and Oluwadare, T. 2016. Assessment of food safety practices in a rural community in Southwest Nigeria. *Food and Public Health* 6(3):59–64.

FDA (US Food and Drug Administration) 2002. Isolation and enumeration of *Enterobacter sakazakii* from dehydrated infant formula. *Food and Drug Administration.* www. cfsanfdagov/~comm/mmesakaz.html.

FDA (US Food and Drug Administration) 2012. *Bad bug book: Foodborne pathogenic microorganisms and natural toxins.* Center for Food Safety and Applied Nutrition (CFSAN). College Park, MD, United States.

FDA (US Food and Drug Administration) 2013. *Voluntarily recalls greek yogurt because of product concerns 2013.* http://www.fda.gov/Safety/Recalls/ucm367298. html.

FDA (US Food and Drug Administration) 2015. Foodborne illness-causing organisms in the U.S. Factsheet. http://www.fda.gov/downloads/Food/FoodborneIllnessConta minants/M187482.pdf.

Finkelstein, R. A. 1996. *Cholera, Vibrio cholerae* O1 and O139, and other pathogenic vibrios. In: Baron, S. (Ed.), *Medical Microbiology.* University of Texas Medical Branch at Galveston, Galveston, TX.

Fleckenstein, J. M., Bartels, S. R., Drevets, P. D., Bronze, M. S. and Drevets, D. A. 2010. Infectious agents of food- and water-borne illnesses. *American Journal of Medical Sciences* 340(3):238–46.

Food Advisory Consumer Service 2009. Food poisoning. http://www.foodfacts.org.za/Articles/Food Poisoning.

Food Safety News 2013. *Low-income children face higher risk of foodborne illness, study says.* http:// www. Food safety news .com/2013/income and foodborne illness/.

Food Safety Newsletter 2015. From farm to plate, make food safe. *Newsletter Food Safety* 1(1):3–8.

Forbes, L. B., Parker, S. and Scandret, W. B. 2003. Comparison of a modified digestion assay with trichinoscopy for the detection of *Trichinella* larvae in pork. *Journal of Food Protection* 66(6):1043–6.

Francis, T., Korns, R., Voight, R., et al. 1955. An evaluation of the poliomyelitis vaccine trials. *American Journal of Public Health and the Nation's Health* 45(5 Pt 2):1–63.

Fratamico, P. M., Bhunia, A. K. and Smith, J. L. 2005. *Foodborne pathogens: Microbiology and molecular biology.* Caister Academic Press, Poole, UK.

Gadaga, T. H., Mutukumira, A. N. and, Narvhus, J. A. 2001. Growth characteristics of *Candida kefyr* and two strains of *Lactococcus lactis* subsp. *lactis* isolated from Zimbabwean naturally fermented milk. *International Journal of Food Microbiology* 70(1–2):11–9.

Galindo, C. L., Rosenzweig, J. A., Kirtley, M. L. and Chopra, A. K. 2011. Pathogenesis of *Y. enterocolitica* and *Y. pseudotuberculosis* in human yersiniosis. *Journal of Pathogens*:1–16.

Garcia, J., Bennett, D. H., Tancredi, D., et al. 2013. Occupational exposure to particulate matter and endotoxin for California dairy workers. *International Journal of Hygiene and Environmental Health* 216(1):56–62.

Gkogka, E., Reij, M. W., Havelaar, A. H., Zwietering, M. H. and Gorris, L. G. H. 2011. Risk-based estimate of effect of foodborne disease on public health, Greece. *Emerging Infectious Diseases* 17(9):1–9.

Gonzales, I., Rivera, J. T., Garcia, H. H. and Cysticercosis Working Group in Peru 2016. Pathogenesis of *Taenia solium* taeniasis and cysticercosis. *Parasite Immunology* 38(3):136–46.

Greenough, W. B., Rosenberg, I. S., Gordon, R. S., Davies, B. I. and Benenson, A. S. 1964. Tetracycline in the treatment of cholera. *Lancet* 41:355–7.

Gul, H. C. and Erdem, H. 2015. Brucellosis (*Brucella* species). In: Bennett, J. E., Dolin, R. and Blaser, M. J. (eds.), *Mandell, Douglas, and Bennett's principles and practice of infectious diseases.* Elsevier Saunders, Philadelphia, PA, 2584–90.

Gurgui, M., Sanchez, F., March, F., et al. 2011. Nosocomial outbreak of *Blastoschizomyces capitatus* associated with contaminated milk in a haematological unit. *Journal of Hospital Infection* 78(4):274–8.

Haghdoost, A. A. 2008. Molecular epidemiology, concepts and domains. *Journal of Kerman University of Medical Sciences* 15(1):97–104.

Hajar, R. 2013. Food poisoning in Morocco: Evolution and risk factors. *International Journal of Scientific and Engineering Research* 4(11):25–7.

Harrell, L. W. and Deardorff, T. L. 1990. Human nanophyetiasis: Transmission by handling naturally infected coho salmon (*Oncorhynchus kisutch*). *Journal of Infectious Diseases* 161(1):146–8.

Hawksworth, D. L. 2001. The magnitude of fungal diversity: The 1.5 million species estimate revisited. *Mycological Research* 105(12):1422–32.

Hennekinne, J. A., De Buyser, M. L. and Dragacci, S. 2012. *Staphylococcus aureus* and its food poisoning toxins: Characterization and outbreak investigation. *FEMS Microbiology Reviews* 36(4):815–36.

Hong, S. T. and Fanq, Y. 2012. *Clonorchis sinensis* and clonorchiasis, an update. *Parasitology International* 61(1):17–24.

Hong, S., Kim, K., Yoon, S., Park, W. Y., Sim, S. and Yu, J. R. 2014. Detection of *Cryptosporidium parvum* in environmental soil and vegetables. *Journal of Korean Medical Sciences* 29(10):1367–71.

Hossain, M., Uddin, M. B., Al Hassan, A., Islam, R. and Cho, H. S. 2014. Potential risk factors analysis of dairy cattle management against brucellosis. *Veterinary Research International* 2:96–102.

Hui, Y. H., Pierson, M. D. and Gorham, J. R. 2001. *Foodborne disease handbook*, Vol 1. *Bacterial pathogens*. Marcel Dekker, New York.

Hunt, R. H., Camilleri, M., Crowe, S. E., et al. 2015. The stomach in health and disease. *Gut* 64(10):1650–68.

Hunter, C. J., Petrosyan, M., Ford, H. R. and Prasadarao, N. V. 2008. *Enterobacter sakazakii*: An emerging pathogen in infants and neonates. *Surgical Infections* 9(5):533–9.

Hussain, M. A. and Dawson, C. O. 2013. Economic impact of food safety outbreaks on food businesses. *Foods* 2(4):585–9.

Iriart, X., Fior, A., Blanchet, D., Berry, A., Neron, P. and Aznar, C. 2010. *Monascus ruber*: Invasive gastric infection caused by dried and salted fish consumption. *Journal of Clinical Microbiology* 48(10):3800–2.

Ito, A. and Budke, C. M. 2014. Culinary delights and travel? A review of zoonotic cestodiases and metacestodiases. *Travel Medicine and Infectious Disease* 12(6 Pt A):582–91.

Iwamoto, M., Ayers, T., Mahon, B. E. and Swerdlow, D. L. 2010. Epidemiology of seafood-associated infections in the United States. *Clinical Microbiological Review* 23(2):399–409.

Janda, J. M. and Abbott, S. L. 2010. The genus *Aeromonas*: Taxonomy, pathogenicity, and infection. *Clinical Microbial Review* 23(1):35–73.

Johler, S., Giannini, P., Jermini, M., Hummerjohann, J., Baumgartner, A. and Stephan, R. 2015. Further evidence for Staphylococcal food poisoning outbreaks caused by egc-encoded enterotoxins. *Toxins* 7(3):997–1004.

Josefson, P., Strålin, K., Ohlin, A., et al. 2011. Evaluation of a commercial multiplex PCR test (SeptiFast) in the etiological diagnosis of community-onset bloodstream infections. *European Journal of Clinical Microbiology and Infectious Diseases : Official Publication of the European Society of Clinical Microbiology* 30(9):1127–34.

Kaper, J. B., Nataro, J. P. and Mobley, H. L. T. 2004. Pathogenic *Escherichia coli*. *Nature Review in Microbiology* 2(2):123–40.

Karam, A., Eveillard, J. R., Ianoto, J. C. et al. 2005. Disseminated cutaneous and visceral fusariosis in an aplastic patient: An unusual digestive entry. *Annales de Dermatologie et de Venereologie* 132(3):255–8.

Kariuki, S., Gordon, M. A., Feasey, N. and Parry, C. M. 2015. Antimicrobial resistance and management of invasive *Salmonella* disease. *Vaccine* 33(S3):21–9.

Kaye, K. S., Anderson, D. J., Choi, Y., Link, K., Thacker, P. and Sexton, D. J. 2008. The deadly toll of invasive methicillin-resistant *Staphylococcus aureus* infection in community hospitals. *Clinical Infectious Diseases : An Official Publication of the Infectious Diseases Society of America* 46(10):1568–77.

Khaleej Times 2014. *518 cases of food borne diseases recorded in Dubai.* http://www. khaleejtimes.com/lifestyle/health-fitness/518-cases-of-foodborne-diseases/

Kilbourne, E. D. 1973. The molecular epidemiology of influenza. *Journal of Infectious Diseases* 127(4):478–87.

Kim, Y. B., Okuda, J., Matsumoto, C., Takahashi, N., Hashimoto, S. and Nishibuchi, M. 1999. Identification of *Vibrio parahaemolyticus* strains at the species level by PCR targeted to the toxR gene. *Journal of Clinical Microbiology* 37(4):1173–7.

Kirk, M. 2014. Foodborne disease surveillance needs in Australia: Harmonization of molecular laboratory testing and sharing data from human, animal and food sources. *New South Wales Public Health Bulletin* 15(2):4–5.

Kirk, M., Ford, L., Glass, K. and Hall, G. 2014. Foodborne illness, Australia, circa 2000 and circa 2010. *Emerging Infectious Diseases* 20(11):1857–62.

Kirk, M. D., Pires, S. M., Black, R. E., et al. 2015. World Health Organization estimates of the global and regional disease burden of 22 foodborne bacterial, protozoal, and viral diseases, 2010: A data synthesis. *PLoS Medicine* 2:100–92.

Koopmans, M., von Bonsdorff, C. H., Vinjé, J., de Medici, D. and Monroe, S. 2002. Foodborne viruses. *FEMS Microbiology Reviews* 26(2):187–205.

Kumar, M. S., Ghosh, S., Nayak, S. and Das, A. P. 2016. Recent advances in biosensor based diagnosis of urinary tract infection. *Biosensor and Bioelectronics* 80:497–510.

Lacey, J. A., Johanesen, P. A., Lyras, D. and Moore, R. J. 2016. Genomic diversity of necrotic enteritis associated strains of *Clostridium perfringens*: A review. *Avian Pathology : Journal of the W.V.P.A* 45(3):302–7.

Lazar, S. P., Lukaszewicz, J. M., Persad, K. A. and Reinhardt, J. F. 2014. Rhinocerebral *Mucor circinelloides* infection in immune compromised patient following yogurt ingestion. *Delaware Medical Journal* 86(8):245–8.

Lazcka, O., Del Campo, F. J. and Munoz, F. X. 2007. Pathogen detection: A perspective of traditional methods and biosensors. *Biosensor and Bioelectronics* 22(7):1205–17.

Lee, K. W., Lillehoj, H. S., Jeong, W., Jeoung, H. Y. and An, D. J. 2011. Avian necrotic enteritis: Experimental models, host immunity, pathogenesis, risk factors, and vaccine development, host immunity, pathogenesis, risk factors, and vaccine development. *Poultry Sciences* 90(7):1381–90.

Lee, S. C., Billmyre, R. B., Li, A., et al. 2014. Analysis of a foodborne fungal pathogen outbreak: Virulence and genome of a *Mucor circinelloides* isolate from yogurt. *mBio* 5:13–4.

Leong, D., Alvarez-Ordonez, A., Jooste, P. and Jordan, K. 2016. *Listeria monocytogenes* in food: Control by monitoring the food processing environment. *African Journal of Microbial Research* 10:1–14.

Liu, Y., Tam, Y. H., Yuan, J., et al. 2015. A foodborne outbreak of gastroenteritis caused by *Vibrio parahaemolyticus* and Norovirus through non-seafood vehicle. *PLoS ONE* 10(9):137–48.

Löfmark, S., Edlund, C. and Nord, C. E. 2010. Metronidazole is still the drug of choice for treatment of anaerobic infections. *Clinical Infectious Diseases : An Official Publication of the Infectious Diseases Society of America* 50(S1):16–23.

Lopez, L., Roos, R., Cressey, P., Horn, B. and Lee, J. 2016. Foodborne disease in New Zealand 2015. *Ministry for Primary Industries.* http://16020-Annual-Foodborne-Disease-Report-2015.

Lydyard, P., Cole, M., Holton, J., et al. 2010. *Case studies in infectious disease.* Taylor and Francis Group, New York.

Mafu, A. A., Pitre, M. and Sirois, S. 2009. Real-time PCR as a tool for detection of pathogenic bacteria on contaminated food contact surfaces by using a single enrichment medium. *Journal of Food Protection* 72(6):1310–4.

Mairiang, E. and Mairiang, P. 2003. Clinical manifestation of opisthorchiasis and treatment. *Acta Tropica* 88(3):221–7.

Majowicz, S. E., Meyer, S. B., Kirkpatrick, S. I., et al. 2016. Food, health, and complexity: Towards a conceptual understanding to guide collaborative public health action. *BMC Public Health* 16:487.

Majowicz, S. E., Musto, J., Scallan, E., et al. 2010. The global burden of nontyphoidal *Salmonella* gastroenteritis. *Clinical Infectious Diseases : An Official Publication of the Infectious Diseases Society of America* 50(6):882–9.

Malangu, N. 2016. Risk factors and outcomes of food poisoning in Africa. *Science, Technology and Medicine*:14–40.

Marini, E., Magi, G., Vincenzi, C., Manso, E. and Facinelli, B. 2016. Ongoing outbreak of invasive listeriosis due to serotype 1/2a *Listeria monocytogenes*, Ancona province, Italy, January 2015 to February 2016. *EURO Surveillance : Bulletin Europeen sur les Maladies Transmissibles = European Communicable Disease Bulletin* 21(17):1–4.

Marroquin-Cardona, A. G., Johnson, N. M., Phillips, T. D. and Hayes, A. W. 2014. Mycotoxins in a changing global environment-a review. *Food and Chemistry Toxicology* 69:220–30.

Martin, S. 2001. Congenital toxoplasmosis. *Neonatal Network* 20(4):23–30.

Martinello, M., Nelson, A., Bignold, L. and Shaw, D. 2012. "We are what we eat!" Invasive intestinal mucormycosis: A case report and review of the literature. *Medical Mycology Case Reports* 1(1):52–5.

Martinez, A., Dominguez, A., Torner, N., et al. 2008. Epidemiology of foodborne *Norovirus* outbreaks in Catalonia, Spain. *BMC Infectious Diseases* 8(47):2–7.

McKenna, M. 2012. Food poison's hidden legacy. *Scientific American.* https://www.sci entificamerican.com/article/food-poisonings-hidden-legacy/.

Miyagishima, K., Abela-Riddder, B. and Savelli, C. J. 2013. *Planning for rapid response to outbreaks of animal diseases transmissible to humans via food.* US National Library of Medicine. http://www.ncbi.nlm.nih.gov/pubmed/24.

Moore, D. A., Girdwood, R. W. and Chiodini, P. L. 2002. Treatment of anisakiasis with albendazole. *Lancet* 360(9326):54.

Mossel, D. A. A., Jansen, J. T. and Struijk, C. B. 1999. Microbiological safety assurance applied to smaller catering operations world-wide: From angst through ardour to assistance and achievement the facts. *Food Control* 10(3):195–211.

Musher, D. M., Manhas, A., Jain, P., et al. 2007. Detection of *Clostridium difficile* toxin: Comparison of enzyme immunoassay results with results obtained by cytotoxicity assay. *Journal of Clinical Microbiology* 45(8):2737–9.

Narsaiah, K., Jha, S. N., Bhardwaj, R., Sharma, R. and Kumar, R. 2012. Optical biosensors for food quality and safety assurance: A review. *Journal of Food Science and Technology* 49(4):383–406.

Newman, K. L., Leon, J. S., Rebolledo, P. A. and Scallan, E. 2015. The impact of socio-economic status on foodborne illness in high-income countries: A systematic review. *US National Library of Medicine, National Institute of Health* 143(12):81.

Niyogi, S. K. 2005. Shigellosis. *Journal of Microbiology* 43(2):133–43.

Noordhout, C. M., Devleesschauwer, B., Angulo, F. J., et al. 2014. The global burden of listeriosis: A systematic review and meta-analysis. *Lancet Infectious Disease* 14(11):1073–82.

Okojie, P. W. and Isah, E. C. 2014. Sanitary conditions of food vending sites and food handling practices of street food vendors in Benin City, Nigeria: Implication for food hygiene and safety. *Journal of Environmental and Public Health* 701316:1–6. https://doi.org/10.1155/2014/701316.

Oliver, M. R., Van-Voorhis, W. C., Boeckh, M., Mattson, D. and Bowden, R. A. 1996. Hepatic mucormycosis in a bone marrow transplant recipient who ingested naturopathic medicine. *Clinical Infectious Diseases: An Official Publication of the Infectious Diseases Society of America* 22(3):521–4.

Omernik, A. and Płusa, T. 2015. Toxins of *Clostridium perfringens* as a natural and bioterroristic threats. *Pol Merkur Lekarski* 39(231):149–52.

Oppenheimer, G. M. 2005. Becoming the Framingham study 1947–1950. *American Journal of Public Health* 95(4):602–10.

Pankhurst, L. J., Deacon, L. J., Liu, J., et al. 2011. Spatial variations in airborne microorganism and endotoxin concentrations at green waste composting facilities. *International Journal of Hygiene and Environmental Health* 214(5):376–83.

Park, M. S., Kim, Y. S., Lee, S. H., Kim, S. H., Park, K. H. and Bahk, G. J. 2015. Estimating the burden of foodborne disease, South Korea, 2008–2012. *PubMed Journal* 12(3):207–13.

Pelzer, K. D. 2011. *Emerging infectious diseases*. http://www.courses .iddl.vt.edu/AEID_I/pdf/web/4Pelzer_NandQ.html.

Pineda, C., Kaushik, A., Kest, H., Wickes, B. and Zauk, A. 2012. Maternal sepsis, chorioamnionitis, and congenital *Candida kefyr* infection in premature twins. *Pediatric Infectious Disease Journal* 31(3):320–2.

Pitt, J. I. and Hocking, A. D. 2009. *Fungi and food spoilage*. Springer, New York.

ProMED 2015. *Food/waterborne illness outbreaks*. Africa. http://regionalnews.safef oodinternational.org/page/Africa%3A.

Radosavljevic, M., Koenig, H., Letscher-Bru, V., et al. 1999. *Candida catenulata* fungemia in a cancer patient. *Journal of Clinical Microbiology* 37(2):475–7.

Rajapakse, S., Chrishan Shivanthan, M., Samaranayake, N., Rodrigo, C. and Deepika Fernando, S. 2013. Antibiotics for human toxoplasmosis: A systematic review of randomized trials. *Pathogens and Global Health* 107(4):162–9.

Ramani, S., Atmar, R. L. and Estes, M. K. 2014. Epidemiology of human noroviruses and updates on vaccine development. *Current Opinion in Gastroenterology* 30(1):25–33.

Rasooly, A. and Herold, K. E. 2006. Biosensors for the analysis of food and waterborne pathogens and their toxins. *Journal of AOAC International* 89(3):873–83.

Rebagliati, V., Philippi, R., Tornese, M., Paiva, A., Rossi, L. and Troncoso, A. 2009. Foodborne botulism in Argentina. *The Journal of Infection in Developing Countries* 3(4):250–3.

Reddy, S. P., Wang, H., Adams, J. K. and Feng, P. C. 2016. Prevalence and characteristics of *Salmonella* serotypes isolated from fresh produce marketed in the United States. *Journal of Food Protection* 79(1):6–16.

Rey Matias, C. A., Pereira, I. A., Santos de Araújo, M., Mercês Santos, A. F., Lopes, R. P. and Christakis, S. 2016. Characteristics of *Salmonella* spp. isolated from wild birds confiscated in illegal trade markets, Rio de Janeiro, Brazil. *BioMed Research International*:1–7.

Riemann, H. P. and Cliver, D. O. 2006. *Foodborne infections and intoxications.* Academic Press, San Diego, CA.

Ritter, A. C. and Tondo, E. C. 2014. Foodborne illnesses in Brazil: Control measures for 2014 FIFA World Cup travelers. *The Journal of Infection in Developing Countries* 8(3):254–7.

Roberts, J. A. 2000. Economic aspects of foodborne outbreaks and their control. *British Medical Bulletin* 56(1):133–9.

Rothman, L. 2015. French retirees have contributed to a foodborne illness crisis. https://munchies.com/en us/article/French-retirees.

Ryan, M. P., Dillon, C. and Adley, C. C. 2011. Nalidixic acid-resistant strains of *Salmonella* showing decreased susceptibility to fluoroquinolones in the midwestern region of the Republic of Ireland due to mutations in the gyrA gene. *Journal of Clinical Microbiology* 49(5):2077–9.

Saadatnia, G. and Golkar, M. 2012. A review on human toxoplasmosis. *Scandinavian Journal of Infectious Diseases* 44(11):805–14.

Safe Food International 2011. *Africa food and waterborne illness outbreaks.* Safe Food International, Washington, 1–76.

Saha, D., Karim, M. M., Khan, W. A., Ahmed, S., Salam, M. A. and Bennish, M. L. 2006. Single-dose azithromycin for the treatment of cholera in adults. *The New England Journal of Medicine* 354(23):2452–62.

Sanders, S. Q., Boothe, D. H., Frank, J. F. and Arnold, J. W. 2007. Culture and detection of *Campylobacter jejuni* within mixed microbial populations of biofilms on stainless steel. *Journal of Food Protection* 70(6):1379–85.

Scharff, R. L. 2012. Economic burden from health losses due to foodborne illness in the United States. *Journal of Food Protection* 75(1):123–30.

Smulders, F. J. M., Norrung, B. and Budka, H. 2013. *Foodborne Viruses and Prions and Their Significance for Public Health (Food Safety Assurance and Veterinary Public Health).* Wageningen Academic Publishers, Wageningen.

Sobel, J. 2005. Botulism. *Clinical Infectious Diseases* 41(8):1167–73.

Sokheng, V. 2014. More troops injured in Mali. http://www.phnompenhpost.com/national/more-troops-injured-mali.

Soon, J. M., Singh, H. and Baines, R. 2011. Foodborne disease in Malaysia: A review. *Food Control* 22(6):824.

Sridhar, S., Lau, S. K. and Woo, P. C. 2015. Hepatitis E: A disease of reemerging importance. *Journal of the Formosan Medical Association* 114(8):681–90.

Stephan, R., Schumacher, S. and Zychowska, M. A. 2003. The VIT® technology for rapid detection of *Listeria monocytogenes* and other *Listeria* spp. *International Journal of Food Microbiology* 89(2–3):287–90.

Susser, M. 1998. Does risk factor epidemiology put epidemiology at risk? Peering into the future. *Journal of Epidemiology and Community Health* 52(10):608–11.

Sutherland, J. C. and Jones, T. H. 1960. Gastric mucormycosis: Report of case in a Swazi. *South African Medical Journal = Suid-Afrikaanse Tydskrif vir Geneeskunde* 34:161.

Switt, A. I., Soyer, Y., Warnick, L. D. and Wiedmann, M. 2009. Emergence, distribution, and molecular and phenotypic characteristics of *Salmonella enterica* serotype 4,5,12:i:-. *Foodborne Pathogen and Disease* 6(4):407–15.

Taege, A. 2010. *Foodborne disease.* Cleveland Clinic, Center for Continuing Education, 121.

Talebreza, A., Memariani, M., Memariani, H., Shirazi, M. H., Shamsabad, P. E. and Bakhtiari, M. 2015. Prevalence and antibiotic susceptibility of *Shigella* species isolated from pediatric patients in Tehran. *Archives of Pediatric Infectious Diseases* 4(1):32–9.

Tauxe, R. V. 2015. *Treatment and prevention of Yersinia enterocolitica and Yersinia pseudotuberculosis Infection.* UpToDate. http://www.uptodate.com/contents/treatment-and-prevention-of-yersinia-enterocolitica-and-yersinia-pseudotuberculosis-infection.

Thomas, M. K., Murray, R., Flockhart, L. et al. 2013. Estimates of the burden of foodborne illness in Canada for 30 specified pathogens and unspecified agents, circa 2006. *Foodborne Pathogens and Disease* 7(10):639–40.

Times of Israel 2016. *Most Israeli poultry infected with bacteria tied to food poisoning.* http://www.timesofisrael.com/most-israeli-poultry-infected-with-bacteria-tied-to-food-poisoning.

Tomsikova, A. 2002. Risk of fungal infection from foods, particularly in immunocompromised patients. *Epidemiology Microbial Immunology* 51(2):78–81.

Torgerson, P. R., Devleesschauwer, B., Praet, N., et al. 2015. World Health Organization estimates of the global and regional disease burden of 11 foodborne parasitic diseases, 2010: A data synthesis. *PLoS Medicine* 12(12):100–92.

Tsao, C. H., Chen, C. C., Tsai, S. J., et al. 2013. Seasonality, clinical types and prognostic factors of *Vibrio vulnificus* infection. *The Journal of Infection in Developing Countries* 7(7):533–40.

Valero, M. A., Perez-Crespo, I., Periago, M. V., Khoubbane, M. and Mas-Coma, S. 2009. Fluke egg characteristics for the diagnosis of human and animal fascioliasis by *Fasciola hepatica* and *F. gigantica. Acta Tropica* 111(2):150–9.

Vallabhaneni, S., Mody, R. K., Walker, T. and Chiller, T. 2015a. The global burden of fungal diseases. *Infectious Disease Clinics of North America* 30(1):1–11.

Vallabhaneni, S., Walker, T. A., Lockhart, S. R., et al. 2015b. Notes from the field: Fatal gastrointestinal mucormycosis in a premature infant associated with a contaminated dietary supplement-Connecticut, 2014. *MMWR. Morbidity and Mortality Weekly Report* 64(6):155–6.

Velusamy, V., Arshak, K., Korostynska, O., Oliwa, K. and Adley, C. 2010. An overview of foodborne pathogen detection: In the perspective of biosensors. *Biotechnology Advances* 28(2):232–54.

Vermorel-Faure, O., Lebeau, B., Mallaret, M. R., et al. 1993. Food-related fungal infection risk in agranulocytosis. Mycological control of 273 food items offered to patients hospitalized in sterile units. *La Presse Médicale* 22(4):157–60.

Vineis, P. 2007. Commentary: First steps in molecular epidemiology: Lower et al. 1979. *International Journal of Epidemiology* 36(1):20–6.

Wain, J., Hendriksen, R. S., Mikoleit, M. L., Keddy, K. H. and Ochiai, R. L. 2015. Typhoid fever. *Lancet* 385(9973):1136–45.

Weagant, S. D. 2008. A new chromogenic agar medium for detection of potentially virulent *Yersinia enterocolitica. Journal of Microbiology Methods* 72(2):185–90.

Wells, C. L. and Wilkins, T. D. 1996. *Clostridia: Sporeforming anaerobic bacilli. Medical microbiology.* University of Texas Medical Branch at Galveston, Galveston, TX.

WHO 2000. WHO surveillance programme for control of foodborne infections and intoxications in Europe 8th Report 1999-2000 Country Reports: Iceland. http://www.bfr.bund.de/internet/8threport/.

WHO 2005. *International travel and health, vaccine-preventable diseases and vaccines.* World Health Organization, Geneva, Switzerland. http://www.who.int/ith/ITH _chapter_6.pdf.

WHO 2006. *Five keys to safer food manual.* World Health Organization, Geneva, 5–20.

WHO 2006a. *Preventive chemotherapy in human helminthiasis: coordinated use of anthelminthic drugs in control interventions: A manual for health professionals and programme managers.* World Health Organization, Geneva.

WHO 2006b. *Brucellosis in humans and animals.* World Health Organization, Geneva.

WHO 2008a. *Foodborne disease outbreaks: Guidelines for investigation and control.* World Health Organization, Geneva.

WHO 2008b. Fact sheet on fascioliasis. In: *Action against worms, World Health Organization,* Geneva (December 2007). Newsletter 1–8.

WHO 2013. *Salmonella (Non-Typhoidal).* World Health Organization, Geneva. http:// www.who.int/mediacentre/factsheets/fs139/en.

WHO 2015. *Estimates of the global burden of foodborne diseases, 2015.* World Health Organization, Geneva. http://www.who.int/foodsafety/areas_work/foodborne-dise ases/infographics_combined_en.pdf?ua=1.

WHO 2015. *World health statistics.* World Health Organization, Geneva, 13–120.

WHO 2016. *Taeniasis/cysticercosis.* World Health Organization, Geneva. http://www. who.int/mediacentre/factsheets/fs376/en/.

Williams, M. S., Golden, N. J., Ebel, E. D., Crarey, E. T. and Tate, H. P. 2015. Temporal patterns of *Campylobacter* contamination on chicken and their relationship to campylobacteriosis cases in the United States. *International Journal of Food Microbiology* 208:114–21.

Winters, D. R. H. 2012. Not sick yet: Food-safety-impact litigation and barriers to justifiability. *Brooklyn Law Review* 77(3):914–27.

Wiwanitkit, V. 2018. Important emerging and reemerging tropical foodborne diseases. In: Holban, A. M. and Grumezescu, A. M. (eds.), *Foodborne diseases.* Academic Press, London, 33–55.

World Health Organization 2016. *WHO estimates of the global burden of foodborne diseases.* 2015. http://apps.who.int/iris/bitstream/10665/199350/1/9789241565165_ eng.pdf.

World Health Organization Report 2016. *World health statistics.* World Health Organization, Geneva, 52–110.

Xiao, L., Morgan, U. M., Fayer, R., Thompson, R. C. and Lal, A. A. 2000. *Cryptosporidium* systematics and implications for public health. *Parasitology Today* 16(7):287–92.

Yugo, D. M. and Meng, X. J. 2013. Hepatitis E virus: Foodborne, waterborne and zoonotic transmission. *International Journal of Environmental Research and Public Health* 10(10):4507–33.

Zarain, P. L., Lopéz-Téllez, G., del Carmen Rocha-Gracia, R., et al. 2012. Nested-PCR as a tool for the detection and differentiation of gram-positive and gram-negative bacteria in patients with sepsis-septic shock. *African Journal of Microbiology Research* 6:4601–7.

Zhang, H. N., Hou, P. B., Chen, Y. Z., et al. 2016. Prevalence of foodborne pathogens in cooked meat and seafood from 2010 to 2013 in Shandong Province, China. *Iranian Journal of Public Health* 45(12):1577–85.

Zollner-Schwetz, I. and Krause, R. 2015. Therapy of acute gastroenteritis: Role of antibiotics. *Clinical Microbial Infection* 21(8):744–9.

CHAPTER 4

Nucleic Acid-Based Molecular Techniques in Microbial Food Safety

Benssan K. Varghese, V.J. Rejish Kumar,
and Radhakrishnan Preetha

CONTENTS

4.1 INTRODUCTION

Fermented foods with high nutritional quality are produced by using different types of microorganisms. Various beneficial microbes are involved in producing cheese, wine, milk products, and many alcoholic beverages. The useful microbes include different strains of *Lactobacillus, Bifidobacterium, Bacillus, Enterococcus, Acetobacter, Aspergillus*, and Yeast (Mohania et al. 2008). However, some microbes can also cause spoilage of foods and cause rancidity, color change, and foul smell formation. Foodborne pathogens can enter into the system due to unhygienic practices, which should be prevented to ensure food safety (Mohania et al. 2008).

Different analytical techniques are available to detect the microbial activity in food, including biochemical, molecular, and immunological techniques (Ndoye et al. 2011). Spoilage microbes can enter into the system at any stage of production. It can be through stocks such as raw meat, fish, and vegetables. Improper storage and handling are other important factors that cause contamination of food products. Other than microbes, physical and chemical contaminants also cause a severe threat to food safety. These contaminants can enter the food during processing, harvesting, post-harvest treatment, and storage. Molecular techniques are used for the identification of the pathogens that cause contamination. These techniques will help to ensure the quality, safety, and traceability of food products. The application of these techniques provides a better idea about microbial ecology, which is useful for assessing and improving the quality of food components (Lauri and Mariani 2009). It also helps to select a good quality starter culture for different food products. Molecular techniques help scientists to identify and address the risk present in food items (Rizo et al. 2018). Moreover, it is a rapid and sensitive method for microbial detection and identification, and it also controls biogenic amines that cause unpleasant flavor and toxicity (Heerthana and Preetha 2019).

Molecular techniques are fundamental in identifying the microorganisms that are difficult to culture in vitro, for example, the foodborne viral pathogen norovirus. The molecular detection method is the only reliable diagnosis. Molecular techniques are used to detect the microorganisms classified as culture-dependent and culture-independent (Sheikha and Hu 2018). Culture dependent techniques include arbitrarily primed-polymerase chain reaction (AP-PCR), repetitive extragenic palindromic-polymerase chain reaction (Rep-PCR), ribotyping, amplified fragment length polymorphism (AFLP), amplified ribosomal DNA restriction analysis (ARDRA), random amplified polymorphic DNA (RAPD), restriction fragment length polymorphism (RFLP), and pulsed-field gel electrophoresis (PFGE), multilocus sequence typing (MLST) (Sheikha and Hu 2018).

Culture-independent techniques include automated ribosomal intergenic spacer analysis (ARISA), temperature gradient gel electrophoresis (TGGE), single-stranded conformation polymorphism-polymerase chain reaction (SSCP-PCR), length heterogeneity -polymerase chain reaction (LH-PCR), pyrosequencing, fluorescent in situ hybridization (FISH), quantitative polymerase chain reaction (qPCR), flow cytometry (FC), multiplex PCR, real-time PCR, microarray, denaturing gradient gel electrophoresis (DGGE), and terminal restriction fragment length polymorphism (T-RFLP) (Sheikha and Hu 2018). Classification of different molecular techniques are shown in Figure 4.1. The culture-independent molecular technique is not dependent on the physiological state of microorganisms, therefore viable but non-culturable, and stressed and injured microorganisms can also be detected. Modern molecular methods need less time to complete the analysis compared to conventional culture-based methods. Molecular techniques can also be applicable when typical strains or strains from a discriminative medium are absent

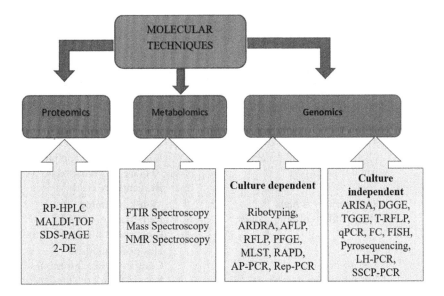

FIGURE 4.1 Different molecular techniques used in food safety analysis.

or underperforming (Fricker et al. 2008). The DNA-based detection method is highly specific and will effectively detect or differentiate microorganisms. Nucleic acid-based molecular techniques can identify specified serotypes, genotypes, or pathotypes by identifying virulence genes or other marker genes (Ceuppens et al. 2014). The main drawback of the molecular method is identifying bare DNA and RNA from dead cells of microorganisms can cause low clarity results, mainly in food safety and quality control (Ceuppens et al. 2014). Recently, different types of omic-based analyses are in use in food quality analysis. It includes genomics, transcriptomics, proteomics, metabolomics, and multi-omics such as metagenomics, metatranscriptomics, and metaproteomics.

Metagenomics is a nucleic acid-based technique that provides the picture of the microbial community in a sample, including the non-culturable organism. At the same time, multi-omics help to identify the different kinds of pathogens present in the food sample and also help to determine the mode of action of a pathogen and its by-products. However, modern molecular techniques are costly compared to traditional molecular methods due to the requirements of adequate laboratory infrastructure, expensive reagents, and specialized equipment (Sheikha and Hu 2018). The advantages and disadvantages of molecular techniques are given in Table 4.1. This chapter discusses nucleic acid-based techniques such as PCR, multiplex PCR, Real-time PCR, RAPD, RFLP, ARDRA, T-RFLP, FISH, LH-PCR, SSCP, DGGE, TGGE, ARISA, Microarrays, DNA-based biosensors, and omic technologies for assuring microbial food safety.

4.2 NUCLEIC ACID-BASED TECHNIQUES IN FOODBORNE PATHOGEN IDENTIFICATION

4.2.1 Polymerase Chain Reaction (PCR)

PCR is a widely used method for detection of genetically modified organisms, authentication of food products, and identification of foodborne pathogens. PCR produces amplified

TABLE 4.1   Advantages and Disadvantages of Molecular Techniques

Method	Advantages	Disadvantages
ARDRA	Rapid Has high reproducibility Easy to perform Medium cost of setup	Has less discriminating power
RFLP	Has good repeatability The detection limit is low	High active to single mutation and false negative
MLST	The differentiating capacity of closely similar genotypes	Requires special equipment Requires expertise to perform Reagents are costly
PFGE	Repeatability is high Higher differentiating power	Time-consuming Requires more labor It depends on a specific strain
RAPD	Rapid Easy to perform Much less time-consuming	Has poor stability Consists of less information Requires a lot of standardization
ARISA	Less expensive High detection capacity	It may underestimate sample richness
TGGE/TGGE	Rapid Low cost Multiple analysis of a sample at the same time Shows the precise representation of the evolution of microbial population	Same migration behavior for heterologous sequence
T-RFLP	High sensitivity Good throughput Helps to determine the appropriate difference in genotypes	Qualitative Needs a clone library for identification Equipment is costly The sequence needs to be known for selection of the enzyme
qPCR	Efficient and robust Highly sensitive Stable DNA biomarkers	It is an expensive method Requires expertise to perform
FCM	Experiment on the viability of microorganisms Quantitative and qualitative	It is labor-intensive Very expensive
FISH	Each microbial cell can be easily visualized, identified, enumerated, and localized	It requires a probe design It is a labor-intensive Challenging to understand the simultaneously massive quantity of various targets
SSCP	Doesn't need gradient gel Automated sequencer performs it	It is reliant on the accessibility of a dependable database

targeted DNA for various analyses. PCR includes isolation of DNA, amplification of target genes, and observation of the amplified product. It relies on thermally stable DNA polymerases, and it needs specifically designed DNA primers of the target region of the DNA. The PCR reaction can repeatedly cycle via a sequence of different temperatures for denaturation, annealing, and extension. PCR is considered the primary method to make

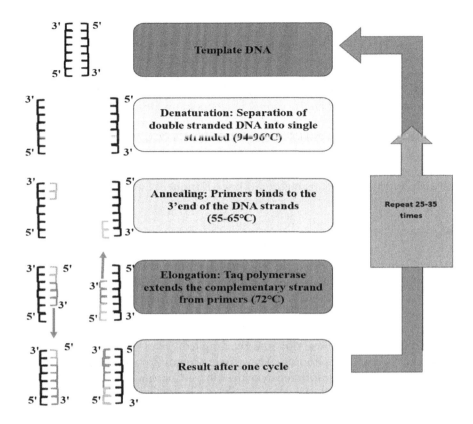

FIGURE 4.2   PCR flow chart.

amplicons of focused parts of DNA fragments. The working principle of PCR in the form of a flow chart is given in Figure 4.2. Amplified PCR products are observed using agarose gel electrophoresis. After observation, the products are confirmed by sequencing. PCR is a powerful and essential tool used in various food safety applications, including the rapid detection of adulterants present in the food products. PCR is also applied for detecting GMOs and in the authentication of food products. Another significant area is the detection of foodborne pathogens (Boldura and Popescu 2016).

4.2.2 Real Time PCR (RT-PCR)

RT-PCR is a modern form of standard PCR. It is a quantitative PCR (qPCR) that gives a picture of the quantity of the targets in the sample. qPCR is generally used to estimate the number of nucleic acid outputs at the time of PCR, and it will measure cDNA and DNA. It also determines the genes that are present inside the various samples (Lobert et al. 2010). This method is rapid and allows lesser chance for contamination. qPCR is highly specific, easy to handle, and easy to perform (Mackay et al. 2002; Maibach and Altwegg 2003). This assay analyzes the amplicons throughout the reaction by quantifying the fluorescent signal formed by the target-specific DNA probes (TaqMan) labeled with fluorescent reporters or by imprecise fluorescent dyes (SYBR green) can present with double-stranded DNA. The quantity of PCR amplicons corresponds to the strength of the fluorescence (Omiccioli et al. 2009; Zhao et al. 2014).

This method detects *Campylobacter* and *Salmonella* in the poultry industry. *Salmonella enterica* in liquid egg, peanut butter, and artificially contaminated chicken can be detected using qPCR (Chen et al. 2010). It can also be used for the detection of *Listeria monocytogenes*, *Salmonella* spp., and *Escherichia coli* O157 in contaminated chicken, pork, beef, and turkey (Suo et al. 2010); RT-PCR is also used for identification of *Salmonella* spp. and *Listeria monocytogenes* in unnaturally and biologically contaminated vegetables, dairy products, meat, fruits, fish, eggs, omelet, chocolate bars, and a wide range of other cooked dishes (Ruiz-Rueda et al. 2011). It is also used to detect *Staphylococcus aureus*, *Salmonella*, and *Shigella* in raw pork (Ma et al. 2014). *Clostridium tyrobutyricum* spores in artificially contaminated raw milk and heat-treated milk can also be detected by this method (Lopez-Enriquez et al. 2007).

4.2.3 Multiplex PCR

This technique helps in the detection of multiple pathogens using a single PCR. The principle of mPCR is the same as the conventional PCR. But the difference is in the use of multiple specific primers in mPCR to detect multiple pathogens. Still, in the conventional PCR we only use a single specific primer. The design of the primers is the key for mPCR. All the primers should have the same annealing temperature to perform properly within a particular reaction and the amplicon sizes should be distinct enough to distinguish in gel electrophoresis. The primers' quantity and composition are also essential to overcome primer–primer hybridization and primer dimer formation (Zhao et al. 2014). Factors such as buffer concentration, the concentration of reagents, DNA template quantity, and processing temperature are also essential. This method's main advantage is the lower cost, short time duration, and fewer reagent requirements. One of the significant drawbacks of this method is that it cannot detect amplified fragments that have the same length, and it shows less clarity in lower quantities of the product (McPherson and Moller 2000).

Earlier, this method was used for the detection of limited pathogens. But now, with the advanced primer designing tools, mPCR can detect more than five organisms at a time (Chen et al. 2012). This method is generally used for the recognition of microorganisms like *Staphylococcus aureus, Salmonella enteritidis, Shigella flexneri, Escherichia coli, Listeria* spp., and *Listeria monocytogenes*. Multiplex PCR is also used for the detection of *Salmonella* spp., and *Salmonella Enteritidis* in contaminated cheese, pork sausages, and chicken (Silva et al. 2011), *Listeria monocytogenes, Escherichia coli* O157:H7, *Staphylococcus aureus*, *Salmonella* spp., and *Yersinia enterocolitica* in infected pork (Guan et al. 2013).

4.2.4 Random Amplified Polymorphic DNA (RAPD)

RAPD DNA is a method which uses the basic principles of PCR techniques for the analysis and identification of genetic variation. The reaction includes utilizing arbitrary primers in a PCR that leads to the amplification of random fragments of DNA. This technique was independently developed by two different laboratories and known as Arbitrary primed PCR (AP-PCR) and RAPD (Welsh and McClelland 1990; Williams et al. 1990). Here, the reaction identifies the sequence polymorphism of nucleotides in PCR using only a single primer sequence of random nucleotides. Then, the nucleotide's unique species (primer) is attached to two different DNA sites on opposite strands. There is a production of discrete DNA through thermocycling amplification if the primers attach within an amplifiable distance. The individuals show polymorphism that results from the differences of sequence in one or both of the primer binding sites resulting in a specific RAPD band. These polymorphisms are taken as genetic markers for the identification of a species.

In RAPD, prior understanding of the DNA sequence of the required gene is not necessary, as arbitrary primers want to attach anywhere in the strand sequence (Anolles et al. 1991). By this, the RAPD method is well known for differentiating the DNA of an organism whose genome composition is unknown. RAPD markers failed to distinguish whether the DNA amplified segment was from a locus of two copies (homozygous) or one copy (heterozygous). The dissimilarity between the template and primer can lead to a decrease in PCR product or, sometimes, a total loss of the product, making the RAPD results difficult to interpret. RAPD is used for differentiating spoilage microorganisms, specifically yeasts. *Monascus* strains are a type of fungus in fermented food products that is identified by using this technique. This analysis method is useful for distinguishing *cremoris*, and it helps in the detection of genetic variations within the species.

4.2.5 Polymerase Chain Reaction-Restriction Fragment Length Polymorphism (PCR-RFLP)

PCR-RFLP or cleaved amplified polymorphic sequence (CAPS) is a technique used to detect intraspecies variations and interspecies organisms. It is a fact that multi nucleotide polymorphisms (MNPs), single nucleotide polymorphisms (SNPs), and microindels can be a recognition site for restriction enzymes (Narayanan 1991). In RFLP, the primary step is the amplification of samples containing variations. The amplified fragment is then treated with suitable restriction enzymes to identify the restriction fragments of various sizes by the presence or absence of the recognition site. After restriction digestion, the samples can be observed using gel electrophoresis. The advantages of RFLP include lower cost and limited requirement of modern advanced instruments. However, it requires specific endonucleases, and it can be difficult to find suitable enzymes for differentiating the various SNPs. The RFLP method is time-consuming due to the inclusion of steps like gel electrophoresis separation. This technique is not convenient for the multiple investigations of vast SNPs due to the need for a specified pair of primers and restriction enzymes for each SNP (Rasmussen 2012).

PCR-RFLP includes several steps, such as primers' design, selection of suitable restriction enzymes, amplification, restriction digestion, and gel electrophoresis for the differentiation of restriction fragments. Now there are many programs that are available for the design of primers, selection of suitable restriction enzymes, and for analyzing the restriction patterns. The PCR-RFLP is a quick, reliable tool for identifying the meat of various animals (Kesmen et al. 2013). This method helps in the differentiation of the lineage groups of *L. monocytogenes* (Rip and Gouws 2020). PCR-RFLP is also used for the identification of sheep, beef, and chicken in heterogenous meat mixtures. It is a rapid and simple tool for the identification and profiling of microbial populations in meat products (Martya et al. 2012). It is used to detect SNP and evaluate the protein-coding genes of *Lactobacillus delbrueckii* in milk products (Giraffa et al. 2003; Ota et al. 2009).

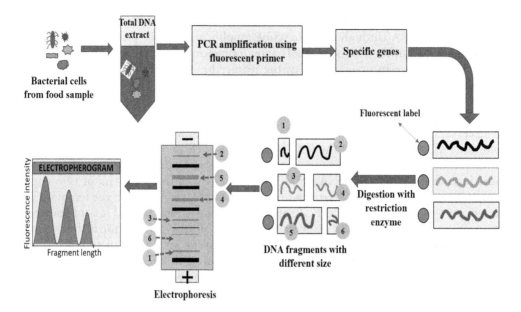

FIGURE 4.3   Mechanism of T-RFLP.

4.2.6 Terminal-RFLP (T-RFLP)

T-RFLP is an independent culture technique for microbial population assessment in which only terminal fragments of an amplified gene are detected and identified. This method is used to analyze highly complex populations of microorganisms by their variations in 16S rRNA (ribosomal RNA). It helps to explain the complex communities in the absence of any genomic sequence information. This method is swift and highly sensitive (Moesender et al. 1999). The reaction allows us to understand the microbial populations in diverse environments and to understand the organism's behavior toward the changes in physiological or environmental parameters. Fundamental T-RFLP analysis mainly consists of five steps outlined in (Figure 4.3) and listed here:

- Isolation and purification of the DNA
- Amplification by PCR and digestion using restriction enzyme
- Detection of the digested products and separation using electrophoresis
- Analysis of the results to produce the fragment units for different samples
- Congregate analysis based on the profile of the sample

The use of multiplex T-RFLP has been reported for the detection of *Listeria* and *Salmonella* spp. in milk. T-RFLP is also used for the identification of lactic acid bacteria (LAB) in fermented milk products (Yu et al. 2009) and to study the *S. aureus* coagulase gene diversity in various food products.

4.2.7 Single Strand Conformation Polymorphism (SSCP)

Single-stranded conformation polymorphism (SSCP) is a commonly employed technique that describes the genomic variants in many samples and a range of different organisms.

This method identifies the difference, such as mutations at a single point and other small-scale changes, detecting through electrophoresis. If a DNA has a sequence of mutations (we can also consider even a single change in base pair), it can be measured by differences in mobility related to wild type DNA when observed under partially denaturing or non-denaturing conditions (Tahira et al. 2009).

In SSCP, the first step is the isolation of DNA. The total quality, accuracy, and DNA sequence length can be affected by the sample that we take and the technique chosen for the extraction of nucleic acid (Masato et al. 1989). The DNA extraction methods will differ according to tissue type or source type, how it was collected from the source, and how we handled or sorted the sample before the extraction. The second step of SSCP is the design of the primer and performing the PCR reaction. Thereafter, amplify the region of interest by PCR reaction using fluorescently labeled primers. After completing the PCR, the PCR products are denatured and processed for capillary electrophoresis. Then, the fluorescent-labeled DNA fragments are separated according to size and discovered by using a laser or cameras. Then the chart of the SSCP profile is analyzed (Tahira et al. 2009). The steps involved in SSCP are shown in Figure 4.4.

SSCP is used for the investigation of microorganisms in many food industries. For example, in a cheese manufacturing company, SSCP monitors *Staphylococcus* spp. (Delbes and Montel 2005); SSCP is used to confirm the authenticity of canned tuna (Rehbein et al. 1999); and in fresh milk, it differentiates and classifies the bacterial variety (Verdier-Metz et al. 2009). *Aureobasidium pullulans*, *Alternaria alternata*, *Candida rugosa*, *Eurotium amstelodami*, *Pichia membranifaciens*, and C. *tropicalis*, and species such as *Itersonilia perplexans*, *Botrytis aclada*, *Neofusicoccum parvum*, *Pleospora herbarum*, and *Lasiodiplodia theobromae* are the different types of fungal DNA identified by SSCP in contaminated meat products and heat-processed meat products (Dorn et al. 2013).

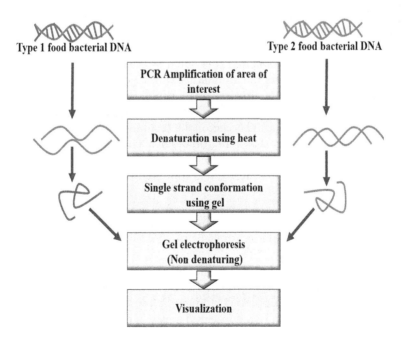

FIGURE 4.4   Mechanism of SSCP.

4.2.8 Amplified Ribosomal DNA Restriction Analysis (ARDRA)

This method is an extended version of restriction fragment length polymorphism (RFLP) of the gene encoding the 16S ribosomal subunit of bacteria. This method comprises the amplification of the preserved regions of the 16S genes supported by restriction digestion (Sklarz et al. 2009). The digested products are analyzed for the presence or absence of bands, and these data are used to create a phylogram or cladogram (Vaneehoutte et al. 1993). A comparison of the microbial communities present in the food products can be analyzed using ARDRA. *Bifidobacteria* are the common bacteria that are analyzed by ARDRA in dairy products such as yogurt, ice cream, sour cream, and cheese desserts (Mazo et al. 2010).

ARDRA is a species-level identification tool for various foodborne organisms and is used to identify *Lactobacillus* (Roy et al. 2001) and *Weissella* (Jang et al. 2002) species in multiple food items. Molecular differentiation of *Bifidobacterium* spp. can also be studied using ARDRA (Roy and Sirois 2000). *Bifidobacterium* spp. is reported to have various health benefits for humans that are abundant in fermented products; ARDRA helps in the efficient detection of *Bifidobacterium* spp. in those products (Krizovà et al. 2006; Youn et al. 2008).

4.2.9 Length Heterogeneity-PCR (LH-PCR)

Using this method, a primer that is fluorescently labeled examines the relative quantity of sequences in an amplified form that originates from various microorganisms. These fluorescently labeled strains are then isolated with the help of gel electrophoresis and examined using a combination of automated gene sequencer and a laser-induced fluorescence. The extracted DNA from various samples are magnified by PCR using an unlabeled reverse primer and fluorescently labeled forward primer. After amplification, the products are denatured chemically using formamide. The amplicons are separated by polyacrylamide gel electrophoresis (Nancy et al. 2000). The data are formed in the form of fluorescence data into electropherograms. Each peak denotes distinct genotypes, and the height of the peak represents the abundance of the genotype. LH-PCR methods are highly reproducible and efficient. It is an easy, rapid, and suitable method to analyze microbial population formation in different fields such as food safety, soil analysis, and pollution control. The main disadvantage of the technique is the limited availability of the LH database. LH-PCR is used as a microbiological characterization tool for studying the effects of soybean meal (SBM) (Heikkinen et al. 2006). This methodology is also useful for multiple species detection in fermented food items (Sardaro et al. 2018).

4.2.10 Automated Ribosomal Intergenic Spacer Analysis/
Ribosomal Intergenic Spacer Analysis (ARISA/RISA)

This low-cost technique helps to compare microbial communities. The prokaryotic DNA encodes for 16S rRNA and highly conserved 23S rRNA genes. These encode small and large subunit genes in the rRNA operon (Madigan 2009). There is an internal transcribed spacer (ITS) present in between two conserved genes, which is a noncoding region, and it displays a considerable difference in nucleotide sequences and length between species. In ARISA/RISA, after DNA isolation, the spacer region is amplified using PCR

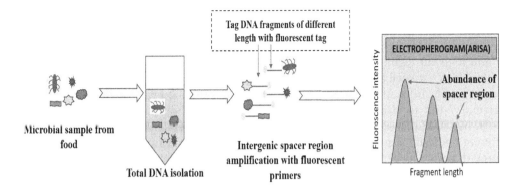

FIGURE 4.5 Mechanism of ARISA.

(Marsh 2005). The formed fragments detected using gel electrophoresis (RISA), or using primers with fluorescent labels, can convert various fragment lengths into peaks on an electropherogram (ARISA). The working mechanism of ARISA is shown in Figure 4.5. It is a sensitive and rapid fingerprinting method (Osborn et al. 2000). One of the main disadvantages of ARISA is confusion caused by the presence of more than one peak in the community profile obtained from a single organism; unrelated microorganisms have similar spacer lengths that can result in the wrong interpretation of community diversity. ARISA assesses microbial diversity in cheese (Porcellato et al. 2014); examines the effect of salt content and the cation category on biological pathogens in cheddar cheese (Porcellato et al. 2014); and detects the presence of fungus during wine production in countries such as South Africa and Austria.

4.2.11 Multilocus Sequence Typing (MLST)

MLST refers to a technique used to characterize the bacterial species isolates by using internal fragments (450–500 bp) of multiple housekeeping genes of both strands by an automated DNA sequence. The various sequences that appear inside a bacterial species marked as different alleles. For every isolate, the alleles present in each locus denote the sequence type (ST) or allelic profiles (Urwin and Maiden 2003). In multilocus sequence typing, nucleotide quantity variation among the alleles should be avoided. The sequences exist in various allele numbers when they differ at a site of nucleotide or different locations. MLST analyzes the DNA sequence differences in a group of housekeeping genes and distinguishes the strains by using their allelic profiles. In MLST various sequences will be selected for every housekeeping gene as alleles, and these alleles at the loci give an allelic profile (Maiden et al. 1998). The chain of profiles then becomes the identification marker for strain typing. Sequences that are dissimilar at even a single nucleotide had arranged as different alleles.

The workflow of MLST includes collection of data, data analysis, and analysis of multilocus sequence. In the first step, the variations are identified by determining the nucleotide sequence of gene fragments. In the second step, allele numbers are assigned to all characteristic sequences and create an allelic profile and a sequence type (ST). When new alleles and sequence types are obtained, they are collected for verification and kept in the database. Finally, the similarities of the isolates are analyzed by comparing the allelic profile. Bioinformatics methods are used to analyze, arrange, and combine the enormous amount of data created during the sequencing and identification procedures.

The entire MLST process is automated and merges the advances in bioinformatics and high throughput sequencing with genetics methods. This method can analyze relationships between bacteria, and provide high separating power for different isolates (Urwin and Maiden 2003). MLST has application in public health, scientific research, veterinary, and food safety. Although it is suitable for population studies, it is very costly. By the sequence preservation of genes, MLST occasionally shows low clarity for different bacterial strains. MLST is useful for ensuring and improving food safety, as it provides more in-depth phylogenetic information about particular species. This technique is used to detect *Campylobacter* in undercooked poultry and unpasteurized milk (Jolley et al. 2018).

4.3 GEL ELECTROPHORESIS

4.3.1 Pulsed-Field Gel Electrophoresis (PFGE)

PFGE is applied to identify big DNA molecules on the gel using an electric field that periodically changes its direction. PFGE helps to resolve DNA of more than 15kb size. The working principle of PFGE is shown in Figure 4.6. Many instruments utilize the pulsed electrophoresis effect to improve clarity in small and large molecules of DNA (Basim and Basim 2001), including field-inversion gel electrophoresis (FIGE), orthogonal-field alternation gel electrophoresis (OFAGE), rotating gel electrophoresis (RGE), contour-clamped homogeneous electric fields (CHEF), and transverse-alternating field gel electrophoresis (TAFE) (Nsofor 2016).

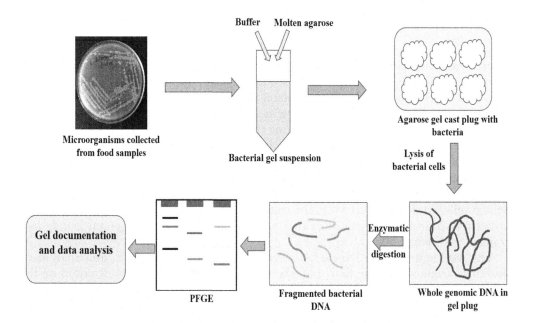

FIGURE 4.6 Mechanism of PFGE.

PFGE has a gel box and a provision for regulating the temperature, a cooler, switching unit, and a power supply (Basim et al. 1999). The gel box consists of an immobilized gel present inside the electrode arrangement and an electrophoresis buffer. The heat exchange mechanism used to control the buffer temperature and the inlet and outlet ports in the gel box help recirculate it throughout the box (Lai et al. 1989). The power supply output should meet the current and voltage requirements for the gel box. In a typical PFGE, the gap between electrodes is 25–50 cm. PFGE separation depends on different molecular factors such as the configuration of DNA, electric field strength, sample concentration, temperature, pulse time, electrical field shape, gel composition, reorientation angle, switch interval, agarose concentration, and restriction enzymes. PFGE is useful for examining biological pathogens in food products (Pleitner et al. 2014) and for identifying and analyzing a variety of foodborne pathogens, for example, *Listeria monocytogenes* (Fugett et al. 2007). PFGE can conduct further outbreak analysis of foodborne diseases caused by different microorganisms such as *Campylobacter, E. coli, Listeria, Salmonella,* and *Shigella.*

4.3.2 Temperature Gradient Gel Electrophoresis and Denaturing Gradient Gel Electrophoresis (TGGE and DGGE)

TGGE and DGGE are electrophoresis methods used to heat chemical material for denaturing the sample as it travels through the gel (acrylamide gel). TGGE and DGGE are used for the analysis of RNA and DNA and some proteins. TGGE depends on the temperature gradient alterations in the composition that separates the nucleic acids. DGGE will discriminate between different genes of a similar size according to their different denaturing capacities, which in turn depends on their base-pair sequence (Fischer and Lerman 1979).

When an electric field is applied, the negatively charged DNA travel toward the positive electrode. The gel becomes a web with pores of similar size to that of the DNA strand's diameter. In TGGE, a temperature gradient allows the spread of DNA across the gel. At room temperature, the DNA strand remains in a stable double-stranded form (Viglasky 2013). When the temperature rises, the strands start to melt, and their rate of movement through the gel will decrease. The temperature at which separation of strands occur depends on the sequence. Therefore, the TGGE is a size-dependent and sequence-dependent method for the separation of DNA molecules. It provides information regarding the melting characteristics and heat stability that depends on the composition of the nucleotides. The specialty of DGGE is the high level of denaturing conditions that the DNA is subjected to, changing the (double-stranded fragment into single-stranded fragments (Tabatabaei et al. 2009). In denaturing gradient gel electrophoresis (DGGE), the gel comprises a denaturing agent. The denaturing agents (formamide/urea) cause the DNA to melt at different stages. Thereby the DNA will spread through the gel, allowing for measurement of single components.

The denaturation process in the gel is very sharp and specific (Bordy and Kern 2004). Most of the fragments separate as a step-wise process, and domains of the fragments rapidly change to single-stranded inside a small field of the denaturing condition (Barasinski and Garnweitner 2020). DGGE help us to understand the variations in DNA sequence and mutations in genes. There are disadvantages to this technique. One is the low reproducibility of results obtained with the denaturing agents (chemical gradients) used in DGGE. This can be addressed in the TGGE method, where temperature is acting as the denaturing agent. DGGE and TGGE are useful for the analysis of variations in a microbial population (Bordy and Kern 2004). Also, the microbial origin and their geography can be evaluated by this method. DGGE is used for the investigation of microbial variations of *Lactobacillus* present in sausages and cheese. Besides prokaryotes, this method can also be used in the identification of airborne myxomycetes.

4.4 FLUORESCENCE IN SITU HYBRIDIZATION (FISH)

FISH is a cytogenic technique in which fluorescent probes will bind to the chromosome parts, which show a high degree of sequence complementarity. The method consists of binding fluorescence tagged target-specific probes of nucleic acid to their relative DNA or RNA pairs inside the cell in the tissue of interest (Gerami and Zembowicz 2011). FISH is a commonly applied microbial ecosystem for identifying, quantifying, and studying the spatial structure of the microbes using 16S rRNA or functional gene-based fluorescent probes. The selection of the FISH probe is crucial for an accurate result. The FISH processes include fixation of the tissue/cells, denaturation, hybridization, washing, and observation under a fluorescent microscope. Fluorescence labeling of the probes can be direct or indirect. The detection mechanism of FISH is shown in Figure 4.7.

FISH can detect pathogens in many processed food products. *Listeria* and *Salmonella* are the most detected species through this method. FISH can identify *Salmonella* spp. in carrots, lettuce, spinach, sweet corn roots, and tomato (Kljujev et al. 2018), *Yersinia* spp. in minced pork meat (Rohde et al. 2017), *Escherichia coli* O157 in beef (Almeida et al. 2013), *Enterobacteriaceae* and *Pseudomonas* spp. in milk (Yamaguchi et al. 2012), *Salmonella* spp. in powdered infant formula (AlmeIda et al. 2010), *Salmonella* spp. in

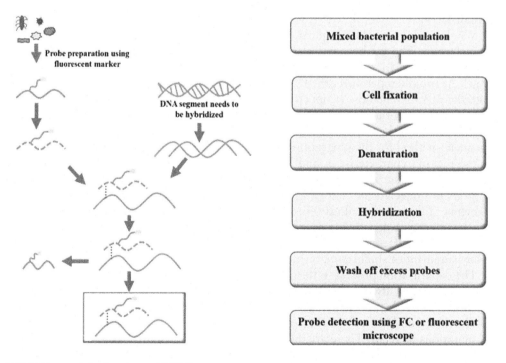

FIGURE 4.7   Mechanism of FISH.

tomato and pork samples (Bisha and Brehm-Stecher 2009; Rathnayaka 2011). *Listeria monocytogenes* in mozzarella cheese, julienne cabbage, and milk (Shimizu et al. 2009), *Escherichia coli* in ikura (seafood from Japan), and Chopped chicken meat (Ootsubo et al. 2003).

4.5 DNA MICROARRAY

DNA microarray is a fluorescent probe-based recent technique used in the food industry to characterize microbes. Here, small nucleic acid probes are used for identification. This method consists of multiple probes, such as cDNA or oligonucleotides, assigned to a hard frame, such as specially treated glass or silicon slide. Many modifications are introduced to this array method to perform further analysis. The hybridization of DNA fragments to probes that are fluorescently labeled are identified by advanced instrumentations and analyzed using modern software. Short oligonucleotides, long oligonucleotides, and PCR amplicons are the different kinds of probes used in microarrays. Microarray can identify food pathogens from samples in lab conditions (Borucki et al. 2005); it is also used for the molecular analysis of *Campylobacter* spp. and *E. coli* in food causing foodborne disease (Wu et al. 2003); microarray (or biochip) methods are suitable for the identification of GMOs in food. Microarray is used to examine the ultra-high temperature (UHT) treated skim milk for the analysis of the expression of gene profiles of *L. monocytogenes* strain F2365. The microarray was also applied to study probiotic microbes in food, such as *Lactobacillus acidophilus* in milk (Peril et al. 2009). *Lactobacillus sakei* genes produced during the fermentation of meat have also been analyzed using this method (Hufner et al. 2007).

4.6 NUCLEIC ACID-BASED BIOSENSOR

A typical biosensor has a bio receptor, a transducer, and a signal processor. Biosensors are categorized into nucleic acid biosensors, antibody biosensors, whole-cell biosensors, phage-based biosensor, and enzyme biosensors depending on the type of bio receptor. Nowadays, DNA biosensors are gaining widespread attention in foodborne pathogen detection (Wu et al. 2019). In the nucleic acid biosensor, a complementary strand of DNA is used as the bio receptor. Compared to other molecular techniques, DNA biosensors are a highly specific and sensitive method for foodborne pathogen detection. DNA biosensors are based on nucleic acid hybridization. DNA biosensors are mainly electrochemical or optical or piezoelectric. Electrochemical DNA biosensor is the best conventional method for foodborne pathogen identification due to its low detection limit and excellent reproducibility. In the case of the electrochemical biosensor, the single-stranded nucleic acid is used as a bio receptor. When a reaction occurs between the bio receptor and target, it produces or consumes ions or electrons, causing an alteration in potential or electric current or other electric properties of the solution. Thereby biological signals are converted into a detectable electrical signal. The commonly used transducers are gold electrodes, pencil graphite, carbon ionic liquid electrode, glassy carbon electrode, and screen-printed electrodes. The advantages of electrochemical DNA biosensors include its sensitivity, lifetime, low cost, and specificity. Electrochemical impedance spectroscopy is another technique applied to electrochemical DNA biosensors (Wu et al. 2019).

The different types of optical DNA biosensors are based on fiber optics, surface plasmon resonance (SPR), bio molecular interaction analysis, and Raman spectroscopy

(Junhui et al. 1997). The DNA biosensors which use fiber optic to transduce signal after target receptor interaction are called fiber optic DNA biosensors. Fiber optics is the device that transmits light from one end to another end through a sequence of internal reflection (Byoungho 2003). There are two types of fiber optic DNA biosensors–labeled and label-free. In the labeled fiber optic biosensor, the transducer converts the signal emitting from the fluorescent-labeled DNA probe after hybridization with the target DNA strand. In the case of a label-free fiber optic DNA biosensor, the DNA strand immobilized on its surface hybridizes with target DNA. Hence, it will alter the total internal reflection of an optical fiber. The change in the intensity of light can be recorded (Ashok et al. 2012). SPR biosensor is based on changes in the surface optical properties, which change the resonance angle due to changes in the refractive index. Piezoelectric DNA Biosensor operation is centered on a quartz crystal that oscillates with specific frequency at a given oscillating voltage. The oscillating frequency will change after hybridization, allowing for monitoring. Another piezoelectric DNA sensor was made by fixing DNA probes on a quartz crystal microbalance (QCM) (Kim and Choi 2014). The QCM is a delicate mass measuring system that reflects a change in mass after hybridization. The advantages of piezoelectric DNA biosensors are durability, low-cost mass production, and chemical inertness.

The usage of biosensors was reported in food industries for the identification of microorganisms such as *Listeria, Campylobacter, Escherichia coli, Salmonella,* and many more (Taylor et al. 2006); electrochemical DNA biosensor is used to analyze *Bacillus cereus* in infant formula and milk (Sheikhzadeh et al. 2016). DNA biosensors are used to determine *S. typhimurium* in apple juice. The DNA biosensor used for the detection of bacteria belongs to Enterobacteriaceae, including *Salmonella* in milk (Wu et al. 2019).

4.7 OMIC-BASED ANALYSIS

With the advent of advanced next-generation sequencing (NGS) techniques, it is now easy to analyze large numbers of sequences in a short period. It propelled the development of various omic technologies, including genomics, transcriptomics, proteomics, metabolomics, and multi-omics such as metagenomics, metatranscriptomics, and metaproteomics (Rizo et al. 2018). Multi-omics will help identify the different kinds of pathogenic groups present in food samples and determine the action and by-products produced by the organisms. Metagenomics gives a quick picture of the microbial community in a sample, including the non-culturable organism. This technique is utilized in the study of viruses and phages. Metatranscriptomics deals with gene expression-related studies of microorganisms present in nature. Metaproteomics helps us understand the metabolic and physiological state of microorganisms and the environmental effects on proteins and gene expressions (Horgan and Kenny 2011). It also helps to monitor post-translational modification and its role in metabolism. In metabolomics, the diversity of metabolites produced are analyzed for safety, nutrition, and sensory evaluation. Metabolomics is also used for the selection of biomarkers for fermentation.

DNA sequencing permits the confirmation of nucleotide sequences in a DNA molecule. In food safety analysis, various sequencing techniques are used for the identification, detection, and examination of foodborne microorganisms that cause serious illness, so that the disease can be controlled. DNA sequencing techniques categorize into three domains–advanced DNA sequencing, basic DNA sequencing, and NGS. Basic DNA sequencing is the earliest technique of DNA sequencing. It includes the Sanger method of DNA sequencing, which describes chain-termination, used to identify foodborne pathogens. Advanced DNA sequencing involves Shotgun sequencing. In food testing, Shotgun sequencing is employed for the analysis of food items. It provides precise validation of ingredients and detection of impurities. By this technique, the identification of the organism can be made even at the strain level. It can detect multiple strains of an organism (Haiminen et al. 2019). This technique describes the identification of food-related microorganisms in beef production (Yang et al. 2016). Next-generation sequencing is also called "high-throughput sequencing." It involves sequencing techniques such as Illumina sequencing and Pyrosequencing, used for the analysis of lactic acid bacteria during Kimchi fermentation (Jung et al. 2013). This method is useful in evaluating the fungal and bacterial diversity of yogurts (Xu et al. 2015). In another study, this technique was used for the characterization of microbes present in the rumen fluid of dairy cows (Thoetkiattikul et al. 2013).

4.8 CONCLUSION

Many safety issues have been reported in the food sector due to the intervention of unwanted microbes in processed foods. Molecular methods are a handy tool to assure the safety of fermented foods, dairy products, meat/poultry products, aquaculture products, probiotics, and single-cell proteins. This chapter provides a detailed description of the molecular techniques used in food safety. There are many methods for identifying and understanding the microorganism characteristics that can cause food spoilage or foodborne diseases. Moreover, molecular techniques are the most suitable and effective methods of understanding the organism behavior and mode of action toward a host. In short, better understanding of molecular techniques such as PCR, Multiplex PCR, Real-time PCR, RAPD, RFLP, ARDRA, T-RFLP, FISH, LH-PCR, SSCP, DGGE, TGGE, ARISA, omic technologies, Microarrays, and DNA-based biosensors described in this chapter will be of much help in ensuring the quality and safety of processed food.

REFERENCES

Almeida, C., Azevedo, N. F., Fernandes, R. M., Keevil, C. W. and Vieira, M. J. 2010. Fluorescence in situ hybridization method using a peptide nucleic acid probe for identification of *Salmonella* spp. in a broad spectrum of samples. *Appl Environ Microbiol* 76(13):4476–85.

Almeida, C., Sousa, J. M., Rocha, R., et al. 2013. Detection of *Escherichia coli* O157 by peptide nucleic acid fluorescence in situ hybridization (PNA-FISH) and comparison to a standard culture method. *Appl Environ Microbiol* 79(20):6293–300.

Anolles, C. G., Bassam, B. J. and Gresshoff, P. M. 1991. High resolution DNA amplification fingerprinting using very short arbitrary oligonucleotide primers. *Biotechnology N Y* 9(6):553–7.

Ashok, K., Sandip, K. D., Sharma, P. and Suman 2012. DNA based biosensors for detection of pathogens. In *Molecular approaches for plant fungal disease management*, 31–35. https://www.researchgate.net/publication/235997982.

Barasinski, M. and Garnweitner, G. 2020. Restricted and unrestricted migration mechanisms of silica nanoparticles in agarose gels and their utilization for the separation of binary mixtures. *J Phys Chem C* 124(9):5157–66.

Basim, E. and Basim, H. 2001. Pulsed-field gel electrophoresis (PFGE) technique and its use in molecular biology. *Turk J Biol* 25:405–18.

Basim, H., Stall, R. E., Minsavage, G. V. and Jones, J. B. 1999. Chromosomal gene transfer by conjugation in the plant pathogen *Xanthomonas axonopodis* pv. *vesicatoria. Am Phytopathol Soc* 89(11):1044–9.

Bisha, B. and Brehm-Stecher, B. F. 2009. Simple adhesive-tape-based sampling of tomato surfaces combined with rapid fluorescence in situ hybridization for *Salmonella* detection. *Appl Environ Microbiol* 75(5):1450–5.

Boldura, O. M. and Popescu, S . 2016. PCR: A powerful method in food safety field. *Polymerase Chain React Biomed Appl* 135–58. doi:10.5772/65738.

Borucki, M. K., Reynolds, J., Call, D. R., Ward, T. J., Page, B. and Kadushin, J. 2005. Suspension microarray with dendrimer signal amplification allows direct and high-throughput subtyping of *Listeria monocytogenes* from genomic DNA. *J Clin Microbiol* 43(7):3255–9.

Brody, J. R. and Kern, S. E. 2004. History and principles of conductive media for standard DNA electrophoresis. *Anal Biochem* 333(1):1–13.

Byoungho, L. 2003. Review of the present status of optical fiber sensors. *Opt Fiber Technol* 9(2):57–79.

Ceuppens, S., Hessel, C. T., Rodrigues, R. Q., Bartz, S., Tondo, E. C. and Uyttendaele, M. 2014. Microbiological quality and safety assessment of lettuce production in Brazil. *Int J Food Microbiol* 18194:67–76.

Chen, J., Tang, J., Liu, J., Cai, Z. and Bai, X. 2012. Development and evaluation of a multiplex PCR for simultaneous detection of five foodborne pathogens. *J Appl Microbiol* 112(4):823–30.

Chen, J., Zhang, L., Paoli, G. C., Shi, C., Tu, S. I. and Shi, X. 2010. A real-time PCR method for the detection of *Salmonella enterica* from food using a target sequence identified by comparative genomic analysis. *Int J Food Microbiol* 137(2–3):168–74.

Delbes, C. and, Montel, M. C. 2005. Design and application of a *Staphylococcus* specific single strand conformation polymorphism-PCR analysis to monitor *Staphylococcus* populations diversity and dynamics during production of raw milk cheese. *Lett Appl Microbiol* 41(2):169–74.

Dorn, S., Holzel, C. S., Janke, T., Schwaiger, K., Bauer, J. and Bauer, J. 2013. PCR-SSCP-based reconstruction of the original fungal flora of heat-processed meat products. *Int J Food Microbiol* 162(1):71–81.

Fischer, S. G. and Lerman, L. S. 1979. Length-independent separation of DNA restriction fragments in two-dimensional gel electrophoresis. *Cell* 16(1):191–200.

Fricker, C. R., DeSarno, M., Warden, P. S. and Eldred, B. J . 2008. False-negative β-D-glucuronidase reactions in membrane lactose glucuronide agar medium used for the simultaneous detection of coliforms and *Escherichia coli* from water. *Lett Appl Microbiol* 47(6):539–42.

Fugett, E. B., Schoonmaker-Bopp, D., Dumas, N. B., Corby, J. and Wiedmann, M. 2007. Pulsed-Field Gel electrophoresis (PFGE) analysis of temporally matched *Listeria monocytogenes* isolates from human clinical cases, foods, ruminant farms, and urban and natural environments reveals source-associated as well as widely distributed PFGE types. *J Clin Microbiol* 45(3):865–73.

Gerami, P. and Zembowicz, A. 2011. Update on fluorescence in situ hybridization in melanoma: State of the art. *Arch Pathol Lab Med* 135(7):830–37.

Giraffa, G., Lazzi, C., Gatti, M., Rossetti, L., Mora, D. and Neviani, E. 2003. Molecular typing of *Lactobacillus delbrueckii* of dairy origin by PCR-RFLP of protein-coding genes. *Int J Food Microbiol* 82(2):163–72.

Guan, Z. P., Jiang, Y., Gao, F., Zhang, L., Zhou, G. H. and Guan, Z. J. 2013. Rapid and simultaneous analysis of five foodborne pathogenic bacteria using multiplex PCR. *Eur Food Res Technol* 237(4):627–37.

Haiminen, N., Edlund, S., Chambliss, D., et al. 2019. Food authentication from shotgun sequencing reads with an application on high protein powders. *NPJ Sci Food* 3(24):1–11. https://doi.org/10.1038/s41538-019-0056-6.

Heerthana, V. R. and Preetha, R . 2019. Biosensors: A potential tool for quality assurance and food safety pertaining to biogenic amines/volatile amines formation in aquaculture systems/products. *Rev AQUA Cult* 11(1):220–33.

Heikkinen, J., Vielma, J., Kemiläinen, O. et al. 2006. Effects of soybean meal based diet on growth performance, gut histopathology and intestinal microbiota of juvenile rainbow trout (*Oncorhynchus mykiss*). *J Aquacult* 261(1):259–68.

Horgan, R. and Kenny, C. L. 2011. 'Omic' technologies: Genomics, transcriptomics, proteomics and metabolomics. *Obstet Gynaecol* 13(3):189–95.

Hufner, E., Markieton, T., Chaillou, S., Crutz-Le, C. A. M., Zagorec, M. and Hertel, C. 2007. Identification of *Lactobacillus sakei* genes induced during meat fermentation and their role in survival and growth. *Appl Environ Microbiol* 73(8):2522–31.

Jang, J., Kim, B., Lee, J., Kim, J., Jeong, G. and Han, H. 2002. Identification of *Weissella* spp. by the genus-specific amplified ribosomal DNA restriction analysis. *FEMS Microbiol Lett* 212(1):29–34.

Jolley, K. A., James, E., Bray., and Maiden, M. C. J. 2018. Open-access bacterial population genomics: BIGSdb software, the PubMLST.org website and their applications. *Wellcome Open Res* 3:124.

Jung, J. Y., Lee, S. H., Jin, H. M., Hahn, Y., Madsen, E. L. and Jeon, C. O. 2013. Metatranscriptomic analysis of lactic acid bacterial gene expression during kimchi fermentation. *Int J Food Microbiol* 163(2–3):171–9.

Junhui, Z., Hong, C. and Ruifu, Y. 1997. DNA based biosensors. *Biotechnology Advances* 15(1):43–58.

Kesmen, Z., Yasemin, celebi, Y., Güllüce, A. and Yetim, H. 2013. Detection of seagull meat in meat mixtures using real-time PCR analysis. *Food Control* 34(1):47–9.

Kim, S. and Choi, S. J. 2014. A lipid-based method for the preparation of a piezoelectric DNA biosensor. *Anal Biochem* 458:1–3.

Kljujev, I., Raicevic, V., Vujovic, B., Rothballer, M. and Schmid, M. 2018. *Salmonella* as an endophytic colonizer of plants - A risk for health safety vegetable production. *Microb Pathog* 115:199–207.

Krizovà, J., Spanovà, A. and Rittich, B. 2006. Evaluation of amplified ribosomal DNA restriction analysis (ARDRA) and species-specific PCR for identification of *Bifidobacterium* species. *Syst Appl Microbiol* 29(1):36–44.

Lai, E., Birren, B. W., Clark, S. M., Simon, M. I. and Hood, L. 1989. Pulsed field gel electrophoresis. *BioTechniques* 7(1):34–42.

Lauri, A. and Mariani, P. O. 2009. Potentials and limitations of molecular diagnostic methods in food safety. *Genes Nutr* 4(1):1–12.

Lobert, S., Hiser, L. and Correia, J. J. 2010. Expression profiling of tubulin isotypes and microtubule-interacting proteins using real-time polymerase chain reaction. *Methods Cell Biol* 95:47–58.

Lopez-Enriquez, L., Rodriguez-Lazaro, D. and Hernandez, M. 2007. Quantitative detection of *Clostridium tyrobutyricum* in milk by real-time PCR. *Appl Environ Microbiol* 73(11):3747–51.

Ma, K., Deng, Y., Bai, Y., et al. 2014. Rapid and simultaneous detection of *Salmonella*, *Shigella*, and *Staphylococcus aureus* in fresh pork using a multiplex real-time PCR assay based on immunomagnetic separation. *Food Cont* 42:87–93.

Mackay, I. M., Arden, K. E. and Nitsche, A. 2002. Real-time PCR in virology. *Nucleic Acids Res* 30(6):1292–305.

Madigan, M. M. 2009. Brock biology of microorganisms, 12th edition. *Int Microbiol* 11:65–73.

Maibach, R. C. and Altwegg, M. 2003. Cloning and sequencing an unknown gene of *Tropheryma whipplei* and development of two LightCycler PCR assays. *Diagn Microbiol Infect Dis* 46(3):181–7.

Maiden, M. C., Bygraves, J. A., Feil, E., et al. 1998. Multilocus sequence typing: A portable approach to the identification of clones within populations of pathogenic microorganisms. *Proc Natl Acad Sci USA* 95(6):3140–5.

Marsh, T. L. 2005. Culture-independent microbial community analysis with terminal restriction fragment length polymorphism. *Methods Enzymol* 397:308–29.

Marty, E., Buchs, J., Eugster-Meier, E., Lacroix, C. and Meile, L. 2012. Identification of *staphylococci* and dominant lactic acid bacteria in spontaneously fermented swiss meat products using PCR–RFLP. *Food Microbiol* 29(2):157–66.

Masato, O., Hiroyuki, I., Hiroshi, K., Kenshi, H. and Takato, S. 1989. Detection of the polymorphisms of human DNA by gelelectrophoresis as single-strand conformation polymorphism. *Proc Natl Acad Sci USA* 86(8):2766–70.

Mazo, J. Z., Dinon, A . Z., Tagliari,C., Ilha, E. C., Sant'Anna, E. S. and Arisi, A. C. M. 2010. Amplified ribosomal DNA restriction analysis (ARDRA) of new isolated strains of *Bifidobacteria* from newborn babies microbiota. *Afr J Microbiol Res* 4(22):2339–42.

McPherson, M. J. and Moller, S. G. 2000. *Polymerase chain reaction.* Oxford: BIOS Scientific Publishers Ltd, 1–18.

Moeseneder, M. M., Arrieta, J. M., Muyzer, G., Winter, C. and Herndl, G. J. 1999. Optimization of terminal-restriction fragment length polymorphism analysis for complex marine bacterioplankton communities and comparison with denaturing gradient gel electrophoresis. *Appl Environ Microbiol* 65(8):3518–25.

Mohania, D., Nagpal, R., Kumar, M., et al. 2008. Molecular approaches for identification and characterization of lactic acid bacteria. *J Dig Dis* 9(4):190–8.

Nancy, J. R., Mary, E., Schutter, R., Dick, P., David, D. and Myrold 2000. Use of length heterogeneity PCR and fatty acid methyl ester profiles to characterize microbial communities in soil. *Appl Environ Microbiol* 66(4):1668–75.

Narayanan, S. 1991. Applications of restriction fragment length polymorphism. *Ann Clin Lab Sci* 21(4):291–6.

Ndoye, B., Rasolofo, E. A., LaPointe, G. and Roy, D. 2011. A review of the molecular approaches to investigate the diversity and activity of cheese microbiota. *Dairy Sci Technol* 91(5):495–524.

Nsofor, C. A. 2016. Pulsed-Field Gel electrophoresis (PFGE): Principles and applications in molecular epidemiology: A review. *Int J Curr Res Sci* 2(2):38–51.

Omiccioli, E., Amagliani, G., Brandi, G. and Magnani, M. 2009. A new platform for real-time PCR detection of *Salmonella* spp., *Listeria monocytogenes* and *Escherichia coli* O157 in milk. *Food Microbiol* 26(6):615–22.

Ootsubo, M., Shimizu, T., Tanaka, R., Sawabe, T., Tajima, K. and Ezura, Y. 2003. Seven-hour fluorescence in situ hybridization technique for enumeration of *Enterobacteriaceae* in food and environmental water sample. *J Appl Microbiol* 95(6):1182–90.

Osborn, A. M., Moore, E. R. B. and Timmis, K. N. 2000. An evaluation of terminal-restriction fragment length polymorphism (T-RFLP) analysis for the study of microbial community structure and dynamics. *Environ Microbiol* 2(1):39–50.

Ota, M., Asamura, H., Oki, T. and Sada, M. 2009. Restriction enzyme analysis of PCR products. *Methods Mol Biol* 578:405–14.

Peril, M. A., Tallon, R. and Klaenhammer, T. R. 2009. Temporal gene expression and probiotic attributes of *Lactobacillus acidophilus* during growth in milk. *J Dairy Sci* 92(3):870–86.

Pleitner, A. M., Hammons, S. R., McKenzie, E., Cho, Y. and Oliver, H. F. 2014. Introduction of molecular methods into a food microbiology curriculum. *J Food Sci Educ* 13(4):68–76.

Porcellato, D., Brighton, C., McMahon, D. J., et al. 2014. Application of ARISA to assess the influence of salt content and cation type on microbiological diversity of Cheddar cheese. *Lett Appl Microbiol* 59(2):207–16.

Rasmussen, H. B. 2012. Restriction fragment length polymorphism analysis of PCR-Amplified Fragments (PCR-RFLP) and Gel electrophoresis – Valuable tool for genotyping and genetic fingerprinting. In: *Gel Electrophor Princ Basics*, 315–34.

Rathnayaka, R. 2011. Effect of sample pre-enrichment and characters of food samples on the examination for the *Salmonella* by plate count method and fluorescent in situ hybridization technique. *Am J Food Technol* 6(9):851–6.

Rehbein, H., Mackie, I. M., Pryde, S., et al. 1999. Fish species identification in canned tuna by PCR-SSCP: Validation by a collaborative study and investigation of intraspecies variability of the DNA-patterns. *Food Chem* 64(2):263–8.

Rip, D. and Gouws, P. A. 2020. PCR-restriction fragment length polymorphism and pulsed-field gel electrophoresis characterization of *Listeria monocytogenes* isolates from ready-to-eat foods, the food processing environment, and clinical samples in South Africa. *J Food Prot* 83(3):518–33.

Rizo, J., Guillén, D., Farres, A., Díaz-Ruiz, G., Wacher, S. S. C. and Rodríguez-Sanoja, R . 2018. Omics in traditional vegetable fermented foods and beverages. *Crit Rev Food Sci Nutr* 60(5):791–809.

Rohde, A., Hammerl, J. A., Appel, B., Dieckmann, R. and Dahouk, S. A. 2017. Differential detection of pathogenic *Yersinia* spp. by fluorescence in situ hybridization. *Food Microbiol* 62:39–45.

Roy, D. and Sirois, S. 2000. Molecular differentiation of *Bifidobacterium* spp. with amplified ribosomal DNA restriction analysis and alignment of short regions of the ldh gene. *FEMS Microbiol Lett* 191(1):17–24.

Roy, D., Sirois, S. and Vincent, D. 2001. Molecular discrimination of *Lactobacilli* used as starter and probiotic cultures by amplified ribosomal DNA restriction analysis. *Curr Microbiol* 42(4):282–9.

Ruiz-Rueda, O., Soler, M., Calvó, L. and García-Gil, J. L. 2011. Multiplex realtime PCR for the simultaneous detection of *Salmonella* spp. and *Listeria monocytogenes* in food samples. *Food Anal Methods* 4(2):131–8.

Sardaro, M. L. S., Perin, L. M., Bancalari, E., Neviani, E. and Gatti, M. 2018. Advancement in LH-PCR methodology for multiple microbial species detections in fermented foods. *Food Microbiol* 74:113–9.

Sheikha, A. F. E. and Hu, D. 2018. Molecular techniques reveal more secrets of fermented foods. *Crit Rev Food Sci Nutr* 60(1):11–32.

Sheikhzadeh, E., Chamsaz, M., Turner, A. P. F., Jager, E. W. H. and Beni, V. 2016. Label-free impedimetric biosensor for *Salmonella typhimurium* detection based on poly [pyrrole-co-3-carboxyl-pyrrole] copolymer supported aptamer. *Biosens Bioelectron* 80:194–200.

Shimizu, S., Ootsubo, M., Kubosawa, Y., Fuchizawa, I., Kawai, Y. and Yamazaki, K. 2009. Fluorescent in situ hybridization in combination with filter cultivation (FISHFC) method for specific detection and enumeration of viable *Clostridium perfringens*. *Food Microbiol* 26(4):425–31.

Silva, D. S. P., Canato, T., Magnani, M., Alves, J., Hirooka, E. Y. and Oliveira, T. C. R. M. 2011. Multiplex PCR for the simultaneous detection of *Salmonella* spp. and *Salmonella enteritidis* in food. *Int J Food Sci Tech* 46(7):1502–7.

Sklarz, M. Y., Angel, R., Gillor, O. and Soares, M. I. M. 2009. Evaluating amplified rDNA restriction analysis assay for identification of bacterial communities. *Antonie Leeuwenhoek* 96(4):659–64.

Suo, B., He, Y., Tu, S. I. and Shi, X. 2010. A multiplex real-time polymerase chain reaction for simultaneous detection of *Salmonella* spp., *Escherichia coli* O157, and *Listeria monocytogenes* in meat products. *Foodborne Pathog Dis* 7(6):619–28.

Tabatabaei, M., Zakaria, M. R., Rahim, R. A., et al. 2009. PCR-Based DGGE and FISH analysis of methanogens in anaerobic closed digester tank treating palm oil mill effluent. *Electron J Biotechnol* 13. http://www.ejbiotechnology.info/content/vol12/issue3/full/4/.

Tahira, T., Kukita, Y., Higasa, K., Okazaki, Y., Yoshinaga, A. and Hayashi, K. 2009. Estimation of SNP allele frequencies by SSCP analysis of pooled DNA. *Methods Mol Biol* 578:193–207.

Taylor, A. D., Ladd, J., Yu, Q., Chen, S., Homola, J. and Jiang, S. 2006. Quantitative and simultaneous detection of four foodborne bacterial pathogens with a multi-channel SPR sensor. *Biosens Bioelectron* 22(5):752–8.

Thoetkiattikul, H., Mhuantong, W., Laothanacharoen, T., et al. 2013. Comparative analysis of microbial profiles in cow rumen fed with different dietary fiber by tagged 16S rRNA gene pyrosequencing. *Curr Microbiol* 67(2):130–7.

Urwin, R. and Maiden, M. C. J. 2003. Multi-locus sequence typing: A tool for global epidemiology. *Trends Microbiol* 11(10):479–87.

Vaneechoutte, M., Beenhouwer, H. D., Claeys, G., et al. 1993. Identification of *Mycobacterium* spp. by using amplified ribosomal DNA restriction analysis. *J Clin Microbiol* 31(8):2061–5.

Verdier-Metz, I., Michel, V., Delbès, C. and Montel, M. C. 2009. Do milking practices influence the bacterial diversity of raw milk? *Food Microbiol* 26(3):305–10.

Viglasky, V. 2013. Polyacrylamide temperature gradient gel electrophoresis. *Methods Mol Biol* 1054:159–71.

Welsh, J. and McClelland, M. 1990. Fingerprinting genomes using PCR with arbitrary primers. *Nucleic Acids Res* 18(24):7213–8.

Williams, J. G., Kubelik, A. R., Livak, K. J., Rafalsk, J. A. and Tingey, S. V. 1990. DNA polymorphisms amplified by arbitrary primers are useful as genetic markers. *Nucleic Acids Res* 18(22):6531–5.

Wu, C. F., Valdes, J. J., Bentley, W. E. and Sekowski, J. W. 2003. DNA microarray for discrimination between pathogenic O157. H7 EDL933 and non-pathogenic *Escherichia coli* strains. *Biosens Bioelectron* 19(1):1–8.

Wu, Q., Zhang, Y., Yang, Q., Yuan, N. and Zhang, W. 2019. Review of electrochemical DNA biosensors for detecting food borne pathogens. *Sensors (Basel)* 19(22):4916.

Xu, H., Liu, W., Gesudu, Q ., et al. 2015. Assessment of the bacterial and fungal diversity in home-made yoghurts of Xinjiang, China by Pyrosequencing. *J Sci Food Agric* 95(10):2007–15.

Yamaguchi, N., Kitaguchi, A. and Nasu, M. 2012. Selective enumeration of viable *Enterobacteriaceae* and *Pseudomonas* spp. in milk within 7 h by multicolor fluorescence in situ hybridization following microcolony formation. *J Biosci Bioeng* 113(6):746–50.

Yang, X., Noyes, N. R., Doster, E., et al. 2016. Use of metagenomic shotgun sequencing technology to detect foodborne pathogens within the microbiome of the beef production chain. *Appl Environ Microbiol* 82(8):2433–43.

Youn, S. Y., Seo, J. M. and Ji, G. E. 2008. Evaluation of the PCR method for identification of *Bifidobacterium* spp. *Lett Appl Microbiol* 46(1):7–13.

Yu, J., Sun, Z., Liu, W., et al. 2009. Rapid identification of lactic acid bacteria isolated from home-made fermented milk in Tibet. *J Gen Appl Microbiol* 55(3):181–90.

Zhao, X., Lin, C. W., Wang, J. and Oh, D. H. 2014. Advances in rapid detection methods for foodborne pathogens. *J Microbiol Biotechn* 24(3):297–312.

CHAPTER 5

Exploring the Landscape of the Genome and Transcriptome in Microbial Food Safety

Adil Hussain and Amjad Iqbal

CONTENTS

5.1 INTRODUCTION

All organisms living on Earth consume food, though the quality and quantity of the food consumed are highly variable. Living things include animals, plants, fungi, protozoa, and bacteria. As these have all been divided into distinct kingdoms of life, each of the kingdoms includes organisms that can consume organisms from other kingdoms as food. In that sense, all living organisms are food for some other organisms. For humans, consumption of particular food and beverage may sometimes cause disease or injury due to foodborne hazards that may either be of a chemical, physical or biological origin. Food may be contaminated with a variety of microbes within the field, transportation, distribution, storage, and from the field to the processing plant and ultimately to the consumer. There are thousands of types of such microbes in the soil, water, and air that cause unwanted food spoilage resulting in food safety problems. Consumption of such food may cause minor to major health-related problems in humans and other organisms. The number of cases reporting microbial food poisoning may range in the millions, especially in the less developed areas of the world, thereby causing losses worth millions of dollars annually. Such organisms include various types of fungi, bacteria, viruses, and other parasites. The armada of toxins, enzymes, and other chemical cocktails secreted by these microbes not only spoil the food but are also one of the major reasons for food poisoning or foodborne illness. Symptoms often include vomiting, body pain, fever, and diarrhea. In short, microbial food safety problems have a major impact on the world economy, and it is important to contain them; and to do that, it is important to detect them.

Detection of microbes in food has in the past relied on culture-based and microscopy techniques; however, new and sophisticated methods have been developed recently that have made detection of different microbes in food more rapid and precise. These methods use a blend of microscopic, physical, biochemical, serological, genetic, and molecular biology-based approaches. Yet scientists have always been looking for more sensitive, precise, specific, rapid, and cost-effective detection methods. This chapter focuses on the use and prospects of the next-generation sequencing (NGS) technologies in microbial food safety and quality. During the past decade, NGS techniques including genomics, transcriptomics, proteomics, metabolomics, and lipidomics (commonly called "omics") revolutionized food science and technology and food microbiology, giving birth to *foodomics*, a discipline that studies the integrated application of omics in Food Science (Cifuentes 2009) and integrates other fields of study such as food chemistry, biology, computational biology, bioinformatics, and data mining (Capozzi and Bordoni 2013).

5.2 WHOLE GENOME SEQUENCING

NGS tools, such as whole-genome sequencing (WGS) and powerful bioinformatics techniques have played a key role in the evolution of modern-day food safety and quality. Today, WGS allows detection and the most detailed comparison of individual strains of different food microbes via single nucleotide polymorphism (SNP) and genomic multi-locus sequence typing (MLST) (Jagadeesan et al. 2019). These techniques are now applied routinely in fields ranging from common diagnostics to investigations of disease outbreak, antimicrobial resistance, and food authenticity (Goodwin et al. 2016; Allard et al. 2018; Quainoo et al. 2017, 2018). Today, WGS is the preferred method of detection of a single type or strain of microbes such as fungi, bacteria, viruses, or others rather than traditional culture-based methods. An important limitation of culture-based methods is that a many microbes cannot be cultured on an artificial substrate (e.g. all types of viruses and several fungi that are obligate parasites and grow only inside a living host). On the other hand, metagenomics applies NGS to generate sequences of multiple types of microbes in a single biological sample making detection and identification easier in cases of multiple microbial infections.

WGS-based approaches were introduced into the diagnostics and public health system about a decade ago (Köser et al. 2012) after which at least four countries (United States, United Kingdom, Denmark, and France) adopted WGS-based microbial pathogen analysis for surveillance of foodborne microbes (Allard et al. 2016; Ashton et al. 2016). This led to the early detection of a significantly higher number of outbreak cases in the following years in the United States (US) alone (Jackson et al. 2016).

Furthermore, WGS is equally important and beneficial for the food industry. For example, an obvious benefit of WGS for the food industry is the identification and analysis of the root cause during microbial spoilage or contamination events. WGS also helps identify and differentiate between new and recurring microbes in production lines as well as predict the virulence and spoilage capacities of various microbes and strains. Food supply chains are a key component of the food industry, and it is crucial to have stringent and efficient mechanisms of tracking and tracing pathogens along the food chain.

WGS-based approaches not only help track and trace the source of the pathogen to the food handlers or food service environments but can also supply key information, such as genetic changes across the entire genome of pathogens over time.

The US began real-time surveillance for *Listeria monocytogenes* in 2013 in environmental, clinical, and food samples via Pulsed-Field Gel Electrophoresis (PFGE); WGS was also used in parallel. Within a couple of years, PFGE and WGS identified multiple disease clusters associated with specific seasonal foods. The PFGE-based system produced three clusters of which two were associated with seasonal food products and one with both seasonal food products as well as ready-to-eat (RTE) food products. However, on the other hand, WGS results indicated that the third single cluster could be excluded, as the genome sequence from these samples indicated that these *L. monocytogenes* isolates were unrelated. This helped public health and food regulatory authorities to confirm the source of contamination as an implicated food vehicle, thereby accelerating the regulatory response to this contamination event and preventing additional listeriosis cases. It is clear that a similar approach can also be applied to other foodborne pathogens of economic importance, such as *Salmonella*, *E. coli*, *Campylobacter*, and others and, in this regard, the initiatives taken by the US Department of Agriculture (USDA), Food and Drug Administration (FDA), and Centers for Disease Control and Prevention (CDC) are especially commendable and could be a case study for other developing countries.

Another related case study is that of real-time *L. monocytogenes* surveillance of human infections in Denmark where WGS replaced PFGE first, because of high discriminatory typing and subtyping by PFGE and second, because WGS is species-independent and fits much better into real-time surveillance. With this, the authorities successfully investigated several cases between 2014 and 2015 and tracked and traced their source to commercial production lines associated with a variety of RTE products. Other similar case studies can be found where WGS-based approaches are revolutionizing food safety and quality.

5.3 TRANSCRIPTOMICS

Another important and interesting NGS technique is transcriptomics. It is basically the aggregate of all the messenger RNA (mRNA) molecules expressed in an organism. Transcriptomics-based techniques promise efficient, fast, and reliable measures to monitor microbial and chemical contaminants in food. An interesting fact about the transcriptome of an organism is its spatial and temporal variation which simply means that the transcriptome of an organism varies with the types of cells and/or tissues and with time. The transcriptomic response of different tissues to different stimuli is also variable at different times. This opens up an opportunity of obtaining a great deal of information about the response of food material to infection by microbes at the transcript level. Transcriptomic analysis of infected food materials not only gives detailed information about the number and types of genes activated in response to infection but also shows the magnitude of up-regulation and/or down-regulation of expression profiles of these genes in different tissues, local or systemic, and over a certain period. On the other hand, transcriptomic analysis of microbes outside and inside food materials causing spoilage can help identify key microbial genes required for infection and those responsible for encoding enzymes and toxins that kill the host resistance machinery and cause spoilage. The genetic information so obtained can be further utilized for functional genomics studies and for engineering plant varieties or cultivars that are resistant to infection by various organisms. Transcriptomics include DNA microarrays and the recently developed NGS technique known as RNA-Seq (Wang et al. 2009).

Wang et al. (2018) performed the transcriptomic analysis of *Pseudomonas fragi* NMC25 subjected to modified atmospheric packaging (MAP). Results indicated that MAP induced differential expression of 559 different genes in *P. fragi* involving down-regulation of genes involved in the electron transport chain, ATP-binding cassette transporters, and flagellar and fimbrial proteins and those involved in DNA replication and repair; this resulted in the inhibition of *P. fragi* aerobic respiration, nutrient uptake, motility, and growth. However, results also indicated that *P. fragi* NMC25 rerouted their pathways for energy production, amino acid synthesis, membrane lipid composition, and other metabolic processes to adapt to MAP-induced stress. This indicates the NMC25 can survive MAP, but their spoilage ability is compromised. In another study, Illikoud et al. (2019) performed transcriptome and volatilome analysis during the growth of *Brochothrix thermosphacta* in beef and shrimp juices. *B. thermosphacta* is one of the main meat and seafood spoilers that cause spoilage by producing malodorous volatile organic chemicals. Variations in host-dependent expression profiles of different genes indicated that the type of chemicals produced depends on the type of food and storage conditions, especially in packaging, and on the type of strain, indicating differential metabolic functioning depending on the type of substrate and strain capacities.

5.4 PROTEOMICS

Proteomics deals with the comprehensive study of proteins, involving identification, structural elucidation, and biological functions. In a broad sense, the word proteome also involves any changes in the native protein of an organism exposed to a particular set of conditions (Anderson and Anderson 1998). Historically, the study of proteomics was started in 1975 with the extraction of proteins from the *Escherichia coli*, a gram-negative rod-shaped bacterium, but the scientists were unable to identify the proteins. Later on, the proper idea of proteomics was introduced by Australian scientist, Mark Wilkins, in the 1990s. The idea came after the introduction of genomics and genome to represent the complete set of genes in a living system. Since then, the field of proteomics has seen robust growth because of rigorous technological advancement in the isolation, fractionation, structural characterization, and role of proteins. The study of proteins/proteomics depends on the isolation and fractionation of proteins from the complex mixture by chromatographic and electrophoretic techniques, elucidation by advanced mass spectrometry (MS), and analysis of MS data by bioinformatics.

Recently, proteomics has been employed with great success in the field of food science and technology to maintain the quality and safety of the food. Proteomics has made it possible to identify GMOs, allergens, and foodborne pathogens in a variety of meat, milk, and plant-based food products. The technique has revolutionized the food world as food safety and food quality is of great concern not only to food marketability but also to consumers' health. An underdeveloped or unreliable method for ensuring quality and safety can diminish consumers' trust, causing economic setbacks for the industry. It is indeed, important to establish a foolproof method to analyze food products during the whole process of the food chain.

Proteomics can serve this purpose very well through quantitative and qualitative analysis of allergens and biological toxins/contaminants (such as bacteria, fungi, and other pathogens) in various foods accurately (Table 5.1). Though human food is a

TABLE 5.1 Technologies Used to Ensure Food Safety and Quality

Technique	Food Item	Detection	Reference
2D Electrophoresis	Milk	Milk protein characteristic for infant nutrition	D'auria et al. (2005)
2D Electrophoresis	Milk	Phosphorylated A_{s1}-casein and glycosylated K-casein for cheese making	Jensen et al. (2012)
LC-ESI-MS/MS and RP-HPLC	Milk	Bioactive peptides	Holder et al. (2014)
2D Electrophoresis coupled with mass spectrometry	Meat	Proteins involved in glycolytic metabolism and chaperon proteins	Hamelin et al. (2006)
2D-LC-MS/MS	Meat	Soybean proteins as adulterant	Leitner et al. (2006)
MALDI-TOF and TOF MS/MS	Fish	Peptides responsible for the flavor	Zhang et al. (2012)
SDS-PAGE, MALDI-TOF, and MALDI-TOF/TOF	Fish	Functional proteins and peptides	Sanmartín et al. (2012)
MALDI-TOF-TOF and MS/MS	Food matrices	Shiga toxin by *E. Coli* O157:H7	Fagerquist et al. (2014)
MS/MS	Food matrices	Staphylococcal enterotoxin B	Callahan et al. (2006)
ELISA	Peanut	Peanut allergens	Iqbal and Ateeq (2013)

complex mixture of biological compounds, proteomics, accompanied by high-throughput separation techniques along with high-resolution characterization, can ably determine the native as well as extraneous proteins and observe the changes that occur in protein conformation during processing. The separation techniques include electrophoresis (one-dimensional or two-dimensional) and chromatographic techniques (one-dimensional or two-dimensional), whereas the characterization technique includes mass spectrometry (Gašo-Sokač et al. 2010).

5.4.1 Proteomics and Food Quality

Food quality can be judged by taste, odor, consistency, and nutritional value of the food product. Proteomics is one of the best tools that can accurately judge the quality of foods and food products. The best example that can determine the role of the proteomics in food is the assessment of meat and milk quality. Through proteomics, one can judge the color, taste, odor, and consistency as quality parameters of these foods. As meat is rich in proteins, the changes in the proteins can be influenced by genetics, environment, and post-mortem processing. A specific trait can be examined for a certain quality parameter; for example, tenderness of cow, chicken, or goat meat can be evaluated through proteomics (Cui et al. 2013). Color is one of the important properties of the meat, as it affects the consumers' choice to buy the meat. Meat redness and color stability are linked with glycolytic enzymes, such as phosphoglucomutase-1, glyceraldehyde-3-phosphate dehydrogenase, and pyruvate kinase (Canto et al. 2015; Gao et al. 2016). Processing time from slaughter until sale of the meat significantly influences the meat tenderness, which is closely linked to the differential expression of the proteins (actin, myosin heavy chain, and troponin T) (Lametsch et al. 2002). The quality of farmed fish can easily be evaluated through recent proteomics techniques. Flavor peptides have been identified in fish through proteomics, using an electronic tongue and MALDI-TOF/TOF MS/MS (Zhang et al. 2012). The changes that occur in the texture and flavor of the fish during cold storage can also be monitored through proteomic analysis. Milk protein information has also been improved by the introduction of proteomics. For example, the protein composition of milk from various species has been evaluated through proteomics to find the proteins with the best characteristics for infant milk production. Moreover, a 2-D gel electrophoretic profiling of milk proteins from various species has provided the foundation to understand the allergic proteins from non-human species (D'auria et al. 2005). Proteomic analysis has greatly contributed to the production of quality cheese. Milk has optimal coagulating properties for cheese making; and this property influenced by the presence of phosphorylated αs_1-casein and glycosylated κ-casein (Jensen et al. 2012).

5.4.2 Proteomics and Food Safety

Food safety deals with the identification and rejection of food that is unfit for human consumption. Food can be contaminated with a variety of agents (adulterants, allergens), but the most important is the foodborne microorganisms (bacteria, fungi, and other pathogens). Thermal processing of the food can also bring conformational changes in the native proteins that can be detected by the proteomics techniques. In the past, protein separation was achieved by 2-D polyacrylamide gel electrophoresis (2D-PAGE), but the method was laborious, time-consuming, and sometimes failed to separate very low or high molecular weight proteins. Multidimensional chromatographic techniques have the advantage in protein isolation and separation over previous techniques, with MS proving to be the best technique for protein identification. In practice, two methods have been employed in microbial protein analysis, (i.e. bottom-up and top-down). In the bottom-up technique, proteins are digested to small peptides and then analyzed by MS, while in the top-down technique, intact proteins are assessed by MS. In the last few years, protein chip technology has been used regularly for proteins and peptides analysis. This high-throughput method based on a microfluidic chip can do protein extraction, separation, and identification in one go and is recognized as a rigorous, fast, and cheap method (Sedgwick et al. 2008).

Bacteria are the main threat to food safety that causes food poisoning. *Clostridium botulinum, Listeria monocytogenes, E. coli* O157:H7, *Staphylococcus aureus, Campylobacter jejuni* are some of the common foodborne pathogens that pose a constant threat to food. *Listeria monocytogenes* is a gram-positive rod, non-spore-forming, facultatively anaerobic, and the most common universal foodborne pathogen that causes illness and death all over the world. In the past, detection of *L. monocytogenes* was cumbersome and time-consuming, but MALDI-TOF MS has revolutionized *L. monocytogenes* detection. *E. coli* O157:H7, a Shiga toxin producer, is a filamentous rod-shaped pathogen that is easily transmitted from cattle or cattle products. The Shiga toxin-producing *E. coli* can simply be characterized by a top-down approach through advance proteomic techniques, such as MALDI-TOF-TOF and tandem mass spectrometry (MS/MS) (Fagerquist et al. 2014). *S. aureus* is another toxin-producing non-motile, gram-positive, facultative anaerobic cocci that mainly contaminates milk and dairy products. *S. aureus* is an indicator microorganism for milk hygiene and quality that can be detected by 2D electrophoresis and shotgun MS analysis. Proteins, such as serotransferrin, fibrinogen b-chain, and antimicrobial polypeptide (cathelicidin) are known biomarkers for the presence of S. *aureus* contamination. Also, the disparity in protein expression at the bovine milk fat globular membrane may confirm the presence of S. *aureus*. The presence of staphylococcal enterotoxin B in the food matrices can also be detected by a very sensitive tryptic fragment-based detection by employing tandem MS (MS/MS), which verify the existence of S. *aureus* in that food (Callahan et al. 2006). Protein Standard Absolute Quantification (PSAQ) with the aid of immunocapture is one of the best and most advanced techniques for qualifying staphylococcal toxins to absolute level (Piras et al. 2016). In the area of pathogen detection, mass spectroscopy (especially MALDI-TOF MS) is the most reliable and fastest way of pathogen detection in complex food matrices. Through MALDI-TOF MS, the profiling of the complete bacterial proteome is possible subject to the availability of fingerprints of the tested bacteria. The fingerprints can further help in the classification of subspecies, strains, and serovar. In addition to the pathogens, MALDI-TOF MS has also been used successfully for the identification and characterization of lactic acid bacteria in fermented food products. (Nguyen et al. 2013).

The presence of mycotoxins, including aflatoxins, trichothecenes, zearalenone, fumonisins, ochratoxins, and patulin, is a key food safety issue all over the world. The mycotoxins are mainly produced in foods and food products by *Aspergillus*, *Penicillium*, and *Fusarium*. The fungi from these genera can multiply in foods and produce a variety of mycotoxins under set conditions (temperature and humidity). The identification of the mycotoxin-producing fungi in foods is very important because the consumption of mycotoxins can cause cancer, mutagenicity, gastrointestinal and renal failure, and immunosuppression (reducing the efficiency to combat infectious diseases) in humans. Proteomics is one of the best methods to identify the released proteins (Doyle 2011). Immunochemical and chromatographic techniques are the best for routine analysis of the mycotoxins. Liquid chromatography with the aid of tandem MS can be used for the separation and characterization of mycotoxins in food matrices. But proteomics is the best tool to study the changes in the proteome of causative organisms during mycotoxin production under favorable conditions (Choi et al. 2012).

Besides the bacteria and fungi, *Giardia intestinalis*, *Toxoplasma gondii*, and *Taenia saginata* are also constant threats to food safety. These food-hazardous protozoans and helminths, which can cause various intestinal diseases in humans, can be detected by using proteomics. Chromatographic coupled with MS techniques may be the best option for monitoring these food-related pathogens (Piras et al. 2016).

Food allergens, which are believed to be protein or glycoproteins in nature, are responsible for food-related allergies. The most common allergenic food items are peanuts, tree nuts, milk, soybeans, crustaceans, egg, fish, and wheat. The term 'food allergy' has been defined by the European Academy of Allergology and Clinical Immunology (EAACI) as "Allergy is a hypersensitivity reaction initiated by immunological mechanisms," whereas food allergens are the foods that intensify the immunological reactions. Food allergies pose a great threat to human health in the industrialized areas of the world due to life-threatening anaphylaxis (a condition in which the subject has difficulty breathing). The main symptoms of allergy that are stimulated by ingesting offending substances are rhinitis, asthma, cramps, diarrhea, vomiting, eczema, angioedema, and anaphylactic shock. Because it is common for various food products to be processed on the same production line (e.g., peanut biscuits, and glucose biscuits), it is difficult to know whether a food is contaminated with allergens or not (Iqbal and Ateeq 2013; Iqbal et al. 2016; Iqbal et al. 2018). As discussed earlier, food allergens are protein in nature; thus proteomics is the best method for exploring the presence, composition, and nature of food allergens. The presence of allergens in food can be detected in two ways–gel-based and non–gel-based. 2D electrophoresis and 2D immunoblotting are gel-based, whereas high performance liquid chromatography (HPLC)-MS and IgE binding assay of the trypsinized proteome are non–gel-based assays. Other proteomic methods include 2D electrophoresis combined with MS for identifying existing and novel allergic proteins in food; ELISA and immunoblotting (IgE-based assays) are very good methods for detecting food allergens, but reliability falls off at very low concentrations (approx. 5 ppm); multiple reaction monitoring (MRM) is a highly sensitive and robust technique in proteomics that can screen the allergic proteins in food, even in minute quantities (Piras et al. 2016).

5.5 LIPIDOMICS

Lipids are non-polar components of the food that can be simple or complex and provide more energy than carbohydrates and proteins per gram. Simple lipids consist of fatty acids, monoglycerides, diglycerides, and triglycerides. The saturated fatty acid (single-bonded between carbon to carbon) in a lipid molecule results in semi-solid fat, whereas the unsaturated fatty acid (double bond) results in the liquid oil. Additionally, the trans-fatty acid has hydrogen atoms on the opposite side of the double bond, having a high melting point, but is less healthy, compared to cis-fatty acid with a hydrogen atom on the same side of the double bond, low melting point, and is healthier. The fatty acid that contains one double bond is known as a monounsaturated fatty acid (MUFA), while those that contain two or more double bonds are known as a polyunsaturated fatty acid (PUFA). Most of the fatty acids are produced inside our body except a few–for example, linoleic acid and alpha-linolenic acid. Foods that are rich in linoleic acid and alpha-linolenic acid should be part of a healthy, balanced diet. Complex lipids consist of moieties other than lipids, such as phospholipids, glycolipids, lipoproteins, cholesterol, etc. Lipids are responsible for the integrity of cellular membranes in living organisms and play an important role in hormones, metabolism, energy storage, and controlling physiological and biochemical responses. However, a high intake of lipids can lead to various health issues, such as obesity, high cholesterol, hypertension, and cardiovascular diseases. The chemical composition and structure of the lipids can affect the sensory quality of the foods. Lipidomics is the study of the functional and structural properties of lipids (lipidome) in a biological system. Spener et al. (2003) defined lipidomics as "the full characterization of lipid molecular species and their biological roles concerning the expression of proteins involved in lipid metabolism and function, including gene regulation." Lipidomics is applicable in the fields of disease biomarker discovery, food nutrition, and plant research. In food science, it is widely used for food quality assessment, authenticity, and chemical safety (i.e., the presence of growth-promoting substances in meat). Moreover, food processing industries are often using edible oils composed of free fatty acids, diglycerides, phospholipids, and sterols to achieve superior characteristics in their products. The edible oils from various sources (animals, plants, and microorganisms) are different from each other in terms of physical, chemical, and nutritional properties because of their composition. To select edible oil with supreme quality from the above-mentioned sources, the study of the lipidome is necessary. However, owing to the large number of molecular lipids with different structures, functions, and presence in significantly different concentrations in a system, it is difficult to analyze all by one technique at the same time. The two most common approaches in lipidomics are hypothesis-driven targeted analysis, and a comprehensive, hypothesis-generating, non-targeted profiling method. In the former approach, specific lipids from the same class can be analyzed. The method is sensitive and robust but gives limited information concerning the total lipidome. The latter one is a very comprehensive, yet semi-quantitative, approach. The most common techniques used in lipidomics are chromatography, mass spectroscopy (MS), and nuclear magnetic resonance (NMR) (Marchand et al. 2018).

Traditionally, thin layer chromatography (TLC) was used for the separation of different classes of lipids (steroids, triglycerides, free fatty acids, and phospholipids) with limitations related to resolution and sensitivity. The method was replaced by gas chromatography (GC) and liquid chromatography (LC). GC proved to be a very efficient method in the separation of various lipids after derivatization (methyl esterification). Thermal stability and volatility of analytes limit the performance of GC. High-performance liquid chromatography (HPLC) coupled with UV or refractive index detector was not suitable for the identification of lipids due to low sensitivity. NMR is not widely used in lipidomics because it is impossible to identify lipids in the complex mixture at the molecular level by this method. The method of choice in lipidomics is MS combined with chromatographic separation (Hyötyläinen et al. 2013). UPLC-Q-Exactive Orbitrap MS is the best method for the profiling of milk lipids obtained from various animals; in fact, this method, was used to identify 14-lipid biomarkers of purity in milk (Li et al. 2017). Similarly, derivatization Gas chromatography-electron impact-mass spectrometry (GCEI)-MS has been named the best method for identifying diacylglycerols (DAG) positional isomers for the confirmation of extra virgin oil purity (Zhu et al. 2013). The determination of fatty acid concentration and polar to non-polar fatty acids ratio is used to verify beef quality (Legako et al. 2015). The selected ion monitoring electrospray ionization-tandem mass spectrometry has shown lower levels of plasmalogens in the juices contaminated with anaerobic bacteria (Řezanka et al. 2015).

5.6 METABOLOMICS

Metabolomics is the newly emerging and fast-growing field dealing with the comprehensive analysis (qualitative and quantitative) of metabolites by sophisticated technologies in a biological system. Metabolites are the chemical compounds produced during the process of metabolism. The two types of metabolites produced during the metabolic process are: primary metabolites–produced for the growth of the organism and include carbohydrates, fats, proteins, and vitamins, etc.; and secondary metabolites–produced from the primary metabolites, but they do not form the basic molecular skeleton of an organism. Metabolomics fills a vital role in food safety, as it identifies the extraneous and hazardous metabolites in foods and food products. Foods and food products may contain desirable metabolites that are required to improve the taste and nutritional quality. Some microorganisms partially digest the food, making it tasty and beneficial to health. Probiotics, sauerkraut, tofu, yogurt, and pickles are examples that are produced by beneficial microorganisms. On the contrary, other microorganisms produce toxic metabolites in foods and food products, causing illness and sometimes death in humans. The food metabolome is multifaceted in nature with great variability, and depends on the type of food products and the raw materials used. The metabolites present in food items have different solubility/polarity and molecular weights (carbohydrates, lipids, proteins, etc.). Since the concentration of various metabolites differs depending on the food type, it is impossible to analyze all metabolites using the same test. There is no instrument available to date that can analyze the entire range of metabolites in one go. Besides the variability in the metabolites themselves, food matrices increase the difficulty of analyzing metabolites in food samples and are accountable for unreliable data that affects the precision and accuracy of the method used. To overcome this difficulty, samples can be pre-treated using phase extraction methods, such as liquid extraction and solid-phase extraction. These pre-treatment steps are time-consuming, but they are useful for producing clean samples for metabolome analysis. In recent years, several methods have been developed to isolate, detect, quantify, and characterize as many metabolites as possible in one run. Another challenge in the field of metabolomics is the interpretation of the large set of data produced by high-throughput techniques. This hurdle can be overcome by using high-tech statistical and multi-variant data analysis tools, like cluster analysis, pathway mapping, comparative overlays, and heat maps (Roessner and Bowne 2009). Metabolomics is done for target analysis, metabolite profiling, metabolome analysis, and metabolic fingerprinting. In target analysis, one best technology is used for the analysis of a specific set of metabolites. It is a quantitative approach through which a group of metabolites are quantitated with the help of labeled isotopes such as 15N or 13C (internal or external). Additionally, by using internal standard, the analysis can be done by semi-quantitative or quantitative methods. In metabolite profiling, a large set of compounds (both known and unknown) are analyzed using GC/MS. Metabolome analysis means determining and identifying as many metabolites as possible using state of the art techniques, such as LC-MS/MS, GC-MS, and/or NMR. In metabolic fingerprinting, the profile of the sample of interest is developed, which is then used to compare with the data of the other samples to separate the differences among the samples. The unique signals are then separated and the metabolites corresponding to the signals are identified and characterized (Roessner and Bowne 2009).

5.7 PROS AND CONS OF NGS IN MICROBIAL FOOD SAFETY MANAGEMENT

Some of the key benefits of using NGS-based approaches in microbial food safety:

- NGS-based methods have significantly higher performance compared to traditional methods. These allow the detection and characterization of microbes up to the strain level even in samples with multiple infections, and the results are highly precise, specific, and sensitive.
- NGS analyses such as WGS and transcriptomics are becoming cheaper and cost-effective as compared to a plethora of traditional subtyping techniques that mostly vary according to the target pathogen. Furthermore, a single NGS run provides researchers with the entire genetic code of a pathogen, which can be recorded, stored, and analyzed for a myriad of other purposes, such as serotyping, identification of SNPs and virulence factors, characterization of the chemical or antibiotic resistance of microbes, and comparison with data sets from other sources over space and time (Joensen et al. 2014) resulting in better value for the money.
- WGS and transcriptomics results are available within a few weeks, are quickly analyzed via bioinformatics approaches, can be electronically shared across the network, and are easily analyzed for different purposes.

- Other benefits of NGS include the ease of sharing, universality in terms of methodology and analyses, ease of learning and flexibility, and compatibility with many other food safety assessment systems.
- The cost of NGS applications is of concern, especially in countries where genetic and molecular biology equipment and reagents are not manufactured locally. Furthermore, estimated costs on the establishment of a real-time surveillance system with a coordinated network of NGS facilities in related public and private sectors may play a prohibitive role in the adoption of NGS based technology.
- A potential limitation is that of data storage. WGS, transcriptomics, and metabolomics experiments generate a large amount of data, usually in giga- or terabytes, that need physical as well as virtual or cloud space in national and/or international repositories and databases, such as the National Center for Biotechnology Information (NCBI), European Nucleotide Archive (ENA), DNA Data Bank of Japan (DDBJ), European Bioinformatics Institute (EBI), and others.
- Lack of high-speed internet, and the specialized, well-trained bioinformaticians, required to handle and interpret NGS data.

Furthermore, as described earlier, NGS methods cannot stand alone. They must be used in combination with routine epidemiological and food monitoring and testing systems that include surveillance and analysis of clinical, food, industrial, and environmental samples.

5.8 CONCLUSION

Food safety is a global phenomenon, and consumers have the right to have access to safe and nutritious food (FAO 1996). Although in many countries of the world, considerable progress has been made in making food systems safe, about 600 million foodborne illnesses and 420,000 deaths from 31 major food safety hazards occurred worldwide in 2010 alone (WHO 2015), not to mention the inestimable, related economic and social costs in cases of epidemic or pandemic situations (FAO 1996). Food safety is also an important element in achieving global food security. Poor food safety diminishes the economic as well as health aspects of food security. WGS offers great potential in the way investigations and assessments are conducted to manage microbial food safety issues and illnesses. Together with the rapidly declining cost and the high degree of precision in identification and characterization of microbes at the strain level, NGS-based approaches have significantly increased the public trust in food technicians, food authorities, and industries. However, despite all its merits or advantages, the application of NGS-based approaches in food science and technology is limited, particularly in developing countries. It is important for these countries to describe and understand the potential merits, demerits, technical challenges, and requirements of this technology and to make appropriate decisions about when and where to apply NGS technology in existing food regulation and control networks and sectors. It is, however, also important to mention that NGS technologies cannot suffice alone. The technology needs to be integrated with epidemiological information, routine laboratory testing, inspections, surveillance, supply chain infrastructure, public health information, and others. This indicates that the establishment of an integrated food control system comprised of a multi-sector network is necessary at the national level as the same has been recommended for the Food and Agriculture Organization (FAO) especially for developing countries (FAO 2016).

REFERENCES

Allard, M. W., Bell, R., Ferreira, C. M., et al. 2018. Genomics of foodborne pathogens for microbial food safety. *Current Opinion in Biotechnology* 49:224–9. doi:10.1016/j.copbio.2017.11.002.

Allard, M. W., Strain, E., Melka, D., et al. 2016. Practical value of food pathogen traceability through building a whole-genome sequencing network and database. *Journal of Clinical Microbiology* 54(8):1975–83. doi:10.1128/jcm.00081-16.

Anderson, N. L. and Anderson, N. G. 1998. Proteome and proteomics: New technologies, new concepts, and new words. *Electrophoresis* 19(11):1853–61.

Ashton, P. M., Nair, S., Peters, T. M., et al. 2016. Identification of *Salmonella* for public health surveillance using whole genome sequencing. *Peer Journal* 4:e1752. doi:10.7717/peerj.1752.

Callahan, J. H., Shefcheck, K. J., Williams, T. L. and Musser, S. M. 2006. Detection, confirmation, and quantification of staphylococcal enterotoxin B in food matrixes using liquid chromatography–mass spectrometry. *Analytical Chemistry* 78(6):1789–800.

Canto, A. C. V. C. S., Suman, S. P., Nair, M. N., et al. 2015. Differential abundance of sarcoplasmic proteome explains animal effect on beef *Longissimus lumborum* color stability. *Meat Science* 102:90–98.

Capozzi, F. and Bordoni, A. 2013. Foodomics: A new comprehensive approach to food and nutrition. *Genes and Nutrition* 8(1):1–4. doi:10.1007/s12263-012-0310-x.

Choi, Y. E., Butchko, R. A. and Shim, W. B. 2012. Proteomic comparison of *Gibberella moniliformis* in limited-nitrogen (fumonisin-inducing) and excess-nitrogen (fumonisin-repressing) conditions. *Journal of Microbiology and Biotechnology* 22(6):780–7. doi:10.4014/jmb.1111.11044.

Cifuentes, A. 2009. Food analysis and foodomics. *Journal of Chromatography A* 1216(43):7109. doi:10.1016/j.chroma.2009.09.018.

Cui, Y.-F., Zhao, G.-M., Li, M.-Y. and Huang, X.-Q. 2013. Application of proteomics to understand the molecular mechanisms behind meat quality. *Packaging and Food Machinery* 2.

D'auria, E., Agostoni, C., Giovannini, M., et al. 2005. Proteomic evaluation of milk from different mammalian species as a substitute for breast milk. *Acta Paediatrica* 94(12):1708–13.

Doyle, S. 2011. Fungal proteomics: From identification to function. *FEMS Microbiology Letters* 321(1):1–9.

Fagerquist, C. K., Zaragoza, W. J., Sultan, O., et al. 2014. Top-down proteomic identification of Shiga toxin 2 subtypes from shiga toxin-producing *Escherichia coli* by matrix-assisted laser desorption ionization–tandem time of flight mass spectrometry. *Applied and Environmental Microbiology* 80(9):2928–40.

FAO. 1996. *World food summit.* Rome. http://www.fao.org/wfs/index_en.htm.

FAO. 2016. *Application of whole genome sequencing (WGS) in food safety management.* http://www.fao.org/documents/card/en/c/61e44b34-b328-4239-b59c-a9 e926e327b4/

Gao, X., Wu, W., Ma, C., et al. 2016. Postmortem changes in sarcoplasmic proteins associated with color stability in lamb muscle analyzed by proteomics. *European Food Research and Technology* 242(4):527–35.

Gaso-Sokac, D., Kovac, S. and Josic, D. 2010. Application of proteomics in food technology and food biotechnology: Process development, quality control and product safety. *Food Technology and Biotechnology* 48(3): 284–395.

Goodwin, S., McPherson, J. D. and McCombie, W. R. 2016. Coming of age: Ten years of next-generation sequencing technologies. *Nature Reviews Genetics* 17(6):333–51. doi:10.1038/nrg.2016.49.

Hamelin, M., Sayd, T., Chambon, C., et al. 2006. Proteomic analysis of ovine muscle hypertrophy. *Journal of Animal Science* 84(12):3266–76.

Holder, A., Thienel, K., Klaiber, I., Pfannstiel, J., Weiss, J. and Hinrichs, J. 2014. Quantification of bio- and techno-functional peptides in tryptic bovine micellar casein and β casein hydrolysates. *Food Chemistry* 158:118 24.

Hyötyläinen, T., Bondia, P., Isabel and Orešič, M. 2013. Lipidomics in nutrition and food research. *Molecular Nutrition and Food Research* 57(8):1306–18.

Illikoud, N., Gohier, R., Werner, D., et al. 2019. Transcriptome and volatilome analysis during growth of *Brochothrix thermosphacta* in food: Role of food substrate and strain specificity for the expression of spoilage functions. *Frontiers in Microbiology* 10:2527. doi:10.3389/fmicb.2019.02527.

Iqbal, A. and Ateeq, N. 2013. Effect of processing on the detectability of peanut protein by ELISA. *Food Chemistry* 141(3):1651–4.

Iqbal, A., Shah, F., Hamayun, M., et al. 2016. Allergens of *Arachis hypogaea* and the effect of processing on their detection by ELISA. *Food and Nutrition Research* 60 (1):28945.

Iqbal, A., Shah, F., Jamal, Y., et al. 2018. Detection of food allergens by ELISA and other common methods. *Fresenius Environmental Bulletin* 27:8340.

Jackson, B. R., Tarr, C., Strain, E., et al. 2016. Implementation of nationwide real-time whole-genome sequencing to enhance listeriosis outbreak detection and investigation. *Clinical Infectious Diseases* 63(3):380–6. doi:10.1093/cid/ciw242.

Jagadeesan, B, Gerner-Smidt, Peter, A, Marc W., et al. 2019. The use of next generation sequencing for improving food safety: Translation into practice. *Food Microbiology* 79:96–115. doi:org/10.1016/j.fm.2018.11.005.

Jensen, H. B., Poulsen, N. A., Andersen, K. K., et al. 2012. Distinct composition of bovine milk from jersey and holstein-friesian cows with good, poor, or noncoagulation properties as reflected in protein genetic variants and isoforms. *Journal of Dairy Science* 95(12):6905–17.

Joensen, K. G., Scheutz, F., Lund, O., et al. 2014. Real-time whole-genome sequencing for routine typing, surveillance, and outbreak detection of verotoxigenic *Escherichia coli*. *Journal of Clinical Microbiology* 52(5):1501–10. doi:10.1128/JCM.03617-13.

Köser, C. U., Ellington, M. J., Cartwright, E. J. P., et al. 2012. Routine use of microbial whole genome sequencing in diagnostic and public health microbiology. *PLoS Pathogens* 8(8):e1002824. doi:10.1371/journal.ppat.1002824.

Lametsch, R., Roepstorff, P. and Bendixen, E. 2002. Identification of protein degradation during post-mortem storage of pig meat. *Journal of Agricultural and Food Chemistry* 50(20):5508–12.

Legako, J. F., Dinh, T. T. N., Miller, M. F. and Brooks, J. C. 2015. Effects of USDA beef quality grade and cooking on fatty acid composition of neutral and polar lipid fractions. *Meat Science* 100:246–55.

Leitner, A., Castro-Rubio, F., Marina, M. L. and Lindner, W. 2006. Identification of marker proteins for the adulteration of meat products with soybean proteins by multidimensional liquid chromatography– tandem mass spectrometry. *Journal of Proteome Research* 5(9):2424–30.

Li, Q., Zhao, Y., Zhu, D., et al. 2017. Lipidomics profiling of goat milk, soymilk and bovine milk by UPLC-Q-exactive orbitrap mass spectrometry. *Food Chemistry* 224:302–9.

Marchand, J., Martineau, E., Guitton, Y., Le Bizec, B., Dervilly-Pinel, G. and Giraudeau, P. 2018. A multidimensional ^{1}H NMR lipidomics workflow to address chemical food safety issues. *Metabolomics* 14(5):60.

Nguyen, D. T. L., Van Hoorde, K., Cnockaert, M., De Brandt, E., Aerts, M. and Vandamme, P. 2013. A description of the lactic acid bacteria microbiota associated with the production of traditional fermented vegetables in Vietnam. *International Journal of Food Microbiology* 163(1):19–27.

Piras, C., Roncada, P., Rodrigues, P. M., Bonizzi, L. and Soggiu, A. 2016. Proteomics in food: Quality, safety, microbes, and allergens. *Proteomics* 16(5):799–815.

Quainoo, S., Coolen, J. P. M., van Hijum, S., et al. 2017. Whole-genome sequencing of bacterial pathogens: The future of nosocomial outbreak analysis. *Clinical Microbiology Reviews* 30(4):1015–63. doi:10.1128/CMR.00016-17.

Quainoo, S., Coolen, J. P. M., van Hijum, S., et al. 2018. Correction for Quainoo et al., "whole-genome sequencing of bacterial pathogens: The future of nosocomial outbreak analysis". *Clinical Microbiology Reviews* 31(1). doi:10.1128/CMR.00082-17.

Řezanka, T., Matoulková, D., Benada, O. and Sigler, K. 2015. Lipidomics as an important key for the identification of beer-spoilage bacteria. *Letters in Applied Microbiology* 60(6):536–43.

Roessner, U. and Bowne, J. 2009. What is metabolomics all about? *Biotechniques* 46(5):363–5.

Sanmartín, E., Arboleya, J. C., Iloro, I., Escuredo, K., Elortza, F. and Moreno, F. J. 2012. Proteomic analysis of processing by-products from canned and fresh tuna: Identification of potentially functional food proteins. *Food Chemistry* 134 (2):1211–9.

Sedgwick, H., Caron, F., Monaghan, P. B., Kolch, W. and Cooper, J. M. 2008. Lab-on-a-chip technologies for proteomic analysis from isolated cells. *Journal of the Royal Society Interface* 5(S2):S123–30.

Spener, F., Lagarde, M., Géloên, A. and Record, M. 2003. What is lipidomics? *European Journal of Lipid Science and Technology* 105(9):481–2.

Wang, G., Ma, F., Chen, X., et al. 2018. Transcriptome analysis of the global response of *Pseudomonas fragi* NMC25 to modified atmosphere packaging stress. *Frontiers in Microbiology* 9(1277). doi:10.3389/fmicb.2018.01277.

Wang, Z., Gerstein, M. and Snyder, M. 2009. RNA-seq: A revolutionary tool for transcriptomics. *Nature Reviews. Genetics* 10(1):57–63. doi:10.1038/nrg2484.

WHO. 2015. WHO estimates of the global burden of foodborne diseases: Foodborne disease burden epidemiology reference group 2007–2015. Geneva. http://apps.who.int/iris/bitstream/10665/199350/1/9789241565165_eng.pdf).

Zhang, M.-X., Wang, X.-C., Liu, Y., Xu, X.-L. and Zhou, G.-H. 2012. Isolation and identification of flavour peptides from puffer fish (*takifugu obscurus*) muscle using an electronic tongue and MALDI-TOF/TOF MS/MS. *Food Chemistry* 135 (3):1463–70.

Zhu, H., Clegg, M. S., Shoemaker, C. F. and Wang, S. C. 2013. Characterization of diacylglycerol isomers in edible oils using gas chromatography–ion trap electron ionization mass spectrometry. *Journal of Chromatography A* 1304:194–202.

CHAPTER **6**

Proteomics Applications in Postharvest and Food Science

S. Prasanna Raghavender, C. Anoint Yochabedh,
and Radhakrishnan Preetha

CONTENTS

6.1 INTRODUCTION

Proteomics is the study of proteins present in an organism, which also involves largescale identification, characterization, and analysis of the proteins. As proteins are involved in all biological functions, a comprehensive analysis provides information regarding their function, structure, and interaction within the cell (Twyman 2005). Proteomics applies in different fields such as drug development, plant breeding, characterization of microbes, extraction of metabolites from plants, analysis of genetically modified organisms, and so on. Proteomics has a wide range of applications in various fields for the following purposes: posttranslational modification, protein–protein interactions, protein sequencing, profiling of protein, and proteome mining (Rain et al. 2001; Graves and Haystead 2002).

Postharvest processing is performed to extend the shelf life of harvested food crops immediately after harvest. Processing of fruits and vegetables immediately after harvest helps in to reduce losses, as they have limited shelf life (Pedreschi et al. 2009). Postharvest loss occurs mainly due to inadequate storage and poor management, which in turn causes a decline in production. Proteomics is an effective method as it not only increases production but also assures food security. Protein biomarkers are useful to detect the disorders in the crops and prevent loss during storage due to microbial diseases (Pedreschi et al. 2009). These biomarkers also predict the shelf life and other characters of the food crops during storage.

Food spoilage has been a common issue since olden times. With the help of these proteomic tools, the quality of the food material can be defined by identifying the food pathogen. It is also used as a labeling tool to prevent microbial spoilage. Proteomics plays an essential role in food processing, food authentication, quality control, traceability, food safety, and nutritional aspects. Demand for healthier and safer foods is always increasing; through proteomics, safe food products can be developed. The food product quality depends on the raw materials. Hence raw materials can be tracked before processing, and safe food products can be prepared by this method. Foodborne diseases caused by microorganisms and the toxins produced by them can also be detected through this technique. The applications of proteomics are further explained in Figure 6.1.

Proteomics finds its importance throughout food processing, starting from the validations of the received raw materials. Specialized quantitative tools such as isotope labeling (Clifton et al. 2009) are also used as quality markers for determining freshness in meat and diagnosing disease in livestock. Different types of proteins present in milk can be characterized and studied using this method. The addition of natural additives to increase growth in farm animals can also be detected through proteomic technology.

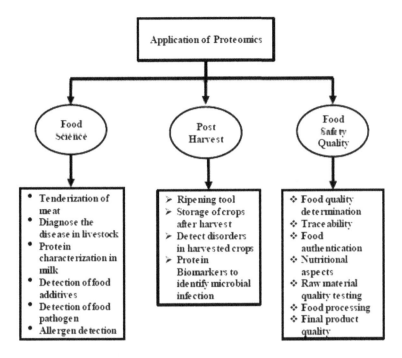

FIGURE 6.1   Applications of proteomics in food science, postharvest, food safety, and quality. (Reproduced, adapted, and modified from Gaso-Sokac et al. 2010).

Genomic expressions of genetically modified crops can also be viewed through this method. The quality of seafood can be determined using proteomics, and, similarly, allergens in seafood can be identified using proteomics. In wine processing, spoilage caused in the grapes can be identified by these methods. The use of proteomics the field of food science is explained in detail in this chapter.

6.2 PROTEOME ANALYSIS IN FOOD SAMPLES

A group of proteins that an organism expresses is called the proteome. It is used to define the various types of proteins formed at a selected time (Twyman 2005). A proteome is defined in words of the assembly, affluence, sequence, localization, modification, interlinkage, and biological function. The analysis requires a broad spectrum of technologies and methods for data integration and data mining (Twyman 2005).

6.2.1 Extraction of Food Protein

The extraction of proteins from plant sources is a huge undertaking, as plants contain a considerable wall of fibrous material. On the other hand, plant cells consist of vacuoles with watery contents which makes the extraction process difficult compared to bacterial or animal cells, and the resultant yield of protein is also meager (O'Farrell et al. 1977). There are different methods for extracting protein from the plant source without disturbing the other components (phenols, terpene, carotene, and other phytochemical compounds) (Carpentier et al. 2008). Those methods include 2D gel electrophoresis, liquid chromatography, and mass spectrometry (MS). In MS, there are two ways–one is gel-based and the other is a gel-free approach (Ho et al. 2006).

6.2.2 Proteomics Tools

6.2.2.1 Chromatography

Chromatography is used to separate proteins in the given food sample based on the differential distribution between the mobile phase and stationary phase. The proteins are separated using the chromatographic method based on various properties such as affinity, molecular weight, and ionic interactions (Ly and Wasinger 2011). Commonly used chromatographic techniques for protein purification are ion exchange, affinity chromatography, size exclusion chromatography, and reverse phase high-performance chromatography.

6.2.2.2 Enzyme-Linked Immunosorbent Assay (ELISA)

ELISA is one of the proteomic tools used to identify protein, peptide, hormones, antibodies, and antigens (Raghavendra and Pullaiah 2018). It is helpful in the detection of antigens and allergens in the food (Watabe et al. 2016). The ELISA technique is based on antigen–antibody interaction, and the detection is usually done with the help of an enzyme and a substrate. The two common enzymes used are alkaline phosphatase (AP) and horse radish peroxidase (HRP). Among these, HRP is commonly used because of its smaller molecular size. Enzyme–substrate interactions produce fluorescence, which can be detected (Buyukkoroglu et al. 2018). There are 4 varieties of ELISA, and they are indirect ELISA, sandwich ELISA, competitive ELISA, and reverse ELISA (Buyukkoroglu et al. 2018).

6.2.2.3 Western Blot Analysis

Western blot analysis can be used to identify and track definite protein or peptide from a tissue. The proteome which has to be determined should specifically bind with a particular antibody. This tool can detect a particular protein from a complex mixture because of its specificity to antibodies. Western blot analysis provides details about the expression and the size of the isolated protein (Hirano 2012). The working procedure for Western blot includes separation, transfer, and antibody interaction.

Separation is carried out using SDS-PAGE. This is a gel electrophoresis procedure that converts peptides into smaller fragments. Initially, the peptides or proteins should be denatured, and then treated with SDS, which provides a net negative charge to the whole peptide. Then this is allowed to pass through the PAGE, which causes the entrapment of these fragments (Nicholas and Nelson 2013). Then, the entrapped fragments are transferred to a membrane (usually nitrocellulose). It is treated with the primary antibody first, which is specific to the peptide or the sample. This, in turn, is treated with a secondary antibody, which binds with the primary antibody, which is used to detect the bands (Tuzmen et al. 2018).

6.2.2.4 Protein Microarray

Protein characterization can be quickly done through protein microarray. Protein microarray is otherwise called protein chips, where assays are performed with the help of a small number of purified proteins (Hall et al. 2007). These assays are classified under different categories favoring high throughput protein studies within a single experiment. They are obtained by immobilizing the protein sample onto a microscopic slide with the help of a spotter. The immobilized protein is used as a probe to identify different functions and activities. The fluorescence or radiolabeling methods are commonly used to measure the resulting signals. There are different types of microarrays available for the study of biochemical activities of the protein. Parameters, such as protein expression levels, specificities, and their binding affinity, can be determined with the help of an analytical microarray (Shruthy et al. 2016). Today these microarrays are primarily used for clinical diagnosis, as selected target proteins alone can be detected in antibody microarray. The relation between type of diet and gene expression in individuals is an unexploited area of research in human nutrition (Hall et al. 2007). Functional protein microarray is helpful in various fields, as it is used as a marker for the identification of multiple diseases, screening of pharmaceutical products, and toxicants identification. In reverse phase protein microarray, protein samples are immobilized. Secondary antibodies like fluorochrome are added to achieve high fluorescence signals, and they are detected by fluorescence or colorimetric assays (Hall et al. 2007; Krishnan and Davidovitch 2015). This fluorescence defines the binding affinity of the protein sample against the target protein. In food safety, microarray-based sensors are developed for detection of foodborne pathogens and food contaminants (Hall et al. 2007).

6.2.2.5 Gel-Based Approaches

In gel-based methods for protein purification, protein extract is collected, and the separation process is done using gel electrophoresis. When the electric field is passed, negatively charged particles travel toward the anode, and the process ends when they reach the end of the gel. Sodium dodecyl sulfate polyacrylamide gel electrophoresis (SDS-PAGE) is the most widely used approach in gel-based proteomics for protein disjunction (Buyukkoroglu et al. 2018). In this method, the protein is denatured, and a uniform negative charge is provided with help of an anionic detergent (SDS) (Cerny et al. 2013). Sodium dodecyl gel electrophoresis eliminates the intrinsic nature of the protein. During SDS-PAGE the primary structure of the protein alone is retained and is separated based on molecular mass. Then it is visualized by staining with Coomassie blue (Neuhoff et al. 1990), silver stain (Blum et al. 1987), radiolabeling (Patton 2002), and fluorescence (Chevalier et al. 2004). This method can determine the molecular weight of the protein sample. Protein samples with unspecified molecular weight can also found by comparison with known molecular weight.

The above method can be done in a single dimension (1D) or in two dimensions (2D), the latter is the most common technique in proteomics. In the gel-based method, isoelectric point PI and molecular weight are the two significant factors for the separation process (Buyukkoroglu et al. 2018). Isoelectric focusing separates protein based on the difference in their charges and surrounding pH (Grog et al. 2000). When 2D gel electrophoresis is used along with MS, more than 4000 proteins can be identified (Imin et al. 2001).

6.2.2.6 Mass Spectrometry (MS)

Before being injected into the mass spectrometer, the protein should be digested using protease enzymes. Each protease enzyme has its own specific cleaving site through which the desired peptide can be digested according to the specification (Figure 6.2). MS consists of the following components: an inlet device, to allow the analyte to the ionizing source; an ion source is used to produce ions; after ionization the analytes are segregated

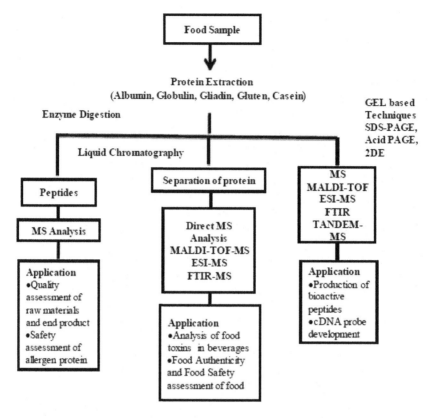

FIGURE 6.2 Outline of protein identification through MS analysis method. (Reproduced, adapted, and modified from O' Farrel (1975) and Klose (1975).

according to mass to charge ratio (m/z) with the help of an analyzer; and a detector to quantify ions.

6.2.2.6.1 Ionization Techniques Electron Spray Ionization (ESI) and Matrix-Assisted Laser Desorption Ionization (MALDI) are the two easy ionizing techniques that are employed primarily for ionization of protein (Fenn et al. 1989). The solvent mixture is used to dissolve the sample; under atmospheric pressure, the solution is atomized, and a charge on the droplet's surface is generated by providing a high electrical field. Droplets instantly become much smaller by solvent vaporization and introduced into an analyzer. Since it is generating multiple charged ions, electrospray is tremendously utilized for precise mass determination, especially for heat sensitive, larger molecular mass compounds such as proteins (Steen and Mann 2004).

In the MALDI method, a dry matrix mixture with the protein sample is taken; the radiation produced by a laser is absorbed by a matrix and behaves like a receptacle for the laser energy; and the ablation of the sample is reduced, and the sample is ionized. Further, the ions pass to an analyzer (Andersen et al. 1996; Steen and Mann 2004). Other soft ionization methods, such as chemical ionization (CI), are also used for ionization of protein. In CI a reagent ion X^+ reacts with analyte A and ionizes analyte ($X^+ +A \rightarrow [A+H]^+ + [X-H]^+$). In this reaction, X^+ is derived from ionized reaction gases such as methane, water, and isobutane.

6.2.2.6.2 Mass Analyzer The next step is the mass analyzer with different specifications. Commonly used mass analyses in proteomics include quadrupole (Q), ion trap (quadrupole ion trap, QIT; linear ion trap, LIT or LTQ), time-of-flight (TOF) mass analyzer, Fourier-transform ion cyclotron resonance (FTICR) mass analyzer, and so on (Han et al. 2008). The path of ions entering the radio frequency quadrupole field is selectively stabilized or destabilized using an alternating electrical field in quadrupole mass analyzer (Lane 2005). In ion traps, the ion gets trapped in a dynamic electric field, and, with respect to their m/z value, they are sequentially ejected into the detector (Steen and Mann 2004). Time of flight, as the name infers, uses time and as well as electric field required to speed up the ions at the said capability. The required time to attain the sensor is noted. Since the particles are identical, their kinetic energy is identical. So, the mass of the ions determines the velocities and reaches the detectors accordingly (Wollnik 1993). In MALDI-TOF the molecules which are separated reach the detector at different time intervals. This combined analyzer is a sensitive method which helps to detect a low quantity of sample and short measurement time. This helps to identify posttranslational modifications and their by-products (Pedreschi et al. 2010). In the Magnetic sector analyzer, the ions are generated, accelerated, and are passed through a curved track (sector) that reaches the detector. The heavier ions with high momentums are allowed to pass through the sector by increasing the magnetic field. The ions having equal centrifugal and centripetal forces will pass through the sector (Pedreschi et al. 2010). In addition to that, different hybrid mass analyses are used in proteomics (Han et al. 2008).

6.2.2.6.3 Detector A detector is vital to determine the fragmentation results. It is a transducer which records the current generated when an ion hits a surface (Lane 2005). One of the methods used commonly is electron multiplier. Individual energetic particles, such as electrons, protons, or ions, are detected with this because of its high sensitivity. In an electron multiplier before amplification, the particles should be converted into electrons. The electrodes called dynodes are used for amplification. An incident ion beam ejected out 2 electrons from the earliest dynode and they were expedited to the subsequent dynode where each caused 2 more electrons (four in all) to be expelled, which in turn are expedited to a third dynode, and so on. Thus, the signal gets amplified (Pedreschi et al. 2010).

6.2.2.7 Edman Degradation

The amino acid sequence in a peptide or protein can be determined by Edman degradation, developed by Pehr Edman. In the Edman degradation reaction, amino acids are isolated separately and recognized as their phenylthiohydantoin derivatives (Liu et al. 2016). Phenyl isocyanate is one of the reagents used in Edman degradation. The process further proceeds by the incubation of the sample in anhydrous acids, such as trifluoroacetic acid, which breaks the linkage between successive amino acids. The initial amino acid is split as anilinothiazolinone derivative (ATZ-amino acid), and the remaining peptide chain can be injuncted and directed to the subsequent degradation cycle. Then, ATZ amino acid is converted to a phenylthiohydantoin derivative (PTH-amino acid) by extracting it with ethyl acetate. HPLC techniques are used to further separate them (Shively 2000; Berg et al. 2002; Nelson and Cox 2005). This method is not very popular since the introduction of MS, but it is still used in proteomics (Speicher et al. 2001; Miyashita et al. 2001).

6.2.2.8 Isotope-Coded Affinity Tag (ICAT)

ICAT is one of the proteomic tools used for the quantitative analysis of the proteome or peptides (Buyukkoroglu et al. 2018). Proteins or peptides containing cysteine residue are labeled with a biotin tag, and, hence, they are a chemical method of labeling. This tool is generally used to detect allergens in food (Kirsch et al. 2009). Initially, the mixture of protein is isolated, and further, it is purified as it contains cysteine residue peptides. Then, these peptides which contain cysteine residue, are labeled with ICAT reagent with light and heavy reagents. Next, these tagged peptides are separated using affinity chromatography with avidin. ICAT works by first purifying the peptides according to the target, then tagging them with ICAT reagents holding two isotopes, C12 and C13, where both are identical chemically. In the final step, LC-MS/MS is used to measure the C12 and C13 labeled-cysteine holding peptides (Gygi et al. 1999).

6.2.2.9 Stable Isotope Labeling by Amino Acids in Cell Culture (SILAC)

SILAC is one of the proteomics tools for the quantification of proteins. This is carried out by metabolic labeling of the protein with a specific isotope; then it is examined using MS (Ong and Mann, 2007). The working principle includes the complete replacement of amino acid, which is present in the peptide with the nucleus of a stable isotope and this is called a "labeling amino acid" (Buyukkoroglu et al. 2018).

6.2.2.10 Isobaric Tag for Relative and Absolute Quantification (iTRAQ)

iTRAQ is a quantification tool that helps to quantify the peptide by labeling them with isobaric tags (Wiese et al. 2007). In this technique the primary amines of peptides and proteins from samples are initially treated with one of the isobaric tags (iTRAQ reagents) and labeled. There are at present two commonly used reagents, 4-plex and 8-plex. These reagents have the same mass, but they differ from each other, having a scattering of heavy isotopes surrounding its structure. Initially, the samples are treated with tags, and then they are pooled together. Then the pooled samples are quantified with LC-MS/MS (Unwin 2010).

6.3 APPLICATION OF PROTEOMICS IN POSTHARVEST TECHNOLOGY

Proteomics help in the reduction of loss of fruits and vegetables that occurs during postharvest processing in the following ways Sharma (2010):

1. Identification and verification of gene product of a specific trait
2. Identifying the markers to verify initial quality and determination of the remaining shelf life
3. Diagnosing physiological and postharvest pathogen issues

6.3.1 Ripening of Fruits

Ripening is a critical end step in fruit processing. This involves several variations such as aggregated biosynthesis of ethylene, smoothening of the cell wall, deterioration of chlorophyll, and synthesis of pigments (Pedreschi et al. 2013). As mentioned above, climacteric fruit ripens after picking from trees, causing increased respiration rate and synthesis of ethylene. A proteome study was performed to explain the development and ripening of pericarps of the cherry tomato by Faurobert et al. (2007). It was asserted that when the stress metabolism increased, the protein levels related to above-mentioned carbon compounds also increased. On the other hand, in the case of non-climacteric fruit ripening they are independent of ethylene production. Grapes are one of the essential categories of fruit in wine production, and proteomics studies were reported on grape skin during ripening (Deytieux et al. 2007). This study proved that the defense proteins like chitinase and thaumatin-like proteins, increased during ripening.

6.3.2 Maize Development

Maize (*Zea mays*) is a commonly produced crop because of its high nutritive value (Pechanova et al. 2017). The maize leaf's chloroplasts are transformed as specific bundle sheath (BS) and mesophyll. A proteomics study helped to differentiate photosynthetic activity between mesophyll and BS cells (Majeran et al. 2005). This research proved that enzymes from the Calvin cycle and lipid metabolism were seen in mesophyll stroma were the same as enzymes that help in the synthesis of starches that are found in BS cells. Another application of proteomics is in maize kernels, as it is the main component of maize. Endosperm in the maize kernel has starch as the main constituent and plays a role in germinating seeds. They play a role in the release of actins tubulins and other cell division/detoxification dependent proteins. The high expression of protease indicates an increase in protein, causing a switch from growth to storage (Mechin et al. 2007). The reproduction process involves pollen germination, and pollen tube growth in flowers was studied using proteomics. In mature maize pollen, the proteins present are mainly for tube wall modification, energy metabolism, posttranslational modifications, and degradations. At the same time, enzymes, such as pectin methyl esterase and inorganic pyrophosphates, are involved in the growth of the pollen tube, hence the stages of reproduction were differentiated in maize using proteomic study (Zhu et al. 2007).

6.3.3 Perishable Fruits

Fruits which are undergoing natural decay are called perishable fruits. There are different types of stresses which cause deterioration in fruits. Low and high temperature plays a vital role in postharvest changes, and high-temperature treatment is usually used before low-temperature storage for protection (Pedreschi 2017). The albedo effect of *Murcott tangor* after storing it at 4°C for two weeks showed an increase in the cysteine proteinase and a reduction in ascorbate peroxides (Lliso et al. 2007). Cai et al. (2014) examined the changes in *Vitis labruscana* when stored for 50 days at 95% relative humidity and 2°C. The enzymes which are associated with glycolysis and the citric acid cycle were down-regulated, whereas, heat shock proteins, antioxidant proteins, cell wall degrading proteins, and proteasomes were up-regulated. Dehydration is another major cause of during transpiration. It is essential to minimize this loss of water to maintain the commodity (Pedreschi 2017). One of the most important regulators for dehydration in plants is abscisic acid (Danquah et al. 2013). The changes of proteome in the albedo of *Murcott tangor* at a high relative humidity (99%) and low humidity (60%) for 15 days were determined (Lilso et al. 2007). Synthesis of cysteine protease was found because of water stress. The proteomics study in the above report concluded that changes in the protein content of albedo of the citrus fruit under storage is related to the programmed cell death.

6.3.4 Induced Resistance (IR)

Synthetic fungicides have traditionally been used to protect fruits from postharvest diseases. Though they were helpful, fungicides cause several issues, such as resistance and toxicity. This issue is generally overcome by the plant's defense system, called induced resistance, in combination with fungicides. Two types of IR have been identified. They are systemic acquired resistance (SAR), induced by specific pathogens and infection and induced systemic resistance (ISR), caused by certain anabolic nonpathogenic microbes. ISR defense in fruits is jasmonate and ethylene dependent whereas it is independent of salicylic acid (SA). ISR is not correlated with pathogen related plant protein (PR) gene expression and this is induced by microbial, physical, and chemical agents (Van Loon 1997).

6.3.4.1 Microbial Agents

Application of the antagonistic property of yeast to prevent fungi disease has gained attention in postharvest disease control (Chan and Tian 2006). The role of antagonistic yeast in inducing resistance in fruit was also reported in the literature (Chan and Tian 2005). Using quadrupole TOF tandem mass spectrometer, 19 proteins were identified in harvested peaches after treating them with antagonistic *Pichia membranefaciens*. The identified proteins included antioxidant related proteins, stress responsive, and energy pathway proteins. It was suggested that resistance in peach fruits was induced by *Pichia membranefaciens* and also proved the involvement of antioxidants, PR proteins, and enzymes related with sugar metabolism (Chan et al. 2007).

6.3.4.2 Heat Treatment

Heat is one of the important postharvest treatments. Heat storage was reported to decrease the occurrence of disease and its severity (Terry and Joyce 2004). In a previous study, the analysis of 2D-DIGE revealed that the peach fruits expressed 52 proteins after heat treatment or after transfer to 20°C when compared to the fruits kept at 20°C. Moreover, most of the expressed proteins (around 93%) have an important function in plant metabolisms such as defense mechanisms. In addition to that, few proteins were identified as heat shock proteins (HSP) in heated peach fruits which gained resistance across some chilling injury symptoms (Lara et al. 2009). Zhang et al. (2011) spotted thirty protein spots after heat treatment in peach fruits. These spots included proteins with different roles.

6.3.4.3 Chemical Agents

Hormones such as salicylic acid (SA) and oxalic acid (OA) are released and cause IR in plants and fruit. SA is released as a signaling component and activates several plant defense mechanisms (Durner et al. 1997). This further protects from pathogens and activates system acquired resistance (SAR) in certain plants and harvested fruits (Yao and Tian 2005). Research conducted by Chan et al. (2007), in which thirteen SA-induced proteins were recognized, indicates that their roles were predominantly involved in energy pathway and they acted as an antioxidant in the peach fruits. OA is vital in the activation of SAR, apoptosis, and stress response (Kim et al. 2008). It is also involved in the extension of shelf life after harvest. Reduced ethylene production and delayed softening of plums were found when they are subjected to OA. After OA treatment, during storage, inhibition of flesh reddening and anthocyanin production were evident in plum fruits (Wu et al. 2011) OA treatment influenced metabolism, and it was studied using proteomics tool. For example, after OA treatment, twenty-five proteins which were expressed differentially were identified in jujube fruit (Wang et al. 2009).

6.4 PROTEOMICS IN FOOD SCIENCE

Through proteomics, it is possible to alter specifications without introducing any other effects. Food pathogens can be detected using proteomics technology, thereby decreasing the rate of foodborne illness. The quality of both animal and plant-origin foods can be found out quickly with the help of specific biomarkers. This helps in assuring the quality and freshness of the food product. It has particular applications in different types of food products (Colgrave 2017).

6.4.1 Proteomics in the Meat Industry

The necessary and essential steps for proteomics are isolation and fast analysis of the protein source. 2-DGE is the most widely used method to separate protein samples according to their isoelectric point and molecular weight (Vercauteren et al. 2007). DIGE and tools such as MS are used for sample labeling and accurate identification of proteins, respectively. MALDI and ESI are used for detection based on ionization techniques. Mass spectrum-based methods are used in the quantification of meat and its tenderness, and this method is also used to identify biomarkers in order to diagnose possible diseases in them (Schiess et al. 2009).

6.4.1.1 Proteomics in Meat Quality Detection

The common factors affecting the meat-producing industries include the breed, genetic traits, feed provided, and sex of the animal. Alterations in those factors might result in the deterioration of the quality of meat, which can be identified using proteomics tools (Paredi et al. 2012). The meat quality to be tested is analyzed, and the respective biomarkers are used to predict and validate quality. If any deterioration in quality is detected or if the quality level is shallow, controlling is done with newly developed tools. Meat tenderness can be identified with the help of biomarkers. These markers are involved in the identification of different biological processes like apoptosis, the structure of myofibril, heat shock proteins, breaking down of proteins, etc. As of now, tenderness is difficult to calculate by comparison through many results due to the metabolic reactions taking place within the products (Wu et al. 2015).

6.4.2 Proteomics in Poultry

Heat stress is a commonly faced problem in the poultry fields, where heat makes the roosters and chickens go infertile. 2D electrophoresis analysis revealed acute heat stress, which impairs protein translation. Proteomics has brought in recent findings that can help in the breeding of chickens even under high heat stress. Biomarkers are used to detect the natural additives, which are added to increase growth, muscle metabolic rate, to select the feed efficiency for the herd, and also to protect against microbial infections (Zhang et al. 2015).

6.4.3 Proteomics in Milk

Biomarkers help to increase the levels of lactation in cows, thereby increasing dairy production. The success of reproduction in the farms has been achieved either naturally or artificially through the help of MS-based investigation of neuropeptide patterns (Colgrave 2017). Proteomics also helps in the detection of the addition of adulterants in milk either by quantitative method (Di Girolamo et al. 2014; Bernardi et al. 2015) or by semi-quantitative methods (Camerini et al. 2016). Early lactation of cows is one of the drawbacks as it is characterized by negative energy and availability of less nutrition, which fails to meet the energy requirement of the lactating cow. Proteomics can solve this because plasma proteomics showed that ketosis was due to inflammation, and it was analyzed through 2DE and iTRAQ (Yang et al. 2012). Many other diseases, including fatty liver and difficulties in lactating animals, were detected and treated to increase production.

6.4.4 Proteomics in Genetically Modified (GM) Crops

Genetic modification of crops is one of the recent advancements in the field of food science that produces food crops using specific traits that are resistant to insects, tolerant to herbicides, and other environmental conditions. There are various methods to characterize GM crops, for example transcriptome profiling; where changes in genomic expressions in GM crops such as maize, rice, and barley were identified (Coll et al. 2008; Kogel et al. 2010; Montero et al. 2011). Proteomic analysis methods such as 2DE and MALDI-TOF MS were used for the transgenic proteins (Cry IA(b)/cry IA(b)) analysis in GM crops. Proteomics is an advanced tool, besides transcriptome, immobilized pH gradients and MS, that is used to study the protein–protein interaction and evaluate the effect of proteome in cash crops. Proteins play a major role in metabolism and cell development. In addition to that, they have played a role as toxins, antinutrients, and allergens in plants. So, proteomics study of GM crops ensures the safety of the product for consumption. Maize crop, which is genetically modified, is the world's most commercialized crop (Albo et al. 2007). An insect-resistant maize, MON810, was produced by introducing gene (cry1Ab) from *Bacillus thuringiensis*. Six differentially expressed proteins in GM maize were identified due to transgene.

6.4.5 Applications in Seafood

Seafood is capable of causing anaphylaxis reactions, as the human immune system identifies the proteins present in those foods as foreign particles (Almeida et al. 2018). Marine foods, such as crabs, lobsters, shrimps, oysters, sea cucumber, and squid, can cause such allergic reactions in humans. Proteomics helps to detect allergens and to evaluate the quality of fish. In aquaculture industries, there are chances of health hazards; to prevent them, FAO, along with the United Nations, has stabilized codes for practice (FAO 2012) for handling fresh and frozen food. In recent years, a seafood protein separation study was done using the 1DE or 2DE method for seafood. Immunoblotting is used for processing the sample of allergic patients and N-terminal sequencing by the Edmund degradation process is also reported (Almeida et al. 2018). Quantification of allergens can be done with the help of ELISA with specific antibodies, and limited fish allergens were quantified by Triple quadrupole mass spectrometer. In aquaculture, the cultured products are affected by bacteria, fungi, and other parasites. The gel-based techniques are the best methods for health management in aquaculture (Figure 6.3).

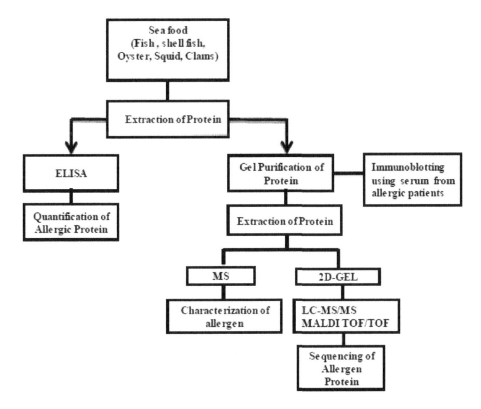

FIGURE 6.3 Proteomics tools in allergen identification in aquaculture products (Reproduced, adapted, and modified from Di Girolamo et al. 2015).

6.4.6 Applications in Wine Processing

Various approaches have been used to understand the process of wine and beer processing. Proteomics has made this simpler as it helps in studying the genome of yeast involved in fermentation and this study helps meet the quality demands of consumers. Proteomics also helps in the analysis of spoilage (Cilindre et al. 2008) caused in grapes through gel-based electrophoresis. Wine processing is affected by the seasonal changes, and a proteomic study was used to correlate *Saccharomyces* strains used for fermentation, in which results were promising (Fleet and Heard, 1993; Esteve-Zarzoso et al. 1998; Fernandez et al. 2000). Through proteomics study, one can link the proteome of yeast which is involved in the fermentation and sensory properties of wine.

6.4.7 Applications in Fermented Foods

Fermentation is one of the major metabolic processes that cause chemical changes in food substrate with the help of enzymes or microbes. The changes carried out for food material with the help of microorganisms help to enhance its acceptability and shelf life (Doyle et al. 2013). The anaerobic process through which the carbohydrate in food is catabolized into alcohol and carbon dioxide and energy is released in form of ATP is called fermentation. In research directed by Cardenas et al. (2014), they aimed to develop a method of protein extraction from high starchy foods. Pozol is an example of fermented food produced from maize, which contains high starch and low protein content. Several microbes, such as yeast, fungi, lactic, and non-lactic bacteria, have been extracted from this food. In this study 15-day-old fermented samples were used to extract the proteomes, including fungi proteins, mainly of *Aspergillus*, proteins of the *Lactobacillus* genus. In another study, authors fermented probiotic with compound wheat embryo Chinese medicine (CWECM). Then the functional proteins from the sample were isolated and analyzed using proteome-based methods. The results concluded that the macromolecules were converted into small peptides with increasing functional proteins. This made Chinese medicine more easily absorbable in the human body (Huang et al. 2019). In a work performed by Kim et al. (2017), *Weissella* species were identified from a few fermented Korean foods. *Weissella* is a hetero-fermentative lactic acid gram-negative bacterium. Korean fermented food, such as Kim chi and jeotgal, are fermented with the help of these bacteria. The authors have developed a novel rapid and accurate method of MALDI-TOF MS basis in-house database for the extraction of this species.

6.4.8 Application for Identification of Food Allergen

Food allergens are the components found in food that cause allergic responses in humans upon consumption. They are usually present in food or could be added during processing. These allergens are proteins in most cases but, in some instances, they could be other components and chemicals such as additives (Piras et al. 2016). The symptoms can be mild or severe according to the exposure and the immunity of individuals. Mild symptoms include itchiness, vomiting, diarrhea, and low blood pressure. The severe symptom is called anaphylaxis. Allergies in the human body are caused when immunoglobulin E (Ig E) detects the food allergen and binds with it. The most common food allergens are peanuts, eggs, selfish, fish, tree nuts, and so on (Marzano et al. 2020).

MS has helped to overcome several proteome assays for detecting and quantification of food allergens. This could be performed with a simple single-stage LC-MS/MS (Piras et al. 2016). Recently, multiple reaction monitoring (MRM) and parallel reaction monitoring have shown the path in the identification and quantification of food allergens (Marzano et al. 2020). Various proteomics tools have been used to detect the presence, domain, and composition of allergens. The detection of allergens can be done only by the detection of IgE levels in the bloodstream. Two types of determinants for IgE are linear, which corresponds to the 1° structure of the protein, and conformational, which corresponds to the 2° and 3° structure of the proteins (Picariello et al. 2013). The allergens can be detected in two ways with the help of proteomics tools, gel-based approaches, and gel-free approaches. As discussed previously in this chapter, gel-based methods, such as 2DGE combined with MS, are quite useful for the detection of allergens. Gel-free methods include chromatography techniques combined with MS. One such example is HPLC-MS/MS, which detects allergens by trypsinizing the proteome by IgE assay. The authors reported that the combination of 2D-MS has proven to be the best technique for the detection of allergens with proteomics tools (Piras et al. 2016).

6.4.9 Application for Food Pathogen/Spoilage Organism Detection

Proper nourishment is very much essential to maintain human health. Spoilage of food is one of the factors which deplete the quality of the food that is to be consumed. The microbes which spoil the food are called spoilage microbes, whereas the ones that cause disease in human are called pathogens. This either causes infection (where the pathogen enters the body and causes illness) or intoxication (where the toxin produced from the pathogen in food causes illness upon consumption) (Bintsis 2017).

Food spoilage pathogens can be detected easily with the help of MALDI-TOF/MS. This is commonly used because of its rapid nature, easy handling, and co-effectiveness. The advantage of this method is that even if the sample is a large molecule, it could be analyzed without fragmentation of molecules (Bohme et al. 2017). Foodborne pathogens have recently been detected by a combination of ESI-MS with simple chromatography techniques, more precisely liquid chromatography. LC-ESI-MS is more specific and efficient in the detection of specific biomolecules and pathogens. The results were comparatively efficient compared with the results obtained from MALDI-TOF/MS (Bohme et al. 2017).

6.4.10 Single Cell Proteomics

Single-cell proteins are used for various studies because of their heterogeneous nature. Proteomic technology helps in viewing more data within an individual cell. Single-cell proteomic tools are classified into two types, by measuring particular data in thousands of cells or by measuring more data in the same cell (Marr et al. 2016).

Single proteomic assay methods are classified as qualitative methods, semi qualitative methods, and quantitative methods. Qualitative methods are used to show the variation between the positive and negative target proteins (Czerkinsky et al. 1983). The enzyme linked immune spot (ELISpot) assay can be used for the detection of cytokines in a cell; it is a most responsive technique for secreted protein detection. Semi qualitative methods are used to measure protein in relative units as all the parameters in a single cell cannot be analyzed through biophysical or biochemical methods. User friendly methods such as flow cytometry are suggested when the sample availability is less and proteins that are localized are in larger amounts (Wu and Singh 2013). In the quantitative method, ELISA, which relies on the calibration curve for transforming protein signals, is used. Single cell proteomics is very efficient as large number of proteins from thousands of cells can be analyzed, toxins in food can be detected, and foodborne diseases can be controlled (Wu and Singh 2013).

6.4.11 Applications in Food Processing

Food processing is one of the essential steps in the product development. It is essential to subject the raw material to several processing treatments to obtain a finished product. Most commonly, all products require thermal treatment, at least for sterilization, either during processing or for the finished product. This has been carried out to increase the quality and nutritional attributes and lengthen the shelf life of the product. Even though this helps in several ways, the main components of the food undergo several chemical reactions (Gallardo et al. 2013). Proteins are the nutritional, functional, integral, and also fundamental components of the food. Upon heat treatment, the proteins and enzymes in food tend to misfold or unfold, causing the proteins to precipitate (Pedreschi et al. 2010).

This is overcome by implementing non-thermal treatments in food processing. A few techniques, such as High Hydrostatic pressure (HPP), electric pulse field, UV, and radiations, cause the food to be fresh with increased nutritive value and safety to consume (Pedreschi et al. 2010). High pressure, when combined with temperature, helps inactivate several enzymes. HPP causes the protein unfolding, which causes deterioration of food quality as it affects the non-covalent bonds (Rastogi et al. 2007).

6.4.12 Proteomics and Food Safety

Food safety is one of the basic requirements and major concerns for consumers. Today, more customers are demanding food which meets both nutritional and safety parameters. Several innovative techniques, such as "omics," have been implemented to maintain food safety and security. The combination of such methods with food is called "foodomics" (Jagadeesh et al. 2017). The main aim of this technique is to provide safe food for good health and to prevent disease risk factors in foods (Gallardo et al. 2013). The food industries must be authentic to provide safe food.

6.4.12.1 Food Authenticity

Food authentication is a tool that is used to verify the authenticity and originality of the food product as per the label. The authenticity of the food is maintained by several regulations that prevent makers of food products from misleading consumers. Traditionally, it was carried out morphologically, but it was challenging to differentiate species of fewer differences (Gallardo et al. 2013). With recent advancements, proteomics paved a way to overcome such issues and difficulties. Tools such as MS and others help in authenticating at a faster rate along with specificity. This is usually carried out in two phases (Gallardo et al. 2013):

Discovery phase: Here, the protein or peptide which is specific to each species is isolated. This is also called species-specific biomarkers. After isolating, this is characterized using simple tools such as 2D gel electrophoresis then further, with MS.

Target driven phase: The second step in product authenticity is the target driven phase. Here the biomarkers, isolated in the previous phase, are further analyzed using monitoring tools, such as Selective Reaction Monitoring (SRM), MRM, and Scanning Microwave Impedance Microscopy (SMIM), which are combined with MS to detect the authenticity. The advantage is it includes a combination of faster sample digestion with high intensity focused ultrasound (HIFU) and SMIM/MS (Gallardo et al. 2013).

6.4.12.2 Food Traceability

Food traceability is a field of food science that has several definitions and dimensions (Mo-Xi and Xin 2019). According to Moe (1998), food traceability is defined as the ability to track the product from the supply chain to production, storage, and sales. It provides the ability to track the site of production and intended use with the recorded data and the details on the product label. There are several contexts in which traceability of food could be used. There are two types of food traceability according to Swink et al. (2007). Manufacture–supplier food traceability is one which includes the data collection and tracking from the manufacturer and principal suppliers. This covers the raw material place of origin, grade, and technology changes. In contrast, the process of collecting as well as tracking the details of customer's needs, new trends in the market, grade, and origin of the food product is called Manufacture–buyer (MB) food traceability. It is essential to use both manufacture-supplier and MB to have updated technology and market changes, respectively.

The application of proteomics tools in MS approaches has shown the detection of variations such as species, quality, and geographical variations (Piras et al. 2016). In research conducted by Bellgard et al. (2013), the fish product evaluation was carried out to find the various species of fish of the same genus. Here, the MALDI-TOF-TOF tool was implemented with a screening. The traceability of the volatile compounds in olive oil was studied using GC-MS and GC/GC-MS (Cajka et al. 2010). Products such as honey were classified according to the protected geographical origin with the help of ICP-MS and statistical tools (Chudzinska and Baralkiewicz 2011).

6.4.12.3 Application in Food Adulteration

Food adulteration is a vast problem found throughout the world. Usually, food adulteration is something where the main ingredient of the food product which determines the quality is either removed or replaced with a lesser quality ingredient (Piras et al. 2016). The health of the consumer is ensured by providing safe and quality food. This was carried out using fat analysis with high sensitivity and specificity. In processed foods, one of the primary and commonly found adulterations lately is cross-species contamination (Piras et al. 2016).

The adulteration of horse and pork meat with beef was one of the adulterations, and a new technique was developed for the detection of specific peptides from the meat. It was reported that about 0.24% of horse or pork meat peptide was detected using MRM and MRM3 techniques (Von Bargen et al. 2014). Proteomics tools serve as a tool for the detection of chicken meat from mixed meat. Here the authors included off-gel fractionation to develop the fractions then detected using LC-ESI-MS/MS. The peptides were then measured using AQUA labeling. Through this method, the authors were able to identify about 0.5% of chicken peptides from mixed meat samples (Sentandreu et al. 2010). In another study, the difference in the expression of myosin in individual species was studied. This was carried out for meat products such as cattle, pig, turkey, chicken, goose, and duck species. The peptides were digested, and then 2D-MALDI-TOF was carried out (Montowska and Pospiech 2012). Milk and milk products are recently reported for adulteration, most commonly with low-quality milk substitutes (Piras et al. 2016). In a comparative study performed by Hinz et al. (2012), milk of different species such as bovine, caprine, equine, camel, and buffalo were studied. ß-lactoglobulin, which is one of the major whey protein components in milk, was found in all samples except camel milk. An iTRAQ-based approach was used to acquire about 211 proteins from milk whey of cow, yak, buffalo, goat, and camel. This paved a path for the identification of adulteration of good quality milk with low quality or bovine milk (Yang et al. 2013).

6.5 CONCLUSION

To summarize, proteomics has broad applications in the area of food science and postharvest technology. This chapter gives the reader an idea about different techniques in proteomics and their applications in food science and postharvest technology. Proteomics finds usage in the determination of quality of milk, seafood, meat, and poultry. Recently, proteomics has been widely used to study GM crops. Other applications of proteomics are in fermented foods, single-cell proteins, food processing, and beverages, all described in this chapter. Providing safe food to the consumer is one of the main challenges for food technologists. Proteomics helps in providing safe food by detecting food pathogens, allergens, and adulterates. In short, this chapter explains the role of proteomics in different areas of food science including food safety and quality. Moreover, it provides a window for better understanding of proteomics applications for food technologists, thereby providing foolproof services to the consumer.

REFERENCES

Albo, A. G., Mila, S., Digilio, G., Motto, M., Aime, S. and Corpillo, D. 2007. Proteomic analysis of a genetically modified maize flour carrying the Cry1Ab gene and comparison to the corresponding wild-type. *Maydica* 52(4):443–55.

Almeida, A. M. D., Eckersall, D. and Miller, I. 2018. *Proteomics in domestic animals: From farm to systems biology*, 489. Switzerland: Springer International Publishing.

Andersen, J. S., Svensson, B. and Roepstorff, P. 1996. Electrospray ionization and matrix assisted laser desorption/ionization mass spectrometry: Powerful analytical tools in recombinant protein chemistry. *Nature Biotechnology* 14(4):449–57.

Bellgard, M., Taplin, R., Chapman, B., et al. 2013. Classification of fish samples via an integrated proteomics and bioinformatics approach. *Proteomics* 13(21):3124–30.

Berg, J. M., Tymoczko, J. L. and Stryer, L. 2002. *Biochemistry*, 1050. New York: WH Freeman.

Bernardi, N., Benetti, G., Haouet, N. M., et al. 2015. A rapid high-performance liquid chromatography-tandem mass spectrometry assay for unambiguous detection of different milk species employed in cheese manufacturing. *Journal of Dairy Science* 98(12):8405–13.

Bintsis, T. 2017. Foodborne pathogens. *AIMS Microbiology* 3(3):529–63.

Blum, H., Beir, H. and Gross, H. J. 1987. Improved silver staining of plant proteins, RNA, and DNA in polyacrylamide gels. *Electrophoresis* 8(2):93–9.

Bohme, K., Fernandez-No, I. C., Mata, P. C. and Velazquez, J. B. 2017. Proteomics of food spoilage pathogens. In: *Proteomics in food science*, eds. M. L. Colgrave, 417–31. Cambridge: Academic Press.

Buyukkoroglu, G., Dora, D. D., Ozdemir, F. and Hazel, C. 2018. Techniques for protein analysis. In: *Omics technologies and bioengineering*, eds. D. Barh and V. Azeveo, 317–51. Cambridge: Academic Press.

Cai, H., Yuan, X., Pan, J., Li, H., Wu, Z. and Wang, Y. 2014. Biochemical and proteomic analysis of grape berries (*Vitis labruscana*) during cold storage upon postharvest salicylic acid treatment. *Journal of Agricultural and Food Chemistry* 62(41):10118–25.

Cajka, T., Riddellova, K., Klimankova, E., Cerna, M., Pudil, F. and Hajslova, J. 2010. Traceability of olive oil based on volatiles pattern and multivariate analysis. *Food Chemistry* 121(1):282–9.

Camerini, S., Montepeloso, E., Casella, M., Crescenzi, M., Marianella, R. M. and Fuselli, F. 2016. Mass spectrometry detection of fraudulent use of cow whey in water buffalo, sheep, or goat Italian ricotta cheese. *Food Chemistry* 197(B):1240–8.

Cardenas, C., Barkla, B. J., Wacher, C., Olivares, L. D. and Sanoja, R. R. 2014. Protein extraction study for the proteomic study of a Mexican traditional starchy food. *Journal of Proteomics* 111:139–47.

Carpentier, S., Panis, B., Vertommen, A., et al. 2008. Proteome analysis of non-model plants: A challenging but powerful approach. *Mass Spectrometry Reviews* 27(4):354–77.

Cerny, M., Skalak, J., Cerna, H. and Brzobohaty, B. 2013. Advances in purification and separation of post-translationally modified proteins. *Journal of Proteomics* 92:2–27.

Chan, Z. and Tian, S. 2005. Interaction of antagonistic yeasts against postharvest pathogens of apple fruit and possible mode of action. *Postharvest Biology and Technology* 36(2):215–23.

Chan, Z. and Tian, S. 2006. Induction of H_2O_2-metabolizing enzymes and total protein synthesis by antagonistic yeast and salicylic acid in harvested sweet cherry fruit. *Postharvest Biology and Technology* 39(3).314–20.

Chan, Z., Qin, G., Xu, X., Li, B. and Tian, B. 2007. Proteome approach to characterize proteins induced by antagonist yeast and salicylic acid in peach fruit. *Journal of Proteome Research* 6(5):1677–88.

Chevalier, F., Rofidal, V., Vanova, P., Bergoin, A. and Rossignol, A. 2004. Proteomic capacity of recent fluorescent dyes for protein staining. *Phytochemistry* 65(11):1499–506.

Chudzinska, M. and Baralkiewicz, D. 2011. Application of ICP-MS method of determination of 15 elements in honey with a chemometric approach for the verification of their authenticity. *Food and Chemical Toxicology* 49(11):2741–9.

Cilindre, C., Jegou, S., Hovasse, A., et al. 2008. Proteomic approach to identify champagne wine proteins as modified by *Botrytis cinerea* infection. *Journal of Proteome Research* 7(3):1199–208.

Clifton, J. G., Huang, F., Kovac, S., Yang, X., Hixson, D. C. and Josic, D. 2009. Proteomic characterization of plasma-derived clotting factor VIII-von Willebrand factor concentrates. *Electrophoresis* 30(20):3636–46.

Colgrave, M. L. 2017. *Proteomics in food science: From farm to fork*, 538. Cambridge: Academic Press.

Coll, A., Nadal, A., Palaudelmàs, M., et al. 2008. Lack of repeatable differential expression patterns between MON810 and comparable commercial varieties of maize. *Plant Molecular Biology* 68(1–2):105–17.

Czerkinsky, C. C., Nilsson, L. A., Nygren, H., Ouchterlony, O. and Tarkowski, A. 1983. A solid-phase enzyme-linked ImmunoSpot (ELISPOT) assay for enumeration of specific antibody-secreting cells. *Journal of Immunological Methods* 65(1–2):109–21.

Danquah, A., de Zelicourt, A., Colcombet, J. and Hirt, H. 2013. The role of ABA and MAPK signaling pathways in plant abiotic stress responses. *Biotechnology Advances* 32(1):40–52.

Deytieux, C., Geny, L., Lapaillerie, D., Claverol, S., Bonneu, M. and Doneche, B. 2007. Proteome analysis of grape skins during ripening. *Journal of Experimental Botany* 58(7):1851–62.

Di Girolamo, F., Masotti, A., Salvatori, G., Scapaticci, M., Muraca, M. and Putignani, L. 2014. A sensitive and effective proteomic approach to identify she-donkey's and goat's milk adulterations by MALDI-TOF MS fingerprinting. *International Journal of Molecular Sciences* 15(8):13697–719.

Di Girolamo, F., Muraca, M., Mazzina, O., Lante, I. and Dahdah, L. 2015. Proteomic applications in food allergy: Food allergenomics. *Current Opinion in Allergy and Clinical Immunology* 15(3):259–66.

Doyle, M., Steenson, L. R. and Meng, J. 2013. Bacteria in food and beverage production. In: *The prokaryotes: Applied bacteriology and biotechnology*, eds. E. Rosenberg, E. F. DeLong, S. Lory, E. Stackebrandt and F. Thompson, 241–56. Heidelberg: Springer.

Durner, J., Shah, J. and Klessig, D. F. 1997. Salicylic acid and disease resistance in plants. *Trends in Plant Science* 2(7):266–74.

Esteve-Zarzoso, B., Manzanares, P., Ramon, D. and Querol, A. 1998. The role of non-*Saccharomyces* yeasts in industrial winemaking. *International Microbiology* 1(2):143–8.

Faurobert, M., Mihr, C., Bertin, N., et al. 2007. Major proteome variations associated with cherry tomato pericarp development and ripening. *Plant Physiology* 143(3):1327–46.

Fenn, J. B., Mann, M., Meng, C. K., Wong, S. F. and Whitehouse, C. M. 1989. Electrospray ionization mass spectrometry: Protein structure. *Mass Spectrometry in Molecular Sciences* 246(4926): 64–71.

Fernandez, M., Ubeda, J. F. and Briones, A. I. 2000. Typing of non-*Saccharomyces* yeasts with enzymatic activities of interest in winemaking. *International Journal of Food Microbiology* 59(1–2):29–36.

Fisheries, F. A. O. 2012. Aquaculture department. *The State of World Fisheries and Aquaculture* 1:1–53.

Fleet, G. H. and Heard, G. M. 1993. Yeast-growth during fermentation, 27–54. Harwood Academic, Lausanne.

Gallardo, J. M., Ortea, I. and Carrera, M. 2013. Proteomics in food science. In: *Foodomics: Advanced mass spectrometry in modern food science and nutrition*, ed. A. Cifuentes, 125–65. New Jersey: John Wiley and sons.

Gallardo, J. M., Ortea, I. and Carrera, M. 2013. Proteomics and its application for food authentication and food-technology research. *TrAC Trends in Analytical Chemistry* 52:135–41.

Gaso-Sokac, D., Kovac, S. and Josic, D. 2010. Application of proteomics in food technology and food biotechnology: Process development, quality control and product safety. *Food Technology and Biotechnology* 48(3):284–95.

Graves, P. R. and Haystead, T. A. 2002. Molecular biologist's guide to proteomics. *Microbiology and Molecular Biology Reviews* 66(1):39–63.

Grog, A., Obermaier, C., Boguth, G., et al. 2000. The current state of two-dimensional electrophoresis with immobilized pH gradients. *Electrophoresis* 21(6):1037–53.

Gygi, S. P., Rist, B., Gerber, S. A., Turecek, F., Gelb, M. H. and Aebersold, R. 1999. Quantitative analysis of complex protein mixtures using isotope-coded affinity tags. *Nature Biotechnology* 17(10):994–9.

Hall, D. A., Ptacek, J. and Snyder, M. 2007. Protein microarray technology. *Mechanism of Aging and Development* 128(1):161–7.

Han, X., Aslanian, A. and Yates, J. R. 2008. Mass spectrometry for proteomics. *Current Opinion in Chemical Biology* 12(5):483–90.

Hinz, K., O'Connor, P. M., Huppertz, T., Ross., R. P. and Kelly, A. L. 2012. Comparison of the principal proteins in bovine, caprine, buffalo, equine, and camel milk. *Journal of Dairy Research* 79(2):185–91.

Hirano, S. 2012. Western blot analysis. In: *Nanotoxicity*, ed. J. Reineke, 87–97. New Jersey: Humana Press.

Ho, E., Hayen, A. and Wilkins, M. R. 2006. Characterization of organellar proteomes: A guide to subcellular proteomic fractionation and analysis. *Proteomics* 6(21):5746–57.

Huang, J., Lyu, X., Liao, A., et al. 2019. Proteomics-based analysis of functional proteins after fermentation of compound wheat embryo Chinese medicine. *Grain and Oil Science and Technology* 2(3):57–61.

Imin, N., Kerim, T., Weinman, J. J. and Rolfe, B. G. 2001. Characterisation of rice anther proteins expressed at the young microspore stage. *Proteomics* 1(8):1149–61.

Jagadeesh, D. S., Kannegundla, U. and Reddy, R. K. 2017. Application of proteomic tools in food quality and safety. *Advances in Animal and Veterinary Sciences* 5(5):213–25.

Kim, E., Cho, Y., Lee, Y., et al. 2017. A proteomics approach for rapid identification of *Weissella* species isolated from Korean fermented foods on MALDI-TOF MS supplemented with an in-house database. *International Journal of Food Microbiology* 243:9–15.

Kim, K. S., Min, J. Y. and Dickman, M. B. 2008. Oxalic acid is an elicitor of plant programmed cell death during *Sclerotinia sclerotiorum* disease development. *Molecular Plant–Microbe Interactions* 21(5):605–12.

Kirsch, S., Fourdrilis, S., Dobson, R., Scippo, M. L., Rogister, G. M. and Pauw, E. D. 2009. Quantitative methods for food allergens: A review. *Analytical and Bioanalytical Chemistry* 395(1):57–67.

Klose, J. 1975. Protein mapping by combined isoelectric focusing and electrophoresis of mouse tissues. *Humangenetik* 26(3):231–43.

Kogel, K.-H., Voll, L. M., Schäfer, P., et al. 2010. Transcriptome and metabolome profiling of field-grown transgenic barley lack induced differences but show cultivar-specific variances. *Proceedings of the National Academy of Sciences of the United States of America* 107(14):6198–203.

Krishnan, V. and Davidovitch, Z. 2015. *Biological mechanisms of tooth movement, 312.* New Jersey: John Wiley and Sons.

Lane, C. S. 2005. Review mass spectrometry-based proteomics in the life sciences. *Cellular and Molecular Life Sciences* 62(7–8):848–69.

Lara, M. V., Borsani, J., Budde, C. O., et al. 2009. Biochemical and proteomic analysis of 'Dixiland' peach fruit (*Prunus persica*) upon heat treatment. *Journal of Experimental Botany* 60(15):4315–33.

Liu, Y., Wang, Z., Zhang, H., et al. 2016. A photo thermally responsive nanoprobe for bioimaging based on Edman degradation. *Nanoscale* 8(20):10553–7.

Lliso, I., Tadeo, F. R., Phinney, B. S., Wilkerson, C. G. and Talon, M. 2007. Protein changes in the albedo of citrus fruits on post-harvesting storage. *Journal of Agricultural and Food Chemistry* 55(22):9047–53.

Ly, L. and Wasinger, V. 2011. Protein and peptide fractionation, enrichment, and depletion: Tools for the complex proteome. *Proteomics* 11(4):513–34.

Majeran, W., Cai, Y., Sun, Q. and Wijk, K. J. 2005. Functional differentiation of bundle sheath and mesophyll maize chloroplasts determined by comparative proteomics. *The Plant Cell* 17(11):3111–40.

Marr, C., Zhou, J. X. and Huang, S. 2016. Single-cell gene expression profiling and cell state dynamics: Collecting data, correlating data points, and connecting the dots. *Current Opinion in Biotechnology* 39:207–14.

Marzano, V., Tilocca, B., Fiocchi, A. G., et al. 2020. A perusal of food allergens analysis by mass spectrometry-based proteomics. *Journal of Proteomics* 215: 103636.

Mechin, V., Thevenot, C., Guilloux, M. L., Prior, J. L. and Damerval, C. 2007. Developmental analysis of maize endosperm proteome suggests a pivotal role for pyruvate orthophosphate dikinase. *Plant Physiology* 143(3):1203–19.

Miyashita, M., Presley, J. M., Buchholz, B. A., et al. 2001. Attomole level protein sequencing by Edman degradation coupled with accelerator mass spectrometry. *Proceedings of the National Academy of Sciences of the United States of America* 98(8):4403–8.

Moe, T. 1998. Perspectives on traceability in food manufacture. *Trends in Food Science and Technology* 9(5):211–4.

Montero, M., Coll, A., Nadal, A., Messeguer, J. and Pla, M. 2011. Only half the transcriptomic differences between resistant genetically modified and conventional rice are associated with the transgene. *Plant Biotechnology Journal* 9(6):693–702.

Montowska, M. and Pospiech, E. 2012. Myosin light chain isoforms retain their species-specific electrophoretic mobility after processing, which enables differentiation between six species: 2DE analysis of minced meat and meat products made from beef, pork, and poultry. *Proteomics* 12(18):2879–89.

Mo-xi, S. and Xin, Y. M. 2019. Leveraging core capabilities and environmental dynamism for food traceability and firm performance in a food supply chain: A moderated mediation model. *Journal of Integrative Agriculture* 18(8):1820–37.

Nelson, D. and Cox, M. 2005. *Lehninger principles of biochemistry*, 1119. New York: W.H. Freeman and Company.

Neuhoff, V., Stamm, R., Pardowitz, I., Arold, N., Ehrhardt, W. and Taube, D. 1990. Essential problems in the quantification of proteins following colloidal staining with Coomassie brilliant blue dyes in polyacrylamide gels and their solution. *Electrophoresis* 11(2):101–17.

Nicholas, M. W. and Nelson, K. 2013. North, south, or east? Blotting techniques. *Journal of Investigative Dermatology* 133(7):1–3.

O'Farrell, P. H. 1975. High resolution two-dimensional electrophoresis of proteins. *Journal of Biological Chemistry* 250(10):4007–21.

O'Farrell, P. Z., Goodman, H. M. and O'Farrell, P. H. 1977. High-resolution two-dimensional electrophoresis of basic as well as acidic proteins. *Cell* 12(4):1133–42.

Ong, S.-E. and Mann, M. 2007. Stable isotope labeling by amino acids in cell culture for quantitative proteomics. *Quantitative Proteomics by Mass Spectrometry*, 37–52.

Paredi, G., Raboni, S., Bendixen, E., de Almeida, A. M. and Mozzarelli, A. 2012. "Muscle to meat" molecular events and technological transformations: The proteomics insight. *Journal of Proteomics* 75(14):4275–89.

Patton, W. F. 2002. Detection technologies in proteome analysis. *Journal of Chromatography. Part B* 771(1–2):3–31.

Pechanova, O. and Pechan, T. 2017. Proteomics as a tool to understand maize biology and to improve maize crops. In: *Proteomics in food science*, ed. M. L. Colgrave, 35–56. Cambridge: Academic Press.

Pedreschi, R. 2017. Postharvest proteomics of perishables. *Proteomics in Food Science* 3–16.

Pedreschi, R., Hertog, M., Lilley, K. and Sand Nicolai, B. 2010. Proteomics for the food industry: Opportunities and challenges. *Critical Reviews in Food Science and Nutrition* 50(7):680–92.

Pedreschi, R., Hertog, M., Robben, J., Lilley, K., et al. 2009. Differential protein expression of conference pear slices submitted to extreme gas concentrations. *Journal of Agricultural and Food Chemistry* 57:6977–7004.

Pedreschi, R., Lurie, S., Hertog, M., Nicolai, B., Mes, J. and Woltering, E. 2013. Postharvest proteomics and food security. *Proteomics* 13(12–13):1772–83.

Picariello, G., Mamone, G. and Addeo, F. 2013. Proteomic-based techniques for the characterization of food allergens. In: *Foodomics: Advanced mass spectrometry in modern food science and nutrition*, eds. C. Nitride and P. Ferranti, 69–99. New Jersey: John Wiley and Sons.

Piras, C., Roncada, P., Rodrigues, P. M., Bonizzi, L. and Soggiu, A. 2016. Proteomics in food: Quality, safety, microbes, and allergens. *Proteomics* 16(5):799–815.

Raghavendra, P. and Pullaiah, T. 2018. Cellular and molecular diagnostics: An introduction. In: *Advances in cell and molecular diagnostics*, eds. P. Raghavendra and T. Pullaiah, 1–32. Cambridge: Academic Press.

Rain, J. C., Selig, L., De Reuse, H., Battaglia, V., Reverdy, C., Simon et al. 2001. The protein–protein interaction map of *Helicobacter pylori. Nature* 409(6817):211–5.

Rastogi, N. K., Raghavarao, K., Balasubramaniam, V. M., Niranjan, K. and Knorr, D. 2007. Opportunities and challenges in high-pressure processing of foods. *Critical Reviews in Food Science and Nutrition* 47(1):69–112.

Schiess, R., Wollscheid, B. and Aebersold, R. 2009. Targeted proteomic strategy for clinical biomarker discovery. *Molecular Oncology* 3(1):33–44.

Sentandreu, M. A., Fraser, P. D., Halket, J., Patel, R. and Bramley, P. M. 2010. A proteomic-based approach for the detection of chicken in meat mixes. *Journal of Proteome Research* 9(7):3374–83.

Sharma, S. K. 2010. *Postharvest management and processing of fruits and vegetables: Instant notes*, 416. New Delhi: NIPA.

Shively, J. E. 2000. The chemistry of protein sequence analysis. *EXS* 88:99–117.

Shruthi, B. S., Vinodhkumar, P. and Selvamani. 2016. Proteomics: A new perspective for cancer. *Advanced Biomedical Research* 5:67.

Speicher, K. D., Gorman, N. and Speicher, D. W. 2001. N-terminal sequence analysis of proteins and peptides. *Current Protocols in Protein Science* 11(10):141.

Steen, H. and Mann, M. 2004. The ABC's (and XYZ's) of peptide sequencing. *Molecular and Cellular Biology* 5(9):699–711.

Swink, M., Narasimhan, R. and Wang, C. 2007. Managing beyond the factory walls: The effects of four types of strategic integration on manufacturing plant performance. *Journal of Operations Management* 25(1):148–64.

Terry, L. A. and Joyce, D. C. 2004. Elicitors of induced disease resistance in postharvest horticultural crops; a brief review. *Postharvest Biology and Technology* 32(1):1–13.

Tuzmen, S., Baskin, Y., Nursal, A. F., et al. 2018. Techniques for nucleic acid engineering: The foundation of gene manipulation. In: *Omics technologies and bio-engineering*, eds. D. Barh and V. Aevedo, 247–315. Cambrige: Academic Press.

Twyman, R. M. 2005. *Principles of Proteomics*, 273. Milton Park: BIOS Scientific Publishers.

Unwin, R. 2010. Quantification of proteins by iTRAQ. *Methods in Molecular Biology* 658:205–15.

Van Loon, L. C. 1997. Induced resistance in plants and the role of pathogenesis-related proteins. *European Journal of Plant Pathology* 103(9):753–65.

Vercauteren, F. G. G., Arckens, L. and Quirion, R. 2007. Applications and current challenges of proteomic approaches, focusing on two-dimensional electrophoresis. *Amino Acids* 33(3):405–14.

Von Bargen, C., Brockmeyer, J. and Humpf, H. -U. 2014. Meat authentication: A new HPLC-MS/MS-based method for the fast and sensitive detection of horse and pork in highly processed food. *Journal of Agricultural and Food Chemistry* 62(39):9428–35.

Wang, Q., Lai, T., Qin, G. and Tian, S. 2009. Response of jujube fruits to exogenous oxalic acid treatment based on proteomic analysis. *Plant and Cell Physiology* 50(2):230–42.

Watabe, S., Morikawa, M., Kaneda, M., et al. 2016. Ultrasensitive detection of proteins and sugars at the single-cell level. *Communicative and Integrative Biology* 9(1):1–30.

Wiese, S., Reidegeld, K. A., Meyer, H. E. and Warscheid, B. 2007. Protein labeling by iTRAQ: A new tool for quantitative mass spectrometry in proteome research. *Proteomics* 7(3):340–50.

Wollnik, H. 1993. Time of flight mass analyzers. *Mass Spectrometry Reviews* 12(2):89–114.

Wu, F., Zhang, D., Zhang, H., et al. 2011. Physiological and biochemical response of harvested plum fruit to oxalic acid during ripening or shelf-life. *Food Research International* 44(5):1299–305.

Wu, M. and Singh, A. K. 2013. Single-cell protein analysis. *Current Opinion in Biotechnology* 23(1):83–8.

Wu, W., Fu, Y., Therkildsen, M., Li, X. M. and Dai, R. M. 2015. Molecular understanding of meat quality through application of proteomics. *Food Reviews International* 31(1):13–28.

Yang, Y., Bu, D., Zhao, X., Sun, P., Wang, J. and Zhou, L. 2013. Proteomic analysis of cow, yak, buffalo, goat, and camel milk whey protein: Quantitative differential expression patterns. *Journal of Proteome Research* 12(4):1660–7.

Yang, Y. X., Wang, J. Q., Bu, D. P. et al. 2012. Comparative proteomic analysis of plasma proteins during the transition period in dairy cows with or without subclinical mastitis after calving. *Czech Journal of Animal Science* 57(10):481–9.

Yao, H. and Tian, S. 2005. Effects of pre- and postharvest application of salicylic acid or methyl jasmonate on inducing disease resistance of sweet cherry fruit in storage. *Postharvest Biology and Technology* 35(3):253–62.

Zhang, J., Gao, Y., Lu ,Q., Sa, R. and Zhang, H. 2015. iTRAQ-based quantitative proteomic analysis of longissimus muscle from growing pigs with dietary supplementation of non-starch polysaccharide enzymes. *Journal of Zhejiang University-(Science)* 16(6):465–78.

Zhang, L., Yu, Z., Jiang, L., Jiang, J., Luo, H. and Fu, L. 2011. Effect of post-harvest treatment on proteome change of peach fruit during ripening. *Journal of Proteomics* 74(7):1135–49.

Zhu, J., Alvarez, S., Marsh, E. L., et al. 2007. Cell wall proteome in the maize primary root elongation zone. II. Region-specific changes in water-soluble and lightly ionically bound proteins under water deficit. *Plant Physiology* 145(4):1533–48.

CHAPTER 7

Impact of Sequencing and Bioinformatics Tools in Food Microbiology

Ramachandran Chelliah, Eric Banan-Mwine Daliri,
Fazle Elahi, Imran Khan, Shuai Wei, Su-Jung Yeon,
Kandasamy Saravanakumar, Inamul Hasan Madar,
Sumaira Miskeen, Ghazala Sultan, Marie Arockianathan,
Shanmugarathinam Alagarsamy, Thirumalai Vasan,
Myeong-Hyeon Wang, Usha Antony,
Devarajan Thangadurai, and Deog Hwan Oh

CONTENTS

7.1 INTRODUCTION—ROLE OF BIOINFORMATICS IN FOOD SCIENCE

In today's world, market demands for healthy, safe, and nutritional food products with minimal synthetic preservatives are rising regularly. World Health Organization (2019) reports that about 600 million people worldwide suffer from diarrheal diseases, associated with 420,000 deaths worldwide (Hoffmann et al. 2017). Therefore, maintaining food safety and quality is an evolving challenge for food scientists, regulators, and the food industry globally. Nevertheless, before reaching the consumer's plate, food commodities must pass through various stages such as harvesting, transportation, storage, refining, and dissemination through which they are vulnerable to attack by a variety of pests, microbes, and oxidative degradation that significantly deteriorate product quality and also have a negative impact on consumer health (Holton et al. 2017). The numerous endogenous factors (bioactive compounds, nutrients, pH, water interaction) and external factors (gas composition, temperature, and microorganisms) of food ecosystems impact on pathogenic microbe development (Holton et al. 2017).

Present methods of protection, detection of biological pollutants, and their related threats are inadequate, expensive, and have their own limitations. In this sense, recent developments in omics sciences (genomics, transcriptomics, proteomics, and metabolomics) may be used to identify, avoid, and regulate foodborne microbes and pathogens and recognize the mode of action of phytochemicals at the cellular level. Thus, in recent years, "foodomics" (an integration of various omics technologies), the omics definition in food science, has been given substantial attention by industry and law enforcement authorities to address major issues related to food bioactivity, health, quality, and traceability (Cifuentes 2017). This offers a detailed description of product efficiency and nutritional attributes. Food strategies, such as neuroscience, transcriptomics, genomics, and therapeutics, will reveal the molecular mechanisms of gene expression, RNA, protein, and antioxidants linked to nutritional bioactivity, safety, and quality (Cifuentes 2017).

Bioinformatics, which uses a statistical approach to analyze biological data, has made significant strides in the field of food science and nutrition in recent years. Bioinformatics software can successfully implement and interpret data from foodomics (combining food chemistry, biological science, and data analysis) systems related to taste, flavor, health, and food quality (Holton et al. 2013). Knowledge produced by the successful implementation of these methods will evaluate and characterize operational genes, proteins, and metabolite-related responses to unique molecules' biological functioning. This also enhances information regarding the role of different food ingredients (preservative and nutritional agents) at the molecular level, deciphering their association with specific genes and their impact on proteins and metabolic products that could be useful in developing successful products with preservative and therapeutic potential (Holton et al. 2013).

Sequence knowledge feature classification is one of the basic functions of bioinformatics. The wide range of sequencing strategies produces comprehensive genomics results. The influence of these data involves thorough identification of structural elements in some of these data, and correlation of sequence information with function, for example by matching predicted protein sequences with recognized mechanisms (Walsh et al. 2018). This type of research will classify gene activities (critical knowledge for metabolic modeling) as anticipated functions for most genes in the gene encoding; and it will recommend characteristics for different bacterial strains through imposing anticipated roles among all genes on receptor databases, predictive characteristics of, for example, *Bifidobacteria* in the gut ecosystem, or even predictive advanced features of diverse microbial ecosystems (Maldonado-Gómez et al. 2016). For genes where a genetic resemblance query does not produce a reasonable assessment, the function can be concluded by correlating the existence and abstinence of the gene in organisms with both the presence and absence of a certain phenotypic trait in the same group of organisms (also pointed to as genome-trait matching [GTM]). For example, a collection of proteins was anticipated to require plant degradation (oligosaccharides by binding bacterial source of differentiation to presence/absence of genes) (Martens et al. 2011). Comparative study of the gene sequences of a species in which certain strains have a beneficial effect, for example, taste enhancement, while others are negative (bacterial contamination) can also be used to distinguish the molecular factors that may underlie these variations, as was done with the *Brettanomyces bruxellensis* (yeast) (Woolfit et al. 2007). Tools to connect omics to genotypes are PhenoLink and DuctApe. Both methods include a genome sequence, which could be fairly difficult to acquire for culturally complex microbiota. Strategies such as multiple displacement amplification can be applied to amplify DNA from a single cell, and a number of genetic alignment methods can also be applied to construct reads from single-cell sequencing (Wu et al. 2014). Mobile components like transposons, plasmids, or phages may transfer functionality across one to another strain. An example is the transition of galactose-using operations among *Lactococcus lactis* strains analyzed by next-generation sequencing (NGS) and computational biology (Thum 2015). To this end, finding possible transposon induction points is critical and can be promoted by bioinformatics methods such as Transposon Insert Finder.

7.1.1 Sequence Technology Overview

Microbial gene sequencing has become the main stream of food microbiology due to increased accessibility and enhanced data sequencing acceleration and reliability. It is a result of developments in sequencing tools commonly called next-generation sequencing. NGS comprises parallel processing and single-molecule sequencing, thereby delivering short and long sequencing reads (Buermans et al. 2014). Short-read transcription is highly accurate and generates 100–300 bp read lengths, which are then organized into imperfect or so-called draft genomes. Due to the difficulties in integrating recurring regions and broad genomic refinements, such as insertions, deletions, and inversions, full genomes could not be constructed from short texts in a single sequence process (Sudmant et al. 2015). It is not a problem for several implementations, particularly descriptive genomics and phylogeny, but also where full genomes and lengthy readings are needed to determine complicated genomic regions. Lengthy-read sequencing generates reads from 10 to 50 Kb in length, but at higher error levels (Loman and Pallen 2015). Microbial DNA sequencing might generally be implemented on platforms such as Illumina, Ion Torrent, PacBio, and Nanopore. Although more comprehensive technical explanations and correlations are very well represented in a variety of recent reports (Figure 7.1), such as Deurenberg et al. (2017), Sekse et al. (2017), and Slatko et al. (2018).

The specific tools selection is based on what technology is applied and depends on whether to apply that sequencing software as well as on the sequencing performance. Maximizing high performance can lead to reduced sequencing expenses for each sample. However, the number of samples sequenced in a single run is a function of the desired performance and coverage, focusing on the method. Single nucleotide polymorphism (SNP) research of bacterial genomes likewise can be done with fairly low coverage, meaning more DNA samples can be processed in a single sequencing cycle (Nielsen et al. 2011). In comparison, metagenomic research aimed at identifying certain microbial genes involved

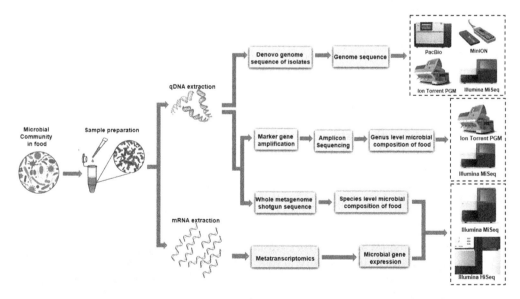

FIGURE 7.1 Schematic overview of the different high-throughput sequencing approaches applicable to food microbiology and suggestions for the sequencing platforms most suitable for each approach (Deurenberg et al. 2017).

in a culture demands much better coverage, reducing the number of samples that can be included in a single test, typically raising the sequencing expense for each sample (Tringe et al. 2005).

7.2 FIRST GENERATION DNA SEQUENCING

Recent implementations of whole genome shotgun sequencing (WGSS) have been adopted in public health monitoring fairly quickly compared to previous technological developments, with records of their use from early investors around 2011 (Lienau et al. 2011; Koser et al. 2012). Although primarily applied for comparative analysis of foodborne disease outbreaks identified by typing technologies including pulsed field gel electrophoresis (PFGE), microbial pathogen WGSS has now been used for retrospective monitoring of foodborne pathogens in a minimum four countries: United Kingdom, Denmark, France, and the United States (Allard et al. 2016; Ashton et al. 2016). The year after the introduction of WGSS for prospective evaluation surveillance of listeriosis in the United States, more and smaller outbreaks were observed, outbreaks were reported earlier, the source of outbreaks was identified more regularly, and the overall number of confirmed infection-related cases improved (Jackson et al. 2018). In the field of global health, WGSS is implemented as a substitute tool, replacing most existing detection and characterization approaches in the microbiology laboratory such as serotyping, pathogenic screening, assessment of antimicrobial resistance, and former molecular typing techniques (Jackson et al. 2018). In a public health environment, replacement of conventional microbiological detection and typing processes with a single optimized analytical WGSS workflow makes implementation price-effective as well as offering more reliable, substantive data to public health than previously collected (Quan et al. 2018). Despite global health leadership, WGSS is gradually been accepted to be applied in food manufacturing. This is not really due to the need to consider human health solutions, but also attributed to the tremendous benefits and opportunities this technology provides for enhancing nutritional quality and efficiency. In a pathogen or spoilage contamination case, risk analysis is a crucial and significant advantage for the food industry (Oliver et al. 2005). For example, WGSS can easily differentiate among both unique and recurrent organism implementation into the manufacturing environment. It could also be applied to forecast characteristics, such as lethality or pathogen antibiotic resistance, or a bacterial contamination organism's potential to crack a particular product preservation barrier (Laxminarayan et al. 2013). Although industrial food safety research may not involve the comprehensive microbial characterization needed by research laboratories, WGSS is increasingly investigating the source of microbial contamination (Rantsiou et al. 2017; Van Hoorde and Butler 2018). As the cost of sequencing decreases with technological advances, it encourages industry to consider adopting its use.

7.2.1 WGSS-Based Monitoring and Mapping Concepts

Molecular subtyping techniques have proven beneficial for detecting pathogens across the food chain, seeking to recognize origins of outbreak and transmission path (Gerner-Smidt et al. 2019). The mode of transmission can show that isolates from cases, food handler, or food service environment came from a common source. The additional knowledge provided via WGSS significantly improves our ability to identify contamination sources.

Over time, bacteria change and their DNA can be used to calculate their progression. Although prior molecular subtyping techniques observed sequence variations in a minor portion of the microbiome genome, WGSS recognizes them in the entire genome and thus more precisely explains strain genetic similarity (Brown et al. 2019). In monitoring and mapping, the relatedness of outbreak bacterial sequences as well as the food supply chain is evaluated to decide whether they may be part of the same transmission chain. However, as discussed in WGSS data, observational proof is necessary to validate and define a dissemination chain.

Presently, there are two major methods for analyzing genomic data to establish the relationship among strains, respectively SNP-based and gene-based strategies. WGSS data analysis by either methodology is a dynamic process in which several steps are incorporated to obtain final results, such as SNP or allele matrices and phylogenetic trees (Timme et al. 2017). The enormous quantity of data produced by WGSS threatens its evaluation (Wyres et al. 2014; Deurenberg et al. 2017). This contributed to the development of numerous technology solutions, predominantly through scholastic efforts, which usually require professional knowledge and skills to implement and operate. Moreover, recently professionally established software has become accessible, offering a viewer-friendly interface, enabling non-bioinformatics expertise to perform analyses with adequate training for both bioinformatics software and final WGSS result interpretation. Consumer software can be costly, but as limited bioinformatics knowledge is required, it may still be more cost-effective for actors in the food industry (Moskowitz et al. 2005).

7.2.2 Phylogenetic Analysis

The genetic diversity observed by SNP, or gene-by-gene review, can also be used to predict phylogenetic relationships among bacterial isolates, generally seen as a phylogenetic tree. The tree represents the measured evolutionary structure (acquired through various possible tree inference algorithms such as parsimony, maximum likelihood, and Bayesian or distance methods) of isolates as a sequence of root or evolutionary divergence branches. Grouped isolates within tree leaves are closer to each other than certain isolates somewhere else in the tree. Ajawatanawong (2017) and Hedge and Wilson (2016) are suggested for more in-depth study of the concepts behind phylogenetic trees.

7.3 NEXT GENERATION SEQUENCING HIGH THROUGHPUT SEQUENCING

Foods, especially fermented foods, may harbor large populations of microbes, both pathogenic and beneficial (Rezac et al. 2018). However, apart from the time-consuming nature of culturing microbes, only a small percentage of microbes are effectively culturable by traditional culture-based approaches (Schoustra et al. 2013). Indeed, viable but non-culturable (VBNC) bacteria include those, such as enteropathogenic *Vibrio cholerae*, *Escherichia coli*, *Legionella pneumophila*, *Helicobacter pylori*, *Vibrio alginolyticus,* and *Vibrio vulnificus* (types 1 and 2) (Fakruddin et al. 2013). This therefore calls for the application of more efficient methods for enumeration and identification of food microbes. Application of DNA-based methods makes identification and classification of bacteria, yeast, and even viruses more effective and this overcomes the challenges posed by VBNC microorganisms (Lee and Bae 2018). Over the years, metagenomics (a culture-independent and sequencing-based technique) has been useful in studying the levels and diversity of various microbial populations in various samples (Shakia et al. 2019). In the past decade, NGS methods have provided a cheaper and faster genome sequencing alternative for microbial identification and characterization (Ward et al. 2018). These high-throughput sequencing methods focus on sequencing the DNA of the entire microbial communities using targeted approaches like PCR-amplicon sequencing of marker genes (such as 16S rRNA genes) or shotgun sequencing of all the DNA in a given sample (Ezponda et al. 2020). Platforms for NGS include the Genome Sequencer FLX+ System (454 Life Sciences), Genome Analyzer System (Illumina), Ion Semiconductor Sequencing, and PacBio Single Molecule Real Time Sequencing (Daliri et al. 2017).

Though 16S rRNA studies are useful in capturing the biodiversity of many samples using minimal sequencing, many studies are currently using shotgun metagenomics as it allows a complete capture of most members of the microbiome and elucidate potential genes and functional pathways (Shakaya et al. 2019). A key limitation of shotgun metagenomics, however, is it does not differentiate between active and inactive members of a microbiome (Shakaya et al. 2019). This therefore implies that shotgun metagenomics cannot distinguish between microbes that are actively contributing to the observations in an ecosystem from those that are dormant. Meanwhile, the use of RNA sequencing (RNASeq) methods to record expressed RNA transcripts within an ecosystem at a given time under a given condition provides better information about active members (Orellana et al. 2019). Application of a combination of proteomics and mass spectrometry provides a powerful insight into actively expressed proteins, but must be paired with reference genomes or metagenomes from which targeted peptides can be matched (Figure 7.2). However, when only RNASeq is applied, less expressed genes as well as the entire meta-transcriptome including non-coding RNAs can be detected, annotated, and mapped to metabolic pathways (Anamika et al. 2016). With the advent of NGS technologies to RNA, it is possible to measure known transcript targets (Li et al. 2019) as well as discover unknown transcripts and transcript variants directly from a sequence data. This field has seen an exponential increase in the number of meta-transcriptomic projects, and most of them represent differential gene expression analysis, which aims to identify active members, genes, and pathways within a given microbiome (Grattepanche et al. 2020). However, the presence of inadequate reference genomes could result in ineffective functionality or taxonomic characterization (Albin et al. 2019). For this reason, it is important to assemble meta-transcriptomic data from the same or similar samples, if possible. More so, since the relative members and the gene expression of a given microorganism are dynamic, there is the need for very large meta-transcriptomic data points (i.e., reads) (Shakya et al. 2019). This implies the need for long read technologies. Longer reads would be helpful for assembly, taxonomy determination, and functional analysis. This will also provide better resolution of transcript isoforms, polycistronic operons, and different genes with high similarity.

7.4 SEQUENCING TECHNIQUES APPLIED TO CHARACTERIZE FOOD-RELATED MICROBIOMES—16S rDNA SEQUENCING

Specifically, 16S rDNA gene is contained in most bacteria and one part of nine hypervariable regions flanked by conserved sequences, which afford specificity to design primer for PCR to amply and sequence specific loci to identify and discriminate bacterial species associated with the food matrix (Jean-Marc et al. 1993; Cao et al. 2017). 16S rDNA

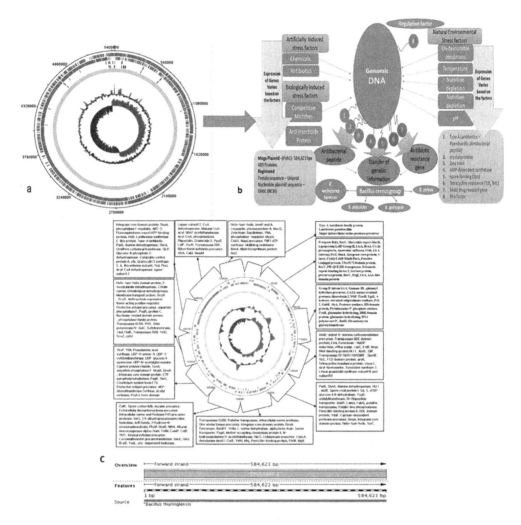

FIGURE 7.2   (a–c) Schematic overview of the de-novo assembly approaches applicable to metagenomic sequencing platforms most suitable for each approach (Chelliah et al. 2019).

sequencing is one of the most important methods other than culture methods for analyzing a microbiome.

The 16S rDNA sequencing is a form of metagenomics which can be used for predictive metagenomics profiling, which is a cost-effective and straightforward method and in line with the multi-omics approach in food microbiology (Ferrocino and Cocolin 2017). Marios et al. (2018) proposed a bioinformatics workflow for analysis of NGS-based 16S rDNA derived from food metagenomics that showed possibilities in applying in the investigation of the ecology of foodborne pathogens encountered in various food products (Marios et al. 2018). Keller et al. (2014) described a proof-of-concept of computational approaches to retain functional information about microbial communities assessed through 16S rDNA (meta) barcoding and developed an automatic pipeline that might infer preliminary or supplementary genomic content of a community. Chauhan et al. (2013) found that the combination of ~70 bp V3 and ~100 bp V6 regions in the 16S rDNA are promising targets for rapid identification of at least six common dairy aerobic spore-forming bacterial species, and *Geobacillus* isolates could be identified at a genus level.

16S rDNA high-throughput sequencing-based molecular-based methods are useful new technologies to study microbial communities including both culturable and nonculturable populations, which have offered a powerful and economic way to gain insight into the bacterial community composition in large numbers of samples (Table 7.1) (Zhang et al. 2019). It may also be applicable for bacterial species identification in various fields with its highly discriminative power, and through analysis of 16S rDNA sequences, 99% or 100% similarity for all *Pseudomonas* species associated with freshwater fish were obtained (Kačániová et al. 2019). The efficient detection of similar species based on 16S rDNA pathogen specific sequences was developed with high specificity (Eom et al. 2007). However, methods developed based on 16S rDNA will always be unable to differentiate closely related species that share identical same region sequences, for example, *Geobacillus stearothermophilus* and *Geobacillus thermoleovorans* are indistinguishable in their 16S rDNA sequences (Weng et al. 2009). Thus, housekeeping genes can be investigated to differentiate bacteria with similar region solely or combined with 16S rDNA.

7.5 IMPROVING METABOLITE PRODUCTION AND BIOMASS

Improvement of the food production process by optimizing biomass has gained immense interest around the globe. Microbial production of fermented food products can be improved through bioinformatics tools (Alkema et al. 2016). Genome-scale metabolic modeling (GMM) techniques have been widely used to rationally improve fermentation yield (Maarleveld et al. 2013). In GMM process, the genome sequence of the organism is used as an inventory of the metabolic potential of the strain of interest. Metabolic models have been made for many microbes, including several of food-relevant microorganisms (Heavner et al. 2013). Importantly, GMMs for modeling organisms have been updated several times over the years since their introduction because more relevant biological information became available (Gu et al. 2019). Complete GMMs along with algorithms such as flux balance analysis, which uses linear programming, allow the *in silico* simulation of growth of organisms under the metabolic constraints to help in optimization of growth conditions for an organism of interest (Alkema et al. 2016; Wegkamp et al. 2010). Moreover, these models can suggest cost-effective or alternative substrates for fermentation (Teusink et al. 2009) while they increase the yield of metabolites such as amino acids (Park et al. 2014) or succinic acid (Otero et al. 2013).

7.6 BOOSTING THE TEXTURE AND FLAVOR PROFILE OF FOOD

Food texture is one of the important characteristics used by people to judge food quality. It can be depicted as hard, soft, rough, liquid, solid, crispy, smooth, creamy, etc. This can be related to physical properties like density, surface tension, viscosity, etc. (Bourne 2002). Sensory assessment tests are used to measure the rheological and mechanical properties of food structures in which the interactions of proteins, lipids, and polysaccharides takes place. The texture also plays an important role in release of flavor and its perception (Bourne 2002). This is achieved when the food is broken down in the mouth, changing the texture of food and releasing and allowing flavors from the food to reach the taste receptors. In the mouth, water dissolves most of the hydrophilic aroma

TABLE 7.1 A Comprehensive List of Publications in Recent 10 Years Using 16S rDNA Sequencing to Study Food Microbiomes

Microbiomes	Food Samples	Country	Conclusions	References
Lactobacillus	Goat's milk	Northwest Algeria	16S rDNA sequence distinguishes the closely related species *L. plantarum* and *L. pentosus*	(Marroki et al. 2011)
Bacillus sp.,	Iranian Kefir type drink	Iran	16S rDNA-based method is a fast and efficient method to track contaminations in the Kefir drink	(Hosseini et al. 2012)
Acinetobacter baumannii, Stenotrophomonas Maltophilia, and *Ralstonia pickettii*	Stored jaggery	India	The amplified 16S rDNA and BLAST analysis showed 98%–99 % homology with pathogenic bacteria	(Verma et al. 2012)
Bacillus cereus	Bottled water	Denmark and Sweden	16S rDNA gene sequencing was possible to separate 2 *B. cereus* strains by the principal component plot, despite the close sequence resemblance	(Hansen et al. 2013)
Gram-positive catalase-negative cocci	Sheep bulk tank milk	Spain	16S rDNA sequencing identified 23 species of GPCNC, *Enterococcus, Streptococcus, Lactococcus, Aerococcus,* and *Trichococcus* genera	(de Garnica et al. 2014)
Lactococcus lactis	Dairy industry	Turkey	PFGE results were confirmed by 16S rDNA sequence analysis	(Gunay-Esiyok et al. 2014)
Vibrio spp.	Seafood, meat, and meat products	Libya	Retrieved *Vibrio* isolates were confirmed using partial amplification of 16S rDNA by universal oligo-nucleotides primers	(Azwai et al. 2016)
Enterohemorrhagic *Escherichia coli* O157	Raw milk and locally made dairy products	Libya	A number of 11 out of 27 isolates were confirmed to be *E. coli* by partial sequencing of 16S rDNA	(Garbaj et al. 2016)
Pseudomonas spp.	Crisp grass carp	China	Through RFLP patterns and 16S rDNA sequencing from the clones, *Pseudomonas* spp. was found dominant in CGC fillets	(Li et al. 2018)

(Continued)

TABLE 7.1 (CONTINUED) A Comprehensive List of Publications in Recent 10 Years Using 16S rDNA Sequencing to Study Food Microbiomes

Microbiomes	Food Samples	Country	Conclusions	References
Clostridium species	Sheep and cattle carcasses	Iran	16S rDNA gene sequencing analysis could be used as a complementary method for MALDI-TOF MS to verify the identification of the organism	(Bakhtiary et al. 2018)
Bacillaceae, Lactococcus, Lactobacillus, Streptococcus, and *Staphylococcus*	Pasteurized and unpasteurized gouda cheese samples	USA	Gouda cheeses assessed were comprised of the same organisms although with different population levels	(Salazar et al. 2018)
Erwinia, Serratia, Arcobacter, Chryseobacterium, and *Pseudomonas*	Whole bird carcass rinses	USA	A more complete characterization of the microbial communities was identified using 16S rDNA next sequencing	(Wages et al. 2019)
Pseudomonas genus	Raw cow milk	New Zealand	Sequencing of the 16S rDNA gene and MALDI-TOF offer a powerful and economic way to gain insight into the bacterial community composition in large numbers of samples	(Zhang et al. 2019)

compounds whereas oils and fats dissolve hydrophobic aroma compounds (Kramer and Szczesniak 2012). Thus, fundamental physical principles are followed in the interaction between aroma and odor activity with the texture and structure properties of food.

Flavor constitutes both aroma and taste of the food material (Kramer and Szczesniak 2012). These flavors and aromas are volatile low molecular weight organic molecules which are very sensitive to air, heat, light, and moisture. Flavors are mainly perceived through the nose, whereas taste is experienced in mouth through taste receptors when the food is chewed. Flavors will make a great impact on the consumer's decision to consume it or not. The flavors are classified as natural, processed, compounded, taste modifier, and abnormal flavors. Thus flavor arises in natural conditions or during cooking or is intentionally added to enhance the property or additives or as a result of degradation (Kramer and Szczesniak 2012). The main criteria to improve the flavor of the food substance: to convert a bland one to a new flavored one, convert weak to strong ones, to complement the existing one, to cover undesirable flavor, to overcome seasonal variation in natural flavors, if the existing one is limited and poses toxic effects (Bourne 2002).

7.6.1 Need of Flour Enhancement in Foods

The use of flavorings in meat products is vast but care should be taken while adding any seasonings or flavorings so that it should not affect the original taste. Ground herbs and spices are used traditionally to flavor meat products. These will enhance the natural flavor of meat products by modifying them to suit to individual palates, but it should not slog their taste buds. For industrial purposes, seasonings must be in the form of dry powders with appropriate herbs and spices along with other permitted additives which include flavor enhancers, hydrolyzed vegetable protein, yeast extracts, phosphates, and colorant. The prime additive used in savory foods is salt and the flavor of these foods can be enriched with the addition of spices and herbs with flavor enhancers (Kramer and Szczesniak 2012).

The fermentation process often influences the quality and taste properties of the food item. Such properties are mainly due to a specific microorganism, and this can be altered through fermentation (bioconversion process), for example, through introducing adjunct strains to cheese fermentations or by introducing extra cellular polysaccharide (ECP) producing organisms to enhance yogurt texture (de Paula et al. 2015). Wine flavor compounds can also be altered by changing fermentation parameters or through altering wine-fermenting starter microbial culture. Although enhancements can be made by evaluating a range of test environments, computational biology and data analysis could be applied to optimize prototype designs. A microorganism's performance through distinct fermentation parameters can be inferred from the gene expression of certain microbes. Using a biochemical model, the development of *Lactobacillus lactis* MG1363 flavor was hypothesized and then scientifically confirmed. Similarly, *Lactobacillus delbrueckii* subsp. *bulgaricus* showed how this organism adapts for milk fermentation and yogurt processing (Hao et al. 2011). Identical analyses were performed for *Oenococcus oeni* and yeast genomes, and their correlation to wine fermentation (Bartowsky et al. 2011). Due to the extreme complexity of the yeast genome system, this analysis is a highly competitive task.

In genes where a genome resemblance search does not provide a reliable predictor, their activity can be concluded by correlating the existence and abstinence of the genes in organisms with the involvement and disappearance of certain morphological traits in the same group of organisms (also known as Global Translation Model [GTM]). GTM growth in different carbohydrates can be projected comparatively accurately based on gene information, for instance *Lactobacillus lactis*, *Lactobacillus plantarum*, *Bifidobacterium breve,* and *Lactobacillus paracasei* (Alkema et al. 2016). It has become evident that some studies which predicted a more complex genotype, such as stress tolerance, became less determined when based only on genetic material. For better prediction of these phenotypes, details on the transcript levels of the genes should be taken into account. Likewise, transcriptome trait matching (TTM) can be used to combine the expression of microorganism genes with a product's texture and flavor proportion, such as enhancing organic acid synthesis by modifying fermentation parameters (Kramer and Szczesniak 2012).

The impact on flavor and texture are induced primarily by the metabolites formed or transformed throughout fermentation. Peptide patterns may be used explicitly to predict the final sensory properties rather than aligning gene material with effects on taste and structure performance. The universal standard check of a fermented product's sensory properties is a quantitative descriptive study conducted by a professional sensory jury (Kramer and Szczesniak 2012). Such tests are complex and require large product quantities to be generated. The finding depends on the expertise of the jury and the characteristics used to define the properties of the product. With proteomics profiling techniques, hundreds of metabolites can now be identified in food samples concurrently. That, in combination with the advancement of limited scale product inspection techniques, has contributed to the introduction of many new statistical methods for associating instrumental data, such as gas chromatography and Matrix Assisted Laser Desorption/Ionization Time of Flight Mass Spectrometry (MALDI TOF-MS), with sensory data (Soeryapranata et al. 2002).

7.6.2 Non-Thermal Textural Modification of Foods—3D Printing

Currently, 3D printing technology is used to produce a variety of food textures from different sources with respect to visual appeal. The latest technology is used to produce a variety of products like cereal-based foods, chocolates, fish gel, protein-fiber diet, fruit-based snacks, and juice gel. For this technology, the starting material must possess suitable rheological properties. Hence, it can be extruded and it must also hold its shape (Sun et al. 2015). Gels, micro-gels, and emulsion gels which are soft, stable, and have smaller particles can be used as thickening agents for texture-modified foods (Sun et al. 2015). To this, nutrients can be added, like fiber, whey protein, milk protein, and fish oil to make more attractive health products.

7.6.3 Bioinformatics in Food Augmentation

Bioinformatics plays a prominent role in the improvement of the quality and processing of foods. The field of bioinformatics can be used efficiently to access all the discovered genomics, proteomics, and metabolomics data available in the databank so as to supply the necessary information to the industries in order to improve the nutrition, taste, and quality of the food (Ashinoff 2000).

7.6.4 Food Taste

The new approaches in bioinformatics play a key role in ensuring the taste, quality, and safety of food. The different taste receptors discovered and submitted to the database were Degenerin-1 ion channel (sour taste), G protein-coupled receptors (bitter taste), Tas 1r3 receptor (sweet taste), ENaC-epithelial ion channel (salt taste), and mGluR4-brain glutamate receptor (umami) (Singh 2011). 'Toward the submission of genetic data of receptors, it is easy for informatics to study the receptors' evolution. These molecular receptor data help to create new generations of taste modifiers for foods. New developments in the algorithm and software based on the existing receptor data can help to create many simulation models. These model compounds can be developed into great food additives (Carocho et al. 2014). This will also pave the way to understanding the basis of interaction between molecules which show strong, complementary, and antagonist effects. Sequence similarity algorithms were also developed to determine the homology between receptors.

7.6.5 Food Flavor

In dairy products, lactic acid bacteria play a pivotal role in the formation of flavor. Exploring the genetic sequence of these bacteria can bring out many potential flavors in the fermented foods. Moreover, the flavors released from the food stuffs are not due to a single compound but due to the interaction of many molecules/compounds. In this aspect, bioinformatics plays a key role in bringing out new-flavored products by permutation and combination of various flavors from the data obtained from existing knowledge, consumer preferences, and needs (Kramer and Szczesniak 2012).

7.6.6 Food Quality

For better quality of food, the crops must be of good quality. With bioinformatics, the new genes can be identified using genomics to produce transgenic crops for better quality and quantity. This technique can also be applied for identifying specific targets and compounds for making the crops pest-or herbicide-resistant. This 'omics' also helps in the production of high-yield and disease-resistant crops (Kaul et al. 2016).

Spoilage is one of the important aspects of food. It is mainly caused by foodborne pathogens (Xu et al. 2010). The tremendous data available in the database can be assessed to predict both positive and negative effects of microbes on food and can also be used to improve the production, quality, and nutritional aspects of food. So, it is best to develop molecular markers to identify the occurrence of spoilage and pathogens in the food samples. The US FDA developed a bioinformatics tool for detecting and identifying foodborne pathogens using microarray (Fang et al. 2010).

The bioactive peptides in food exerted various biological properties like antioxidant, immunomodulating, antimicrobial, etc. Using the protein databases UniProtKB, SwissProt, and TreMBL, many bioactive peptides have been discovered (Udenigwe et al. 2013). The occurrence and frequency of these crypt-bio peptides in the primary structure is calculated using the formula a/N where "a" is the number of peptides exhibiting bioactivity of particular type present in sequence and "N" is the total number of amino acids present in the proteins. This bioinformatics approach not only saves time but also screens the molecules in a fast manner. The software programs developed to produce various *in silico* peptide profiles created by different enzymes' proteolytic action are BIOPEP, ExPASY, and PoPS (Minkiewicz et al. 2008).

7.6.7 Metabolomics

Understanding various microbial metabolic pathways helps us to design compounds which can contribute to the texture, flavor, structure, and safety of food products (Alkema et al. 2016). The metabolic pathways of microbes can be understood and controlled using modern bioinformatics tools in order to obtain desired products. Metabolomics is the high-throughput characterization of complex mixture of small molecules. The metabolites obtained in metabolic pathways have substantial impact on the nutrition and taste of food stuffs (Kramer and Szczesniak 2012). Metabolomics, a promising bioinformatic tool, will provide insight to the peculiar trait of each metabolite in order to get targeted and to enhance its function. Apart from this, it also has the potential to improve crop cultivation and food processing methods to meet the desires and concerns of consumers (Deliza et al. 2005). This bioinformatics branch has great impact on the various fields of food science like identification of compounds, sensory characteristics, safety, and processing.

7.6.8 Food Databases

(i) *FooDB*: This database provides information about food constituents, chemistry, and biology including information about the texture, color, flavor, taste, and aroma of food. Each entry contains detailed biochemical, compositional, and physiological data (Church 2009).

(ii) *EuroFIR-BASIS*: This database deals with bioactive compounds in plant-based foods. This gives information about the food composition and biological activity of bioactive compounds. It will be a useful resource for disaster management authorities, food regulatory and advisory bodies, epidemiologists, food science researchers, and food industries. The database covers multiple compound classes and 330 major food plants and their edible parts (Church 2009).

(iii) *Food Wiki*: It is the depot for food and nutritional information. The immense quantity of data in this database will be a rich resource for the development of food by improving the quality and nutritional content.

(iv) *Foodomics*: This database integrates omics technologies to different food and nutrition domains to improve the human health. It is a database for food molecular profiles. The National Nutrient Databases for Standard Reference (NNDSR) (Ashley and Snow 2007), a free open resource, features composition of more than 8000 foods, and it is utilized globally for nutritional assessment. Likewise, EuroFIR in Europe provides standard food composition database for scientists and researchers. Another database, the FAO/INFOODS Analytical Food Composition Database (Charrondière et al. 2012) also provides data for the composition of common foods that are consumed globally.

(v) *Additivechem*: This database has created a platform for more than 9000 types of food additives along with their structure, physical and chemical properties, metabolism, synthesis, excretion, and toxicity by linking with 16 other databases. This database will help to explore the interrelationship between the structure and function of food additives (http://www.rxnfinder.org/additivechem/) (Neltner et al. 2013).

(vi) *Allergen Online*: This database provides a list of allergens and their sequence so that the protein can be identified from the database that may present a potential risk of allergenic cross-reactivity. A few other databases dedicated to allergens are FARRP Allergen database, SDAP, AllerMatch, and InformAll (Gendel and Jenkins 2006).

This "omics" technology in food science has made tremendous progress (Herrero et al. 2012). As in other disciplines, food science is also enriched with huge data in their database. These data need to be analyzed and processed with high-throughput "omics" technologies to make it productive. So, the software and algorithms developed along with the information in the database will provide a great platform to explore and decode food composition, flavor, nutrients, chemistry, and biology. As more and more data accumulate in the repository, the development of new bioinformatics technologies will surely create major strikes in the field of food science in the near future.

7.7 ASSESSMENT OF THREATS FROM FOODBORNE PATHOGENS

Instead of determining functionality for all genes in a biosynthetic pathway, specifically testing microbiome genome sequences for genes with different features can be a highly responsive and computationally intense way to classify possible safety and health threats from microbial strains in a culture. A particular bacterium's capacity for antibiotic resistance or toxicity can be explored by evaluating its genetic sequence to a standard massive database, known resistance genes, and virulence factors (Didelot et al. 2012). Different methods were identified to identify bacteria persistence in food products, anaerobic spore-forming organisms in food, and potential pathogens in metagenomics data (Møretrø and Langsrud 2017). This (meta)genomics-based approach can be applied to a broad variety of additional features, for example, antimicrobial peptide synthesis and cleaning resistance frequently applied in food manufacturing systems (Chalker et al. 2002). A prerequisite for obtaining useful results from metagenomics studies is a specialized gene function relationship database and exposure to specialized information on specific features to identify gene roles.

Risk assessment (RA) has been determining the relative degree of the risk posed by the microbial pathogens by which many people become infected. In addition, RA is the scientific assessment of the well-known or potential confrontational effects arising from the food hazard exposure to humans. The WHO and FAO through Codex Alimentarius Commission (CODEX, 2007) encouraged the researchers, especially academic and regulatory agencies, to execute the Microbiological Risk Assessment (MRA) 1. According to Codex (2007) hazard is a biological, chemical, or physical agent in, food with the potential to cause an adverse health effect, while risk is a function of the probability of an adverse health effect and the severity of that effect. The MRA refers determining the risk in relation to biological hazard in food. Generally, MRA included the microbial presence and growth from the manufacture/production to consumer's plate. Each year in Europe, meat alone contributes to 2.3 million foodborne illnesses. Many of these illnesses are attributed to pathogenic bacterial contamination and inadequate operations leading to growth and/or insufficient inactivation occurring along the entire farm-to-fork chain. The MRA comprises hazard identification and characterization, exposure assessment, and risk characterization (Lindqvist et al. 2019). Hazard identification included the identification of microorganism, food, processing, and production in relation to risk assessment. On the other hand, hazard characterization describes the adverse impact on health by foodborne pathogens as regards the type, severity, and duration of the impact (Buchannan et al. 2000). Hazard will be classified based on the pathogenicity: 1. Infectious: first adherence, multiplication, invading to epithelium cells, 2. Intoxication: toxin presence in food, and 3. Toxin release after infection inside the human body. Microorganism has capacity to survive and grow in variable food and create adverse impact on the human (Pouillot et al. 2015). The severity of pathogenicity mainly depends on the concentration of toxin and host. The food plays an important role both in the survival and virulence of the organism, like high gastric pH leads to expression of the specific gene.

The potential WGSS data in MRA has been debated since the beginning of this century. With the introduction of high-throughput DNA sequencing technologies, however, food safety and risk has turned beyond the assessment of microbial exposure (or) presence in different food processes for agents classified at (sub)species and *serovar* level. Moreover, with the rapidly decreasing costs of sequencing, WGSS will soon become a standard surveillance technique for the subtyping of isolates for epidemiological purposes. Although the use of molecular data has proved to be a powerful tool in decision-making during outbreak investigations (Dallman et al. 2014)

7.8 MIXED CULTURE FERMENTATIONS CHARACTERIZATION

Complex fermentations require a (un)defined (wild) starter culture with various microbes (bacteria, yeasts, and fungi) fermenting a substrate into the component together. Sources are fermented cheese, malolactic oil, soy, and sea food. Clear progression of microbes might develop in fermentations, such as the microbes *Saccharomyces cerevisiae* and *Oenococcus oenii*. According to all the above mentioned GTM and TTM strategies to link (transcribe) genes to phenotypes, the occurrence and exclusion (or functionality of) microorganisms can be correlated with the features of the fermentation product.

The initial step in classifying fermentation is to decide which microorganisms are involved at the various fermentation stages and to compare them with other measures, such as metabolomics or phage involvement. Microbial consortia characteristics are described by the operational ability of all bacterial genomes. Metagenomics has a benefit over traditional consortium isolate sequencing as it often exposes DNA in the otherwise un-culturable species. Dependent on the consortium sequences, microorganism features can be anticipated. Because of the progression of microbes in fermentation, omitting DNA from dead microbes is critical before constructing sequence-based computational predication models. One way to draw down 'dead' DNA and hence not sequence it is using propidium monoazide (Varma et al. 2009). NGS strategies that profile, for example, the 16S gene present in all bacteria, have progressively been applied in molecular biology techniques, for example gel-based techniques. The data analysis processing of 16S data through food fermentation is very well established. These characterizations of the taxa at least at the stage of the species in a certain fermentation, however, are difficult to determine.

In addition to the genetic level, a broad biodiversity is not known e.g., 16S sequencing (DeSantis et al. 2007). There is also substantial genetic diversity in a bacterial population. For example, over 14,000 genes are contained throughout the *Lactobacillus* strains (its pan-genome), with a single gene encoded with about 3,000 proteins. A gene family is typically made up of genes that are evolutionary, but that could operate differently based on the particular sequence of proteins. Comparative genome approaches have been applied to discover strained diversity in complex, but fairly well-defined fermentation in general and in particular *Lactobacillus sakei* from meat fermentations, *Lactobacillus sanfranciscensis* in sourdough fermentations (Hüfner and Hertel 2008) and wine yeasts, particularly with *Lactobacillus lactis* and *Leuconostoc mesenteroid* from cheese (Herreros et al. 2003) for combined with molecular strain typing. Shotgun-based metagenomics characterize the DNA in mixed-culture fermentation, but bacterium-level diversity is extremely difficult to conclude from shotgun metagenomics sequence fragments (Bora et al. 2016). Due to the enormous biodiversity, it should be known, in certain mixed culture fermentations, that the actual presence of any stress isolate is thought to be important. The integration of shotgun proteomics and comparative genomics can be particularly useful because shotgun metagenome DNA sequences could be matched with the insulation genomes to provide proof of the metagenomic functionality existing in isolates (Dröge and McHardy 2012).

Metatranscriptomics strategies enable characterization of complicated fermentation sequences generated from mRNA (Jiang et al. 2016). An advantage of metatranscriptomics over metagenomics techniques is that monitoring gene expression allows identifying what genes were strongly expressed in a diverse population. Application of 'metatranscriptomics' using multi-species genomes to assess global expression of genes throughout species was documented for Kim chi (Jung et al. 2013). Previously, metagenome and metatranscriptome sequencing of microbial populations engaged in fermentation of cheese rind, was reported. The advantage of this analysis is that metagenomics and metatranscriptomics profiling were linked to their respective possible origins (genome sequences of rind cheese fermentation isolates). Using laboratory setups such as this one, in conjunction with metabolomics analyses and suitable take-up studies will reinforce this argument by using metagenomics/metatranscriptomics strategies to classify and potentially improve fermentations (Jiang et al. 2016). Bacteriophages serve a major role in industrial fermentations attributable to the pathogen predation trend of sustaining biodiversity, but also that phage surveys interrupt fermentation processes (Dick 2018). Nonetheless, determining the specificities of bacteriophages and the associations among microbes in mixed-cultural fermentation are time-consuming tasks.

Bioinformatics technologies, which monitor the association of microbes and bacteriophages, as well as in the metabolic specifications of microbial consortia existing throughout fermentation may relate to information developments in fermentation stabilization in the future (Bruder et al. 2016). It may be done by research of synthetic consortia. Such consortia are presently being created, and cross-kingdom associations have been studied. In a study where cheese rind bacterial populations were produced based on different omics, fermentation expertise, and devoted follow-up experiments, the capacity for determining characteristics of specific fermentations was illustrated. This research did not specifically identify which specified strains (or close relatives) were present in actual fermentation. This was identified in a given consortium for representative *Lactobacillus lactis* and *Lactobacillus mesenteroides* strains of complicated cheese fermentation and *Lactobacillus lactis* strain (Morea et al. 1999).

7.8.1 Branding, Monitoring, and Detection

Food processing and utilization occur in diverse ecosystems where certain protein, lipid, and carbohydrate sources are accessible next to the microorganisms found in the natural environment (Thomas et al. 2011). The existence of the endogenous flora and the food's macromolecular systems may cause many difficulties in detecting and identifying different microorganisms, such as possible food pathogens or probiotic strains applied to the food product for usability. In addition to conventional DNA-based identification tools such as (q)PCR, novel genomic data-based techniques have been established to allow rapid and precise monitoring or diagnosis of particular species or even strains of natural microflora. Similar amplification and locus sequencing defined as biased between distinct *Lactobacillus plantarum* strains have been shown to measure the relative involvement of various strains across the gastrointestinal tract (Bron et al. 2007). The same technique can also be applied to developing specific primers to differentiate between different species of virulent and non-virulent populations and to identify a strain of significance in food products, enabling unique product packaging (Yap et al. 2014).

In addition to focused monitoring of a single strain, metagenome strategies as defined for analyzing diverse fermented foods, such as cheese (Escobar-Zepeda et al. 2016) and fermented foods (Kannan and Chelliah 2015; Chelliah et al. 2016; Antony et al. 2020) of plant source would also gain in detecting spoilage bacteria. Particularly because these approaches allow independent material profiling and do not involve a culture phase, which could generate bias in the results, they might very well be more precise to determine spoilage bacteria by a product. Culture changes will also have their validity due to restricted expenditures and product necessity. 16S community characterization strategies, particularly in fermented products, will enable diagnosis of low-abundance microbes that may be overgrown in culture-dependent detection systems.

7.9 FOODOMICS: OMICS IN THE FIELD OF FOOD SCIENCE AND NUTRITION

The use of omic sciences in the food industry is commonly known as "foodomics," where this discipline studies food and nutrition domains to improve consumer's well-being, health, and confidence (Alfieri 2018). Therefore, foodomics is presented as a global discipline in which food, nutrition, omics technology, and bioinformatics are combined. In this perspective, nutrigenomics and nutrigenetics can be considered a part of the more general "foodomics" term. The continuous development of transcriptomics, genomics, proteomics, and metabolomics offered widespread prospects for increasing our knowledge and understanding of the different issues, which can now be addressed by foodomics (García-Cañas et al. 2012). For example, foodomics helps to understand the cellular, molecular, and biochemical mechanisms that cause the adverse or beneficial effects of bioactive food components using nutrigenomic approaches (Wittwer et al. 2011). Following the nutrigenetic approaches, one can understand the gene-based differences among individuals in response to a specific dietary pattern (Williams et al. 2008). It helps to identify genes that are involved in the previous stage to the onset of the diseases (Smith et al. 2012). It also helps to determine the effect of food constituents on critical and important molecular pathways (Corella et al. 2011). Moreover, foodomics helps to understand the stress adaptation responses of foodborne pathogens to ensure proper food hygiene, processing, and preservation (Soni et al. 2011) and the complete assessment of food safety, quality, and traceability (O'Flaherty and Klaenhammer 2011). In addition, it is understood that the methodological basis of foodomics are not only to describe the food fingerprint but also to assess the effect of food after ingestion. It is well established that health is highly influenced by genetics. However, diet, lifestyle, and environment also influence the epigenome, gut microbiota and, by association, the transcriptome, proteome, and eventually the metabolome (García-Cañas et al. 2012). Good health needs proper balance in the combination of genetics and nutrition/lifestyle/environment; otherwise poor health is a result. Therefore, foodomics can offer a complete image of the effects of food on the physiological mechanisms critical to human life (Cifuentes 2009). Although foodomics offers huge applications, each omic approach is not easy and requires a high degree of complementary knowledge of researchers working in different areas, typically including biology/medicine, analytical chemistry, bioinformatics, and statistics tools (García-Cañas et al. 2012).

7.10 SELECTION OF LIGAND MOLECULES—MOLECULAR INTERACTION STUDY BASED ON DOCKING MECHANISM (SIMULATION MODELING)

Molecular docking is the study of molecular recognition which aims to predict how two or more molecular structures and binding affinity of a complex are formed by two or more molecules with known structures using widely used computational tools. This study of, for instance, a drug or macromolecule receptor match along to be a perfect fit with each other to make a stable complex (Gane and Dean 2000). Orientation of molecule/drug candidates to the macromolecular targets which binds together in complex predicts the affinity and actions of a given small molecule (O'Hare et al. 2013).

The first molecular docking algorithm was developed in the 1980s by Shoichet and Kun (1993); there are several forces that interact between the two molecules which are used to define various docking scores that measure how good each complex is. These scores take into account towards higher binding strength force on docking. These forces are electrical forces, Van Der Vaals forces, and hydrogen bonds. Docking can be classified into 3 sub-methodologies: (i) Protein–protein docking, (ii) Protein–ligand docking, and (iii) Protein–carbohydrate docking.

Protein–protein docking involves the docking of two protein molecules without any need of experimental constraints (Figure 7.3). It can be further categorized into rigid, semi-rigid, and flexible docking. Docking methods can be divided by the classification according to the degrees of flexibility of the molecules under study (Halperin et al. 2002). *(i) Rigid docking:* The goal of this docking procedure is to generate at least one near-native solution candidate. Both ligand and protein are considered rigid entities, and just the three translational and three rotational degrees of freedom are considered during sampling. The rigid docking should permit steric clashes because unbound proteins may collide in their native position when they are placed in their native interacting position (Andrusier et al. 2008). *(ii) Flexible docking:* Flexible docking strategies require no a priori knowledge of the ligand conformation and hence consist of searches in the space of 6+N translational, rotational, and conformational variables. This type of docking approach becomes a necessity when there is no useful information on the conformation of ligand. Thus, flexibility in design relies heavily on flexible docking procedures with four different strategies currently in use for docking flexible ligands: (a) Monte Carlo or molecular-dynamics docking of complete molecules, (b) in-site combinatorial search, (c) ligand build up, and (d) site mapping and fragment assembly (Figure 7.4).

Protein–ligand docking, the most commonly used method of docking, predicts the position of a ligand and whether it has a protein attached, in this case, to its receptor molecule. The purpose of protein-line array is to predict and classify the structure(s) that

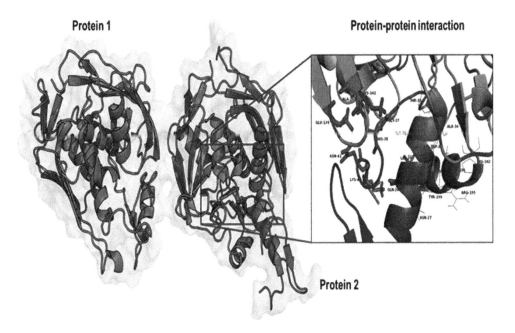

FIGURE 7.3   The graphical representation of protein–protein complex visualized using PyMol.

result from a connection between a particular ligand and a known 3D structure objective protein (Figure 7.5).

Protein-carbohydrate docking was meant to be a key component of numerous biological processes, such as innate immunity, metabolism, and immune response. The docking of carbohydrates involves several important signals. For example, secreted biomolecules including interaction of host cells with disease pathogens, play a crucial part in cell–cell interaction (Figure 7.6). Therefore, choosing a docking program largely depends upon the protein family under consideration. Proteins and carbohydrates interact by combining a genetic conformational search algorithm coupled with an empirical free energy function specific for carbohydrates. The increasing importance of carbohydrate-binding proteins as potential drug targets provides an atomic understanding of carbohydrate recognition revealing some of the unique molecular characteristics of protein–carbohydrate complexes.

Auto Dock is a molecular docking suite consisting of automated docking tools (Figure 7.7). AutoDock consists of two main programs: (i) AutoDock docks the two molecules according to the grid, which is pre-calculated and generated of required dimension around the protein to dock and set by AutoGrid. (ii) AutoGrid is considered one of the best programs when it comes to docking and virtual screening (Wright et al. 2013) (Table 7.2). Various possible problems must be resolved before a protein can be used for AutoDock. That includes missing atoms, breaks in the chain, and alternate locations. Various docking programs utilize potential power grids. These grids represent the calculations of energy, and the grid stores two kinds of potentials in their most basic form: the electrostatic and the van der Waals. The grid was formulated so that the information on the energy contributions of the receptor could be stored at grid points. That allowed it to only be read during ligand scoring.

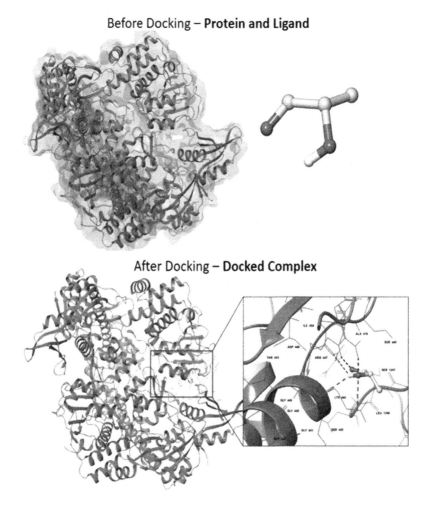

FIGURE 7.4 The graphical representation of protein-ligand complex visualized using PyMol.

7.10.1 Advantages of Docking

(i) There is a huge advantage of the application of docking in a targeted drug delivery system. We can study both the ligand (drug) and receptor (target site) size, shape, charge distribution, polarity, hydrogen bonding and hydrophobic interactions.

(ii) Molecular docking helps in the identification of target sites of the ligand and the receptor molecule.

(iii) Docking also helps in understanding different enzymes and their mechanism of action.

(iv) The "scoring" feature in docking helps in selecting the best fit or the best drug from an array of options.

(v) Molecular docking helps in moving the process of computer-aided drug designing faster and also provides every conformation possible based on the receptor and ligand molecule.

(vi) There are millions of compounds, ligands, drugs, and receptors, the 3D structure of which has been crystallized. Virtual screening of these compounds can be made.

Protein-Ligand Complex
Hydrogen bonds and Hydrophobic interactions

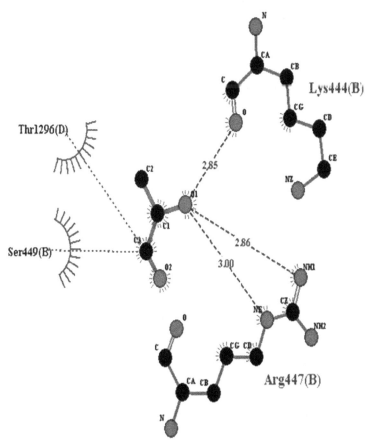

FIGURE 7.5 The protein-ligand complex interaction bonds are visualized using Ligplot+GUI tool.

7.10.2 Limitations of Docking

(i)   The scoring functions used in docking, almost all of them, do not take into account the role played by covalently bound inhibitors or ions (Stephen et al. 2004).

(ii) The methodology and research in protein–protein docking must be greatly increased as success in this field is greatly hampered by many false positives and false negatives (Stephen et al. 2004).

(iii) In protein–small-molecule docking, there can be problems in the receptor structure. A reliable resolution value for small-molecule docking less than the most crystallographic structures have a resolution between 1.5 Å and 2.5 Å. Increasing the use of homology models in docking should be looked at with care as they have even poorer resolution (Vakser et al. 2014). Most applications accept and yield good results for structures below 2.2 Å. All the same, care should be taken while picking a structure.

(iv) In AutoDock, more options can be explored, and the options may vary depending on the complexes being docked, and also the complexity of the problem in hand.

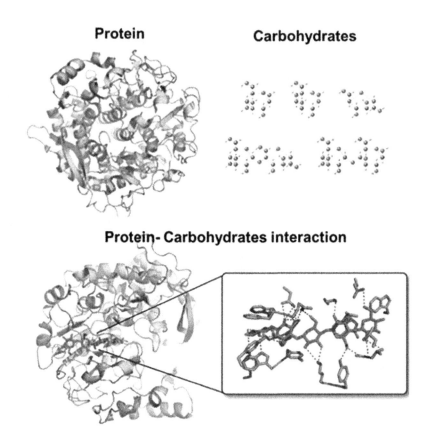

FIGURE 7.6   The protein-carbohydrates complex zoomed in to visualize the shared interaction bonds.

FIGURE 7.7   Steps followed in docking analysis using Autodock.

7.11 AFLATOXIN B1 INHIBITORY POTENTIAL AND THERAPEUTIC CHARACTERIZATION OF SELECTED TEST COMPOUNDS: SELECTION OF THE RECEPTOR MOLECULES

In recent years, worldwide concern over food safety and security has increased due to growing numbers of foodborne illness outbreaks and widespread food contamination. Living organisms, such as bacteria and fungi, cause most of the contaminants in food samples. When compared to bacteria, fungi are one of the predominant contaminants in food samples. It has been estimated that 25% of the world's crops, such as nuts, cereals, and rice, are contaminated by mold and fungal growth (Pandya and Arade 2016). Different types of fungus damage the food materials by secreting various secondary

TABLE 7.2   Most Commonly Used Docking Tools

Docking tool	Description	Website URL
Autodock	Genetic algorithm, Lamarckian genetic algorithm, and simulated annealing	http://autodock.scripps.edu/
Autodock VINA	New generation of Autodock	http://vina.scripps.edu/manual.html
PatchDock	Rigid docking	https://bioinfo3d.cs.tau.ac.il/PatchDock/
GEMDOCK	Generic evolutionary method for docking	http://gemdock.life.nctu.edu.tw/dock/
DOCK	Geometric matching algorithm, docks either small molecules or fragments and includes solvent effect	http://dock.compbio.ucsf.edu/
HADDOCK	Protein–protein docking	http://www.bonvinlab.org/software/haddock2.2/
GLIDE	Exhaustive search and two different scoring function to rank-order compounds	https://www.schrodinger.com/glide
GOLD	Genetic algorithm, flexible ligand, partial flexibility of proteins	https://www.ccdc.cam.ac.uk/solutions/csd-discovery/components/gold/
FlexX	Incremental construction algorithm for protein ligand docking	https://www.biosolveit.de/FlexX/

metabolites, such as toxins and enzymes, that spoil the food sample. Mycotoxins are toxic secondary metabolites produced by different species of fungi, and they represent a serious threat to human and animal health and safety (Bennett et al. 2007; Bircan 2009). Several hundreds of different mycotoxins have been identified, but the most commonly observed mycotoxins that present a concern to human health and livestock include aflatoxins, ochratoxin A, patulin, fumonisins, zearalenone, and nivalenol/deoxynivalenol. They are extremely hazardous to human and animal health, causing a decrease in immunity, nervous system and reproductive system damage, necrosis of tissues, organ lesions, carcinogenesis and mutagenesis, and even death (Koppen et al. 2010).

Aflatoxin is the most poisonous of the mycotoxins and is produced by certain molds (*Aspergillus flavus* and *Aspergillus parasiticus*) which grow in soil, decaying vegetation, and grains. Crops are frequently affected by *Aspergillus* spp. including cereals (corn, sorghum, wheat, and rice), oilseeds (soybean, peanut, sunflower, and cotton), spices (chili peppers, black pepper, coriander, turmeric, and ginger) and tree nuts (pistachio, almond, walnut, coconut, and Brazil nut). It is also found in the milk of animals that are fed contaminated feed, in the form of aflatoxin M1. High doses of aflatoxins can lead to acute poisoning (aflatoxicosis) and it can be life-threatening, usually through damage to the liver. Aflatoxins have also been shown to be genotoxic, meaning they can damage DNA and cause cancer in animal species. There is also evidence that they can cause liver cancer in humans.

Aflatoxin is a difuranocoumarin molecule synthesized through the polyketide pathway (Bennett et al. 2003). Totally 6 out of 18 types of aflatoxins that have been identified are designated as B1, B2, G1, G2, M1, and M2 (Dors et al. 2011). These groups exhibit molecular differences. For example, the B-group aflatoxins (B1 and B2) have a cyclo-pentane ring while the G-group (G1 and G2) contains the lactone ring (Gourama et al. 1995). Whereas the B-group aflatoxins exhibit blue fluorescence, the G-group exhibits yellow-green fluorescence under ultraviolet (UV) light, thus making the use of fluores-cence important in identifying and differentiating between the B and G groups. Aflatoxin B1 is the most common and the most widespread in the world and accounts for 75% of all aflatoxin contaminations of food and feeds (Ayub and Sachan 1997). Aflatoxins M1 and M2 are hydroxylated products of aflatoxins B1 and B2, respectively, and are associ-ated with cow milk upon ingestion of B1 and B2 aflatoxin-contaminated feed. Moreover, once formed from B1 and B2 forms, aflatoxins M1 and M2 remain stable during milk processing (Stroka et al. 2002).

7.12 PHYSICOCHEMICAL CHARACTERIZATION AND DRUG-LIKENESS SCORES OF COMPOUNDS

Drug discovery and development pipeline process is a long-term process which requires huge investments, where the majority of the projects end in non-drug like chemical enti-ties. High throughput screening methods, in-silico analysis, and bioinformatics applica-tions in the drug discovery process reduces the time and cost of investments. Whereby, the lead compound is identified and helps to ease the process of bringing successful drug-like molecules. It is important to understand the complete physicochemical properties of the proposed chemical molecule before terming a molecule as having drug-likeness and proceeding with further process development.

To find out what makes a chemical entity have drug likeness, the available data sets are used to determine the drug likeness (Walters et al. 2012). The empirical rule stated by Lipinski forms the basis of all the new approaches and describes various criteria that qualifies drug-like or non-drug-like chemical molecules, which are known as the "famous rule of five."

7.12.1 Drug Likeness

The term "drug likeness" tries to justify the relationship between physicochemical prop-erties and pharmacological activity. Models or rules predicting drug-likeness are based on physicochemical properties which are calculated from molecular structure of the com-pound of interest. Drug likeness predicts qualitatively the chance for a chemical entity in its initial development phase to become a drug and move to the next phases of trial. Structural features and physicochemical properties of drug-likeness was established by various models (Takaoka et al. 2003); if the compound meets the criteria, it is determined to have drug-likeness, and if it fails the criteria, it is then termed "non-drug" or "non-drug likeness."

7.12.2 Drug Likeness Prediction Methods—Counting Methods

Counting methods involve determining the single dimension molecular descriptor values which influence the drug-likeness molecules of interest. The Lipinski rule of five and its application for predicting drug likeness, along with modified rules of Ghose and Opera, comprise this counting method. The Lipinski rule of five forms the basis, correlating molecular descriptors to drug likeness. A chemical entity passing the rule of five is found to possess better oral bioavailability and has a chance to become a drug-like molecule. Ghose modified the Lipinski parameters and extended the number of compounds to screen, calculated the parameters to qualify more than 80% of the compounds, and proposed certain criteria compared to Lipinski. Pareto analysis was carried out by Opera to cover 80% of compounds to find out parameter limits.

7.12.3 Knowledge-Based Methods

This method uses the scientist's knowledge to predict drug-likeness. Inherent binding energies and scoring of structural fragments of the molecule form the idea behind this method (Walters and Murcko 2002). It focuses on the knowledge of the functional groups. Selecting the appropriate reagent is the prime focus of this method, rather than predicting drug-likeness, because it focuses on intrinsic binding energies.

7.12.4 Functional Group Filters

It is important to identify and eliminate unwanted functional groups because of undesirable qualities, like chemical reactivity and metabolic ability. The rapid elimination of swill method falls into this category, which eliminates unwanted reagents from databases (Walters et al. 1998). Functional group filters find and eliminate reactive, toxic, and undesirable chemical moieties. Reactive and toxic reagents that do not meet the molecular weight criteria are filtered first followed by chemical reactions. In the final step, the compound is screened after taking other physicochemical properties into account.

7.12.5 Multiproperty Optimization

Multiproperty optimization involves envisioning the parameter's influence on the behavior of the chemical molecule by using a combinatorial library (Walters and Murcko 2002). The combinatorial library comprises a database of compounds from the selected subset of reagents, calculates the effects of properties on combinatorial products, and accepts the positive improvements on the drug likeness. The process is iterated until the desirable conditions are attained. Combinatorial libraries were scored by frequency distribution of properties such as logP, molecular weight, hydrogen bond donors, and hydrogen bond acceptor, followed by comparing these values with that calculated by the database. The best match between calculated and obtained from database is selected for further study. The library whose frequency distribution most closely matched that of the database is given the maximum score.

7.12.6 Chemistry Space Methods

Multidimensional analysis of certain parameters of molecules, distinct from the drug likeness to non-drug molecules, is based on the values obtained from calculations. Chemistry space is characterized by calculating a number of descriptors for each molecule and applying the descriptor values as a factor in multidimensional space.

7.12.7 Predicting Models Using Machine Learning Algorithm

To develop consistent models of drug-likeness prediction, numerous molecular descriptors and a number of machine learning approaches have been utilized. This is one of the successful methods for distinguishing drugs from non-drugs. Databases are utilized to statistically analyze the scores obtained and are helpful in learning the criteria that differentiates the drug and non-drug molecules. Recursive partitioning and neural network approaches (Kadam and Roy 2007) are feasible models under machine learning programs.

7.12.8 Recursive Partitioning Approach

The machine learning program known as the recursive partitioning approach is creates a decision tree of one-dimensional descriptors. By working from top to bottom, a decision is made based on the descriptors output obtained from the databases.

7.12.9 Neural Network Approach

Neural network approach strongly reflects the database heritage from which it is selected and behaves similar to a biological nervous system. This approach uses one-dimensional and two-dimensional descriptors to classify the drugs. This approach uses atom-type topological descriptors. The score from a single neuron output layer varies between 0.1 for non-drug to 0.9 for drug likeness.

7.12.10 Physicochemical Parameters Contributing to Drug Likeness Prediction

There are number of efforts to correlate physicochemical parameters with molecular level behavior of drugs. These parameters include but are not limited to molecular weight, lipophilicity measured as the logarithm of the octanol-water partition coefficient, intrinsic aqueous solubility, number of hydrogen bond donors and acceptors, molar refractivity, number of rings, and number of rotatable bonds. Molecular weight is correlated to intestinal and blood brain barrier permeability, where the molecules with greater molecular weight have less permeability. Absorption is well correlated with lipophilicity factors. More number hydrogen bond donors lower permeability across lipid bilayers. Oral drugs have fewer rotatable bonds compared with other classes of drugs.

7.12.11 Existing Criteria for Physicochemical Parameters to Predict Drug Likeness

Drug-likeness prediction models use various aspects of drug physicochemical properties not limited to solubility, permeability, metabolic stability, and transporter effects which are commonly used to describe bioavailability properties. The majority of models (Lipinski et al. 1997; Muegge et al. 2001; Tian et al. 2015) for drug-likeness use physicochemical properties calculated from the molecular structure and compares against drugs. The main aim of the various drug-likeness criteria (or rules) is to give researchers tools to eliminate compounds with a high risk of failure at an earlier stage of the discovery and development process. The following sections describe the various rules or criteria arrived at by various research teams.

(i) Lipinski Rule of Five is one of the first widely used criteria for the basis of all prediction models. Lipinksi states to be a drug a molecule should possess the size (molecular weight≤500 daltons), the hydrophobicity (logP≤5), and polarity (hydrogen-bond acceptors≤10 and donors≤5) of compound, and show good oral bioavailability behavior. If a compound violates one or more parameters of these limits, it may not be orally active. Changes have been made to these limits by various researchers to bring out new prediction models.

(ii) The Opera concept of drug likeness (modified concept of Lipinski), applying more rigorous threshold, limiting the size (molecular weight≤450), hydrophobicity (logP≤4.5), and polarity (hydrogen-bond acceptors≤8 and donors≤5).

(iii) Ghose framed the rule of drug likeness by defining logP between –0.4 and 5.6, molecular weight between 160 and 480, molar refractivity between 40 and 130, and total number of atoms between 20 and 70, including hydrogen bond donor and hydrogen bond acceptor. Found that 80% of the compounds satisfy the qualifying limits.

(iv) Egan rule predicts oral bioavailability by human intestinal absorption stating the logP value should not be more than 5.88 and topological polar surface area should not be more than 131.6.

(v) Varma (2010) rule is one of the recent criteria for determining drug likeness. The rule limits MW≤500 Da; logD (pH depended octanol-water partition coefficient) between –2 and 5 units, hydrogen bond acceptor and hydrogen bond donor ≤9, and number or rotational bonds ≤12. This rule combines the descriptors of the Rule of Five with the one for oral bioavailability.

(vi) Veber rule takes into account the rotatable bonds and polar surface area. Larger molecule to be orally bioavailable should have 10 or fewer rotatable bonds, and polar surface area should not be more than 140. Compact molecule is easier to absorb than extended one.

(vii) Muegge rule states it is a type of pharmacophore like filter to classify as drug like or non-drug like molecule. The limits are molecular weight 200–600 Da, logP between –2 to +5, topological surface area not more than 150, number of rings not more than 7, number of carbon atoms not less than 4, number of heteroatoms more than 1, number of rotatable bonds not more than 15, hydrogen bond donor atoms not more than 5, and hydrogen bond acceptor atoms should be more than 10. This is one of the simplest structural rules.

(viii) Gleeson et al. (2011) rule uses two descriptors size (MW) and hydrophobicity (logP). Introduced ADMET score, to simplify the drug-likeness rule, by using only two physicochemical properties, molecular weight and hydrophobicity (logP), to discriminate between oral and non-oral compounds. Maximum score of 1 for oral compound and 2 for non-oral compounds.

(ix) Hopkins (Bickerton et al. 2012) introduced a quantitative estimate of drug-likeness (QED) score which is a recent score. Hopkins prediction model directly includes aromaticity. It quantifies the compound quality. QED values range from zero (depicts all properties as unfavorable) to one (where all properties are favorable). QED suggests better way to quantify and rank the druggability of molecules according to the merits of chemical attractiveness with their related ligands. QED provides detailed information about drug likeness compared to other approaches. Hopkins QED method is more flexible by replacing the stiff cutoffs with a novel continuous index.

All rules to characterize physicochemical parameters and its impact on drug molecule have a defined threshold for each descriptor. They are relaxed in various ways according to the needs of the medicinal chemist to bring out a conclusion. Sometimes a few violations of set criteria are permissible according to the research projects undertaken.

7.12.12 Toxicity Prediction

It is very important to evaluate chemical safety as early as possible in order to reduce the harm of chemicals to our health. Common toxicity end points are acute toxicity, the adverse effects happening after administration of a single dose of the chemical. Conducting toxicity studies is time-consuming and application is not feasible for large numbers of compounds. Hence developing an *in silico* model to estimate toxicity level is an important approach. Computational methods are more advantageous than experimental approaches, as they are green, fast, cheap, accurate, and, importantly, they can be conducted before synthesizing the compound. US Environmental Protection Agency (US EPA) has established toxicity categories (Li et al. 2014) into 4 types as the dose limit (mg/kg) of oral acute toxicity for Category I (≤ 50), Category II ($>50–\leq 500$), Category III ($>500–\leq 5000$), Category IV (>5000). Category IV is generally regarded as practically nontoxic. Quantitative structure activity relationship (QSAR) models have been established to forecast acute rodent toxicity of chemicals. Toxicity end point of QSAR models were based on relatively small data sets or homologous compounds. Hence it exhibits poor generalization capacity. Multiclass models resulted in high predictive accuracy and have good prediction power for the external validation of data sets. Statistical methods (Horn et al. 2018) such as multilinear regression (MLR) and neural network (NN) were used to build models based on different data sets. Models were built using machine learning programs such as support vector machine (SVM), C4.5 decision tree (C4.5), random forest (RF), κ-nearest neighbor (kNN), and naive Bayes (NB) algorithms. In silico toxicity modules are inbuilt modules available at Discovery Studio's TOPKAT (USA), ADME-Tox Prediction of Advanced Chemistry Development (Canada). Toxicity Estimation Software Tools are free tools provided by US EPA which is mainly based on QSAR. ProTox (Drwal et al. 2014) is a web server tool for the prediction of toxicity in rodents, and it is open source.

7.13 RESOURCES

Evaluating the data sets of known drugs and non-drugs with the available data sets of the predicting molecule is the prime focus. This is generally done by the available drug database resources using web platforms to bring out the desired output. Databases and tools to predict drug likeness include ZINC, PubChem, ChEMBL, DrugBank, Comprehensive Medicinal Chemistry, World Drug Index. The predicting tools on the web and should be explored according to the need of the projects. A few of them are listed in Table 7.3.

TABLE 7.3   Drug Likeness Prediction Resources

Resources	Description
http://admet.scbdd.com/	ADMETLAB. A valuable tool for medicinal chemists in the drug discovery process
http://www.vcclab.org/lab/asnn/	ASNN too. This method uses the correlation between collective responses as a measure of distance amid the analyzed cases for the nearest neighbor technique
http://www.swissadme.ch/	BOILEDEgg, iLOGP and Bioavailability Radar. SwissADME Web tool computes physicochemical, pharmacokinetic, drug-like, and related parameters
http://chembcpp.scbdd.com/	ChemBCPP. Estimates several important chemical properties using a variety of QSAR methodologies
http://www.niper.gov.in/pi_dev_tools/DruLiToWeb/DruLiTo_index.html	DruLiTo. Virtual screening tool uses drug likeness rules
http://molsoft.com/mprop/	MolEdit. High speed Molecular properties prediction tool

The growing number of publications in the drug-like prediction area illustrates that this is one of the key areas of growth in medicinal chemistry. Current commercially available compounds fit the concept of drug-likeness following the above-mentioned rules. The different drug-likeness rules based on physicochemical properties are overlap with one another; the main differences between them are in the cut-off limits of the criteria. The validation of these rules for predicting drug likeness is growing as is the number of compounds of interest.

7.13.1 Monitoring Surface Microbiomes of Ready-To-Eat Foods

As science and technology progresses, so do genome sequencing methods. In 1977, Sanger developed the first sequencing method, and it is still used as a standard method. It is a process of fragmenting DNA, cloning, analyzing the sequence, and assembling it (also known as the shotgun method). However, it is very expensive and time-consuming. For example, human sequencing costs 13 years and 3 billion dollars. Since it is impossible to commercialize, several types of NGS methods have been developed. It is a large-scale sequencing method that reads a large amount of DNA sequences at once as quickly and cheaply as possible. There are high-throughput sequencing, long-read sequencing, and third generation sequencing, and the characteristics are as follows: (i) *High-throughput sequencing*: increases the number to the limit even if the length of the (read) sequence is slightly sacrificed; (ii) *Long-read sequencing*: increases the length of each maximal even if the number of (read) sequencing is slightly sacrificed; (iii) *Third generation sequencing*: reads DNA sequences directly without fragmentation and amplification of genes (Figure 7.8).

In the food industry, the identification of microorganism is critical, as all foods are affected by microorganisms, and this can lead to fermentation or spoilage. Microorganisms are essential for traditionally fermented foods, such as yogurt, cheese, beer, wine, kim chi,

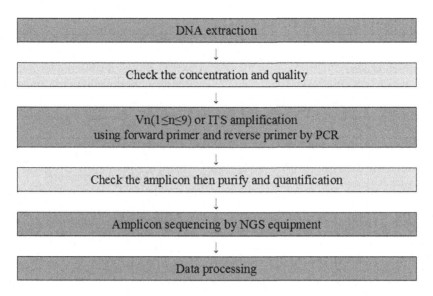

FIGURE 7.8   Investigation procedure of microorganisms by metagenome analysis.

and sufu (tofu). The genomic information can be obtained by NGS, and whole genome information can be used in the epidemiological investigation of pathogens. Metagenomic analysis can confirm the composition of bacteria in the sample without cultivation. Alkema et al. (2016) referenced NGS can directly profile the product, eliminating the need for a culture step which could create bias in the results, and it could well prove to be more specific for detecting spoilage bacteria from a product. Skipping the culturing step can be great benefit in terms of lower cost and material requirements. Meanwhile, in fermented products, 16S rRNA profiling can detect low abundance microbes that can overgrow in culture-dependent detection methods.

To identify the microbiome in a food sample, the following procedures are generally applied as follows. (i) DNA extraction, quality of extracted DNA by kit and concentration of extracted DNA is checked by Nano-Drop. The quality can be assessed by gel electrophoresis. (ii) Amplification by PCR and amplicon sequencing. Since the 16S rRNA in bacteria has 9 variable regions, which is related with the diversity between different species, it can be amplified and analyzed by NGS to obtain the distribution information of the species in sample (Schoch et al. 2012). In fungi, the internal transcribed spacer (ITS) is amplified and analyzed (Schoch et al. 2012). (iii) After amplification by PCR using variable region or ITS, PCR products are detected using gel electrophoresis, then purified (by kit) and quantified using a fluorometer. (iv) Finally, the sample can be sequenced (using Illumina, Thermo Fisher Scientific, Pacific Biosciences, Oxford Nanopore Technologies, etc.) (Table 7.4).

7.13.2 Data Processing

Bioinformatic analyses can be performed by QIIME, Mothur, or MG-RAST. We can analyze how many different types of bacteria are present (α-diversity) and the similarity or heterogeneity between samples (β-diversity) through aligning, clustering into operational taxonomic units (OTUs), and classifying, and obtaining the taxonomy information. The bioinformatics pipeline for analysis is shown in Table 7.5 (Plummer et al. 2015).

TABLE 7.4 Comparison of the Functionality and Features of QIIME, Mothur, and MG-RAST (Adapted from Plummer et al. 2015)

Functionality and Features	QIIME	Mothur	MG-RAST
License	Open-source	Open-source	Open-source
Implemented in	Python	C++	Perl
Current version (at March 13, 2015)	1.9.0	1.34.0	3.5
Cited (according to Scopus at April 8, 2015)	1769	2565	722
Website	http://qiime.org/	http://www.mothur.org/	http://metageromics.anl.gov
Web-based interface	YES (http://www.n3phele.com/) Not supported/maintained by the QIIME team	NO	YES (at website above)
Primary usage	Command line	Command line	GUI (at website above)
Amplicon analysis	YES	YES	YES
Whole metagenome shotgun analysis	YES—experimental only	NO	YES
Sequencing technology compatibility	Illumina, 454, Sanger, Ion Torrent, PacBio	Illumina, 454, Sanger, Ion Torrent, PacBio	Illumina, 454, Sanger, Ion Torrent, PacBio
Quality control	YES	YES	YES
16S rRNA gene Databases searched	RDP, SILVA, Greengenes, and custom databases	RDP, SILVA, Greengenes, and custom databases	M5RNA, RDP, SILVA, and Greengenes
Alignment Method	PyNAST, MUSCLE, INFERNAL	Needleman-Wunsch, blastn, gotoh	BLAT
Taxonomic analysis/ assignment	UCLUST, RDP, BLAST, Mothur	Wang/RDP approach	BLAT
Clustering algorithm	UCLUST, CD-HIT, Mothur, BLAST	mothur, adapts DOTUR and CD-HIT	UCLUST
Diversity analysis	alpha and beta	alpha and beta	alpha
Phylogenetic Tree	FastTree	Clearcut algorithm	YES

(Continued)

TABLE 7.4 (CONTINUED)   Comparison of the Functionality and Features of QIIME, Mothur, and MG-RAST (Adapted from Plummer et al. 2015)

Functionality and Features	QIIME	Mothur	MG-RAST
Chimera detection	UCHIME, chimera slayer, BLAST	UCHIME, chimera slayer, and more	NO
Visualisation	PCA plots, OUT networks, bar plots, heat maps	Dendrograms, heat maps, Venn diagrams, bar plots, PCA plots	PCA plots, heat maps, pie charts, bar plots, Krona and Circos for visualization
User Support	Forum, tutorials, FAQs, help videos	Forum, SOPs, FAWs, user manual	Video tutorials, FAQs, user manual, 'How to' section on website

Note: Where known, the algorithm used by each pipeline is named. The default algorithm, where known, is bolded. GUI: Graphical User Interface; RDP: Ribosomal Database Project; M5RNA: Non-redundant multisource ribosomal RNA annotation; PyNAST: PythonNAST; MUSCLE: MUltiple Sequence Comparison by Log-Expectation; INFERNAL: INFERence of RNA Alignment; BLAST: Basic Local Alignment Search Tool; BLAT: BLAST-Like Alignment Tool; CD-HIT: Cluster Database at High Identity with Tolerance; PCA: Principal Coordinate Analysis; OTU: Operational Taxonomic Unit; FAQ: Frequently Asked Questions; SOPs: Standard Operating Procedures.

TABLE 7.5 High-Throughput Sequencing Studies of Food Fermentations*

Target Gene	Short Description	Publication Year	Food Group	Reference
16S rRNA gene (Bacteria)	Kefir grains and kefir milk	2011	Dairy and fermented milks	Dobson et al.
16S rRNA gene (Bacteria)	Danish raw milk cheese during ripening	2011	Dairy and fermented milks	Masoud et al.
16S rRNA gene (Bacteria)	Kefir grains from different parts of Brazil	2012	Dairy and fermented milks	Leite et al.
16S rRNA gene (Bacteria)	Mozzarella cheese (Italy) and intermediates from two manufactures	2012	Dairy and fermented milks	Ercolini et al.
16S rRNA gene (Bacteria)	Latin style cheese	2012	Dairy and fermented milks	Lusk et al.
16S rRNA gene (Bacteria)	Curd, fresh and smoked Polish cheese (Oscypek)	2012	Dairy and fermented milks	Alegria et al.
16S rRNA gene (Bacteria)	Artisanal soft, semi-hard, and hard cheeses from raw or pasteurized cow, goat, or sheep milk	2012	Dairy and fermented milks	Quigley et al.
16S rRNA gene (Bacteria); ITS1-2 (Fungi)	Kefir grain and kefir milk from different sources	2013	Dairy and fermented milks	Bokulich et al.
16S rRNA gene (Bacteria); ITS1-2 (Fungi)	Swabs from cheesemaking environment and cheese	2013	Dairy and fermented milks	Fuka et al.
16S rRNA gene (Bacteria)	Turkish kefir grains	2014	Dairy and fermented milks	Nalbantoglu et al.
16S rRNA gene (Bacteria)	Whey cultures and cheese curds from water-buffalo mozzarella, Grana Padano, and Parmigiano Reggiano cheeses (Italy) manufacturing	2014	Dairy and fermented milks	De Filippis et al.
16S rRNA gene (Bacteria)	Ewe milk, curd and Canestrato cheese (Italy) during ripening	2014	Dairy and fermented milks	De Pasquale et al.
16S rRNA gene (Bacteria)	Cow milk, curd and Caciocavallo cheese (Italy) during ripening	2014	Dairy and fermented milks	De Pasquale et al.
16S rRNA gene (Bacteria)	Cow milk (from different lactation stages), curd and Fontina cheese (Italy) from three dairies	2014	Dairy and fermented milks	Dolci et al.
16S rRNA gene (Bacteria); ITS1-2 (Fungi)	Bloomy, natural, and washed cheese rinds	2014	Dairy and fermented milks	Wolfe et al.
16S rRNA gene (Bacteria)	Traditional Pico cheese (Portugal) manufactured in three different dairies, monitored during ripening	2014	Dairy and fermented milks	Riquelme et al.

(Continued)

TABLE 7.5 (CONTINUED) High-Throughput Sequencing Studies of Food Fermentations*

Target Gene	Short Description	Publication Year	Food Group	Reference
16S rRNA gene (Bacteria)	Samples of milk, whey, curd, and ripened Poro cheese (Mexico)	2014	Dairy and fermented milks	Aldrete-Tapia et al.
16S rRNA gene (Bacteria); ITS1-2 (Fungi)	Tarag (fermented dairy product) from China and Mongolia	2014	Dairy and fermented milks	Sun et al.
16S rRNA gene (Bacteria)	Samples of core and rind of Herve cheese (Belgium)	2014	Dairy and fermented milks	Delcenserie et al.
16S rRNA gene (Bacteria)	Chinese traditional fermented milk (yond bap) from cow or goat milk	2015	Dairy and fermented milks	Liu et al.
16S rRNA gene (Bacteria); 18S rRNA gene (Fungi)	Naturally fermented cow milks from Mongolia	2015	Dairy and fermented milks	Liu et al.
16S rRNA gene (Bacteria); 26S rRNA gene (Fungi)	Milk kefir grains from different Italian regions	2015	Dairy and fermented milks	Garofalo et al.
16S rRNA gene (Bacteria); ITS1-2 (Fungi)	Matsoni (fermented milk) samples from several geographic areas	2015	Dairy and fermented milks	Bokulich et al.
16S rRNA gene (Bacteria); 26S rRNA gene (Fungi)	Environmental swabs from a dairy plant and cheeses (Italy)	2015	Dairy and fermented milks	Stellato et al.
16S rRNA gene (Bacteria)	Continental cheese produced early and late in the day, at different ripening times	2015	Dairy and fermented milks	O'Sullivan et al.
16S rRNA gene (Bacteria)	Commercial high-moisture mozzarella cheese produced with different acidification methods	2016	Dairy and fermented milks	Guidone et al.
16S rRNA gene (Bacteria)	Undefined strain starters (milk cultures) for high-moisture mozzarella cheese	2016	Dairy and fermented milks	Parente et al.
16S rRNA gene (Bacteria)	Grana-type cheese (Italy) during ripening	2016	Dairy and fermented milks	Alessandria et al.
16S rRNA gene (Bacteria)	Natural whey culture, milk, curd and Caciocavallo cheese (Italy) during ripening	2016	Dairy and fermented milks	De Filippis et al.
16S rRNA gene (Bacteria)	Environmental swabs from a dairy plant and cheeses (Italy)	2016	Dairy and fermented milks	Calasso et al.

(*Continued*)

TABLE 7.5 (CONTINUED)   High-Throughput Sequencing Studies of Food Fermentations*

Target Gene	Short Description	Publication Year	Food Group	Reference
16S rRNA gene (Bacteria)	Spatial distribution of microbiota in Italian ewes' milk cheese	2016	Dairy and fermented milks	De Pasquale et al.
16S rRNA gene (Bacteria)	Rye, durum, and common wheat sourdough	2013	Doughs	Ercolini et al.
16S rRNA gene (Bacteria)	Traditional, sweet leavened doughs	2013	Doughs	Lattanzi et al.
16S rRNA gene (Bacteria)	Rye sourdoughs propagated for two months at 20 and 30°C	2014	Doughs	Bessmeltseva et al.
16S rRNA gene (Bacteria)	Flour and sourdough made of durum wheat grown under organic and conventional farming	2015	Doughs	Rizzello et al.
16S rRNA gene (Bacteria); 18S rRNA gene (Fungi)	Flour, doughs, and related food environments	2015	Doughs	Minervini et al.
16S rRNA gene (Bacteria)	Wheat sourdoughs used for traditional breads in different regions of France	2015	Doughs	Lhomme et al.
16S rRNA gene (Bacteria)	Sourdoughs used for the manufacture of traditional French breads	2015	Doughs	Lhomme et al.
16S rRNA gene (Bacteria)	Dough samples during manufacture of chica (a fermented maize product) in Argentina	2015	Doughs	Elizaquivel et al.
16S rRNA gene (Bacteria)	Rye sourdoughs from four Estonian bakeries	2015	Doughs	Elizaquivel et al.
16S rRNA gene (Bacteria)	Botrityzed wine during fermentation, three vintages, inoculated and uninoculated batches	2012	Fermented beverages–grapes	Bokulich et al.
16S rRNA gene (Bacteria); ITS1-2 (Fungi)	Grape must samples collected in California over two different vintages	2014	Fermented beverages–grapes	Bokulich et al.
16S rRNA gene (Bacteria); 26S rRNA gene and ITS1-2 (Fungi)	Grape must samples collected in different Portuguese regions during fermentation	2015	Fermented beverages–grapes	Pinto et al.
26S rRNA gene (Fungi)	Spanish grape must samples during fermentation	2015	Fermented beverages–grapes	Wang et al.
18S rRNA gene (Fungi)	Italian traditional wine fermentations	2016	Fermented beverages–grapes	De Filippis et al.

(*Continued*)

TABLE 7.5 (CONTINUED)   High-Throughput Sequencing Studies of Food Fermentations*

Target Gene	Short Description	Publication Year	Food Group	Reference
16S rRNA gene (Bacteria); ITS1-2 (Fungi)	Grape must samples collected in three wineries in Northern Italy during fermentation	2016	Fermented beverages–grapes	Stefanini et al.
16S rRNA gene (Bacteria)	Beer (American Coolship Ale) during fermentation and related environment	2012	Fermented beverages–malt	Bokulich et al.
16S rRNA gene (Bacteria)	Barley during malting, two different seasons	2014	Fermented beverages–malt	Juste et al.
16S rRNA gene (Bacteria); ITS1-2 (Fungi)	Traditional Korean alcoholic beverage (Makgeolli), and the starter (Nuruk), during fermentation	2012	Fermented beverages–rice	Jung et al.
16S rRNA gene (Bacteria); ITS1-2 (Fungi)	Kimoto sake during manufacturing and related environmental samples	2014	Fermented beverages–rice	Bokulich et al.
16S rRNA gene (Bacteria); ITS1-2 (Fungi)	Tea fungus (kombucha) samples during fermentation	2014	Fermented beverages–tea	Marsh et al.
16S rRNA gene (Bacteria); ITS1-4 (Fungi)	Pu-erh Japanese traditional tea during fermentation	2015	Fermented beverages–tea	Zhao et al.
16S rRNA gene (Bacteria)	Fermented meat (salami) during ripening	2015	Meat	Greppi et al.
16S rRNA gene (Bacteria)	Fermented meat (salami) during ripening	2015	Meat	Polka et al.
16S rRNA gene (Bacteria)	Fermented seafood	2010	Seafood	Roh et al.
16S rRNA gene (Bacteria)	Narezushi (salted and fermented fish, rice, peppers)	2011	Seafood	Koyanagi et al.
16S rRNA gene (Bacteria)	Traditional fermented sushi (kaburazushi), during fermentation	2013	Seafood	Koyanagi et al.
16S rRNA gene (Bacteria)	Traditional fermented shrimp (Saeu-jeot) during fermentation	2013	Seafood	Jung et al.
16S rRNA gene (Bacteria)	Traditional fermented shrimp (Saeu-jeot) during fermentation at different temperatures	2014	Seafood	Lee et al.
16S rRNA gene (Bacteria)	Korean fish sauce (Myeolchi-Aekjeot)	2015	Seafood	Lee et al.
16S rRNA gene (Bacteria)	Pearl millet slurried with or without groundnuts at the beginning and end of fermentation	2009	Vegetables	Humblot et al.

(Continued)

TABLE 7.5 (CONTINUED)   High-Throughput Sequencing Studies of Food Fermentations*

Target Gene	Short Description	Publication Year	Food Group	Reference
16S rRNA gene (Bacteria)	Vegetable pickle from rice bran (Nukadoko)	2011	Vegetables	Sakamoto et al.
16S rRNA gene (Bacteria)	Baechu (Chinese cabbage) and Chonggak (radish) kimchi prepared with and without starter	2012	Vegetables	Jung et al.
16S rRNA gene (Bacteria)	Fermented Korean soybean paste (Doenjang)	2012	Vegetables	Nam et al.
16S rRNA gene (Bacteria)	Traditional Korean fermented food (Kochujang) made of rice, pepper, soybeans	2012	Vegetables	Nam et al.
16S rRNA gene (Bacteria)	Ten different varieties of kim chi, during fermentation	2012	Vegetables	Park et al.
16S rRNA gene (Bacteria)	Olive surfaces and brine during fermentation	2013	Vegetables	Cocolin et al.
16S rRNA gene (Bacteria)	Kimchi samples during fermentation (100 days)	2013	Vegetables	Jeong et al.
16S rRNA gene (Bacteria)	Fermented Korean soybean lumps (meju)	2014	Vegetables	Jeong et al.
16S rRNA gene (Bacteria)	Started and unstarted Bella di Cerignola table olives	2015	Vegetables	De Angelis et al.

Note: Studies are grouped according to the type of food and ordered by year of publication (adapted from Filippis et al. 2017)

Several researchers have reported the microbiome from fermented food products. Fungi as well as bacteria also have been analyzed more in dairy, fermented milks, and beverages than other food categories (Figure 7.9). Irlinger et al. (2015) reviewed 33 cheese rind studies; they identified 104 bacterial genera and 39 fungal genera by types of cheese, cheese variety, and methods for identification. Wolfe et al. (2014) also analyzed cheese rinds (Figure 7.10). Meanwhile, Xu et al. (2020) characterized the fungal and bacterial communities during the production of sufu, a Chinese fermented food. (Figures 7.11, 7.12, 7.13). Table 7.5 (Filippis et al. 2017) details high-throughput sequencing studies on fermented foods.

Macromolecular structures and the endogenous flora can inhibit detection and tracing of specific microorganisms (Alkema et al. 2016). As seen above, amplicon sequencing using 16S rRNA, or ITS, has been widely used for food-related microbial research. It can be used to identify pathogens and understand the action of microorganisms in the manufacturing of fermented foods. Meanwhile, in order to process a large amount of data, capabilities such as data analysis, storage, and analysis are becoming important, and new data analysis technologies such as deep learning and block chain will be required in combination with the fourth "industrial revolution" (Alkema et al. 2016). Therefore, it is suggested that it is required not only to acquire the NGS technology but also the ability to process data.

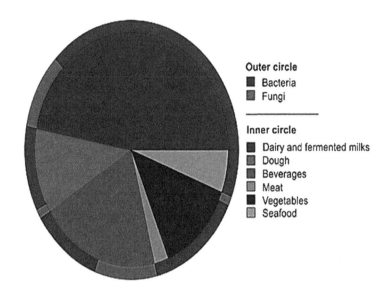

FIGURE 7.9 Pie chart showing the abundance of HTS studies of fermented foods and beverages grouped according to the food matrix. For each food environment, the outer circle shows the proportion of studies analyzing bacterial or fungal communities (adapted from Plummer et al. 2015).

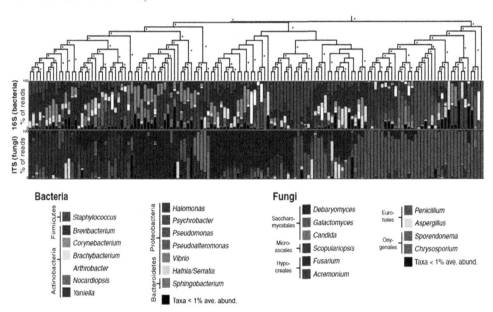

FIGURE 7.10 Distribution of abundant genera across cheese rind communities (adapted from Wolfe et al. 2014). Columns show relative abundance of genera within each cheese. Each column represents averaged data for multiple wheels of an individual cheese. Top row shows bacterial (16S rDNA) data and bottom row shows fungal (internal transcribed spacer or ITS) data. Communities were clustered using a UPGMA tree, and asterisks indicate clusters that were supported with >70% jackknife support. Only those genera that had an average abundance of 1% or greater across all samples are indicated; genera less than 1% abundance are combined and shown in black.

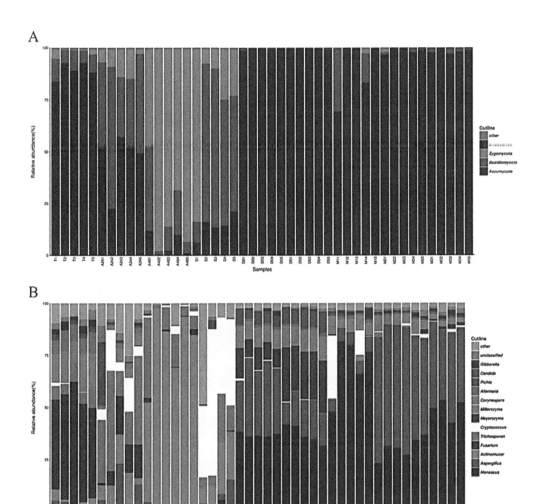

FIGURE 7.11   Relative abundance of fungal composition in sufu samples during the fermentation process at phylum level (A) and genus level (B) (adapted from Xu et al. 2020). Tofu (T), pehtze, inoculated with *A. elegans* for 24 h (A24), 48 h (A48), salt-pehtze (S), fermentation of sufu for 0 day (D0), 5 days (D5), 1 month (M1), 2 months (M2), 3 months (M3).

7.14 CONCLUSION REMARKS AND FUTURE PERSPECTIVES

Bioinformatics is progressively implemented in food fermentation and safety. Below we summarize some innovative new technologies. Sequence-based analysis of microbial features begins. An array of which functionality can be efficiently calculated using sequence data. Uniquely accessible datasets of genotype/phenotype/transcriptome, such as those reported for *Lactobacillus lactis* and *L. plantarum*, might help to develop novel sequence-based operational predictor techniques such as more precisely screened protein domains for sugar-based active enzymes and related promoter or functional binding positions to phenotype.

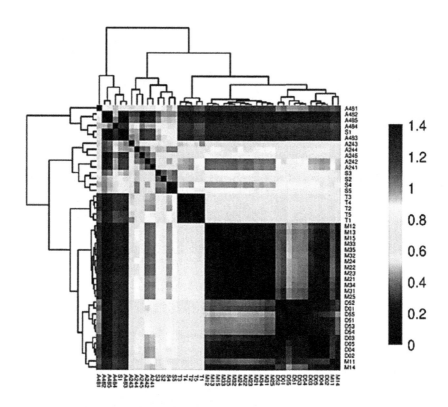

FIGURE 7.12 Analysis of similarity among sufu samples fermented by different times based on the relative abundance of fungal OTUs (operational taxonomic units) (adapted from Xu et al. 2020). Tofu (T), pehtze, inoculated with *A. elegans* for 24 h (A24), 48 h (A48), salt-pehtze (S), fermentation of sufu for 0 day (D0), 5 days (D5), 1 month (M1), 2 months (M2), 3 months (M3).

By diversifying the above details, a silico-based database of culture collections for desired traits can be created. This will involve databases that employ managed vocabulary to incorporate genomics data, systems biology, phenotypes, ingredient information, food sample properties, online parameter measurement during the food cycle, and 'biomarkers' for usability in complex taxonomy (based on GTM). The FAIR (easily accessible, open, integrated, reusable; http:/datafairport.org/) (Wilkinson et al. 2019) concept of storing data must be clearly emphasized. As analysis becomes more structured and computation-intensive, technology and databases can also be set up on a virtual machine that can then be operated on computation networks or the Cloud. Initial steps focused on data integration have been taken in the EU-funded GenoBox program (www.genobox.eu) (Alkema et al. 2016), which aims to build a profile that optimizes genotype and phenotype data in order to scan microbial genomes for versatility and hazard factors.

Likewise, IBM and MARS formed a partnership to sequence the food system (http :/www.research.ibm.com/client/foodsafety/) (Alkema et al. 2016). They seek to evaluate marginal amounts of microbial constituents in several food items worldwide. The subsequent database could be applied to determine the threats of such microbes/functionality in a food product. Since adequate biodiversity was documented in this database, it can also be used for tagging products based on specific microbiota footprints in fermented products or foods containing a microbiome. The major key factor in controlling fermentation

A

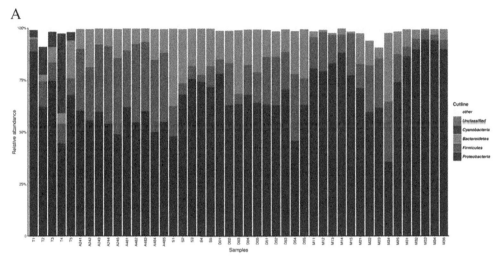

B

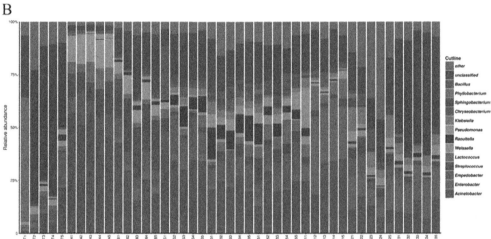

FIGURE 7.13 Relative abundance of bacterial composition in sufu samples during the fermentation process at phylum level (A) and genus level (B) (adapted from Xu et al. 2020). Tofu (T), pehtze which inoculated with *A. elegans* for 24 h (A24), 48 h (A48), salt-pehtze (S), fermentation of sufu for 0 day (D0), 5 days (D5), 1 month (M1), 2 months (M2), 3 months (M3).

efficiency is the associations involving microbes and their ecosystem. For example, this new element of specificity has been analyzed for the microbe–plant relationships of rice and coconut and the applied of systems biology across genome-scale metabolic models by using kinetic models to explain microbe–matrix correlations. Such research involves a comprehensive knowledge and understanding of both microorganism properties and the physical characteristics of the framework in which the organism resides. In summary, the growing quantity of data on food fermentation and sustainability encourages and centralizes the knowledge in datasets that can improve the assessment of fermentation success and sustainability with those of the appropriate research design, software, knowledge, and follow-up experiments.

The field of food microbiology has totally been transformed by high-throughput screening. This has further enabled the deep knowledge of the genomic characterization of starter cultures, probiotics, as well as foodborne pathogens. In addition to this, there is now a better understanding of the culture-independent analysis of mixed microbial communities in foods and food production facilities. Whole genome sequencing of food-related microbial isolates has advanced to the point that it is now commonly used in the verification of the safety of probiotic candidates and also in the detection of outbreaks of foodborne disease. Although these analyses are sequence-based, it could be also be said that culture-independent analyses have also provided certain valuable insights. Typically, the short sequence reads generated by current sequencers results in limited resolution. However, strain-level variations between microorganisms can help in influencing the organoleptic properties of foods, and hence strain-level resolution is far more desirable. We can only anticipate a time in the near future when it will be possible to use metagenomics to provide efforts to fine-tune fermentation processes and eventually provide reliable tests for the presence of foodborne pathogens.

BIBLIOGRAPHY

Ajawatanawong, P., Yanai, H., Smittipat, N., et al. 2019. A novel ancestral Beijing sublineage of *Mycobacterium tuberculosis* suggests the transition site to Modern Beijing sublineages. *Scientific Reports* 9(1):1–12.

Albin, D., Nasko, D., Elworth, R. L., et al. 2019. SeqScreen: A biocuration platform for robust taxonomic and biological process characterization of nucleic acid sequences of interest. In: *2019 IEEE international conference on bioinformatics and biomedicine (BIBM)* (pp. 1729–36). IEEE.

Aldrete-Tapia, A., Escobar-Ramırez, M. C., Tamplin, M. L. and Hernandez-Iturriaga, M. 2014. High-throughput sequencing of microbial communities in Poro cheese, anartisanal Mexican cheese. *Food Microbiology* 44:136–41.

Alegría, Á., Szczesny, P., Mayo, B., Bardowski, J. and Kowalczyk, M. 2012. Biodiversity in Oscypek, a traditional Polish cheese, determined by culture-dependent and -independent approaches. *Applied and Environmental Microbiology* 78(6):1890–8.

Alessandria, V., Ferrocino, I., De Filippis, F., Fontana, M., Rantsiou, K., Ercolini, D. and Cocolin, L. 2016. Micro-biota of an Italian Grana like cheese during manufacture and ripening unraveled by 16S rRNA-based approaches. *Applied and Environmental Microbiology* 82(13):3988–95.

Alfieri, F. 2018. The role of omic sciences in food security and sustainability. *Encyclopedia of Food Security and Sustainability* 8:44.

Alkema, W., Boekhorst, J., Wels, M. and Van Hijum, S. A. 2016. Microbial bioinformatics for food safety and production. *Briefings in Bioinformatics* 17(2):283–92.

Allard, A., Laurens, F. and Costes, E. 2016. Character integration, breeding goals compatibility and selection indexes using genome wide breeding values: A study case. In: *Program/Abstracts of the 8th International Rosaceae Genomics Conference* (pp. 21–4).

Anamika, K., Verma, S., Jere, A. and Desai, A. 2016. Transcriptomic profiling using next generation sequencing - advances, advantages, and challenges. *Next Generation Sequencing-Advances, Applications and Challenges* 9:7355–65.

Andrusier, N., Mashiach, E., Nussinov, R. and Wolfson, H. J. 2008. Principles of flexible protein–protein docking. *Proteins: Structure, Function, and Bioinformatics* 73(2):271–89.

Antony, U., Ilango, S., Chelliah, R., Ramakrishnan, S. R. and Ravichandran, K. 2020. Ethnic fermented foods and beverages of Tamil Nadu. In: *Ethnic fermented foods and beverages of India: Science history and culture* (pp. 539–60). Springer, Singapore.

Anvarian, A. H. P., Cao, Y., Srikumar, S., Fanning, S. and Jordan, K. 2016. Flow cytometric and 16S sequencing methodologies for monitoring the physiological status of the microbiome in powdered infant formula production. *Frontiers in Microbiology* 7:968

Ashinoff, R. 2000. Overview: Soft tissue augmentation. *Clinics in Plastic Surgery* 27(4):479–87.

Ashley, J. and Snow, G. 2007. Computerized nutrient analysis systems. In: *Handbook of nutrition and food* (pp. 103–10). CRC Press, Florida.

Ashton, P. M., Nair, S., Peters, T. M., Bale, J. A., Powell, D. G., Painset, A. and De Pinna, E. M. 2016. Identification of *Salmonella* for public health surveillance using whole genome sequencing. *Peer Journal* 4:e1752.

Ayub, M. Y. and Sachan, D. S. 1997. Dietary factors affecting aflatoxin b1 carcinogenicity. *Malaysian Journal of Nutrition* 3:161–79.

Azwai, S. M., Alfallani, E. A., Abolghait, et al. 2016. Isolation and molecular identification of *Vibrio* spp. by sequencing of 16S rDNA from seafood, meat and meat products in Libya. *Open Veterinary Journal* 6(1):36–43.

Bakhtiary, F., Sayevand, H. R., Remely, M., Hippe, B., Indra, A., Hosseini, H. and Haslberger, A. G. 2018. Identification of *Clostridium* spp. derived from a sheep and cattle slaughterhouse by matrix-assisted laser desorption and ionization-time of flight mass spectrometry (MALDI-TOF MS) and 16S rDNA sequencing. *Journal of Food Science and Technology* 55(8):3232–40.

Bartowsky, E. J. and Borneman, A. R. 2011. Genomic variations of *Oenococcus oeni* strains and the potential to impact on malolactic fermentation and aroma compounds in wine. *Applied Microbiology and Biotechnology* 92(3):441–7.

Benkerroum, N. 2019. Aflatoxins: Production, structure, health issues and incidence in Southeast Asian and sub-Saharan African countries. *International Journal of Environmental Research and Public Health* 17:423.

Bennett, J. W., Kale, S. and Yu, J. 2007. Aflatoxins: Background, toxicology, and molecular biology. In: *Foodborne diseases* (pp. 355–73). Springer Humana Press Inc., Totowa, NJ.

Bennett, J. W. and Klich, M. 2003. Mycotoxins. *Clinical Microbiology Reviews* 16(3):497–516.

Bessmeltseva, M., Viiard, E., Simm, J., Paalme, T. and Sarand, I. 2014. Evolution of bacterial consortia in spontaneously started rye sourdoughs during two months of daily propagation. *PLoS ONE* 9(4):e95449.

Bickerton, G. R., Paolini, G. V., Besnard, J., Muresan, S. and Hopkins, A. L. 2012. Quantifying the chemical beauty of drugs. *Nature Chemistry* 4(2):90.

Bircan, C. 2009. Incidence of ochratoxin A in dried fruits and co-occurrence with aflatoxins in dried figs. *Food and Chemical Toxicology* 47(8):1996–2001.

Bohra, A., Dubey, A., Saxena, R. K., et al. 2011. Analysis of BAC-end sequences (BESs) and development of BES-SSR markers for genetic mapping and hybrid purity assessment in pigeonpea (*Cajanus* spp.). *BMC Plant Biology* 11(1):56.

Bokulich, N. A., Bamforth, C. W. and Mills, D. A. 2012. Brewhouse-resident microbiota are responsible for multi-stage fermentation of American coolship ale. *PLoS ONE* 7(4):e35507.

Bokulich, N. A., Bergsveinson, J., Ziola, B. and Mills, D. A. 2015. Mapping microbial ecosystems and spoilage-gene flow in breweries highlights patterns of contamination and resistance. *eLife* 10:4. https://doi.org/10.7554/eLife.04634.

Bokulich, N. A. and Mills, D. A. 2013. Facility-specific "house" microbiome drives microbial landscapes of artisan cheese making plants. *Applied and Environmental Microbiology* 79(17):5214–23.

Bokulich, N. A., Thorngate, J. H., Richardson, P. M. and Mills, D. A. 2014. Microbial biogeography of wine grapes is conditioned by cultivar, vintage, and climate. *Proceedings of the National Academy of Sciences of the United States of America* 111(1):E139–48.

Bora, S. S., Keot, J., Das, S., Sarma, K. and Barooah, M. 2016. Metagenomics analysis of microbial communities associated with a traditional rice wine starter culture (Xajpitha) of Assam, India. *3Biotech* 6(2):153.

Bourne, M. 2002. *Food texture and viscosity: Concept and measurement.* Elsevier, New York.

Bron, P. A., Meijer, M., Bongers, R. S., De Vos, W. M. and Kleerebezem, M. 2007. Dynamics of competitive population abundance of *Lactobacillus plantarum* ivi gene mutants in faecal samples after passage through the gastrointestinal tract of mice. *Journal of Applied Microbiology* 103(5):1424–34.

Brown, E., Dessai, U., McGarry, S. and Gerner-Smidt, P. 2019. Use of whole-genome sequencing for food safety and public health in the United States. *Foodborne Pathogens and Disease* 16(7):441–50.

Bruder, K., Maiki, K., Cooper, A., Sible, E., Shapiro, J. W., Watkins, S. C. and Putonti, C. 2016. Freshwater metaviromics and bacteriophages: A current assessment of the state of the art in relation to bioinformatic challenges: Supplementary issue: bioinformatics methods and applications for big metagenomics data. *Evolutionary Bioinformatics Online* 12:EBO–S38549.

Brul, S., Bassett, J., Cook, P., Kathariou, S., McClure, P., Jasti, P. R. and Betts, R. 2012. 'Omics' technologies in quantitative microbial risk assessment. *Trends in Food Science and Technology* 27(1):12–24.

Buchanan, R. L. and Appel, B. 2010. Combining analysis tools and mathematical modeling to enhance and harmonize food safety and food defense regulatory requirements. *International Journal of Food Microbiology* 139(S1):S48–56.

Buermans, H. P. J. and Den Dunnen, J. T. 2014. Next generation sequencing technology: Advances and applications. *Biochimica et Biophysica Acta (BBA) - Molecular Basis of Disease* 1842(10):1932–41.

Calasso, M., Ercolini, D., Mancini, L., et al. 2016. Relationships among house, rind and core microbiotas during manufacture of traditional Italian cheeses at the same dairy plant. *Food Microbiology* 54:115–26.

Cao, Y., Fanning, S., Proos, S., Jordan, K. and Srikumar, S. 2017. A review on the applications of next generation sequencing technologies as applied to food-related microbiome studies. *Frontiers in Microbiology* 8:1829.

Carocho, M., Barreiro, M. F., Morales, P. and Ferreira, I. C. 2014. Adding molecules to food, pros and cons: A review on synthetic and natural food additives. *Comprehensive Reviews in Food Science and Food Safety* 13(4):377–99.

Chalker, A. F. and Lunsford, R. D. 2002. Rational identification of new antibacterial drug targets that are essential for viability using a genomics-based approach. *Pharmacology and Therapeutics* 95(1):1–20.

Charrondière, U. R., Rittenschober, D., Nowak, V. and Stadlmayr, B. 2012. FAO/ INFOODS analytical food. Composition database version 1.0–AnFooD1. 0. *International Network of Food Data Systems (INFOODS)*.

Chauhan, K., Dhakal, R., Seale, R. B., et al. 2013. Rapid identification of dairy mesophilic and thermophilic spore forming bacteria using DNA high resolution melt analysis of variable 16S rDNA regions. *International Journal of Food Microbiology* 165(2):175–83.

Chelliah, R., Ramakrishnan, S. R., Prabhu, P. R. and Antony, U. 2016. Evaluation of antimicrobial activity and probiotic properties of wild-strain *Pichia kudriavzevii* isolated from frozen idli batter. *Yeast* 33(8):385–401.

Chelliah, R., Wei, S., Park, B. J. et al. 2019. New perspectives on Mega plasmid sequence (poh1) in *Bacillus thuringiensis* ATCC 10792 harboring antimicrobial, insecticidal and antibiotic resistance genes. *Microbial Pathogenesis* 126:14–8.

Chun, J., Oren, A., Ventosa, A., et al. 2018. Proposed minimal standards for the use of genome data for the taxonomy of prokaryotes. *International Journal of Systematic and Evolutionary Microbiology* 68(1):461–6.

Church, S. M. 2009. EuroFIR synthesis Report no 7: Food composition explained. *Nutrition Bulletin* 34(3):250–72.

Cifuentes, A. 2009. Food analysis and foodomics. *Journal of Chromatography. Part A* 1216(43):7109. https://doi.org/10.1016/j.chroma.2009.09.018.

Cifuentes, A. 2017. Foodomics, foodome and modern food analysis. *TrAC Trends in Analytical Chemistry* 96:1. doi:10.1016/j.trac.2017.09.001.

Cocolin, L., Alessandria, V., Botta, C., Gorra, R., De Filippis, F., Ercolini, D. and Rantsiou, K. 2013. NaOH-debittering induces changes in bacterial ecology during table olives fermentation. *PLoS ONE* 8(7):e69074.

Codex Alimentarius Commission, Joint FAO/WHO Food Standards Programme, and World Health Organization 2007. *Codex alimentarius commission: Procedural manual*. FAO.

Corella, D., Arnett, D. K., Tucker, K. L., et al. 2011. A high intake of saturated fatty acids strengthens the association between the fat mass and obesity-associated gene and BMI. *The Journal of Nutrition* 141(12):2219–25.

Daliri, E. B. M., Wei, S., Oh, D. H. and Lee, B. H. 2017. The human microbiome and metabolomics: Current concepts and applications. *Critical Reviews in Food Science and Nutrition* 57(16):3565–76.

Dallman, T. J., Chattaway, M. A., Cowley et al. 2014. An investigation of the diversity of strains of enteroaggregative *Escherichia coli* isolated from cases associated with a large multi-pathogen foodborne outbreak in the UK. *PLoS ONE* 9(5):e98103.

De Angelis, M., Campanella, D., Cosmai, L., Summo, C., Rizzello, C. G. and Caponio, F. 2015. Microbiota and metabolome of un-started and started Greek-type fermentation of Bella di Cerignola table olives. *Food Microbiology* 52:18–30.

De Filippis, F. and Ercolini, D. 2016. Food microbial ecology in the "omics" era. In: *Reference module in food sciences*. Smithers, G. W. (ed.) (pp. 1–7). Elsevier, USA.

De Filippis, F., La Storia, A. and Blaiotta, G. 2017. Monitoring the mycobiota during Greco di Tufo and Aglianico wine fermentation by 18S rRNA gene sequencing. *Food Microbiology* 63:117–22.

De Filippis, F., La Storia, A., Stellato, G., Gatti, M. and Ercolini, D. 2014. A selected core microbiome drives the early stages of three popular Italian cheese manufactures. *PLoS ONE* 9(2):e89680.

De Filippis, F., Parente, E. and Ercolini, D. 2017. Metagenomics insights into food fermentations. *Microbial Biotechnology* 10(1):91–102.

De Garnica, M. L., Sáez-Nieto, J. A., González, R., Santos, J. A. and Gonzalo, C. 2014. Diversity of gram-positive catalase-negative cocci in sheep bulk tank milk by comparative 16S rDNA sequence analysis. *International Dairy Journal* 34(1):142–5.

De Pasquale, I., Calasso, M., Mancini, L., et al. 2014. Causal relationship between microbial ecology dynamics and proteolysis during manufacture and ripening of protected designation of origin (PDO) cheese Canestrato Pugliese. *Applied and Environmental Microbiology* 80(14):4085–94.

De Paula, A. T., Jeronymo-Ceneviva, A. B., Todorov, S. D. and Penna, A. L. B. 2015. The two faces of *Leuconostoc mesenteroides* in food systems. *Food Reviews International* 31(2):147–71.

Delcenserie, V., Taminiau, B., Delhalle, L., et al. 2014. Microbiota characterization of a Belgian protected designation of origincheese, Herve cheese, using metagenomic analysis. *Journal of Dairy Science* 97(10):6046–56.

Deliza, R., Rosenthal, A., Abadio, F. B. D., Silva, C. H. and Castillo, C. 2005. Application of high pressure technology in the fruit juice processing: Benefits perceived by consumers. *Journal of Food Engineering* 67(1–2):241–6.

DeSantis, T. Z., Brodie, E. L., Moberg, J. P., Zubieta, I. X., Piceno, Y. M. and Andersen, G. L. 2007. High-density universal 16S rRNA microarray analysis reveals broader diversity than typical clone library when sampling the environment. *Microbial Ecology* 53(3):371–83.

Deurenberg, R. H., Bathoorn, E., Chlebowicz, M. A., et al. 2017. Application of next generation sequencing in clinical microbiology and infection prevention. *Journal of Biotechnology* 243:16–24.

Dick, G. 2018. *Genomic approaches in earth and environmental sciences*. John Wiley and Sons, Michigan.

Didelot, X., Bowden, R., Wilson, D. J., Peto, T. E. and Crook, D. W. 2012. Transforming clinical microbiology with bacterial genome sequencing. *Nature Reviews Genetics* 13(9):601–12.

Dobson, A., O'Sullivan, O., Cotter, P. D., Ross, P. and Hill, C. 2011. High-throughput sequence-based analysis of the bacterial composition of kefir and an associated kefirgrain. *FEMS Microbiology Letters* 320(1):56–62.

Dolci, P., De Filippis, F., La Storia, A., Ercolini, D. and Cocolin, L. 2014. rRNA-based monitoring of the micro-biota involved in Fontina PDO cheese production in relation to different stages of cow lactation. *International Journal of Food Microbiology* 185:127–35.

Dors, G. C., Caldas, S. S., Feddern, V., et al. 2011. Aflatoxins: Contamination, analysis and control. *Embrapa Suínos e Aves-Capítulo em Livro Científico (Alice)*, 20: 415-38.

Dröge, J. and McHardy, A. C. 2012. Taxonomic binning of metagenome samples generated by next-generation sequencing technologies. *Briefings in Bioinformatics* 13(6):646–55.

Drwal, M. N., Marinello, J., Manzo, S. G., Wakelin, L. P., Capranico, G. and Griffith, R. 2014. Novel DNA topoisomerase IIα inhibitors from combined ligand-and structure-based virtual screening. *PLoS ONE* 9(12):e114904.

Dugat-Bony, E., Straub, C., Teissandier, A., et al. 2015. Overview of a surface-ripened cheese community functioning by meta-omics analyses. *PLoS ONE* 10(4):e0124360.

Elizaquível, P., Pérez-Cataluña, A., Yépez, A., et al. 2015. Pyrosequencing vs. culture-dependent approaches to analyze lactic acid bacteria associated to chicha, a traditional maize-based fermented beverage from Northwestern Argentina. *International Journal of Food Microbiology* 198:9–18.

Eom, H. S., Hwang, B. H., Kim, D. H., Lee, I. B., Kim, Y. H. and Cha, H. J. 2007. Multiple detection of food-borne pathogenic bacteria using a novel 16S rDNA-based oligonucleotide signature chip. *Biosensors and Bioelectronics* 22(6):845–53.

Ercolini, D. 2013. High throughput sequencing and metagenomics: Moving forward in the culture-independent analysis of food microbial ecology. *Applied and Environmental Microbiology* 79(10):3148–55.

Ercolini, D., De Filippis, F., La Storia, A. and Iacono, M. 2012. "Remake" by high-throughput sequencing of the microbiota involved in the production of water buffalo Mozzarella cheese. *Applied and Environmental Microbiology* 78(22):8142–5.

Escobar-Zepeda, A., Sanchez-Flores, A. and Baruch, M. Q. 2016. Metagenomic analysis of a Mexican ripened cheese reveals a unique complex microbiota. *Food Microbiology* 57:116–27.

Ezponda, T., Alkorta-Aranburu, G., Prósper, F. and Agirre, X. 2020. Genotyping and sequencing. In: *Principles of nutrigenetics and nutrigenomics* (pp. 33–9), Academic Press, London, UK.

Fakruddin, M., Mannan, K. S. B., Chowdhury, A., Mazumdar, R. M., Hossain, M. N., Islam, S. and Chowdhury, M. A. 2013. Nucleic acid amplification: Alternative methods of polymerase chain reaction. *Journal of Pharmacy and Bioallied Sciences* 5(4):245.

Fang, H., Xu, J., Ding, D., et al. 2010. An FDA bioinformatics tool for microbial genomics research on molecular characterization of bacterial foodborne pathogens using microarrays. *BMC Bioinformatics* 11.

Ferrocino, I. and Cocolin, L. 2017. Current perspectives in food-based studies exploiting multi-omics approaches. *Current Opinion in Food Science* 13:10–5.

Franz, R., Clavero, C., Kolbeck, J. and Anders, A. 2016. Influence of ionisation zone motion in high power impulse magnetron sputtering on angular ion flux and NbO x film growth. *Plasma Sources Science and Technology* 25(1). http://www.ncbi.nlm.nih.gov/pubmed/015022.

Fuka, M. M., Wallisch, S., Engel, M., Welzl, G., Havranek, J. and Schloter, M. 2013. Dynamics of bacterial communities during the ripening process of different Croatian cheese types derived from raw ewe's milk cheeses. *PLoS ONE* 8(11):e80734.

Gane and Dean 2000. Recent advances in structure-based rational drug design. *Current Opinion in Structural Biology* 10(4):401–4.

Garbaj, A. M., Awad, E. M., Azwai, S. M., et al. 2016. Enterohemorrhagic *Escherichia coli* O157 in milk and dairy products from Libya: Isolation and molecular identification by partial sequencing of 16S rDNA. *Veterinary World* 9(11):1184.

García-Cañas, V., Simó, C., Herrero, M., Ibáñez, E. and Cifuentes, A. 2012. Present and future challenges in food analysis: Foodomics. *Analytical Chemistry* 84(23):10150–9.

Garofalo, C., Osimani, A., Milanovic, V., et al. 2015. Bacteria and yeast microbiota in milk kefir grains from different Italian regions. *Food Microbiology* 49:123–33.

Gendel, S. M. and Jenkins, J. A. 2006. Allergen sequence databases. *Molecular Nutrition and Food Research* 50(7):633–7.

Gerner-Smidt, P., Besser, J., Concepción-Acevedo, J., et al. 2019. Whole genome sequencing: Bridging one-Health surveillance of foodborne diseases. *Frontiers in Public Health* 7:172.

Gleeson, M., Bishop, N. C., Stensel, D. J., Lindley, M. R., Mastana, S. S. and Nimmo, M. A. 2011. The anti-inflammatory effects of exercise: Mechanisms and implications for the prevention and treatment of disease. *Nature Reviews Immunology* 11(9):607–15.

Gourama, H. and Bullerman, L. B. 1995. *Aspergillus flavus* and *Aspergillus parasiticus*: Aflatoxigenic fungi of concern in foods and feeds: A review. *Journal of Food Protection* 58(12):1395–404.

Grattepanche, J. D. and Katz, L. A. 2020. Top-down and bottom-up controls on micro-eukaryotic diversity (ie, amplicon analyses of SAR lineages) and function (ie, meta-transcriptome analyses) assessed in microcosm experiments. *Frontiers in Marine Science* 6:818.

Greppi, A., Ferrocino, I., La Storia, A., Rantsiou, K., Ercolini, D. and Cocolin, L. 2015. Monitoring of the microbiota of fermented sausages by culture independent rRNA-based approaches. *International Journal of Food Microbiology* 212:67–75.

Gu, C., Kim, G. B., Kim, W. J., Kim, H. U. and Lee, S. Y. 2019. Current status and applications of genome-scale metabolic models. *Genome Biology* 20(1):121.

Guidone, A., Matera, A., Ricciardi, A., Zotta, T., De Filippis, F., Ercolini, D. and Parente, E. 2016. The microbiota of high-moisture Mozzarella cheese produced with different acidification methods. *International Journal of Food Microbiology* 216:9–17.

Gunay-Esiyok, O., Akcelik, N. and Akcelik, M. 2014. Identification of genomic heterogeneity among *Lactococcus lactis* strains by plasmid profiling, PFGE and 16S rDNA sequence analysis. *Polish Journal of Microbiology* 63(2):157–66.

Halperin, I., Ma, B., Wolfson, H. and Nussinov, R. 2002. Principles of docking: An overview of search algorithms and a guide to scoring functions. *Proteins: Structure, Function, and Bioinformatics* 47(4):409–43.

Hansen, T., Skånseng, B., Hoorfar, J. and Löfström, C. 2013. Evaluation of direct 16S rDNA sequencing as a metagenomics-based approach to screening bacteria in bottled water. *Biosecurity and Bioterrorism: Biodefense Strategy, Practice, and Science* 11:S158–65.

Hao, P., Zheng, H., Yu, Y., et al. 2011. Complete sequencing and pan-genomic analysis of *Lactobacillus delbrueckii* subsp. *bulgaricus* reveal its genetic basis for industrial yogurt production. *PLoS ONE* 6(1):e15964.

Heavner, M. E., Gueguen, G., Rajwani, R., Pagan, P. E., Small, C. and Govind, S. 2013. Partial venom gland transcriptome of a Drosophila parasitoid wasp, *Leptopilina heterotoma*, reveals novel and shared bioactive profiles with stinging Hymenoptera. *Gene* 526(2):195–204.

Hedge, J. and Wilson, D. J. 2016. Practical approaches for detecting selection in microbial genomes. *PLoS Computational Biology* 12(2):e1004739.

Herrero, M., Simó, C., García-Cañas, V., Ibáñez, E. and Cifuentes, A. 2012. Foodomics: MS-based strategies in modern food science and nutrition. *Mass Spectrometry Reviews* 31(1):49–69.

Herreros, M. A., Fresno, J. M., Prieto, M. G. and Tornadijo, M. E. 2003. Technological characterization of lactic acid bacteria isolated from Armada cheese (a Spanish goats' milk cheese). *International Dairy Journal* 13(6):469–79.

Hoffmann, S., Devleesschauwer, B., Aspinall, W., et al. 2017. Attribution of global foodborne disease to specific foods: Findings from a World Health Organization structured expert elicitation. *PLoS ONE* 12(9):e0183641.

Holton, A. E., Canary, H. E. and Wong, B. 2017. Business and breakthrough: Framing (expanded) genetic carrier screening for the public. *Health Communication* 32(9):1051–8.

Holton, T. A., Vijayakumar, V. and Khaldi, N. 2013. Bioinformatics: Current perspectives and future directions for food and nutritional research facilitated by a food-wiki database. *Trends in Food Science and Technology* 34(1):5–17.

Horn, B., Esslinger, S., Pfister, M., Fauhl-Hassek, C. and Riedl, J. 2018. Non-targeted detection of paprika adulteration using mid infrared spectroscopy and one-class classification – Is it data preprocessing that makes the performance? *Food Chemistry* 257:112–9.

Hosseini, H., Hippe, B., Denner, E., Kollegger, E. and Haslberger, A. 2012. Isolation, identification and monitoring of contaminant bacteria in Iranian Kefir type drink by 16S rDNA sequencing. *Food Control* 25(2):784–8.

Hüfner, E. and Hertel, C. 2008. Improvement of raw sausage fermentation by stress-conditioning of the starter organism *Lactobacillus sakei. Current Microbiology* 57(5):490–6.

Humblot, C. and Guyot, J. P. 2009. Pyrosequencing of tagged 16S rRNA gene amplicons for rapid deciphering of the microbiomes of fermented foods such as pearl millet slurries. *Applied and Environment Microbiology* 75(13):4354–61.

Irlinger, F., Layec, S., Hélinck, S. and Dugat-Bony, E. 2015. Cheese rind microbial communities: Diversity, composition and origin. *FEMS Microbiology Letters* 362(2):1–11.

Jackson, K. A., Bohm, M. K., Brooks, J. T., et al. 2018. Invasive methicillin-resistant *Staphylococcus aureus* infections among persons who inject drugs-six sites, 2005–2016. *MMWR. Morbidity and Mortality Weekly Report* 67(22):625.

Jean-Marc, N., Yves, V. d. P., Peter, D. R., Sabine, C. and Rupert, D. W. 1993. Compilation of small ribosomal subunit RNA structures. *Nucleic Acids Research* 14(13):3025–49.

Jeong, S. H., Lee, S. H., Jung, J. Y., Choi, E. J. and Jeon, C. O. 2013. Microbial succession and metabolite changes during long-term storage of Kimchi. *Journal of Food Science* 78(5):M763–9.

Jiang, Y., Xiong, X., Danska, J. and Parkinson, J. 2016. Metatranscriptomic analysis of diverse microbial communities reveals core metabolic pathways and microbiome-specific functionality. *Microbiome* 4(1):2.

Jung, J. Y., Lee, S. H., Jin, H. M., Hahn, Y., Madsen, E. L. and Jeon, C. O. 2013. Metatranscriptomic analysis of lactic acid bacterial gene expression during kimchi fermentation. *International Journal of Food Microbiology* 163(2–3):171–9.

Jung, J. Y., Lee, S. H., Lee, H. J. and Jeon, C. O. 2013. Microbial succession and metabolite changes during fermentation of saeu-jeot: Traditional Korean salted seafood. *Food Microbiology* 34(2):360–8.

Jung, M. J., Nam, Y. D., Roh, S. W. and Bae, J. W. 2012a. Unexpected convergence of fungal and bacterial communities during fermentation of traditional Korean alcoholic beverages inoculated with various natural starters. *Food Microbiology* 30(1):112–23.

Kačániová, M., Klūga, A., Kántor, A., Medo, J., Žiarovská, J., Uchalski, C. P. and Terentjeva, M. 2019. Comparison of MALDI-TOF MS biotyper and 16S rDNA sequencing for the identification of *Pseudomonas* species isolated from fish. *Microbial Pathogenesis* 132:313–8.

Kadam, R. U. and Roy, N. 2007. Recent trends in drug-likeness prediction: A comprehensive review of in silico methods. *Indian Journal of Pharmaceutical Sciences* 69(5):609.

Kannan, D., Chelliah, R., Vinolya Rajamanickam, E., Srinivasan Venkatraman, R. and Antony, U. 2015. Fermented batter characteristics in relation with the sensory properties of idli. *Hrvatski Časopis za Prehrambenu Tehnologiju, Biotehnologiju i Nutricionizam* 10(1–2):37–43.

Kaul, S., Sharma, T. and Dhar, M. 2016. "Omics" tools for better understanding the plant–endophyte interactions. *Frontiers in Plant Science* 7:955.

Keller, A., Horn, H., Förster, F. and Schultz, J. 2014. Computational integration of genomic traits into 16S rDNA microbiota sequencing studies. *Gene* 549(1):186–91.

Köppen, R., Koch, M., Siegel, D., Merkel, S., Maul, R. and Nehls, I. 2010. Determination of mycotoxins in foods: Current state of analytical methods and limitations. *Applied Microbiology and Biotechnology* 86(6):1595–612.

Köser, C. U., Holden, M. T., Ellington, M. J., et al. 2012. Rapid whole-genome sequencing for investigation of a neonatal MRSA outbreak. *New England Journal of Medicine* 366(24):2267–75.

Koyanagi, T., Nakagawa, A., Kiyohara, M., et al. 2013. Pyrosequencing analysis of microbiota in Kaburazushi, a traditional medieval. *Bioscience, Biotechnology and Biochemistry* 77(10):2125–30.

Kramer, A. and Szczesniak, A. S. 2012. *Texture measurement of foods: Psychophysical fundamentals; sensory, mechanical, and chemical procedures, and their interrelationships*. Springer Science, Doredrecht.

Lattanzi, A., Minervini, F., Di Cagno, R., et al. 2013. The lactic acid bacteria and yeast microbiota of eighteen sourdoughs used for the manufacture of traditional Italian sweet leavened baked goods. *International Journal of Food Microbiology* 163(2–3):71–9.

Laxminarayan, R., Duse, A., Wattal, C., et al. 2013. Antibiotic resistance - The need for global solutions. *The Lancet Infectious Diseases* 13(12):1057–98.

Lee, S. and Bae, S. 2018. Molecular viability testing of viable but non-culturable bacteria induced by antibiotic exposure. *Microbial Biotechnology* 11(6):1008–16. https://doi.org/10.1111/1751-7915.13039.

Leite, A. M., Mayo, B., Rachid, C. T., Peixoto, R. S., Silva, J. T., Paschoalin, V. M. and Delgado, S. 2012. Assessment of the microbial diversity of Brazilian kefir grains by PCR-DGGE and Pyrosequencing analysis. *Food Microbiology* 31(2):215–21.

Lhomme, E., Lattanzi, A., Dousset, X., et al. 2015a. Lactic acid bacterium and yeast microbiotas of sixteen French traditional sourdoughs. *International Journal of Food Microbiology* 215:1.

Li, F., Hitch, T. C., Chen, Y., Creevey, C. J. and Guan, L. L. 2019. Comparative metagenomic and metatranscriptomic analyses reveal the breed effect on the rumen microbiome and its associations with feed efficiency in beef cattle. *Microbiome* 7(1):6.

Li, G., Shen, M., Yang, Y., et al. 2018. Adaptation of *Pseudomonas aeruginosa* to phage PaP1 predation via O-antigen polymerase mutation. *Frontiers in Microbiology* 9:1170.

Li, Y., Li, J., Li, W. and Du, H. 2014. A state-of-the-art review on magnetorheological elastomer devices. *Smart Materials and Structures* 23(12).

Lienau, E. K., Strain, E., Wang, C., et al. 2011. Identification of a salmonellosis outbreak by means of molecular sequencing. *New England Journal of Medicine* 364(10):981–2.

Lindqvist, R., Langerholc, T., Ranta, J., Hirvonen, T. and Sand, S. 2019. A common approach for ranking of microbiological and chemical hazards in foods based on risk assessment-useful but is it possible? *Critical Reviews in Food Science and Nutrition*:1–14.

Lipinski, C. A., Lombardo, F., Dominy, B. W. and Feeney, P. J. 1997. Experimental and computational approaches to estimate solubility and permeability in drug discovery and development settings. *Advanced Drug Delivery Reviews* 23(1–3):3–25.

Liu, X.-F., Liu, C.-J., Zhang, H., Gong, F., Luo, Y. and Li, X. 2015a. The bacterial community structure of yond bap, a traditional fermented goat milk product, from distinct Chinese regions. *Dairy Science and Technology* 95(3):369–80.

Loman, N. J. and Pallen, M. J. 2015. Twenty years of bacterial genome sequencing. *Nature Reviews in Microbiology* 13(12):787–94.

Lusk, T. S., Ottesen, A. R., White, J. R., Allard, M. W., Brown, E. W. and Kase, J. A. 2012. Characterization of microflora in Latin-style cheeses by next-generation sequencing technology. *BMC Microbiology* 12:254.

Maarleveld, T. R., Khandelwal, R. A., Olivier, B. G., Teusink, B. and Bruggeman, F. J. 2013. Basic concepts and principles of stoichiometric modeling of metabolic networks. *Biotechnology Journal* 8(9):997–1008.

Maldonado-Gómez, M. X., Martínez, I., Bottacini, F., et al. 2016. Stable engraftment of *Bifidobacterium longum* AH1206 in the human gut depends on individualized features of the resident microbiome. *Cell, Host and Microbe* 20(4):515–26.

Marios, M., Valentina, A., Ilario, F., Kalliopi, R. and Luca, C. 2018. A bioinformatics pipeline integrating predictive metagenomics profiling for the analysis of 16S rDNA/rRNA sequencing data originated from foods. *Food Microbiology* 76:279–86.

Marroki, A., Zúñiga, M., Kihal, M. and Pérez-Martínez, G. 2011. Characterization of *Lactobacillus* from Algerian goat's milk based on phenotypic, 16S rDNA sequencing and their technological properties. *Brazilian Journal of Microbiology* 42(1):158–71.

Marsh, A. J., O'Sullivan, O., Hill, C., Ross, R. P. and Cotter, P. D. 2013. Sequencing-based analysis of the bacterial and fungal composition of kefir grains and milks from multiple sources. *PLoS ONE* 8(7):e69371.

Martens, E. C., Lowe, E. C., Chiang, et al. 2011. Recognition and degradation of plant cell wall polysaccharides by two human gut symbionts. *PLoS Biology* 9(12):e1001221.

Masoud, W., Takamiya, M., Vogensen, F. K., Lillevang, S., Al-Soud, W. A., Sørensen, S. J. and Jakobsen, M. 2011. Characterization of bacterial populations in Danish raw milk cheeses made with different starter cultures by denaturing gradient gel electrophoresis and pyrosequencing. *International Dairy Journal* 21(3):142–8.

Minervini, F., Lattanzi, A., De Angelis, M., Celano, G. and Gobbetti, M. 2015. House microbiotas as sources of lactic acid bacteria and yeasts in traditional Italian sourdoughs. *Food Microbiology* 52:66–76.

Minkiewicz, P., Dziuba, J., Iwaniak, A., Dziuba, M. and Darewicz, M. 2008. BIOPEP database and other programs for processing bioactive peptide sequences. *Journal of AOAC International* 91(4):965–80.

Morea, M., Baruzzi, F. and Cocconcelli, P. S. 1999. Molecular and physiological characterization of dominant bacterial populations in traditional Mozzarella cheese processing. *Journal of Applied Microbiology* 87(4):574–82.

Møretrø, T. and Langsrud, S. 2017. Residential bacteria on surfaces in the food industry and their implications for food safety and quality. *Comprehensive Reviews in Food Science and Food Safety* 16(5):1022–41.

Moskowitz, H. R., German, J. B. and Saguy, I. S. 2005. Unveiling health attitudes and creating good-for-you foods: The genomics metaphor, consumer innovative web-based technologies. *Critical Reviews in Food Science and Nutrition* 45(3):165–91.

Muegge, I. and Rarey, M. 2001. Small molecule docking and scoring. *Reviews in Computational Chemistry* 17:1–60.

Nalbantoglu, U., Cakar, A., Dogan, H., Abaci, N., Ustek, D., Sayood, K. and Can, H. 2014. Metagenomic analysis ofthe microbial community in kefir grains. *Food Microbiology* 41:42–51.

Nam, Y. D., Lee, S. Y. and Lim, S. I. 2012a. Microbial community analysis of Korean soybean pastes by next-generation sequencing. *International Journal of Food Microbiology* 155(1–2):36–42.

Neltner, T. G., Alger, H. M., Leonard, J. E. and Maffini, M. V. 2013. Data gaps in toxicity testing of chemicals allowed in food in the United States. *Reproductive Toxicology* 42:85–94.

Nielsen, R., Paul, J. S., Albrechtsen, A. and Song, Y. S. 2011. Genotype and SNP calling from next-generation sequencing data. *Nature Reviews Genetics* 12(6):443–51.

O'Flaherty, S. and Klaenhammer, T. R. 2011. The impact of omic technologies on the study of food microbes. *Annual Review of Food Science and Technology* 2:353–71.

O'Hare, E., Scopes, D. I., Kim, E. M., et al. 2013. Orally bioavailable small molecule drug protects memory in Alzheimer's disease models. *Neurobiology of Aging* 34(4):1116–25.

O'Sullivan, D. J., Cotter, P. D., O'Sullivan, O., Giblin, L., McSweeney, P. L. and Sheehan, J. J. 2015. Temporal and spatial differences in microbial composition during the manufacture of a continental-type cheese. *Applied and Enviromental Microbiology* 81(7):2525–33.

Oliver, S. P., Jayarao, B. M. and Almeida, R. A. 2005. Foodborne pathogens in milk and the dairy farm environment: Food safety and public health implications. *Foodborne Pathogens and Disease* 2(2):115–29.

Orellana, L. H., Hatt, J. K., Iyer, R., et al. 2019. Comparing DNA, RNA and protein levels for measuring microbial dynamics in soil microcosms amended with nitrogen fertilizer. *Scientific Reports* 9(1):1–11.

Otero, J. M., Cimini, D., Patil, K. R., Poulsen, S. G., Olsson, L. and Nielsen, J. 2013. Industrial systems biology of *Saccharomyces cerevisiae* enables novel succinic acid cell factory. *PLoS ONE* 8(1):e54144.

Palittapongarnpim, P., Ajawatanawong, P., Viratyosin, W., et al. 2018. Evidence for host-bacterial co-evolution via genome sequence analysis of 480 Thai *Mycobacterium tuberculosis* lineage 1 isolates. *Scientific Reports* 8(1):1–14.

Pandya, J. P. and Arade, P. C. 2016. Mycotoxin: A devil of human, animal and crop health. *Advanced in Life Science* 5:3937–41.

Parente, E., Cocolin, L., De Filippis, F., et al. 2016a. FoodMicrobionet: A data-base for the visualization and exploration of food bacterial communities based on network analysis. *International Journal of Food Microbiology* 219:28–37.

Park, E. J., Chun, J., Cha, C. J., Park, W. S., Jeon, C. O. and Bae, J. W. 2012. Bacterial community analysis during fermentation of ten representative kinds of kimchi with bar-coded pyrosequencing. *Food Microbiology* 30(1):197–204.

Park, S. H., Kim, H. U., Kim, T. Y., Park, J. S., Kim, S.-S. and Lee, S. Y. 2014. Metabolic engineering of *Corynebacterium glutamicum* for L-arginine production. *Nature Communications* 5(1):1–9.

Pinto, C., Pinho, D., Cardoso, R., et al. 2015. Wine fermentation microbiome: A landscape from different Portuguese wine appellations. *Frontiers in Microbiology* 16:905.

Plummer, E., Twin, J., Bulach, D. M., Garland, S. M. and Tabrizi, S. N. 2015. A comparison of three bioinformatics pipelines for the analysis of preterm gut microbiota using 16S rRNA gene sequencing data. *Journal of Proteomics and Bioinformatics* 8(12):283–91.

Połka, J., Rebecchi, A., Pisacane, V., Morelli, L. and Puglisi, E. 2015. Bacterial diversity in typical Italian salami at different ripening stages as revealed by high-throughput sequencing of 16S rRNA amplicons. *Food Microbiology* 46:342–56.

Pouliot, M., Hoffman, T. J., Stierli, D., Beaudegnies, R., El Qacemi, M. and Pitterna, T. 2019. U.S. Patent Application No. 16/087, 448.

Pouliot, R., Hugron, S. and Rochefort, L. 2015. Sphagnum farming: A long-term study on producing peat moss biomass sustainably. *Ecological Engineering* 74:135–47.

Quan, T. P., Bawa, Z., Foster, D., et al. 2018. Evaluation of whole-genome sequencing for mycobacterial species identification and drug susceptibility testing in a clinical setting: A large-scale prospective assessment of performance against line probe assays and phenotyping. *Journal of Clinical Microbiology* 56(2):e01480-17.

Quigley, L., O'Sullivan, O., Beresford, T. P., Ross, R. P., Fitzgerald, G. F. and Cotter, P. D. 2012a. High-throughput sequencing for detection of subpopulations of bacteria not previously associated with artisanal cheeses. *Applied and Environmental Microbiology* 78(16):5717–23.

Rantsiou, K., Englezos, V., Torchio, F., et al. 2017. Modeling of the fermentation behavior of *Starmerella bacillaris*. *American Journal of Enology and Viticulture* 68(3):378–85.

Rezac, S., Kok, C. R., Heermann, M. and Hutkins, R. 2018. Fermented foods as a dietary source of live organisms. *Frontiers in Microbiology* 9:1785.

Riquelme, C., Camara, S., Dapkevicius Mde, L., Vinuesa, P., da Silva, C. C., Malcata, F. X. and Rego, O. A. 2015. Characterization of the bacterial biodiversity in Picocheese (an artisanal Azorean food). *International Journal of Food Microbiology* 192:86–94.

Rizzello, C. G., Cavoski, I., Turk, J., et al. 2015. The organic cultivation of *Triticum turgidum* ssp. *durum* reflects on the axis flour, sour-dough fermentation and bread. *Applied and Environmental Microbiology* 81(9):3192–204.

Roh, S. W., Kim, K. H., Nam, Y. D., Chang, H. W., Park, E. J. and Bae, J. W. 2010. Investigation of archaeal and bacterial diversity in fermented seafood using barcoded Pyrosequencing. *ISME Journal* 4(1):1–16.

Sakamoto, N., Tanaka, S., Sonomoto, K. and Nakayama, J. 2011. 16S rRNA Pyrosequencing-based investigation of the bacterial community in nukadoko, a pickling bed of fermented rice bran. *Applied and Environmental Microbiology* 144(3):352–9.

Salazar, J. K., Carstens, C. K., Ramachandran, P., et al. 2018. Metagenomics of pasteurized and unpasteurized Gouda cheese using targeted 16S rDNA sequencing. *BMC Microbiology* 18(1):189.

Schoch, C. L., Seifert, K. A., Huhndorf, S., et al. 2012. Nuclear ribosomal internal transcribed spacer (ITS) region as a universal DNA barcode marker for fungi. *Proceedings of the National Academy of Sciences of the United States of America* 109(16):6241–6.

Schoustra, S. E., Kasase, C., Toarta, C., Kassen, R. and Poulain, A. J. 2013. Microbial community structure of three traditional Zambian fermented products: Mabisi, chibwantu and munkoyo. *PLoS ONE* 8(5):e63948.

Sekse, C., Holst-Jensen, A., Dobrindt, U., Johannessen, G. S., Li, W., Spilsberg, B. and Shi, J. 2017. High throughput sequencing for detection of foodborne pathogens. *Frontiers in Microbiology* 8:2029.

Sekse, R. J. T., Hunskår, I. and Ellingsen, S. 2018. The nurse's role in palliative care: A qualitative meta-synthesis. *Journal of Clinical Nursing* 27(1–2):e21–38.

Shakya, M., Lo, C. C. and Chain, P. S. 2019. Advances and challenges in metatranscriptomic analysis. *Frontiers in Genetics* 10:904.

Shoichet, B. K. and Kuntz, I. D. 1993. Matching chemistry and shape in molecular docking. *Protein Engineering, Design and Selection* 6(7):723–32.

Singh, T., Biswas, D. and Jayaram, B. 2011. AADS-An automated active site identification, docking, and scoring protocol for protein targets based on physicochemical descriptors. *Journal of Chemical Information and Modeling* 51(10):2515–27.

Slatko, B. E., Gardner, A. F. and Ausubel, F. M. 2018. Overview of next-generation sequencing technologies. *Current Protocols in Molecular Biology* 122(1):e59.

Smith, C., Ordovas, J., Sanchez-Moreno, C., Lee, Y. and Garaulet, M. 2012. Apolipoprotein A-II polymorphism: Relationships to behavioural and hormonal mediators of obesity. *International Journal of Obesity* 36(1):130–6.

Soeryapranata, E., Powers, J. R., Hill, H. H. Jr, Siems, W. F. III, Al-Saad, K. A. and Weller, K. M. 2002. Matrix-assisted laser desorption/ionization time-of-flight mass spectrometry method for the quantification of β-casein fragment (f 193-209). *Journal of Food Science* 67(2):534–8.

Soni, K. A., Nannapaneni, R. and Tasara, T. 2011. The contribution of transcriptomic and proteomic analysis in elucidating stress adaptation responses of *Listeria monocytogenes*. *Foodborne Pathogens and Disease* 8(8):843–52.

Stefanini, I., Albanese, D., Cavazza, A., Franciosi, E., De Filippo, C., Donati, C. and Cavalieri, D. 2016. Dynamic changes in microbiota and mycobiota during spontaneous "Vino Santo Trentino" fermentation. *Microbial Biotechnology* 9(2):195–208.

Stellato, G., De Filippis, F., La Storia, A. and Ercolini, D. 2015. Coexistence of lactic acid bacteria and potential spoilage microbiota in a dairy-processing environment. *Applied and Environmental Microbiology* 81(22):7893–904.

Stephen, R. C., David, W. G., Sandor, V. and Carlos, J. C. 2004. ClusPro: A fully automated algorithm for protein–protein docking. *Nucleic Acids Research* 32:S1–2.

Stroka, J. and Anklam, E. 2002. New strategies for the screening and determination of aflatoxins and the detection of aflatoxin-producing moulds in food and feed. *TrAC Trends in Analytical Chemistry* 21(2):90–5.

Sudmant, P. H., Rausch, T., Gardner, E., et al. 2015. An integrated map of structural variation in 2,504 human genomes. *Nature* 526(7571):75–81.

Sun, Z., Liu, W., Bao, Q., et al. 2014. Investigation of bacterial and fungal diversity in Tarag using high-throughput sequencing. *Journal of Dairy Science* 97(10):6085–96.

Sun, Z. Y., Xue, L. R. and Zhang, K. 2015. A new approach to finite-time adaptive stabilization of high-order uncertain nonlinear system. *Automatica* 58:60–6.

Takaoka, Y., Endo, Y., Yamanobe, S., et al. 2003. Development of a method for evaluating drug-likeness and ease of synthesis using a data set in which compounds are assigned scores based on chemists' intuition. *Journal of Chemical Information and Computer Sciences* 43(4):1269–75.

Teusink, B., Wiersma, A., Jacobs, L., Notebaart, R. A. and Smid, E. J. 2009. Understanding the adaptive growth strategy of *Lactobacillus plantarum* by in silico optimisation. *PLoS Computational Biology* 5(6):e1000410.

Thomas, F., Hehemann, J. H., Rebuffet, E., Czjzek, M. and Michel, G. 2011. Environmental and gut bacteroidetes: The food connection. *Frontiers in Microbiology* 2:93.

Thum, T. and Condorelli, G. 2015. Long noncoding RNAs and microRNAs in cardiovascular pathophysiology. *Circulation Research* 116(4):751–62.

Tian, S., Wang, J., Li, Y., Li, D., Xu, L. and Hou, T. 2015. The application of in silico drug-likeness predictions in pharmaceutical research. *Advanced Drug Delivery Reviews* 86:2–10.

Timme, R. E., Rand, H., Shumway, M., et al. 2017. Benchmark datasets for phylogenomic pipeline validation, applications for foodborne pathogen surveillance. *Peer Journal* 5:e3893.

Tringe, S. G. and Rubin, E. M. 2005. Metagenomics: DNA sequencing of environmental samples. *Nature Reviews Genetics* 6(11):805–14.

Udenigwe, C. C., Gong, M. and Wu, S. 2013. In silico analysis of the large and small subunits of cereal RuBisCO as precursors of cryptic bioactive peptides. *Process Biochemistry* 48(11):1794–9.

Vakser, I. A. 2014. Protein-protein docking: From interaction to interactome. *Biophysical Journal* 107(8):1785–93.

Van Hoorde, K., Butler, F. and Butler, F. 2018. Use of next-generation sequencing in microbial risk assessment. *EFSA Journal* 16(S1):e16086.

Varma, M., Field, R., Stinson, M., Rukovets, B., Wymer, L. and Haugland, R. 2009. Quantitative real-time PCR analysis of total and propidium monoazide-resistant fecal indicator bacteria in wastewater. *Water Research* 43(19):4790–801.

Verma, A. K., Singh, S., Singh, S. and Dubey, A. 2012. 16S rDNA sequence based characterization of bacteria in stored jaggery in Indian jaggery manufacturing units. *Sugar Tech* 14(4):422–7.

Verma, R. 2010. Customer choice modeling in hospitality services: A review of past research and discussion of some new applications. *Cornell Hospitality Quarterly* 51(4):470–8.

Wages, J. A., Feye, K. M., Park, S. H., Kim, S. and Ricke, S. C. 2019. Comparison of 16S rDNA next sequencing of microbiome communities from post-scalder and post-picker stages in three different commercial poultry plants processing three classes of broilers. *Frontiers in Microbiology* 10:972.

Walsh, A. M. 2018. *High-throughput sequencing-based characterisation of fermented foods and their impacts on host gut microbiota.* PhD Thesis. University College, Cork.

Walters, S. M. 1998. U.S. Patent No. 5,751,661. U.S. Patent and Trademark Office, Washington, DC.

Walters, W. P. 2012. Going further than Lipinski's rule in drug design. *Expert Opinion on Drug Discovery* 7(2):99–107.

Walters, W. P. and Murcko, M. A. 2002. Prediction of 'drug-likeness'. *Advanced Drug Delivery Reviews* 54(3):255–71.

Wang, C., García-Fernández, D., Mas, A. and Esteve-Zarzoso, B. 2015. Fungal diversity in grape must and wine fermentation assessed by massive sequencing, quantitative PCR and DGGE. *Frontiers in Microbiology* 6:1156.

Ward, C. M., To, H. and Pederson, S. M. 2018. ngsReports: An R package for managing FastQC reports and other NGS related log files. *bioRxiv.* http://www.ncbi.nlm.nih.gov/pubmed/313148.

Wegkamp, A., Teusink, B., De Vos, W. and Smid, E. 2010. Development of a minimal growth medium for *Lactobacillus plantarum*. *Letters in Applied Microbiology* 50(1):57–64.

Weng, F. Y., Chiou, C. S., Lin, P. H. P. and Yang, S. S. 2009. Application of recA and rpoB sequence analysis on phylogeny and molecular identification of *Geobacillus* species. *Journal of Applied Microbiology* 107(2):452–64.

WHO. 2019. Nutrition in universal health coverage. No. WHO/NMH/NHD/19.24. World Health Organization, Geneva.

Wilkinson, M. D., Dumontier, M., Aalbersberg, I. J., et al. 2019. Addendum: The FAIR guiding principles for scientific data management and stewardship. *Scientific Data* 6(1):1–2.

Williams, J. G. and Anderson, J. P. 2008. U.S. Patent No. 7,462,452. U.S. Patent and Trademark Office, Washington, DC.

Wittwer, J., Rubio-Aliaga, I., Hoeft, B., Bendik, I., Weber, P. and Daniel, H. 2011. Nutrigenomics in human intervention studies: Current status, lessons learned and future perspectives. *Molecular Nutrition and Food Research* 55(3):341–58.

Wolfe, A. L., Singh, K., Zhong, Y., et al. 2014. RNA G-quadruplexes cause eIF4A-dependent oncogene translation in cancer. *Nature* 513(7516):65–70.

Wolfe, B. E., Button, J. E., Santarelli, M. and Dutton, R. J. 2014. Cheese rind communities provide tractable systems for in situ and in vitro studies of microbial diversity. *Cell* 158(2):422–33.

Woolfit, M., Rozpędowska, E., Piškur, J. and Wolfe, K. H. 2007. Genome survey sequencing of the wine spoilage yeast *Dekkera* (*Brettanomyces*) *bruxellensis*. *Eukaryotic Cell* 6(4):721–33.

Wright, J. D., Sargsyan, K., Wu, X., Brooks, B. R. and Lim, C. 2013. Protein–protein docking using EMAP in CHARMM and support vector machine: Application to Ab/Ag complexes. *Journal of Chemical Theory and Computation* 9(9):4186–94.

Wu, A. R., Neff, N. F., Kalisky, T., et al. 2014. Quantitative assessment of single-cell RNA-sequencing methods. *Nature Methods* 11(1):41.

Wyres, K. L., Conway, T. C., Garg, S., Queiroz, C., Reumann, M., Holt, K. and Rusu, L. I. 2014. WGS analysis and interpretation in clinical and public health microbiology laboratories: What are the requirements and how do existing tools compare? *Pathogens* 3(2):437–58.

Xu, D., Grishin, N. V. and Chook, Y. M. 2012. NESdb: A database of NES-containing CRM1 cargoes. *Molecular Biology of the Cell* 23(18):3673–6.

Xu, D., Wang, P., Zhang, X., Zhang, J., Sun, Y., Gao, L. and Wang, W. 2020. High-throughput sequencing approach to characterize dynamic changes of the fungal and bacterial communities during the production of Sufu, a traditional Chinese fermented soybean food. *Food Microbiology* 86:103340.

Xu, J., Kelly, R., Fang, H. and Tong, W. 2010. ArrayTrack: A free FDA bioinformatics tool to support emerging biomedical research-an update. *Human Genomics* 4(6):1–7.

Yap, K. P., Gan, H. M., Teh, C. S. J., Chai, L. C. and Thong, K. L. 2014. Comparative genomics of closely related *Salmonella enterica* serovar *typhi* strains reveals genome dynamics and the acquisition of novel pathogenic elements. *BMC Genomics* 15(1):1007.

Zhang, D., Palmer, J., Teh, K. H., Biggs, P. and Flint, S. 2019. 16S rDNA high-throughput sequencing and MALDI-TOF MS are complementary when studying psychrotrophic bacterial diversity of raw cows' milk. *International Dairy Journal* 97:86–91.

Zhao, M., Zhang, D. L., Su, X. Q., et al. 2015. An integrated metagenomics/metaproteomics investigation of the microbial communities and enzymes in solid-state fermentation of Puerhtea. *Scientific Reports* 5(1):10117.

PART II

Generation of Sequencing Technologies

CHAPTER 8

Evaluation of Sequencing Technologies in Food Science

Jorianne Thyeska Castro Alves, Mônica Silva de Oliveira,
Gislenne Moia, Rosyely da Silva Oliveira,
Ronilson Santos dos Santos,
Pablo Henrique Caracciolo Gomes de Sá,
and Adonney Allan de Oliveira Veras

CONTENTS

8.1 INTRODUCTION

Genetic sequencing is a technique that includes biochemical processes which allow identification of each nucleotide in a DNA strand. Sequencing methods have undergone several changes over the years, mainly regarding the cost-benefit. Next-generation sequencing (NGS) platforms are currently the standard (Vincent et al. 2017).

NGS platforms generate thousands of base pairs of DNA sequencing quickly. Its speed and affordable price contributed to this technology gaining more visibility in the scientific community. NGS emerged in mid-2005 (Matthijs et al. 2016).

During the last few years, sequencing technologies have made great strides in solving the systematic errors of previous sequencing platforms, achieving a balance between reading length and cost. Thus, the third generation of sequencing technology emerged, such as Oxford Nanopore and PacBio (Winand et al. 2019). In this, the unique DNA molecules are directly sequenced, thus reducing the bias associated with amplification (Cao et al. 2017).

Sequencing techniques have applications in diverse fields of research (biology, genetic diseases, cancer, forensics, pharmacogenomics, pathology, precision medicine, among others). For example, in agriculture, where the system of food production is extremely complex, the genetic information obtained through NGS provides the means to produce food that is nutritious, safe, and accessible (Thottathil et al. 2016).

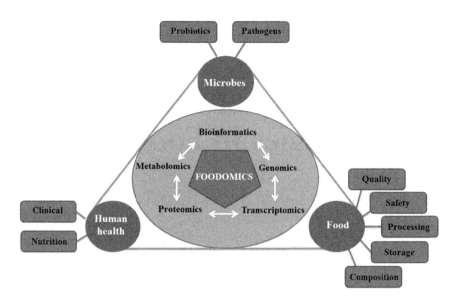

FIGURE 8.1   Flowchart of the main areas and applications of foodomics.

Food science and technology contribute to the success of the modern food system by integrating diverse fields of study, such as biology, chemistry, physics, engineering, microbiology, nutrition, toxicology, biotechnology, genomics, computer science, and other disciplines that provide scientific knowledge for solving problems associated with the food system (Figure 8.1) (Floros et al. 2010).

It is the understanding and application of science that satisfies the needs of society for quality and safe, and sustainable food. The scientific approach applied to understanding food generates solutions that directly impact in the industry and the consumer, improving the food supply, and ensuring that foods are largely safe, tasty, abundant, nutritious, and less costly (Platt 2009).

The impact of modern methods of food production and manufacturing is evident in improved food quality, stability, safety. Other impacts include essential nutrient preservation, addition of vitamins and minerals, removal of toxins, and food design to optimize health and reduce the risk of disease (Floros et al. 2010).

New technologies are constantly being developed, which are essential for the food processing industry as it aims to produce fresh, authentic, convenient, and tasty food products and extend the life of the products it produces (Alfadul and Elneshwy 2010).

This science is still relatively new, being used primarily as a response to the many recurring changes in the developed world. High demands for easily prepared foods represent a major challenge for the food industry since only qualified professionals are able to understand the complexity of food systems.

8.2   CURRENT SEQUENCING PLATFORMS

As previously mentioned, NGS platforms have revolutionized the field of genetics in diverse fields of application. The sequencing platforms most commonly used in research today are discussed below.

Illumina's platform first appeared in 2006 and was quickly adopted by many researchers around the world, due to its cost-benefit, yield, and length of reads (Malla et al. 2019). The platforms follow the principle of sequencing by synthesis (SBS), incorporating reversible chain terminator nucleotides for all four bases; each base is labeled with a different fluorescent marker and the use of a DNA polymerase (Ansorge 2009).

Currently, Illumina has a wide portfolio of platforms that meet different types of needs; among them, the sequencing platforms MiSeq and HiSeqs are the most established. MiSeq was designed as a personal and fast bench sequencer, with short sequencing times of up to 4 hours and sequencing throughput targeted at small genomes. HiSeq, on the other hand, was designed for high performance applications, producing a data quantity of 1 terabyte in 6 days. The HiSeq 2500 model can also be run in fast mode which, despite being less economical, can produce a human genome in just 27 hours (Reuter et al. 2015).

A study by Liu et al. (2020) sought to analyze the metagenomics and metatranscriptomics of the microbial community structure and also the metabolic potential of fermented soybeans. The DNA and RNA of the fermented soybeans were sequenced on an Illumina HiSeq 2500 platform. At the end of the sequencing, it was verified that 92,192,276 raw reads were generated for the metagenomic sequences, with an average of 13.83 Gb and 150 bp in length; and for the metatranscriptomic sequences, 38,798,262 reads were generated with an average of 11 Gb and with an average length of 151 bp (Liu et al. 2020).

Pacific Biosciences (PacBio) is one of the most widely used third generation platforms, launched in 2010. It adopted SBS, like Illumina, but also incorporates fluorescent markers similar to other next generation platforms. The sequencing is done in single-molecule real-time (SMRT) sequencing in which signals are emitted after the incorporation of the nucleotides is detected. PacBio generates a relatively small number of very long reads as opposed to other NGS platforms that normally generate many short reads (Weirather et al. 2017).

Ion Torrent brought the personal genome machine (PGM) to the market in mid-2010. It also works with SBS; however, it detects hydrogen ions, which are released during the polymerization of DNA that occurs when its bases are incorporated into growth chains. Ion Torrent allows for a quick and simple workflow that easily adapts to the needs of researchers in various areas including infectious diseases, reproductive genomics, identification, agrigenomics, among others. PGM can provide approximately 5.5 million readings with an average length of 400 bp, producing a maximum of 2 Gb of data (Adamiak et al. 2018).

Ion Proton, an evolution of Ion Torrent, is a platform based on semiconductors. It can be applied to several research areas such as exome sequencing, transcriptome sequencing, and single-cell sequencing. It generates up to 10 Gb of data with a read length of up to 200 bp, with the execution time varying from 2-4 hours (Zhang et al. 2017).

Oxford Nanopore sequencers emerged as a portable and competitive technology. Through them, readings greater than 150 kb were obtained. The MinION sequencer is a 90 g portable device, which is connected to a computer via USB. This is known for generating ultra-long readings that can reach up to 2 Mb in length, in addition to having a throughput of up to 30 Gb and sequencing time of less than 2 hours (Jain et al. 2016).

8.3 FOOD SCIENCE AND SEQUENCING

Ensuring the quality of food produced is a difficult task. The US Centers for Disease Control and Prevention (CDC), and the Food and Drug Administration (FDA) bear responsibility for overseeing control of food safety, the advances in technology have been instrumental in improving the food safety process and identifying and preventing microbial and chemical risks in food (Brown et al. 2019).

NGS technology has innovated the way DNA sequencing is applied in food safety. This technology is ideal in national and international surveillance systems in support of food security and outbreak responses. NGS has revolutionized the detection of microbiological sources of diseases that are occasionally transmitted by food (Brown et al. 2019).

Although other food safety tests exist, like methods of cultivation based on polymerase chain reaction (PCR) and antigens, the wait for results can take multiple days. In contrast, the results of tests performed through the NGS platforms can be obtained in just a single day–a different but effective methodology (Kumar and Chordia 2017).

Bioinformatics is an important ally in studies of the application of food processing. For example, the study by Waidha et al. points to the processes that help to stabilize food longer through the use of living organisms (usually microorganisms) that optimize safety and conservation. Bioinformatics improves this process by increasing the chance of obtaining positive results through the integration of biomaterials. The study highlights the example of the cocoa plant which takes 3–5 years before starting to form fruits that are widely used in the manufacture of chocolate. Fingerprints are carried out to screen restriction fragment length polymorphism (RFLP) markers in the plants to be used in genotypic detection (their characteristics). It is bioinformatics that makes possible the selection of desired characteristics in the initial stages of the plant (Waidha et al. 2015).

In 2019, Choeisoongnern and colleagues conducted a study to track and identify bacteriocin-like inhibitory substances that produce the lactic acid bacteria found in fermented products. In their work, many methods were used for the processing of samples, such as the isolation of lactic acid bacteria, the characterization of bacteriocin type inhibitors, and random amplification of polymorphic DNA (RAPD), among others. There was also the execution of the 16S rDNA gene sequencing process, in which the representative strains of each RAPD profile group were subjected to genomic DNA isolation and amplification of 16S rDNA using universal primers. The study concludes that fermented products are a rich source of bioactive microbes, and the strains that were screened and identified were heat stable and resistant to pH–essential characteristics for use in the food industry (Choeisoongnern et al. 2019).

8.4 BIOINFORMATICS TOOLS

Today bioinformatics tools, such as databases, are used in food science analysis (Table 8.1). These tools reduce the time for screening bioactive peptides and are illustrative of the direct interaction of bioinformatics in studies related to food allergens (Kumar and Chordia 2017).

The most used databases are UniProtKB, SwissProt, and TreMBL. These databases assist in the discovery and analysis of protein sequences related to bioactive peptides, which are present in foods that have various biological properties, such as antihypertensive, antimicrobial, immunological, and antioxidant (Walther and Sieber 2011).

These databases help to reduce the time required to find bioactive peptides. Another database that is used to identify bioactive peptides is BIOPEP, which also has tools for the evaluation of proteins as precursors of bioactive peptides, including proteolytic processes (Minkiewicz et al. 2019).

TABLE 8.1   Bioinformatics Tools Used in Food Science

Tools	Description	Source
UniProtKB	Central hub of functional information about proteins.	https://www.uniprot.org/
SwissProt	Annotation data with information extracted from the literature and computational analysis evaluated by manual curation.	https://www.uniprot.org/
TreMBL	Computationally analyzed records that require complete manual annotation.	https://www.uniprot.org/
BIOPEP	Tools for the evaluation of protein as precursors of bioactive peptides.	http://www.uwm.edu.pl/biochemia/index.php/pl/biopep
PoPS	Computational toolset to investigate protease specificity.	http://pops.csse.monash.edu.au/
AllerMatch	Web tool in which it is possible to compare the amino acid sequence of a protein with sequences of allergenic proteins.	http://www.allermatch.org/
FARRP Allergen	Provides access to a peer-reviewed list of allergens and a database for identifying proteins that may present a potential risk of allergen cross-reactivity.	http://www.allergenonline.com/
SDAP	A web server that integrates a database of allergenic proteins to various computational tools that can assist in studies of structural biology related to allergens.	http://fermi.utmb.edu/SDAP/sdap_src.html
InformAll	Provides reliable sources of information, including database.	http://research.bmh.manchester.ac.uk/informall/Introduction/
Allergome	An integrated platform that collects information and data for allergens.	http://www.allergome.org/
Immune Epitope Database	A database dedicated to allergenic epitopes.	http://www.immuneepitope.org/
FooDB	Provides information on macronutrients and micronutrients, including many of the constituents that give foods taste, color, flavor, texture, and aroma.	http://www.foodb.ca.
EuroFIR-BASIS	A database that combines the composition of foods and the biological activity of bioactive compounds in foods.	http://www.eurofir.org.

(Continued)

TABLE 8.1 (CONTINUED)   Bioinformatics Tools Used in Food Science

Tools	Description	Source
AutoSNPdb	The AutoSNPdb system was developed for flexible use and allows extension to a wide range of annotations and species.	http://autosnpdb.appliedbioinformatics.com.au/index.jsp?species=
Phenol-Explorer	Database on the effects of food processing on polyphenol content.	http://www.phenol-explorer.eu
ChEMBL	Database with manual curation of bioactive molecules with drug-like properties. It gathers chemical, bioactive, and genomic data to help translate genomic information into effective new drugs.	https://www.ebi.ac.uk/chembldb/
ChemSpider	Chemical structure database that provides quick access to text and structure search for more than 67 million structures from hundreds of data sources.	http://www.chemspider.com/Default.aspx
FEMA GRAS	Presents information on the collection and analysis of data necessary to verify the safety of ingredients.	https://www.femaflavor.org/about
HMDB	Web database containing detailed information on metabolites of small molecules found in the human body.	http://www.hmdb.ca/
MetaComBio	It is a web tool that contains links to chemical databases that describe low molecular weight compounds.	http://www.uwm.edu.pl/metachemibio/index.php/about-metacombio
NutriChem	Provides a basis for mechanically understanding the health consequences of eating behaviors.	http://www.cbs.dtu.dk/services/NutriChem-1.0/
PubChem	Database that provides information on chemical and physical properties, biological activities, safety, toxicity, patents, citations in the literature, among others.	https://pubchem.ncbi.nlm.nih.gov/

The process of proteolysis *in silico* is one that reduces the time needed to track bioactive peptides present in different sources of proteins. It uses proteases that can lead to the discovery of new and reliable precursors of already known bioactive peptides (Waidha et al. 2015). Example of tools used in this process are PoPS (Boyd et al. 2005), EXPASY (Artimo et al. 2012), and BIOPEP (Minkiewicz et al. 2019).

Bioinformatics also has a direct interaction with research on food allergens. Allergens are natural substances that can trigger a hypersensitivity reaction in susceptible people. Most allergens have a certain structural similarity, some even have the same sequence. Therefore, studies of homology and structural bioinformatics can be used to detect possible allergenicity and cross-reactivity of proteins (Kumar and Chordia 2017). There are databases dedicated to the identification and storage of data on food allergens, such as AllerMatch (Fiers et al. 2004), InformAll database (http://research.bmh.manchester.ac.uk/informAll), SDAP (Mari et al. 2006) and FARRP Allergen (https://farrp.unl.edu/).

In addition, there are several food composition databases, such as FooDB, EuroFIR-BASIS, and FoodWikiDB. FooDB is the largest in the world and the most comprehensive resource in food constituents, chemistry, and biology. It provides information on macronutrients and micronutrients, including many of the constituents that give foods taste, color, texture, and aroma. Each chemical entry in FooDB contains more than 100 separate data fields, covering detailed compositional, biochemical, and physiological information. Users can browse or search FooDB by food source, name, descriptors, function, or concentrations. FooDB is available at www.foodb.ca.

Another food composition database is EuroFIR-BASIS. It exclusively combines the composition of foods and the biological activity of bioactive compounds in foods of plant origin. The database covers 17 classes of compounds and 300 European plant foods and their edible parts with data from peer-reviewed and quality-assessed literature. EuroFIR-BASIS is available at https://www.eurofir.org/our-tools/ebasis/ (Gry et al. 2007).

The FoodWikiDB database project is a repository of nutritional and food information. It uses immense amounts of data generated by sequencing and managed by bioinformatics strategies and protocols to improve the quality and nutritional value of food sources. The project is ongoing so that these resources continue to advance and develop information for the food sciences. This comprehensive, centralized database will be important for the progression of bioinformatics in the area of food science (Holton et al. 2013).

Foodomics is a database for molecular food profiles. It released 28 of the USDA's national nutrient databases for standard reference (SR28) and contains data for up to 150 food molecules. Nutrition facts labels (NFLs) on branded foods reflect current information about nutrients but generally contain less than 15 nutrients. The integration of the current NFL and SR28 is necessary to obtain gastronomic profiles of the food intake of a person (or animal), thereby allowing for the implementation of precision nutrition for each person. In this way, the right food molecules are given in the right amount, at the right time, and to the right person, similar to what is done in precision medicine (Allen et al. 2016).

8.5 FUTURE PERSPECTIVES

Bioinformatics combined with advances in "*omic*" sciences, has enabled researchers, regulatory companies, and food industries to guarantee food safety and quality by detailing and accurately identifying microbiological risks in food (Cocolin and Ercolini 2015).

Recently, new proposals have been presented that focus on the analysis of food contamination, which is often due to the lack of testing for pathogens. The study by Lewis et al. (2020) presents improvements in the sequencing technology and bioinformatics pipelines as a solution to this problem. Complementarily, the work of Zhang et al. (2019) proposes the *additiveChem* database for curating food additives, including their molecular structures, biological activities, and accurate toxicological assessments, to explore the relationship between the structure and function of the additives in food.

Another aspect is research focused on the evaluation and control of food quality, seeking to identify contamination in raw materials during food processing before reaching the market. Research is being carried out aimed at developing tools for molecular authentication, which are believed to be the answer to maintaining the quality of food for suppliers and consumers. Considering scientific advances, it is expected that in the coming years, new molecular tools will make use of complex matrices for quick and efficient verification of food quality (Bruno et al. 2019).

Bioinformatics tools are expected to grow exponentially in applications for assessing food quality and safety.

8.6 CONCLUSIONS

In this chapter we present the use of bioinformatics applied in the food sciences, contributing directly to the advancement and expansion of research in this area, through several aspects such as studies related to allergenicity, functionalities, and flavor and highlighting food safety and conservation (Holton et al. 2013). NGS sequencing techniques impact food science, acting directly and quickly in research to prevent and control food problems. In a complementary way, bioinformatics tools grow exponentially and are fundamental to the evolution of the evaluation process, quality, and food safety.

REFERENCES

Adamiak, J., Otlewska, A., Tafer, H., et al. 2018. First evaluation of the microbiome of built cultural heritage by using the ion torrent next generation sequencing platform. *Biodeterioration and Biodegradation* 131:11–8. doi:10.1016/j.ibiod.2017.01.040.

Alfadul, S. M. and Elneshwy, A. A. 2010. Use of nanotechnology in food processing, packaging and safety – Review. *African Journal of Food Agriculture Nutrition Development* 10(6):2719–39.

Allen, H., Lennon, D. J., Lukosaityte, J. and Borum, P. R. 2016. Foodomics database: A new tool for precision medicine and the-omic toolbox. *FASEB Journal* 30:682–713.

Ansorge, W. J. 2009. Next-generation DNA sequencing techniques. *New Biotechnology* 25(4):195–203. doi:10.1016/j.nbt.2008.12.009.

Artimo, P., Jonnalagedda, M., Arnold, K., et al. 2012. ExPASy: SIB bioinformatics resource portal. *Nucleic Acids Research* 40(W1):597–603. doi:10.1093/nar/gks400.

Boyd, S. E., Pike, R. N., Rudy, G. B., Whisstock, J. C. and La Banda, M. G. 2005. POPS: A computational tool for modeling and predicting protease specificity. *Journal of Bioinformatics and Computational Biology* 3(3):551–85.

Brown, E., Dessai, U., McGarry, S. and Gerner-Smidt, P. 2019. Use of whole-genome sequencing for food safety and public health in the United States. *Foodborne Pathogens and Disease* 16(7):441–50. doi:10.1089/fpd.2019.2662.

Bruno, A., Sandionigi, A., Agostinetto, G., et al., 2019. Food tracking perspective : DNA metabarcoding to identify plant composition in complex and processed food products. *Genes Journal*. doi:10.3390/genes10030248.

Cao, Y., Fanning, S. Proos, S., Jordan, K. and Shabarinath, S. 2017. A review on the applications of next generation sequencing technologies as applied to food-related microbiome studies. *Frontiers in Microbiology* 8:1–16. doi:10.3389/fmicb.2017.01829.

Choeisoongnern, T., Sivamaruthi, B. S., Sirilun, S., et al. 2019. Screening and identification of bacteriocin-like inhibitory substances producing lactic acid bacteria from fermented products. *Food Science and Technology* 2061:1–9. doi:10.1590/fst.13219.

Cocolin, L. and Ercolini, D. 2015. ScienceDirect zooming into food-associated microbial consortia · A 'cultural' evolution. *Current Opinion in Food Science* 2. Elsevier Ltd:43–50. doi:10.1016/j.cofs.2015.01.003.

Fiers, M. W., Kleter, G. A., Nijland, H., Peijnenburg, A. A. C. M., Nap, J. P. and Ham, R. C. H. J. V. 2004. Allermatch™, a webtool for the prediction of potential allergenicity according to current FAO/WHO codex alimentarius guidelines. *BMC Bioinformatics* 5:1–6. doi:10.1186/1471-2105-5-133.

Floros, J. D., Newsome, R., Fisher, W., Barbosa-Cánovas, G. V., Chen, H., Dunne, C. P., German, J. B., et al. 2010. Feeding the world today and tomorrow: The importance of food science and technology. *Comprehensive Reviews in Food Science and Food Safety* 9(5):572–99. doi:10.1111/j.1541-4337.2010.00127.x.

Gry, J., Black, L., Eriksen, F. D., et al., 2007. EuroFIR-BASIS - a combined composition and biological activity database for bioactive compounds in plant-based foods. *Trends in Food Science and Technology* 18(8):434–44. doi:10.1016/j.tifs.2007.05.008.

Holton, T. A., Vijayakumar, V. and Khaldi, N. 2013. Bioinformatics: Current perspectives and future directions for food and nutritional research facilitated by a foodwiki database. *Trends in Food Science and Technology* 34(1):5–17. doi:10.1016/j.tifs.2013.08.009.

Jain, M., Olsen, H. E., Paten, B. and Akeson, M. 2016. The Oxford nanopore MinION: Delivery of nanopore sequencing to the genomics community. *Genome Biology* 17(1):1–11. doi:10.1186/s13059-016-1103-0.

Kumar, A. and Chordia, N. 2017. Bioinformatics approaches in food sciences. *Journal of Food: Microbiology, Safety and Hygiene* 2(2):2–5. doi:10.4172/2476-2059.1000e104.

Lewis, E., Hudson, J. A., Cook, N., Barnes, J. D. and Haynes, E. 2020. Next-generation sequencing as a screening tool for foodborne pathogens in fresh produce. *Journal of Microbiological Methods* 105840. doi:10.1016/j.mimet.2020.105840.

Liu, X. F., Liu, C., Zeng, X. G., Zhang, H. Y., Luo, Y. Y. and Li, X. R. 2020. Metagenomic and metatranscriptomic analysis of the microbial community structure and metabolic potential of fermented soybean in yunnan province. *Food Science and Technology* 40(1):18–25. doi:10.1590/fst.01718.

Malla, M. A., Dubey, A., Kumar, A., Yadav, S., Hashem, A. and Allah, E. F. A. 2019. Exploring the human microbiome: The potential future role of next-generation sequencing in disease diagnosis and treatment. *Frontiers in Immunology* 10:1–23. doi:10.3389/fimmu.2018.02868.

Mari, A., Scala, E., Palazzo, P., Ridolfi, S., Zennaro, D. and Carabella, G. 2006. Bioinformatics applied to allergy: Allergen databases, from collecting sequence information to data integration. The allergome platform as a model. *Cellular Immunology* 244(2):97–100. doi:10.1016/j.cellimm.2007.02.012.

Matthijs, G., Souche, E., Alders, M., et al. 2016. Guidelines for diagnostic next-generation sequencing. *European Journal of Human Genetics* 24(1):2–5. doi:10.1038/ejhg.2015.226.

Minkiewicz, P., Iwaniak, A. and Darewicz, M. 2019. BIOPEP-UWM database of bioactive peptides: current opportunities. *International Journal of Molecular Sciences* 20(23). doi:10.3390/ijms20235978.

Platt, G. C. 2009. *Food science and technology.* Vol. 3 (537). Blackwell Publishing Ltd, New Delhi.

Reuter, J. A., Spacek, D. V. and Snyder, M. P. 2015. High-throughput sequencing technologies. *Molecular Cell* 58(4):586–97. doi:10.1016/j.molcel.2015.05.004.

Thottathil, G. P., Jayasekaran, K. and Othman, A. S. 2016. Sequencing crop genomes: A gateway to improve tropical agriculture. *Tropical Life Sciences* 27(1):93–114.

Vincent, A. T., Derome, N., Boyle, B., Culley, A. I. and Charette, S. J. 2017. Next-generation sequencing (NGS) in the microbiological world: How to make the most of your money. *Journal of Microbiological Methods* 138:60–71. doi:10.1016/j.mimet.2016.02.016.

Waidha, K. M., Jabalia, N., Singh, D., Jha, A. and Kaur, R. 2015. Bioinformatics approaches in food industry: An overview. *National Conference on Recent Trends in Biomedical Engineering, Cancer Biology, Bioinformatics and Applied Biotechnology (BECBAB-2015).* 1:1–4. doi:10.13140/RG.2.2.27961.77926.

Walther, B. and Sieber, R. 2011. Bioactive proteins and peptides in foods. *International Journal for Vitamin and Nutrition Research* 81:181. doi:10.1024/0300-9831/a000054.

Weirather, J. L., Cesare, M., Wang, Y., et al. 2017. Comprehensive comparison of pacific biosciences and oxford nanopore technologies and their applications to transcriptome analysis. *F1000Research* 6(1):100. doi:10.12688/f1000research.10571.1.

Winand, R., Bogaerts, B., Hoffman, S., et al. 2019. Targeting the 16S rRNA gene for bacterial identification in complex mixed samples: Comparative evaluation of second (Illumina) and third (Oxford Nanopore Technologies) generation sequencing technologies. *International Journal of Molecular Sciences* 21(1):1–22. doi:10.3390/ijms21010298.

Zhang, D., Cheng, X., Sun, D., et al. 2019. AdditiveChem: A comprehensive bioinformatics knowledge-base for food additive chemicals dachuan. *Food Chemistry* 125519. doi:10.1016/j.foodchem.2019.125519.

Zhang, X., Liang, B., Xu, X., et al. 2017. The comparison of the performance of four whole genome amplification kits on ion proton platform in copy number variation detection. *Bioscience Reports* 37(4):1–10. doi:10.1042/BSR20170252.

CHAPTER 9

First Generation–The Sanger Shotgun Approach

Muhsin Jamal, Sana Raza,
Sayed Muhammad Ata Ullah Shah Bukhari, Saadia Andleeb,
Muhammad Asif Nawaz, Sidra Pervez, Liloma Shah, and Redaina

CONTENTS

9.1 GENOME

The word genome refers to the genetic material of a living organism, specifically the genetic material of organisms within the fields of molecular biology and genetics. In most organisms, the genome is comprised of ribonucleic acid (RNA) and deoxyribonucleic acid (DNA); in viruses, however, the genome is either one or the other. Genomes may have coding and noncoding sequences. All biological information is stored in the DNA of organisms, which include genetic instructions for the reproduction, development, coordination, and cell differentiation in higher organisms. The structure of a DNA molecule is a twist formed by two biopolymer strands. It is also called a polynucleotide chain because it is made up of smaller subunits called nucleotides (Mortier 2016).

DNA is present inside the nucleus, mitochondria, and even in the chloroplast of cells of eukaryotic organisms, whereas it is found in the cytoplasm of prokaryotic cells such as bacteria. A protein coat, called a capsid, encloses the genome (RNA or DNA) of viruses. Normally DNA is transcribed to mRNA and then translated to a specific type of protein (Blomstergren 2003). DNA contains 3 main parts: deoxyribose (pentose sugar), a phosphate group, and an organic base. Nucleotides are made up of nitrogenous bases:

Thymine (T), Adenine (A), Guanine (G), and Cytosine (C). Pyrimidines (T and C) are single ring structures, and Purines (A and G) are double ring structures. Adenine combines with thymine by 2 hydrogen bonds while Cytosine and Guanine bind by 3 hydrogen bonds, whereas RNA have Uracil (U) instead of Thymine (T).

9.2 PURPOSE OF GENOME SEQUENCING

Scientists use genomic sequences to study the basic concept of living organisms in the fields of genetics and molecular biology. A genome or genomic sequence shows the order of DNA nucleotides. Genome sequencing is not related to the "decoding" of the genome, therefore does not directly express the genetic secrets of a whole species. We still have to translate the resulting order of nucleobases into an understanding of how the genome works. The most important steps of genome sequencing are given below:

- Efficiently locate genes within the genome.
- Recognize how the genome works as a whole.
- Identify the different regulatory regions (i.e., regions that control how genes are switched off and on). Genome sequencing enables us to digitally store the genomic information, so that further processing or compression can be performed (Mortier 2016).

9.3 GENOME SEQUENCING TECHNOLOGIES

During the last several decades, a number of genome-sequencing techniques have been developed, ranging from chemical sequencing methods to current next-generation sequencing (NGS). NGS is the easiest, fastest, and least expensive technique (Sanger and Coulson 1975; Sanger et al. 1977). DNA sequencing plays an important role within different fields of molecular biology and genetics. With the passage of time, sequencing technologies are always evolving (Bekel et al. 2009). Through the introduction of whole genome shotgun sequencing and the advent of the first-generation machine for automatic sequencing, the essential stages toward whole genome sequencing were launched thirteen years earlier. (Fleischmann et al. 1995). Sequencing is a continuous process in molecular biology facilitates the exploration of cellular functions, mechanisms of different of pathways, and the role of genes in these pathways. Ewing et al. (1998) stated that about 827 complete genomes have been published comprising about 1842 bacteria whereas 936 eukaryotic cells genomes. In most genome sequencing technologies, the genetic material (i.e., DNA) is broken down into smaller fragments in the very first step. Bioinformatics tools read these short fragments to recombine them into a large single sequence for further analysis (Ewing et al. 1998).

9.4 FIRST-GENERATION DNA SEQUENCING

Sanger and Maxam-Gilbert sequencing technologies were classified as the First-Generation Sequencing Technology. Who initiated the field of DNA sequencing with their publication in 1977.

9.5 HISTORY

"Knowledge of sequences could contribute much to our understanding of living matter"—Frederick Sanger in Heather and Benjamin (2016). In 1943, three scientists, Oswald Avery, Colin MacLeod, and Maclyn McCarty, explored DNA transfer in various strains of *Pneumococcus* (Avery et al. 1944). Since then, subsequent generations of bacteria inherited trait have proven that DNA is the only substance that carries all the genetic information of an organism. Ten years later, Francis Crick and James Watson published a momentous study which showed that DNA was a double helix structure in composition. In 1953, DNA 3-D arrangement was first modeled by Watson and Crick based on crystallographic research of Rosalind Franklin and Maurice Wilkins (Watson and Crick 1953; Zallen 2003). This study contributed to the background for both encoding proteins in nucleic acids and DNA replication. In early 1950, Sanger determined the initial protein sequence that was insulin. His effort demonstrated clearly that proteins were made of well-defined arrangements of amino acid. In late 1960, many proteins were sequenced by the improvement of Edman degradation sequencing, which involves a frequent exclusion of an N-terminal residue from the peptide chain (Shendure et al. 2017).

DNA is a complex long molecule made of smaller subunits, making it difficult to distinguish from one another (Hutchison 2007). There was need for novel strategies. Initially, the efforts of researchers were concentrated on the sequencing of easily accessible pure RNA species, like bacteriophage (single standard RNA), microbial ribosomal, or transfer RNA. However, researchers used RNase enzymes to cut the RNA chain at specific points using available techniques (borrowed from analytical chemistry) to determine nucleotide structure but not sequence order (Holley et al. 1961). In 1965, Robert Holley and classmates succeeded in producing the first entire nucleic acid sequencing of alanine transfer RNA from *Saccharomyces cerevisiae* (Holley et al. 1965).

Similarly, Fred Sanger and coworkers established correlated procedure depend on finding radiolabeled incomplete digestion fragments after 2-D fractionation that permitted scientists to gradually enhance the upward pool of tRNA and rRNA sequences (Brownlee and Sanger 1967; Adams et al. 1969). In 1972, Walter Fiers' laboratory, using the 2-D fractionation method, identified the first complete protein coding gene sequence of "phage MS2" (Jou et al. 1972) followed by its complete genome sequencing 4 years later (Fiers et al. 1976).

In 1974, two techniques were independently developed by the English (Sanger) and an American team (Maxam and Gilbert). Gilbert and Maxam devised a chemical retraction technique while Sanger developed a technique similar to the natural DNA replication phenomenon. Both teams were awarded Nobel Prizes in 1980. Sanger and coworkers used these techniques to sequence bacteriophage φX174 (first DNA genome), or 'PhiX'. Nowadays, it is used as a positive control genome in various sequencing labs (Sanger et al. 1977). The Sanger method became the standard because of its practical application. Their findings opened doors for future researchers to investigate the genetic code of living organisms and inspired them to develop effective and faster sequencing technology. Because of low radioactivity and high effectiveness, the Sanger method of sequencing is the most functional, and it has been automated and commercialized as the "Sanger Sequencing Technology" (Pareek et al. 2011).

TABLE 9.1   Among the Sequencing Strategies Applied are Shotgun Sequencing

Organisms	Genome Size (Mbp)	References
Haemophilus influenzae	1.83	Fleischmann et al. (1995)
Mycoplasma genitalium	0.58	Sterky and Lundeberg (2000)
Methanococcus jannaschii (a)	1.66	Bult et al. (1996)
Borrelia burgdorferi	1.44	Fraser et al. (1997)
Archaeoglobus fulgidus (a)	2.18	Klenk et al. (1998)
Helicobacter pylori (26695)	1.66	Tomb et al. (1997)
Treponema pallidum	1.14	Fraser et al. (1998)
Chlamydia trachomatis	1.04	Stephens et al. (1998)
Rickettsia prowazekii	1.11	Andersson et al. (1998)
Helicobacter pylori (J99)	1.64	Alm et al. (1999)
Methanobacterium thermoautotrophicum (a)	1.75	Smith et al. (1997)
Aquifex aeolicus, Helicobacter pylori	1.50	Deckert et al. (1998)

9.6 MICROBIAL GENOMIC ANALYSIS

Since the 1995 publication of the complete *Haemophilus influenzae* genome sequence (Fleischmann et al. 1995), the genomes of over 400 species of bacteria have been sequenced. Multiple techniques are available for genomic sequence analysis. However, Sanger's shotgun sequencing technique has remained the original genome sequence assemblage for over 25 years (Sanger et al. 1977). The most commonly used approach for microbial genome sequencing is whole genome shotgun sequencing. This method was applied in order to sequence the numerous microorganisms' genomes such as, *Helicobacter pylori, H. influenzae* (Tomb et al. 1997), *Archaeoglobus fulgidus* (Klenk et al. 1998), and *Thermotoga maritima* (Kaiser et al. 2003) (Table 9.1).

9.7 SANGER SEQUENCING METHODS

Sanger's method, also called dideoxy sequencing (or chain termination), is used for single-stranded DNA sequencing. It includes the use of correspondent standard ddNTPs. In ddNTPs, the 3-hydroxyl (OH) group is absent which is essential for chain elongation of DNA; therefore it cannot bind with 5' phosphate of the next dNTP (Chidgeavadze et al. 1984) when these adapted nucleotides (ddNTPs) are inserted into a sequence resulting in chain termination (i.e., the addition of dNTPs are inhibited due to lack of the free 3-OH group). Before performing the DNA sequencing, a double strand DNA is denatured into single strand DNA by using heat. Denaturation is followed by an annealing step in which a specific primer is added. This primer contains a free 3' end and therefore binds to a template strand. Either one of the nucleotides or primers must be radioactively or fluorescently labeled in order to assist in sequence analysis. Then the products with multiple fragments (sizes) are run on gel so that the actual size of the product is observed (Mortier 2016). The solution is categorized into 4 labeled tubes "G," "A," "T," and "C." Once the primer is bound to the DNA strand (Figure 9.1), then chemicals are added to these

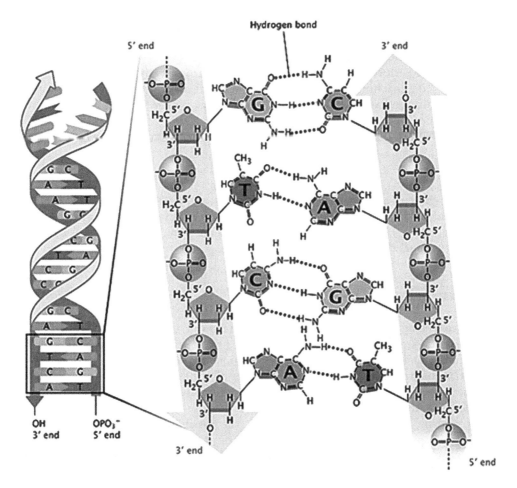

FIGURE 9.1 Chemical formula of dNTPs and ddNTPs: (left) Deoxynucleotides (dNTPs), natural nucleoside triphosphates that get incorporated during DNA polymerase; (right) Reversible dye-terminators, engineered nucleotides used for Illumina sequencing-by-synthesis (Adapted from Schirmer et al. 2016, Creative Commons Attribution 4.0 International).

samples as follows. All four tubes contain DNA polymerase and one of the four dNTPs accordingly: Thymine "T" tube: ddTTP; Cytosine "C" tube: ddCTP; Adenine "A" tube: ddATP; and Guanine "G" tube: ddGTP. Each of the reaction tubes contains a different ddNTP, and the concentration of these typical precursors is about one hundredth of the normal dNTP. The chain of DNA is "grown" by adding a nucleotide to the DNA polymerase and, as a result, a DNA strand is produced. However, this chain elongation event is terminated by ddNTP incorporation in the chain.

The main point of this technique is that all reactions begin from the similar nucleotide but finish with a specified ddNTP (Mortier 2016). In this manner, different lengths of bands are formed. After completion of the reactions, the polymerase chain reaction (PCR) products (DNA) are denatured, followed by electrophoresis. All the reaction tubes are run on separated polyacrylamide gels, and different sized bands are produced separately. DNA bands are visualized by UV light or autoradiography (Sanger and Coulson 1975; Sanger et al. 1977).

Later, significant developments were made to Sanger sequencing. For instance, fluorometric-based detection replaced the "phosphorus- or tritium-radiolabeling." In this modification, reaction can take place in a single tube instead of 4 tubes, and detection of the results is performed using capillary-based electrophoresis. After this, automated DNA sequencing machines were developed (Luckey et al. 1990). With passage of time several modifications were made to Singer's method, making it the most suitable method for DNA sequencing.

9.7.1 Advantages

The Sanger shotgun approach is the most popular sequencing technique because it is the simplest and fastest method. This approach can read a DNA fragment ranging in size from 500 bp to 1.0 kb, and primer annealing and sequencing reactions can be completed within a few hours (Kingsmore and Saunder 2011).

9.7.2 Disadvantages

Sanger sequencing has limitations. First, it can read only small fragments of DNA and electrophoretic separation is required for reading those fragments. Second, despite reliable accuracy and availability, the Sanger technique likewise has application issues due to practical limitations of its workflow (Hert et al. 2008). Even with efficient automation, the Sanger apparatus can only interpret 96 reactions in equivalent, thereby limiting throughput to about 115kb/day (Mardis 2011). Because the sequence is not an original DNA fragment, but rather an enzymatic copy, the possibility of incorporating incorrect bases in the coping sequence exists. Third, the present cost estimate to sequence an entire human genome is 5–30 million USD and could take about 60 years to complete on a single machine (Hert et al. 2008).

9.8 APPLICATION OF DNA SEQUENCING

The discovery of elementary molecular heredity mechanisms, modeling of DNA structure, and other innovative sequencing technologies of DNA are of much significance to genetics research. It was not up until the findings by Maxam and Gilbert (1977), and Sanger et al. (1977) that the first functional techniques of sequencing were established and its implementation was conducted on a larger scale however. Bedbrook and his colleagues isolated and sequenced the plant cDNA for the first time (Bedbrook et al. 1980). The first transgenic plant was developed through *Agrobacterium* through the efficacious incorporation of recombinant DNA and plant biotechnology sequencing methods (Fraley et al. 1983; Herrera et al. 1983).

Sequencing techniques play a key role in plant research, permitting the characterization and alteration of metabolic pathways and genes and the usage of genetic alteration for investigations within marker-assisted selection (MAS), species diversity, seed purity and germplasm characterization. The genome determination of maize, *Arabidopsis thaliana*, and rice were important breakthroughs that assisted in analyzing the entire genome structure and gene characterization of plants (Schnable et al. 2009).

9.9 CONCLUSION

It is concluded here that the first-generation Sanger shotgun approach, developed a half-century ago, is the simplest, easiest, and the most widely used technique in the field of genetics. Although it reads only small fragments of DNA by using ddNTPs precursors, new modifications have been introduced from time to time, leading to second- and third-generation sequencing.

REFERENCES

Adams, J. M., Jeppesen, P. G. N., Sanger, F., et al. 1969. Nucleotide sequence from the coat protein cistron of R17 bacteriophage RNA. *Nature* 223(5210): 1009–14.

Alm, R. A., Ling, L. S. L., Moir, D. T., et al. 1999. Genomic-sequence comparison of two unrelated isolates of the human gastric pathogen *Helicobacter pylori*. *Nature* 397(6715):176–80.

Andersson, S. G., Zomorodipour, A., Andersson, J. O., et al. 1998. The genome sequence of *Rickettsia Prowazekii* and the origin of mitochondria. *Nature* 396(6707):133–40.

Avery, O. T., MacLeod, C. M. and McCarty, M. 1944. Studies on the chemical nature of the substance inducing transformation of pneumococcal types: Induction of transformation by a desoxyribonucleic acid fraction isolated from *Pneumococcus* type III. *The Journal of Experimental Medicine* 79(2):137–58.

Bedbrook, J. R., Smith, S. M. and Ellis, R. J. 1980. Molecular cloning and sequencing of cDNA encoding the precursor to the small subunit of chloroplast ribulose-1, 5-bisphosphate carboxylase. *Nature* 287(5784):692–7.

Bekel, T., Henckel, K., Küster, H., et al. 2009. The sequence analysis and management system–SAMS-2.0: data management and sequence analysis adapted to changing requirements from traditional sanger sequencing to ultrafast sequencing technologies. *Journal of Biotechnology* 140(1–2):3–12.

Blomstergren, A. 2003. *Strategies for de novo DNA sequencing* (Doctoral Dissertation, Bioteknologi).

Brownlee, G. G. and Sanger, F. 1967. Nucleotide sequences from the low molecular weight ribosomal RNA of *Escherichia coli*. *Journal of Molecular Biology* 23(3):337–9.

Bult, C. J., White, O., Olsen, G. J., et al. 1996. Complete genome sequence of the methanogenic archaeon, *Methanococcus jannaschii*. *Science* 273(5278):1058–73.

Chidgeavadze, Z. G., Beabealashvilli, R. S., Atrazhev, A. M., et al. 1984. 2′, 3′-Dideoxy-3'aminonucleoside 5′-triphosphates are the terminators of DNA synthesis catalyzed by DNA polymerases. *Nucleic Acids Research* 12(3):1671.

Deckert, G., Warren, P. V., Gaasterland, T., et al. 1998. The complete genome of the hyperthermophilic bacterium *Aquifex aeolicus*. *Nature* 392(6674):353–8.

Ewing, B., Hillier, L., Wendl, M. C., et al. 1998. Base-calling of automated sequencer traces usingphred. I. Accuracy assessment. *Genome Research* 8(3):175–185.

Fiers, W., Contreras, R., Duerinck, F., et al. 1976. Complete nucleotide sequence of bacteriophage MS2 RNA: Primary and secondary structure of the replicase gene. *Nature* 260(5551):500–7.

Fleischmann, R. D., Adams, M. D., White, O., et al. 1995. Whole-genome random sequencing and assembly of *Haemophilus influenzae* Rd. *Science* 269(5223):496–512.

Fraley, R. T., Rogers, S. G., Horsch, R. B., et al. 1983. Expression of bacterial genes in plant cells. *Proceedings of the National Academy of Sciences USA* 80(15):4803–7.

Fraser, C. M., Casjens, S., Huang, W. M., et al. 1997. Genomic sequence of a lyme disease spirochaete, *Borrelia burgdorferi. Nature* 390(6660):580–6.

Fraser, C. M., Norris, S. J., Weinstock, G. M., et al. 1998. Complete genome sequence of *Treponema pallidum*, the syphilis spirochete. *Science* 281(5375):375–88.

Heather, J. M. and Chain, B. 2016. The sequence of sequencers: The history of sequencing DNA. *Genomics* 107(1):1–8.

Herrera-Estrella, L., Depicker, A., Van Montagu, M., et al. 1983. Expression of chimaeric genes transferred into plant cells using a Ti-plasmid-derived vector. *Nature* 303(5914):209–13.

Hert, D. G., Fredlake, C. P. and Barron, A. E. 2008. Advantages and limitations of next-generation sequencing technologies: A comparison of electrophoresis and non-electrophoresis methods. *Electrophoresis* 29(23):4618–26.

Holley, R. W., Apgar, J., Everett, G. A., et al. 1965. Structure of a ribonucleic acid. *Science* 1462–5.

Holley, R. W., Apgar, J., Merrill, S. H., et al. 1961. Nucleotide and oligonucleotide compositions of the alanine-, valine-, and tyrosine-acceptor "soluble" ribonucleic acids of yeast. *Journal of the American Chemical Society* 83(23):4861–2.

Hutchison III, C. A. 2007. DNA sequencing: Bench to bedside and beyond. *Nucleic Acids Research* 35(18):6227–37.

Jou, W. M., Haegeman, G., Ysebaert, M., et al. 1972. Nucleotide sequence of the gene coding for the bacteriophage MS2 coat protein. *Nature* 237(5350):82–8.

Kaiser, O., Bartels, D., Bekel, T., et al. 2003. Whole genome shotgun sequencing guided by bioinformatics pipelines—An optimized approach for an established technique. *Journal of Biotechnology* 106(2–3):121–33.

Kingsmore, S. F. and Saunders, C. J. 2011. Deep sequencing of patient genomes for disease diagnosis: When will it become routine? *Science Translational Medicine* 3(87):23.

Klenk, H. P., Clayton, R. A., Tomb, J. F., et al. 1998. Erratum: The complete genome sequence of the hyperthermophilic, sulphate-reducing archaeon *Archaeoglobus fulgidus. Nature* 394(6688):101.

Luckey, J. A., Drossman, H., Kostichka, A. J., et al. 1990. High speed DNA sequencing by capillary electrophoresis. *Nucleic Acids Research* 18(15):4417–21.

Mardis, E. R. 2011. A decade's perspective on DNA sequencing technology. *Nature* 470(7333):198–03.

Mortier, T. 2016. *Non-reference-based DNA read compression using machine learning techniques.* Ghent University, Belgium.

Pareek, C. S., Smoczynski, R. and Tretyn, A. 2011. Sequencing technologies and genome sequencing. *Journal of Applied Genetics* 52(4):413–35.

Pray, L. 2008. Discovery of DNA structure and function: Watson and Crick. *Nature Education* 1(1):1–6.

Sanger, F., Air, G. M., Barrell, B. G., et al. 1977. Nucleotide sequence of bacteriophage φX174 DNA. *Nature* 265(5596):687–95.

Sanger, F. and Coulson, A. R. 1975. A rapid method for determining sequences in DNA by primed synthesis with DNA polymerase. *Journal of Molecular Biology* 94(3):441–4.

Sanger, F., Nicklen, S. and Coulson, A. R. 1977. DNA sequencing with chain-terminating inhibitors. *Proceedings of the National Academy of Sciences USA* 74(12):5463–7.

Schirmer, M., D'Amore, R., Ijaz, U. Z., et al. 2016. Illumina error profiles: Resolving fine-scale variation in metagenomic sequencing data. *BMC Bioinformatics* 17(1):125.

Schnable, P. S., Ware, D., Fulton, R. S., et al. 2009. The B73 maize genome: Complexity, diversity, and dynamics. *Science* 326(5956):1112–5.

Shendure, J., Balasubramanian, S., Church, G. M., et al. 2017. DNA sequencing at 40: Past, present and future. *Nature* 550(7676):345–53.

Smith, D. R., Doucette-Stamm, L. A., Deloughery, C., et al. 1997. Complete genome sequence of *Methanobacterium thermoautotrophicum* DeltaH: Functional analysis and comparative genomics. *Journal of Bacteriology* 179(22):7135–55.

Stephens, R. S., Kalman, S., Lammel, C., et al. 1998. Genome sequence of an obligate intracellular pathogen of humans: *Chlamydia trachomatis*. *Science* 282(5389):754–9.

Sterky, F. and Lundeberg, J. 2000. Sequence analysis of genes and genomes. *Journal of Biotechnology* 76(1):1–31.

Tomb, J. F., White, O., Kerlavage, A. R., et al. 1997. Erratum: The complete genome sequence of the gastric pathogen *Helicobacter pylori*. *Nature* 389(6649):412.

Watson, J. D. and Crick, F. H. 1953. Molecular structure of nucleic acids: A structure for deoxyribose nucleic acid. *Nature* 171(4356):737–8.

Zallen, D. T. 2003. Despite Franklin's work, Wilkins earned his nobel. *Nature* 425(6953):15.

CHAPTER 10

Fourth Generation–
In Situ Sequencing

Sadhna Mishra, Arvind Kumar, Shikha Pandhi,
and Dinesh Chandra Rai

CONTENTS

10.1 INTRODUCTION

Sequencing is used to decode the defined array of nucleotides in a polymer of nucleic acids–either deoxyribonucleic acid (DNA) or ribonucleic acid (RNA). Sanger's sequencing method of DNA remained the sole choice for the complete sequencing of genomes from various species, including the human genome for more than two decades. The term "next-generation sequencing" (NGS) was used for these technologies. These methodologies were further developed into second, third, and, newly, fourth-generation sequencing technologies (Ku and Roukos 2013). These technologies can accomplish the millions of sequence reads in a very short period (high-throughput sequencing, massively parallel sequencing) and also include the clonal extension of DNA of interest to produce adequate indication for exposure during the sequencing run (clonal sequencing). A shift from gene to genome investigations across various scientific disciplines allowed researchers to inquire all about the transcriptome, genome, and epigenome of any life form. These NGS technologies are used in numerous basic and applied fields of the scientific community, including drug discovery, forensic science, food science, animal and plant breeding, biotechnology, systematics, and evolutionary biology. The investigated progressive data made NGS an inevitable and global tool for biological research by increasing convenience to high-throughput sequencing methodologies and the simultaneous expansion of inventive bioin-formatics tools. The explosion of this can be verified by the huge amount of scientific publications of NGS. The revolution of genomics occurs by the rapid progression and diminishing cost of NGS technologies. Characterization of the genome is to find genomic regions that are associated with the commercially important traits, profiling of messenger RNA (mRNAs) and micro RNA (miRNAs) to study the control of biological processes and to learn more about evolutionary questions. In addition, these methodologies have also been utilized for ecotoxicological applications. Fast and low-cost DNA sequencing is possible through the improvement of new methods of NGS. NGS technologies are used to find out more about the genome to transcriptome-wide control of biological processes, identification of narrative markers for genetic mapping, phylogenetics, traceability, inhab-itant structure, and exploration of the relationship of loci among traits exaggerated by selection and ecotoxicological applications in fisheries science. As compared with the use of traditional genetic markers, the use of NGS enables investigators to achieve a high degree of resolution and therefore unchain all the information which never was achievable previously (Kumar and Kocour 2017). The period during every sequencing methodology gets decreased while the amount of scientific knowledge has continued growing exponen-tially after the Human Genome Project. By the reduction of production costs and immense sequencing data, the expansion of the NGS technologies has evolved significantly. The different generations of NGS have progressively overcome the restrictions of traditional DNA sequencing techniques and are used for a broad range of applications in molecular biology. Conversely, many technological challenges emerged with these new technologies and remain to be profoundly examined and resolved. Innovations in the third generation technologies are carried out by the removal of bias and copy error coupled with PCR amplification and deploying a particular molecule template approach, also evade the cyclic order approach and thereby enable further massive parallelization (Lee et al. 2015). Genomic analysis can be done directly in the tissue or cell by using fourth-generation sequencing technology. The materialization of fourth-generation sequencing methodolo-gies conserves the spatial synchronization of DNA and RNA sequences up to the subcel-lular level consequently enabling the back-mapping sequencing reads to the original histological perspective. The preference for sequencing nucleic acids straightforwardly in fixed cells by using NGS methods is due to the limitations and challenges of traditional methods being overcome by new methods. This becomes broadly applicable, and the impact of information created by the amalgamation of *in situ* sequencing and NGS meth-ods in research and diagnostics is illustrated by the increase of available data (Ke et al. 2016). The technologies belong from the fourth generation are still investigational. The mutational status and other relationships of any individual or multiple cells can be deter-mined by using fourth-generation sequencing technologies. *In situ* sequencing of a cell in a fixed tissue or organ by using second-generation sequencing technology is possible in fourth-generation sequencing technologies (Mignardi and Nilsson 2014). A study con-ducted by a researcher in breast cancer for multiplex gene expression profiling and analy-ses of point mutations in tissue sections by using *in situ* sequencing (fourth-generation sequencing technologies) has to offer prime concepts for this generation sequencing. The physical limitation during fourth-generation sequencing is determined by the intensity problem, which is caused by the amount of RNA present in the cell for the sequencing progression as it depends on the separation of two diverse points on a layer. This problem is solved by the introduction of the random mismatching primers which decrease the intensity of reading and allow only selected sequencing of specific fragments in the documentation pool in every cycle (Ke et al. 2013). This perspective affords approximately 400 reads at a time which makes it possible to determine thousands of gene expressions of a cell simultaneously with diverse kinds of RNA fragments including mRNA, rRNA (ribosomal RNA), RNAi (RNA interference), single nucleotide polymorphisms (SNPs), microsatellites and noncoding RNA (Mignardi and Nilsson 2014). When compared with other generations of sequencing technologies, fourth generation is predicted to be constructive in certain purposes rather than representing a substitutional option to be used in a wide field of application. The particular advantage of these methodologies includes analysis of a population of cells with single-cell can be done for various applications. Conversely, by using these methods the troubles about practicality, equivalence, cost efficiency, and full incorporation of present sequencing systems require to be solved to enhance efficiency. The single-cell analysis is the most prominent method to know more about the physiological and pathological conditions of the healthy as well as unhealthy tissue. It has been proven in the past few years that *in situ* sequencing technology or scRNA-seq (NGS-based single-cell RNA sequencing) technology is the prevailing implement for various applications including the identification of unusual cells (Grun et al. 2015), to define the lineage of the cell (Blakeley et al. 2015), classification of cell sub-populations (Usoskin et al. 2015), evaluation of tissue composition and their biological insights, appraisal of transcription dynamics and gene regulation (Deng et al. 2014; Shalek et al. 2014; Brennecke et al. 2015; Hanchate et al. 2015). Almost all the single-cell RNA sequencing methods require the separation of the single-cell from the tissue by enzymatic or mechanical methods which results in the loss of endemic information. The direct visualization under the microscope is carried out by using a laser-assisted microdissection method for capturing the interest of cells. After that, these single cells are used for downstream analysis to provide the spatial localization of the cell in the tissue of interest (Ke et al. 2016). Although in such kinds of technologies the same sample requires repeated cycles, this provides endemic resolution in all magnitudes. *In situ* hybridization is another method that uses a computational method that enables the mapping of the scRNA-seq data of information for the tissue of interest (Achim et al. 2015; Satija et al. 2015). The primary knowledge of the endemic pattern of gene expression of the organism of interest is the foremost inadequacy of such techniques. Spatial transcriptomics is the term which refers to the interesting alternative between the *in situ* sequencing and *in situ* hybridization which destines conjunction of *in situ* and *ex situ* mapping and identification by NGS, respectively. The fourth-generation sequencing (*in situ* sequencing) methodology is applicable for many functional areas (Figure 10.1).

10.2 PLATFORMS FOR SEQUENCING

Fourth-generation sequencing technology is an *in situ* sequencing method in which the nucleic acid of individual cells or tissue in a histological part can be sequenced. By using the chemistry of the second generation *in situ*, sequencing can directly examine the composition of nucleic acid in fixed cells or tissues (Reid et al. 2012).

10.3 FOURTH-GENERATION SEQUENCING TECHNOLOGIES

The fourth-generation technologies are very fast and include Oxford Nanopore (or nanopore), Polonator, and Complete Genomics sequencing. Various types of novel sequencing

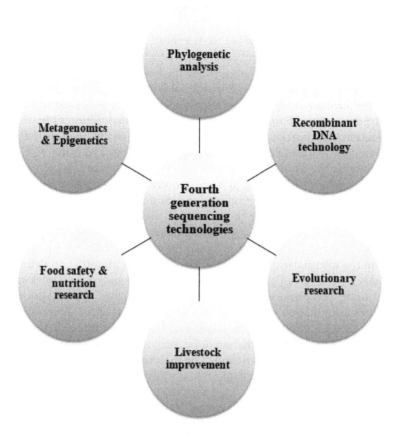

FIGURE 10.1   Functional areas of fourth-generation sequencing (*in situ* sequencing).

chemistries are used in these methodologies which cannot be done by the second or third generation sequencing technologies. Though the chemistry of Polonator resembles–and originally it was a founded for the development of–ABI-SOLiD (a second-generation sequencing technique), its characteristics of being a modular, open, and compatible system make it more innovative, so it is categorized as fourth-generation. The achievements of Nanopore, like sequencing without amplification, synthesis elimination, and RT sequencing with no repeated cycles, make to keep it in the fourth-generation sequencing category. The current non-nanopore DNA sequencing technologies present in the market have elaborate setup requirements for preparation of the sample and need complicated algorithms for processing the data. In 1996, the first publication in PNAS highlighted one of the most powerful sequencing techniques for the detection of single molecules based on nanopore. The nanopore approach is a fast and low-cost option for DNA sequencing technology that belongs to fourth-generation sequencing (Feng et al. 2015).

10.3.1 Sequencing Machinery of Nanopore Technology

The basis of nanopore sequencing is the identification of single DNA molecules by passing them through a petite nanopore compartment. The diameter of nanopores is of nanoscale

and these are tiny biopores that can be divided into a solid-state and biological category (Feng et al. 2015). Nanopore sequencing technology works by way of a little alteration in the electrical current across the nanopore that is absorbed by a conducting fluid
with the voltage applied when a moving nucleotide (or DNA strand) passes through it. A
pore-forming protein with lipid bilayer is needed for the biological nanopore formation;
in solid-state nanopore formation, a synthetic material, like silicon nitride, is required.
Nanopore technology is a recently released technology that offers a profitable nanopore
podium that can attain long lengths read and can also sequence the genome of a phage.
The corporation predicts that nanopore can read 1 billion base pairs of DNA in 6 hours
and costs US $900 (Feng et al. 2015). The nanopore microwell (*cis* and *trans*) is equipped
with electrodes across a single protein that is integrated into a lipid bilayer. An array chip
that is incorporated in microwells is used for the preparation of sample, analysis, and
detection. The nanopore sequencer is mainly adopted for the DNA, but it also has wider
applications for protein and RNA. Inside the nanopore cyclodextrin, an adapter protein
provides a binding site for DNA molecules, whereas the exonuclease united with nanopore breaks the nucleic acid bases from the DNA strand. The bases, which are cleaved
from the DNA strand, are linked to the cyclodextrin and are identified at the speed
of 20 ms/base based on the magnitude differences of current disruption. In the strand
sequencing technique, when the single-stranded DNA is passed through the pore, the
sequencing is potentially quicker with higher accuracy than exonuclease sequencing. The
nano-sized pores in biological membranes are embedded, while in the solid-state film
they are formed and split the reservoir which contains conductive electrolytes into *cis* and
trans microwells. In each chamber in which the electrodes are immersed, the electrolyte
ions are moved in solution through the pore electrophoretically and generate an ionic current signal under a biased voltage. Sometimes, when the nano-sized pore is made barren
by a negatively charged DNA molecule in the *cis* chamber, the flow of current throughout
the nanopore would also be blocked and the current signal interrupted (Venkatesan et al.
2009). By comparison with the biological nanopore, the solid-state nanopore is steadier
and can parallel multiplex on a single device to work. Recently, various companies are
using solid-state nanopore for the sequencing of the whole genome to attain elevated
reads within a short period. Genia nanopore is a similar technique to Oxford Nanopore,
except for extending the growing amplified DNA product; it allows a cleavable label to
enter the pore.

The nanopores have many considerable benefits such as tag-free, ultra-long interpretations (10^4–10^6 bases), throughput is high, and it does not require large samples (Feng
et al. 2015). The nanopore system real-time sequencing of DNA molecules at a low price
(US $25–$40/gigabase of sequence) and can read extremely extended DNA molecules in
one read. The process of sequencing is very fast–that's why it is an up-and-coming selection of choice for many applications. Nanopore sequencing is still in the proof-of-concept
stage and is not available at the commercial level–parallelized or routinized; on the other
hand, the error rate is also high (Gupta and Gupta 2014).

10.3.1.1 Types of Nanopore

On a broad spectrum, nanopore technologies are divided into two categories–biological
and solid-state. It is verified by many groups that both categories of nanopores are used
for the detection of single biological molecules. The biological nanopores have wide
applications in diverse areas including disease diagnosis, single-molecule detection, and
DNA sequencing. Recently advanced nanotechnology has facilitated the development of
solid-state nanopore (Feng et al. 2015). By combining with other devices like transistors

(field-effect), these artificial nanopores can be integrated on a circuit chip and offer the possibility of miniature, transportable sequencing devices for DNA. Recently, another type of nanopore, hybrid nanopores, has been projected with the features of biological and solid-state nanopores together (Derrington et al. 2010). Despite the error rate being very high (over 90%), the sequencing by nanopore technology is rising rapidly (Mikheyev and Tin 2014). Biological nanopores (or transmembrane protein channels) can be inserted into a substrate like liposomes, planar lipid bilayers, etc. The highly reproducible size and well-defined structure are the main advantages of biological nanopores. Cyclodextrin, α-Hemolysin, MspA and Bacteriophage phi29 are some examples of biological nanopores (Gupta and Gupta 2014).

10.3.2 Polonator Technology (Polony-Based Sequencing Technology)

Dr. Church (Harvard Medical School) developed Polonator technology with the vision of decreasing costs and providing an affordable sequencing system with high-quality components for the masses. Polonator is easily upgradable and utilizable for various kinds of applications and is also intended to be modular. The sequencing by Polonator sequencing is carried out by ligation. In this sequencing technology, the two flow cells are connected within the apparatus in which one cell endures the biochemistry and the other is imaged. There are 18 wells in every flow cell, with over 1 billion total beads (streptavidin-coated polystyrene). The reaction with DNA takes place on the surface of the beads. The typical output based on technique and library titration is approximately 8–10 million reads in a single lane and about 150 million reads per run in a dual flow cell. In this technique, the run output is about 4–5 Gb and puts out 26–28 bases read length; on the other hand, the library beads are magnified by eliminating the unamplified beads, increasing the output to 8–10 Gb (Gupta and Gupta 2014).

In the Polonator technology, the four days' duration of the run gives the read length of just 13 + 13 bp paired-end (26 bp). In general, the best mappable reads of 92% gives a mean precision of more than 98%. All the corresponding software is freely accessible for the optimization and interpretation of the data (Gupta and Gupta 2014).

10.3.3 DNA Nanoball Technology (Complete Genomics)

The development and commercialization of this sequencing technology for the whole or complete genome and optimization of this technique is specifically carried out for re-sequencing of human genome applications. This sequencing technology is based on a proprietary DNA range and ligation-based read technology. The technology claims to give sequencing data with 99.99% accuracy which facilitates providing human sequencing services to the scientific community. It provides inclusive elucidation for human genome sequencing at a large-scale, and, by combining various technologies, it asserts to convey low-cost human sequencing with the greatest accuracy–all technological advancements, including sequencing assay, libraries, arrays, instruments, and software at every step of each run. The instrument in this technology is domestic with lofty throughput sequencing. One of the two primary workings of this technology is DNB arrays (DNA nano ball); the other is cPAL (combinatorial l probe-anchor ligation). A silicon chip in which the DNA molecule is efficiently packed is known as DNA nanoball or DNB arrays. The cPAL read unit of this technology gives very accurate data by reading the DNA fragments by using the reagents at low concentrations and cost. The exclusive combination of two units (DNB and cPAL) makes this sequencing technology preferable cost- and quality-wise compared to other commercially available sequencing approaches. On the other hand, the DNB arrays are used for shearing of DNA molecule into fragments that have to be sequenced, and the adopter sequence is used for the circularization of the DNA fragments. Many single-stranded copies of the replicated DNA fragments are formed because the rolling circle replication mechanism occurs for each of the circular fragments. With the help of the compacted DNA nanoball, all the single stranded DNA copies concatenate head-to-tail to form a long strand. The microarray flow-cells can absorb the nanoballs in an extremely arranged model, which allows the sequencing process of DNA nanoballs with high density. Further, ligation with a specific fluorescent probe to the DNA at a specific location of the nucleotide sequence in the nanoball is carried out to unchain the sequencing process. The fluorescent light is recorded for each interrogated position after detecting the fluorescence which determines the base call. The read length for each DNB end is 35 (70 bases per DNB), which is equivalent to the mapping authority of the lengthier reads as compared to other techniques (Gupta and Gupta 2014).

The nanoball technology can pack the highest number of DNA nanoballs that have to be sequenced in an array and also maximize the number of reads per flow cell; on the other hand, the non-progressive cPAL unit can minimize the error rate of the reads. This technology attains preference to sequence the human genome in a large number of samples. On the other hand, this technology also has a few disadvantages, such as this technique is mainly for the sequencing of the human genome–not for other animals. Problems coupled with the sequencing of highly repetitive DNA occurs because of short length read of 35 bp (Gupta and Gupta 2014).

10.3.4 Massively Parallel Spatially Resolved Sequencing

An innovative method developed by Joakim Lundeberg and colleagues is now offered to early access through endemic Transcriptomics, their startup company (Stahl et al. 2016). In this novel technology, a fresh and iced up piece of tissue is accrued at chip which contains 100 µm unique sequence-barcoded oligo-dT capture probes equipped in an ordered manner with sequencing adaptors. The position of the cells is an ordered manner that can be recorded by imaging the tissue; after that, the permeabilization of the sample occurs followed by the mRNA diffusion on the array of the capture probe. Further, these probes are applied as primers for the synthesis of cDNA "on-chip." which the sequencing library generates and can further be retrieved and explored by NGS. Every read can subsequently be mapped back based on special characteristics according to its spatial barcode. Currently, it does not offer a resolution of single-cell but the spatially defined regions in wide transcriptome can be analyzed with high flow capacity.

10.3.5 Single Cell *In Situ* Transcriptomics

To explore the flow capacity of *in situ* RNA unveiling various imaging, chronological and combinatorial tagging approaches that are based on single-molecule fluorescence ISH (smFISH) have been developed. The chronological tagging technique can depend upon single or several fluorophores. The present technique was developed by Lubeck et al. (2014). The hybridization, imagination, and stripping of 24 sets of single dyed labeled recognition probes for a given transcript with DNaseI was carried out. In successive rounds of further cycles, the same probe sets were used but labeled dyes were different under combinatorial design which creates a peerless sequence of labels among hybridization cycles. The highly efficient hybridization reaction can detect more than 95% of all types of mRNA molecules which are present in a particular cell but the rate of colocalization (for precision and efficient multiplexing) is 77.9 ± 5.6 % among primarily two hybridizations (pretty low).

Another method, known as multiplexed error-robust fluorescence in situ hybridization (MERFISH) and developed by Chen et al. (2015), can determine all the information regarding mRNA of a particular single cell including copy number, type, location, and other identities. The binary labeling in MERFISH suggests that the RNAs of interest are either positively or negatively fluorescent for the given cycle of imaging. In this technique, the hybridization sequences with target-specification contained by the encoding probes and comprehended with readout sequences are primarily hybridized to intended RNAs. The fluorescent RNAs of the present cycle are marked as "1," while other RNAs are marked as "0." Among the cycles of imaging, the photobleaching of fluorescence commencing the preceding cycle occurs. The identification of different genes by the exclusive arrangement of readout probes which generates codes of 14–16 bit, after 14–16 cycles of the hybridization process. On the other hand, the rates of calling and errors are decreased and increase, respectively, as the rounds of hybridization cycles increase. The error-robust barcoding scheme, as a result of the introduction of the Hamming distance, was introduced to overcome this issue (also used for the telecommunication), which plays a role in the detection and correction of encoding errors in RNA barcodes. By using this technique, the identification of approximately 1,001 different mRNA species has been done. Additionally, the prediction of the innovative functions of many unannotated genes, mapping of regulatory complexes of many genes can be done by evaluating the expression level instability of various genes. The reason for the limited throughput of the smFISH technique is the requirement of high-resolution optical imaging setups. Consequently, they have limited demonstration at the level of a particular cell and not to the complete section of tissue.

10.3.6 RNA Sequencing by *In Situ* Sequencing Technology

The process of *in situ* sequencing has potential adaptability for diagnostics purposes; the tissue which has to be used is preserved either in fixed formalin or FFPE (embedded on paraffin) by extensive cross-linking which leads to strand cleavage or covalent modification in the nucleic acid molecule (Evers et al. 2011). This can cause length reduction in the integral molecules of RNA which can further be subjected to the extraction from the organisms. Sometimes the cross-links are reversible, but still, it is complicated to attain annotatable reads from such organisms. Interrogation of the approximate length of 40 nts (nucleotides) motifs can be done by *in situ* sequencing because there is no need for nucleic acid isolation from the tissue or organ.

Another method was developed for *in situ* sequencing by which the detection of a particular RNA molecule was carried out by using a single resolution of a nucleotide with padlock probes in the company of rolling circle amplification. In this technique, the RNA molecule is fixed in its natural atmosphere by using paraformaldehyde and subjected to the formation of cDNA from the RNA by using the *in situ* reverse transcription method with specific primers including target-specific lock nucleic acid-modified, or nontarget-specific random primers so that the padlock probes have to hybridize with the target of interest by their flanking ends. The gap-fill approach (the polymerized DNA ligated which circularize the probe) causes the cloning of cDNA into circular DNA which is then subjected to the rolling circle amplification process followed by the generation of substrate for *in situ* chemistry of NGS. In this technique, sequencing by ligation method is used and at complete genomics for the sequencing of the cloned fragment. This approach is applicable for the SNV (single nucleotide variant) differentiation in the mRNA of cultured cells of human and mouse cell β-actin (ACTB). The sequencing of short transcripts of ACTB and HER2 extracted from breast cancer tissue, the 12 and 13 codons of KRAS transcript are sequenced for the detection of rare mutation (KRAS mutation) (Ke et al. 2016). *In situ* technology was used to evaluate the distribution of fused transcript of TMPRSS2-ERG, coupled with a point mutation in somatic cells and the expression of biomarkers present in prostate cancer (Kiflemariam et al. 2014). The expression counterparts of 31 genes and the different expression models among positive cancer and stromal cells (HER2 and VIM) are obtained by the *in situ* technique (Ke et al. 2016). A parallel approach developed by Lee et al. (2015), fluorescent *in situ* sequencing (FISSEQ) for the generation of random libraries, is quite opposite to barcoded probes used in targeted *in situ* sequencing of gene index (Figure 10.2). Fluorescent *in situ* sequencing approach is also used for the tissue sections of the embryo (mouse and drosophila) and fully developed brain section of drosophila. 156,762 reads are procured in 8,102 genes by using the FISSEQ approach with the reading length of 30 bases. The targeted *in situ* sequencing is more sensitive in comparison to FISSEQ at double order of magnitude because the

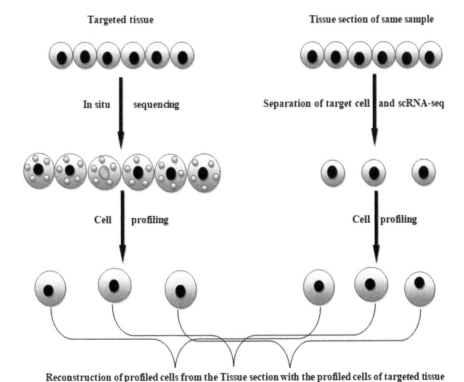

FIGURE 10.2   Gene expression data of a single cell by using *in situ* sequencing (fluorescence) and scRNA-seq for endemic reconstruction.

construction of target-specific library which facilitates the elimination of transcripts with expression level is very high to avoid signal overcrowding. On the other hand, to eliminate the optical extents and signal congestion, partition sequencing approach with target sequencing primers is used. *In situ* sequencing approaches have some limitations, such as in the alternative approach, there must be no spectral overlapping among the used dye and autofluorescence or to use the fluorescent dyes (time-resolved). The dimensional limitation also occurs in the substrate which is subjected to sequencing in terms of cell size. *In situ* sequencing also faces challenges in statistical analysis and elucidation (Ke et al. 2016). Approximately 10 GB of data is generated by the image process in each sequencing cycle which covers 20 mm² at 20X enlargement.

10.4 APPLICATIONS OF FOURTH-GENERATION (*IN SITU*) SEQUENCING TECHNOLOGY

The fourth-generation based technologies are successfully applied in the area of animal biotechnology and epigenetics from the revolution of traditional breeding to evaluate the genomic variations at a very small level. It has a high level of applicability in the field of evolutionary research for the identification of life diversity on the globe. The other fields like metagenomics, ancient DNA analysis, genomic variability, transgenics, food safety, nutrition, and human health (genetic disorder, cancer diagnostic and research, infectious diseases, post- and prenatal diagnostics, the human microbiome, personalized medicines) also have need of fourth-generation sequencing technology (Gupta and Gupta 2014). These NGS based technologies are essential in the areas of tailored medicines and clinical fields. By applying the complete genome sequencing, the evaluation of complex and disease-related genes can be done clinically. The exclusive technology of NGS encourages us to overcome the various technical and ethical challenges which need to be addressed. In clinical research, this application reduces the cost for storage and evaluation of data, plasticity in flow capacity, and elucidation and reporting of data is easier. Some sequencers, like Qiagen gene reader, can be used with a push of a button to evaluate the genomic and molecular investigation (Gupta and Gupta 2014; Feng et al. 2015). *In situ* sequencing technologies are applicable for mutation and genetic variation by interrogation of the defined DNA molecule in the intact cell. Transcript analysis of various cells in correspondence to their cellular circumstances can also be done by using *in situ* sequencing approaches (Larsson et al. 2010; Gut 2013). Fourth-generation sequencing technologies are applicable in the areas of food safety and nutrition to enhance poultry production with economic feasibility through real-time results and also enhance food safety. The microflora of the gut, the attendance of plasmids (genes that are involved in the metabolic pathway) can be determined for the functional screening of the production of nutrients and resistance against antibiotics. The sequencing of the microflora of poultry as well as other animals can be conducted to know the effect on human health and disorders (Gupta and Gupta 2014). *In situ* sequencing technologies are applicable for the sequencing of microbial contaminants; their interpretation of food components could help cure many of the diseases caused by microbial contamination and enlighten our knowledge about food safety and security.

10.5 FUTURE PERSPECTIVE AND RELATED CHALLENGES

Fourth-generation sequencing technology (*in situ* sequencing with the unity of conventional imaging approaches) offers many novel and innovative openings for the recitation of heterogeneity of the tissue. The recent data of scRNA-seq, the various cells have hundreds of similar copies in the brain tissue thus this provides to explore the understanding with neurological turmoils and the workings of the brain. The molecular and cellular mapping of tissue micro atmosphere in array to describe novel interests for therapeutics and anticipation is the splendid challenge in cancer research. Some laboratories are engaged with research that aims to do profiling of endemic (spatial) transcripts of cancer to related treatment. Though *in situ* sequencing corresponds to a potential contrivance, before its broad application many technical issues such as imaging of sample, comparatively low competence of molecular progressions, management and elucidation of data needed to be addressed. The techniques belong from the *in situ* sequencing technology utilize imaging sequencing based on the optical fluorescence which may create trouble with the samples from biological sources that have autofluorescence at an elevated level. This problem can be overcome either by applying light irradiation technique or other protocols for tissue clearing (used in traditional FISH) to the *in situ* sequencing for removal of autofluorescence (Chung and Deisseroth 2013; Yang et al. 2014; Ke et al. 2016). Compression and molecular crowding are the main limitations. That's why the present *in situ* sequencing technologies fail to profile the complete transcriptome; the synergistic combination of this tool with the scRNA-seq methodology can conquer this problem. Identification and definition of cell types between the populations of cells that are isolated from any source (tissue) can be evaluated by the scRNA-seq approach. The similarity and collation of the molecularly defined expression profiles with *in situ* sequencing expression profiles of the same source can be compared. The scRNA-seq defined state and type of the cell can be mapped to their intact location in the tissue which provides a clearer picture of endemic gene expression; on the other hand, *in situ* sequencing gives information about the position of the cells based on biomarker selection which is recognized by scRNA-seq and also gives deep information about the transcriptional state of molecularly defined cells (Figure 10.2).

10.6 CONCLUSION

In conclusion, fourth-generation (*in situ* sequencing) sequencing technology with high throughput are continuously in progression. The transitional gap among the sequencing generations is disordered in the array of changes and the amount and nature of the sequencing approach has resulted in cost reduction. The sequencing of DNA now can be potentially applied for the historical interpretation. There is no displacement by each new generation of the preceding generation technology. The *in situ* sequencing technology endows various factitive illustrations to the genomic and transcriptomic investigations. Further improvement to overcome the technical and physical obstructions can make it possible to be the standard method of sequencing (genomics and transcriptomics).

REFERENCES

Achim, K., Pettit, J. B., Saraiva, L. R., et al. 2015. High-throughput spatial mapping of single-cell rna-seq data to tissue of origin. *Nature Biotechnology* 33:503–9.

Blakeley, P., Fogarty, N. M. E. and Del Valle, I. 2015. Defining the three cell lineages of the human blastocyst by single-cell RNA-seq. *Development* 142:3151–65.

Brennecke, P., Reyes, A., Pinto, S., et al. 2015. Single-cell transcriptome analysis reveals coordinated ectopic gene-expression patterns in medullary thymic epithelial cells. *Nature Immunology* 16:933–41.

Chen, K. H., Boettiger, A. N., Moffitt, J. R., Wang, S. and Zhuang, X. 2015. Spatially resolved, highly multiplexed RNA profiling in single cells. *Science* 348:60.

Chung, K. and Deisseroth, K. 2013. CLARITY for mapping the nervous system. *Nature Methods* 10:508–13.

Deng, Q., Ramsköld, D., Reinius B., et al. 2014. Single-cell RNA-seq reveals dynamic, random monoallelic gene expression in mammalian cells. *Science* 343:193–6.

Derrington, I. M., Butler, T. Z., Collins, M. D., et al. 2010. Nanopore DNA sequencing with MspA. *Proceedings of the National Academy of Sciences USA* 107(37):16060–5.

Evers, D. L., Fowler, C. B., Cunningham, B. R., Mason, J. T. and O'Leary, T. J. 2011. The effect of formaldehyde fixation on RNA: Optimization of formaldehyde adduct removal. *The Journal of Molecular Diagnostics* 13(3):282–8.

Feng, Y., Zhang, Y., Ying, C., Wang, D. and Du, C. 2015. Nanopore-based fourth-generation DNA sequencing technology. *Genomics, Proteomics and Bioinformatics* 13(1):4–16.

Grun, D., Lyubimova, A., Kester, L., et al. 2015. Single-cell messenger RNA sequencing reveals rare intestinal cell types. *Nature* 525(7568):251–5.

Gupta, A. K. and Gupta, U. D. 2014. Next generation sequencing and its applications. In: Verma, A. S., Singh, A., editors. *Animal biotechnology: Models in discovery and translation*. Academic Press: Waltham, MA, 345–67.

Gut, I. G. 2013. New sequencing technologies. *Clinical and Translational Oncology* 15(11):879–81.

Hanchate, N. K., Kondoh, K., Lu, Z., et al. 2015. Single-cell transcriptomics reveals receptor transformations during olfactory neurogenesis. *Science* 350:1251–55.

Ke, R., Mignardi, M., Hauling, T. and Nilsson, M. 2016. Fourth generation of next-generation sequencing technologies: Promise and consequences. *Human Mutation* 37(12):1363–7.

Ke, R., Mignardi, M., Pacureanu, A., et al. 2013. *In situ* sequencing for RNA analysis in preserved tissue and cells. *Nature Methods* 10(9):857–60.

Kiflemariam, S., Mignardi, M., Ali, M. A., Bergh, A., Nilsson, M. and Sjöblom, T. 2014. *In situ* sequencing identifies TMPRSS2–ERG fusion transcripts, somatic point mutations and gene expression levels in prostate cancers. *The Journal of Pathology* 234(2):253–61.

Ku, C. S. and Roukos, D. H. 2013. From next-generation sequencing to nanopore sequencing technology: Paving the way to personalized genomic medicine. *Expert Review of Medical Devices* 10(1):1–6.

Kumar, G. and Kocour, M. 2017. Applications of next-generation sequencing in fisheries research: A review. *Fisheries Research* 186:11–22.

Larsson, C., Grundberg, I., Söderberg, O. and Nilsson, M. 2010. *In situ* detection and genotyping of individual mRNA molecules. *Nature Methods* 7(5):395–7.

Lee, J. H., Daugharthy, E. R., Scheiman, J., et al. 2015. Fluorescent *in situ* Sequencing (FISSEQ) of RNA for gene expression profiling in intact cells and tissues. *Nature Protocols* 10(3):442–58.

Lubeck, E., Coskun, A. F., Zhiyentayev, T., Ahmad, M. and Cai, L. 2014. Single-cell *in situ* RNA profiling by sequential hybridization. *Nature Methods* 11:360–1.

Mignardi, M. and Nilsson, M. 2014. Fourth-generation sequencing in the cell and the clinic. *Genome Medicine* 6(4):31.

Mikheyev, A. S. and Tin, M. M. 2014. A first look at the oxford nanopore MinION sequencer. *Molecular Ecology Resources* 14(6):1097–102.

Reid, D. E., Hayashi, S., Lorenc, M., et al. 2012. identification of systemic responses in soybean nodulation by xylem sap feeding and complete transcriptome sequencing reveal a novel component of the autoregulation pathway. *Plant Biotechnology Journal* 10(6):680–9.

Satija, R., Farrell, J. A., Gennert, D., Schier, A. F. and Regev, A. 2015. Spatial reconstruction of single-cell gene expression data. *Nat Biotechnology* 33: 495–502.

Shalek, A. K., Satija, R. and Shuga, J. 2014. Single-cell RNA-seq reveals dynamic paracrine control of cellular variation. *Nature* 510:363–9.

Stahl, P. L., Salmen, F., Vickovic, S., et al. 2016. Visualization and analysis of gene expression in tissue sections by spatial transcriptomics. *Science* 353:78–2.

Usoskin, D., Furlan, A., Islam, S., et al. 2015. Unbiased classification of sensory neuron types by large-scale single-cell RNA sequencing dmitry. *Nature Neuroscience* 18:145–53.

Venkatesan, B. M., Dorvel, B., Yemenicioglu, S., Watkins, N., Petrov, I. and Bashir, R. 2009. Highly sensitive, mechanically stable nanopore sensors for DNA analysis. *Advanced Materials* 21(27):2771–6.

Yang, B., Treweek, J. B., Kulkarni, R. P., et al. (2014). Single-cell phenotyping within transparent intact tissue through whole-body clearing. *Cell* 158(4):945–58.

CHAPTER 11

Library Preparations and Sequencing Platform for Improving Food Safety

Mahesh Pattabhiramaiah and Shanthala Mallikarjunaiah

CONTENTS

11.1 INTRODUCTION

When dealing with microbial pathogens with regard to food safety, it is of paramount importance that the pathogen under investigation is clearly defined and characterized based on the identification and/or assessment of foodborne outbreaks, assessment of virulence or disease risk, and control technique analyses. Pathogens have predominantly been identified at the species level and beyond the context of the microbial communities in which they exist.

Effective taxonomic awareness of microorganisms and their species is required to promote beneficial food processes, including fermentation, and to minimize adverse events including contamination and spoilage. Conventional methods such as classical gram stain, along with detailed biochemical characteristics, are mainly used for the isolation, identification, and characterization of bacteria of clinical, food, or environmental origin. Although regarded as a "gold standard," culture-dependent techniques can only detect 0.1% of diverse populations, including those present in human intestinal microbiota.

The current approach for the evaluation of foodborne disease outbreaks relies entirely on conventional microbiological techniques that are slow, time and labor-consuming culture-dependent approaches, like enrichment and/or selective steps, and have disadvantages that can contribute to delays which impede the identification of the outbreak. From a public health, policy, and regulatory perspective, it is important to take advantage of any improvements in awareness and understanding of new technologies that may help to monitor foodborne disease outbreaks. Traditional culture-dependent approaches have slowly been augmented by molecular analytical methods over the last decades. The advent of molecular diagnostic methods which focus on nucleic acid detection of the organism has made the detection, identification, and characterization of foodborne pathogens (FBPs) quicker, more accurate, and sensitive (Weile and Knabbe 2009; Dwivedi and Jaykus 2011).

Alternative molecular diagnostics methods, such as polymerase chain reaction (PCR), are fast and boost the outcomes of investigations. While PCR analysis improves the screening efficiency of FBP samples, viable microorganisms still need a mandatory confirmation of positive samples. Furthermore, in circumstances of microbial outbreaks, where traditional approaches such as culture-based methods cannot identify, these approaches are primarily used as additional confirmatory methods. The domain of genomics is developing steadily with the introduction of new generations of technologies, including ultra-high-throughput sequencing, which can provide enhanced insights and further reduce outbreak investigation time scales.

As a result, to enhance awareness of the ecological niche, like food, strategies are needed to classify or characterize microorganisms and forecast the functional dynamics of the diverse microbiological populations involved in the sample. Recent advances in multi-omic innovations have allowed the characterization of numerous microbial species and their variations present in a food sample. For the monitoring of foodborne pathogens and microbial communities, innovation in high-throughput sequencing in the 2000s has facilitated the implementation of accurate subtyping approaches. Today, the deployment of these approaches is significantly transforming the monitoring of FBPs and is anticipated to gradually guide the regulation of FBPs.

Microbial typing is the process of determining or distinguishing between various types of microorganisms in the same microbial community. Although there has been a transition toward genotyping methods using DNA or RNA molecules, which give accurate results, the conventional phenotyping approaches still rely on phenotypic properties such as antimicrobial profiling, chemotaxonomic profiling, phage-typing, serotyping, etc. A blend of methods is often recommended.

For the surveillance and identification of foodborne disease outbreaks, fingerprinting technologies including multi-locus variable number tandem repeat analysis (MLVA), pulsed-field gel electrophoresis (PFGE), multilocus sequence typing (MLST), and DNA sequence-based characterization method have proven powerful (Swaminathan et al. 2001; Joseph and Forsythe 2012; EFSA 2013; Lindstedt et al. 2013). Next-generation sequencing (NGS) technologies have entered the arena, providing incentives to precisely identify and gaining significant access to FBPs (Croucher and Didelot 2015; Franz et al. 2016). In NGS applications, whole-genome sequencing (WGS) and metagenomics have rapidly advanced as concerns microbial hazards and food safety.

Genomes are life architectures that regulate the structural dynamics and efficiency of organisms all across their lifespan via single genes or multiple genes situated in different loci. For delivering both the structural and functional characteristics of genomes, DNA sequencing technologies act as valuable resources (Bai et al. 2012). Due to the advent of next-generation sequencing techniques in 2005, DNA sequencing was accomplished using sequencing methods (Maxam and Gilbert 1977; Sanger et al. 1977; Heather and Chain 2016).

Oswald Theodore Avery confirmed that deoxyribonucleic acid (DNA) was the genetic material. Watson and Francis Crick, in 1953, determined its double-helical structure, attributing to the central dogma of molecular biology. The mysteries of life are decoded by the genomic DNA, which determines the species and individuals that make the DNA sequence crucial to the analysis of the structures and functions of the cells. DNA sequencing innovations may support scientists and health professionals in a wide variety of applications, such as molecular genetic modification, breeding, virulence gene identification, comparative and phylogenetic studies. Ideally, DNA sequencing technologies are simple to operate, rapid, precise, and inexpensive. Over the decades, DNA sequencing technologies and applications have witnessed phenomenal development leading to an explosion of genome data.

Over the past four decades, the swift progress in DNA sequencing technologies has enhanced the capability to precisely identify and characterize the microbiomes of complex food matrices or environmental samples. The universal nature and specificity of nucleic acids makes the molecule an ideal target for bacterial or microbiome characterization. DNA sequencing has steadily evolved from Sanger's low throughput DNA sequencing to next-generation (NGS) and third-generation high-throughput sequencing techniques (Loman and Pallen 2015).

NGS developments have been a central resource in virtually every area of biological science in recent years (Metzker 2010). It allows parallel sequencing of millions of tiny DNA fragments at low per-base costs in a short period. In addition to genome (re)sequencing, NGS offers comprehensive knowledge about the structure of complex DNA (cDNA) organisms, making it the chosen tool for most, though not all, genomic applications, such as transcriptome analysis (RNA-Seq), metagenomics, methylated DNA (MeDip-seq), and DNA-associated protein (ChIP-Seq) profiling. Modern NGS applications are also continually rendered and scrutinized in order to achieve greater and improved data quality.

11.2 LIBRARY PREPARATION

NGS sources require the translation of the nucleic acid source material into normal libraries. There are a broad range of protocols of preparation for NGS libraries, but they have in common that (fragments of) DNA or RNA molecules are fused with adapters containing the elements required for solid surface immobilization and sequencing. Additionally, size selection steps are often carried out and PCR is typically used to amplify libraries. The accuracy of sequencing data relies heavily on the consistency of the sequenced material. The library development method will also ensure a fast molecular recovery of the initial fragments (low bias and high complexity) in order to obtain the best genomic coverage with the least amount of sequencing.

NGS can be generally categorized into the following process steps: sample pre-processing, library preparation, sequencing, and bioinformatics. Notwithstanding the basic principles of this sequencing technique, all modern technology for sequencing requires dedicated sample preparation to yield the sequencing library loaded onto the instrument (Metzker 2010; Goodwin et al. 2016). Sequencing libraries consist of DNA fragments with oligomer adapters at the 5" and 3" end of a given length range and the actual sequencing method. The generated data is analyzed with bioinformatics following sequencing. Many well-established protocols for different sequencing platforms require similar steps and result in the same automation demands. Five potential automation processes identified:

- Mechanical fragmentation
- Enzymatic reactions
- Size selection and clean-up
- Amplification
- Quantification

11.2.1 Mechanical Fragmentation

The concepts of acoustics, sonication, and hydrodynamic fragmentation are applicable to mechanical fragmentation. The most complicated technique is the Illumina sequencing method, which involves the shortest fragment length, roughly 350–550 base pairs (Illumina 2019). Sonic and ultrasonic shearing use acoustic waves to create cavitation bubbles. The bubbles burst after each acoustic stimulus, which induces high local liquid velocity, leading to the breakup of DNA strands. The Diagenode Bioruptor is a machine that deals with sound waves, while the Covaris Focused-ultrasonicator uses ultrasonic waves (Diagenode Megaruptor 2019).

Hydrodynamic fragmentation is desirable when larger fragments are required. Strong shear forces are produced through a small orifice with dimensions of micrometers which require very narrow dividing bands in combination with limited DNA losses. But the lengths of separation are limited to a few thousand base pairs. This technique is used in the Diagenode Megaruptor and the Covaris G-tube that can be used in normal laboratory centrifuge systems.

11.2.2 Enzymatic Reactions

The automated process of enzyme reactions summarizes the protocol steps which involve incubation at a given temperature. This involves enzyme fragmentation, end repair, adenylation , and ligation. This ensures that a given temperature to be maintainable for a specified period is a prerequisite for the automation device.

11.2.3 Size Selection and Cleanup

DNA fragments of specified length should be generated by selection and cleaning of the sizes. This process may involve magnetic beads, columns, or gels. For most developed protocols, the capacity to cope with reversible solid phase immobilization perforations for selection and clean-up is essential (Hawkins et al. 1994; DeAngelis et al. 1995; Xu et al. 2003). But other organizations, including Qiagen, provide the customer with spin-column cleanup kits, and the BluePippin Program advises scale collection by gel for the PACBIO SMRT bell Express Design preparedness package.

11.2.4 Amplification

PCR is the normal mechanism for nucleic acid amplification which involves thermal cycling. The majority of library preparation protocols provide an extension stage for adding sequencing adapters and increasing the concentration of DNA. By comparison, amplification-free processes create more fragments with incomplete adapters (Rohland and Reich 2012).

A disadvantage of amplification during the preparation of the library is a potential PCR prejudice which results in an inadequate representation or even a total absence of certain loci with extreme base compositions (Dohm et al. 2008). Important progress has been made in recent years in minimizing prejudice (Aird et al. 2011), and in library preparation, PCR amplification has become a routine step. However, in order to reliably measure the nucleic acid sequences in the sample, special PCR workflows using unique molecular identifiers (UMIs) may be used to effectively eliminate prejudice. These involve a 2–4-cycle UMI-PCR with a corresponding swap of primers and a second PCR for amplification and amplifier ligation (Ståhlberg et al. 2016; Ståhlberg et al. 2017). The automation phase of these multi-stage PCR workflows should be paid particular attention to avoid contamination throughout the processing of PCR products (Kotrova et al. 2017).

11.2.5 Quantification

Quantifying typically takes the form of a quality check before the sequencing step at the outset of a library preparation protocol. Quantification of samples is highly important when mixing multiple samples in a single series of separate barcodes. Molar differences in concentration among collected libraries may contribute to an inadequate coverage of the less concentrated libraries. Spectrofluorometric methods, real-time quantitative polymerase chain reaction (qPCR) and droplet digital PCR (ddPCR) can be used to quantify the library (Robin et al. 2016) to measure the mass concentration of a sample, whereas spectrofluorometric methods use fluorescent dyes (Mardis and McCombie 2017). By comparing the PCR sample with reference PCRs, a qPCR may give the mass concentrations (Buehler et al. 2010).

Quantifications utilizing qPCR are more reliable than spectrofluorometrical approaches, because only nucleic acids are detected through sequencing adapters. Spectrofluorometric approaches as well as real-time chain reactions to polymerase only result in mass concentrations. The average length of the fragment must be estimated and the molar concentration measured. Commercial systems, such as the Agilent Bioanalyzer or Fragment Analyzer System can be used to determine the average fragment length.

Digital amplification facilitates quantification without reference PCRs in comparison to qPCR (White et al. 2009; Hindson et al. 2011; Hindson et al. 2013). ddPCRs thus automatically demonstrate molar concentration without the need to examine the length of the fragment (Pinheiro et al. 2012; Laurie et al. 2013).

11.3 WHOLE GENOME SEQUENCING

Three major technological revolutions have been accomplished by scientific advances in WGS: first-generation sequencing (whole genome shotgun sequencing), NGS, and third-generation sequencing (single-molecule long-read sequencing (Loman and Pallen 2015).

11.3.1 First Generation DNA Sequencing (Whole Genome Shotgun Sequencing)

The Sanger Chain Termination Process was the first DNA sequencing technique developed by Frederick Sanger and his colleagues in 1977 on a chain-termination process (also known as Sanger sequencing). Another DNA-based sequencing technique was developed by Walter Gilbert based on chemical modification of DNA followed by the cleavage at specific bases. Compared to conventional microbiology, the relatively low performance and high cost of modern sequencing approaches imply that such a technique was primarily employed either as a confirmatory approach or a gleaning knowledge method which conventional methods could not provide. The shotgun approach for Sanger sequencing was used for WGS and assembly, which selectively incorporated chain-terminating dideoxynucleotides by DNA polymerase during DNA replication. A capillary-based; semi-automated variant of traditional Sanger technique is the whole genome shotgun DNA sequencing. To determine its sequence, the first process uses chemicals to break up DNA, while the Sanger method makes copies of DNA strands and determines what nucleotides were added. At this juncture, DNA is fragmented randomly, cloned to a plasmid with a high copy number, and transformed into *E. coli*. The flanking PCR primers are used to amplify the cloned region.

A fluorescently labeled dideoxyribonucleotide (ddNTP) that corresponds to the nucleotide identity at the end position is stochastically terminated in each PCR loop. A high-resolution electrophoretic polymeric gel capillary subsequently separates the DNA fragments, and as the gel exits, an argon laser excites the fluorescent-labeled DNA and the emission spectrum is captured. Read lengths of about 1000 pairs with a precision of 99.99 percent have been obtained using this technique. However, the use of this approach was constrained by low throughput and high operating costs.

Sanger sequencing has been embraced as the "first generation" of laboratory and commercial sequencing applications because of its high performance and lower radioactivity (Sanger et al. 1977). DNA sequencing was challenging and laborious at that time because radioactive materials were required. In 1987, Applied Biosystems introduced the first sequencing device, namely AB370, after years of development, and adopted capillary electrophoresis to speed up and make the sequencing more accurate; it could detect 96 bases at a time, 500 K bases per day, with a reading length of 600 bases. Sanger sequencing technology became the main tool for the completion of the Human Genome Project in 2001 (Smith et al. 1986). This initiative greatly spurred the production of innovative novel sequencing tools to improve speed and precision while minimizing costs and resources at the same time.

11.3.2 Second Generation Sequencing

Owing to the low speed, expense, and time-consuming problems of the first-generation sequencing methods, another new technique was developed with higher throughput from multiple samples and lower cost than the previous techniques; it was known as

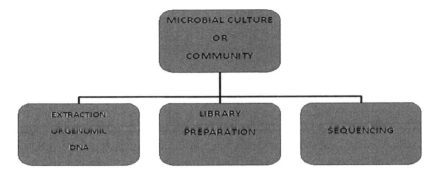

FIGURE 11.1   Next generation sequencing workflow.

"Next Generation Sequencing (NGS) Technologies", or "High Throughput Sequencing Technologies" (Reuter et al. 2015; Kchouk et al. 2017). NGS is a fairly new technology and can be used to retrieve genetic data from multiple candidate genes including single nucleotide polymorphism (SNPs), and is more cost-effective than first-generation sequencing methods (Ansorge 2016).

NGS has groundbreaking impacts on genetic applications such as metagenomics, genomics comparison, high-performance polymorphism detection, RNA analysis, mutation screening, profiling for transcriptomes, methylation, and chromatin remodeling. Sequence (read length), sequence efficiency, high performance, and low cost are the key indicators for the success of the next-generation technology.

NGS techniques involve the extraction of genomic DNA for whole-genome sequencing, and, in metagenomics, the study involves the extraction of genomic DNA from the community. Bioinformatics tools are used for data analysis after a library preparation process that can include PCR and a sequence of the generated library pool (Figure 11.1). NGS technologies include Roche/454 sequencing (https://www.454.com), Illumina (Solexa) (https://www.illumina.com), Applied Biosystems SOLiD™ System (https://www.appliedbiosystems.com), and Ion Torrent sequencing (https://www.iontorrent.com) (Henson et al. 2012; Morey et al. 2013).

Based on their age of appearance, NGS platforms can be categorized into second and third-generation sequencing technologies. Pacific BioSciences and Oxford Nanopore are the latest technologies and characterized by their sequencing technology of the third generation while the rest is the technology of second-generation (Pareek et al. 2011). Third-generation NGS technology is used to sequence individual DNA molecules with no previous amplification level, that is, single long molecule sequencing or clonal amplification; but second-generation NGS platforms rely on PCR, the disadvantage being the creation clusters of a certain DNA template (Khodakova et al. 2016).

11.3.2.1 Roche/454 Sequencing

The KTH Royal Institute of Technology (Stockholm) developed the first next-generation Roche/454 sequencing in 1996 and introduced to the market by 454 Life Sciences in 2005 (https://www.454.com) and subsequently upgraded to GS FLX Titanium series after 3 years of production (Pillai et al. 2017). Roche sequencing is a pyro-sequencing method (sequencing by synthesis [SBS]) that is based on the detection and quantification of DNA polymerase activity performed using the enzyme luciferase. Roche/454 sequencing has been effectively employed for both confirmatory and de-novo sequencing (Fakruddin et al. 2013).

11.3.2.2 Illumina (Solexa) Sequencing

In terms of usability, versatility, and efficiency, Illumina at the forefront of the development of sequence platforms applicable in the field of genomics research for humans and animals. The purpose is to analyze genetic variants and functions using cutting-edge sequencing technologies. Illumina has recently used a sequencing technique that dominates the sequencing industry and generates more than 90% of the world's DNA sequence data. Illumina acquired Solexa, which was published in 2005 (Barba et al. 2014) before the introduction of Mi-Seq and Hi-seq platforms. The Mi-seq platform can sequence upto 15 Gbp, while Hi-seq platforms achieves upto 600 Gbp. In a single run, the (Illumina) Solexa sequencer can sequence up to 1 Gbp. The reversible terminator technique used by Illumina employs an intermediate sequencing principle comparable to the chain termination technique, used during Sanger sequencing, whereby a fluorescent-labeled base prevents the elongation of the strand, and sequence information is unveiled by reading out the label of the incorporated base. To sequence the millions of clusters present on the flow cell surface, Solexa sequencing uses four fluorescently labeled nucleotides and a special DNA polymerase enzyme (Heather and Chain 2016).

11.3.2.3 Applied Biosystems SOLiD Sequencing

Sequencing by oligonucleotide ligation detection (SOLiD) is one of Life Technologies (Thermo Fisher) products (https://www.appliedbiosystems.com). SOLiD employs a clonal amplification by emulsion PCR and detection by optical systems, similar to Illumina's platforms Roche/454, and Thermo Fisher's Ion Torrent (Levy and Myers 2016). Applied Biosystems SOLiD platform uses ligation-mediated synthesis chemistry for sequencing reactions (Valouev et al. 2008; Levy and Myers 2016).

To elongate the complementary strand, the SOLiD sequencing technique involves a series of probe-anchor binding, ligation, imaging, and cleavage cycles. The process starts with the application of amplified DNA fragments to microbeads for bead-based preparation. Beads are often placed on a glass slide onto which the DNA fragments can be fixed. To enable scaling of the number of samples analyzed, the glass slides can be divided up into 8 chambers. The 8-mer oligonucleotides are sequentially bound to DNA fragments with a fluorescent label at the end. The consequential end product is then eliminated and for 5 more cycles the procedure is repeated using hybridized primers; the design enables error checking capacity. The chief drawbacks include short read intervals (50–75 bp), the very long cycles of 7–14 days, and the need for cutting-edge computing technology and advanced data processing knowledgeable staff.

11.3.2.4 Ion Torrent Sequencing: Proton/Personal Genome
Machine (PGM) Sequencing

Life Technologies (now Thermo Fisher), marketed the Ion Personal Genome Machine (PGM), the Ion Torrent sequencing platform (http://www.iontorrent.com). There are four sequencing chips available (Wang et al. 2014). The Ion PI Chip uses a proton sequencer, while the Ion 314, 316, or 318 chips use Ion PGM. The Ion 314 chip offers the lowest reads at 0.5 million reads per chip, while the Ion 318 Chip offers the maximum reads of close to 5.5 million reads per chip. The Proton sequencer offers higher efficiency (10–100 Gb vs. 20 Mb–1 Gb) and higher reads per run (660 Mb vs. 11 Mb) than the PGM chips, but perhaps the read lengths (200–500 bp), run time (4–5 h), and precision (99%) are comparable (Liu et al. 2012; Wang et al. 2014). In the last decade, this sequencing technology has been attributed to comprehensive transcriptomal studies (Yuan et al. 2016).

The sequence network for ion torrent uses an equivalent technology as pyrosequencing but does not use enzyme reactions or optic fluorescent nucleotides branded as other technologies of the second generation (Rothberg et al. 2011; Salipante et al. 2014). It measures the liberation of Hydrogen ion (H^+), a byproduct of incorporation of nucleotide as quantitated changes in pH shifts by means of a new integrated silicon coupled detector (Quail et al. 2012). The consequential alteration in pH is detected by an integrated complementary metal-oxide-semiconductor (CMOS) and an ion sensitive field-effect transistor (ISFET). Detected pH is imprecisely equal to the number of detected and converted into a voltage signal which proportionate with the number of incorporated nucleotides (Goodwin et al. 2016). The system has until now used to detect small genomes and targeted sequencing. However, a new system and new chips that will allow them to push into the high throughput territory of whole genome sequencing has been announced. Ion Torrent launched its follow-on system, that is, Ion Proton in 2012 which allows for larger chips with higher densities needed for exome and whole genome sequencing (https://www.thermofisher.com).

11.4 THIRD GENERATION SEQUENCING

Third-generation technologies have the potential to synthesize extremely precise de novo assemblies of microbial, plant, and animal genomes, permitting novel perceptions into molecular phylogeny. DNA sequencing and mapping technologies in third generation are generating a revolution in the sequencing of high-quality genomes. To address this limitation of PCR amplification used in NGS, third-generation sequencing technology was designed. Contrary to sequencing of the second generation which generates a few hundred pairs of short-reads, the technology from the third generation generates over 100,000 bp molecules or maps. In this case, the sequence of single DNA molecules decreases the risk of low errors by avoiding bias, phasing, or the problems of synchronization associated with amplifications.

Pacific Biosciences (PacBio) Single Molecule Real Time (SMRT), Illumina Tru-seq Synthetic Long-Read technology, and Oxford Nanopore Technologies sequencing platforms are three commercially available third-generation DNA sequencing technologies. The three technologies make long reads ranging between 5000bp and 15,000bp, with some reads exceeding 100,000bp, using a single sequence or clonal amplification and sequences of long molecules.

11.4.1 Pacific Biosciences Single Molecule Real Time (SMRT) Sequencing

PacBio SMRT sequencing was developed and implemented as a third-generation sequencing technique as a single parallel molecular real-time sequencing method (Shin et al. 2013). Contrary to all other sequencing systems, PacBio does not require clonal DNA amplification (Figure 11.2). Pacific Biosciences (https://www.pacificbiosciences.com) markets the PacBio RS II system launched in 2011 (Nakano et al. 2017) and the SMRT real-time sequencing system was the first commercially available long-read single-molecule device with reads of tens of kilobases in length, and complete genomes can be readily assembled. This sequencing method can be applied for whole-genome sequencing, selective sequencing, complex population analysis, RNA sequencing, and epigenetics characterization. In comparison to the most extensively used third-generation sequencing technology, the

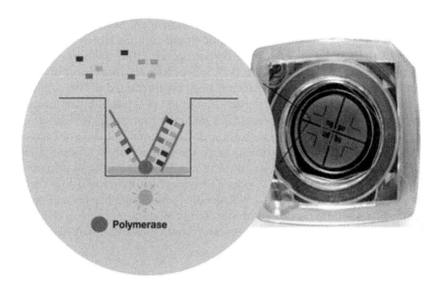

FIGURE 11.2   PacBio sequencing system (https://genetics.thetech.org).

PacBio RS (Ardui et al. 2018) platform has developed into other series of systems, including RSII and Sequel, for the correction of low-quality reads. Features include high read
lengths, uniform coverage, and single-molecule resolution.

The silicon dioxide chips termed zero-mode waveguides (ZMWs) are used in PacBio
SMRT technology for sequencing (Eid et al. 2009; Ambardar et al. 2016). The platform
uses a series of fluorescently labeled nucleotides that are released by a synthesis approach
and in real-time detection. ZMWs harness the properties of light passing through aper-
tures with a diameter smaller than its wavelength. Each ZMW encompasses a DNA polymerase bound to its bottom and the target DNA fragment for sequencing. Throughout
typical speed reverse strand synthesis, the incorporated nucleotide can be identified with
the aid of labeled fluorescent dye.

The main benefit of the SMRT sequencer platform includes long read lengths (for
de novo assemblies of novel genomes), direct evaluation of the individual molecules, and
the ability to prepare templates without PCR amplification (Roberts et al. 2013). The
device documents simplify and improve genomic assembly and perspectives of the disease
heritability including the kinetic of each nucleotide incorporation reaction. PacBio SMRT
technology, however, significantly limits higher error rates (Ardu et al. 2018). SMRT sup-
ports multiple sequence applications which provide genetic data with new standards of
depth and consistency and aids the precise study of sequencing of whole genomes, complex populations, isoforms, or epigenetic modifications.

11.4.2 Nanopore Single Molecule Sequencing

In 2014, Oxford Nanopore Technologies, a UK-based company developed the most recent
third-generation technology called "nanopore sequencing" (Lee et al. 2016). Nanopore
sequencing was implemented to address short-read sequence technology limitations and
permits large DNA molecules to be sequenced from easy-to-read libraries in a short time.
Nanopores are pores of nanometer size and are deemed to be one of the most innovative

technologies for real-time, high-speed, true single-molecule DNA sequencing (Rhee and Burns 2006). It's a single molecule sequencing (SMS) technology that can sequence a single DNA molecule sequencing and of long reads up to 1 Mb and longer without the aid of a replication enzyme (Schadt et al. 2010).

A Nanopore is just a small hole of 1 nm internal diameter. It is incorporated into an engineered membrane which is electrically resistant and is supplied with voltage through the membrane. A library is constructed without the need for PCR amplification steps for high-molecular-weight genomic DNA extraction. The adaptors are linked to the DNA fragments and a processing enzyme is combined during library preparation (Magi et al. 2017). The sequencing initiates when the enzyme-sequence complex interacts with the nanopore. A process enzyme (green) is mixed with DNA. The complex of DNA-enzymes interacts with nanopore (blue). Nanopores serve as the channel between two chambers of an electrophoretic device used for the sequencing technology. The resulting current can be determined by applying a little voltage (~ 100 mV) across the nanopore. The ionic current disruption is caused by the molecules passing through the nanopore, and measuring this disruption allows the molecule to be identified (Derrington et al. 2010). In the context of DNA, each base gives a subtly different reading as it passes through the pore, allowing the direct reading of the sequence. DNA fragments are translocated into pores, allowing for real-time analysis (Loman and Watson 2015; Lu et al. 2016). Nanopore DNA sequence delivers a range of promising benefits over the others, including fast detection of minimal limited content, very long reads, quick results, low cost, and tiny footprints (https:/nanoporetech.com). In addition, the latest MinION ultra-long single-molecule reads called R9.4 were made available (Jain et al. 2017) with a median increase to 92% and a substantially increased performance of 127000 to 217000 reads per flow cell, overcoming previous readers' high error rates.

The Oxford Nanopore MinION is an electronically operated device that sequences DNA through the electronic measurement of minute electric current interruptions as DNA molecules move through a nanopore (Figure 11.3). Small genomes, including *E. coli* (4.5Mbp) or yeast (12Mbp) can be effectively sequenced. The precision per nucleotide of genomes sequenced by MinION has been calculated to > 99.95% using error correctional algorithms close to those used for PacBio reads. The small-scale and low-cost instruments have made it suitable for studying Ebola outbreaks in the field in West Africa and other remote locations.

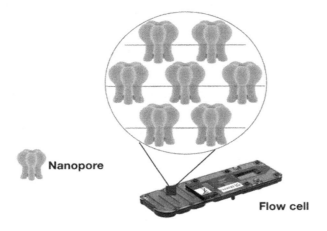

FIGURE 11.3   Nanopore sequencing system (https://genetics.thetech.org).

11.4.3 Helicos/HeliScope Sequencing

The Helicos sequencing method was first commercially introduced by the now-bankrupt Helicos Biosciences for single-molecule fluorescent sequencing (Shendure and Ji 2008; Thompson and Steinmann 2010) Using the Helicos sequencing system and HeliScope single-molecule sequencers, the Seqll sequencing provider (http://seqll.com) sequences genomic DNA and RNA. HeliScope is another example of single DNA molecule sequencing (Braslavsky et al. 2003). In this instance, a synthesis with a highly sensitive fluorescence device is used to do the sequencing of single DNA molecules. Random DNA fragmentation followed by poly-A tailing is used to prepare the DNA library which is then hybridized to surface-hitched poly T oligomers. To synthesize DNA templates of a single molecule, an array of primer is annealed and DNA polymerase complements a single nucleotide resulting in a template-dependent extension. A fluorophore is attached to each nucleotide and is introduced to one nucleotide at a time. To accurately identify the nucleotide being incorporated into the growing strand, recorded images are analyzed and the cycle is repeated for every new nucleotide. The read lengths of HeliScope vary between 25 and 60 bases, with an average of 35 bases; it has efficaciously sequenced the human genome (Pushkarev et al. 2009) to provide insights of disease in a clinical evaluation (Ashley et al. 2010) and sequenced RNA to produce quantitative transcriptomes of tissues and cells (Hickman et al. 2013).

11.5 SEQUENCING TECHNIQUES FOR IDENTIFICATION OF FBPs

Next-generation sequencing (NGS) has potentially been applied in the field of food microbiology which includes disease diagnostics, antimicrobial resistance, forensics, outbreak detection, and food authenticity (Allard et al. 2016; Goodwin et al. 2016; Quainoo et al. 2017). NGS devices are less expensive, more effective, and smaller, and hence many laboratories, including clinical microbial laboratories, have a major impact on food microbiology. Nevertheless, NGS is used to identify new pathogens, which are a basis for the use of NGS in microbial diagnostics. NGS has many advantages compared with conventional diagnostic approaches, such as the ability to detect hostile or non-cultivable species, and co-infections, rather than specific pathogenic protocols. Among the most remarkable benefits of NGS is that unlike many other diagnostic tests, it needs little to no previous knowledge of the pathogen; thus, NGS is a very critical tool to detect FBPs and can potentially be used in two ways: (i) single cultured isolate of a bacterial colony, a virus or any microorganisms whole-genome sequencing determination, and (ii) using a "metagenomics" approach to generate sequences of multiple microorganisms present in the food sample.

WGS can prospectively discriminate pathogens compared to conventional molecular typing methods and is becoming an appropriate potential surveillance tool to control FBPs (Allard et al. 2016; Ashton et al. 2016; Jackson et al. 2016). Application of metagenomics for food safety and quality control is still in its early stages of development and provides a great opportunity to predict the presence or emergence of pathogenic or spoilage microorganisms. These approached are coupled with bioinformatics for the analysis of high-throughput sequences.

Microbial genomic DNA sequencing is technically feasible at platforms such as Illumina, Ion Torrent, PacBio, and Nanopore. Figure 11.4 and Table 11.1 provide a summary of all these extensively used sequencing platforms including robust advances in technology (Deurenberg et al. 2017; Sekse et al. 2017; Slatko et al. 2018).

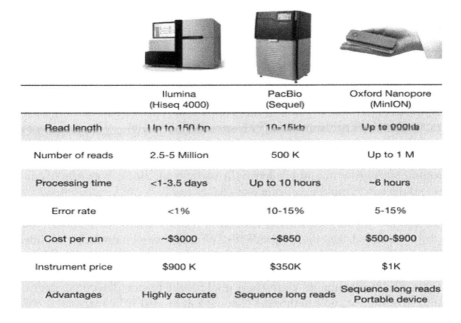

	Ilumina (Hiseq 4000)	PacBio (Sequel)	Oxford Nanopore (MinION)
Read length	Up to 150 bp	10-15kb	Up to 900kb
Number of reads	2.5-5 Million	500 K	Up to 1 M
Processing time	<1-3.5 days	Up to 10 hours	~6 hours
Error rate	<1%	10-15%	5-15%
Cost per run	~$3000	~$850	$500-$900
Instrument price	$900 K	$350K	$1K
Advantages	Highly accurate	Sequence long reads	Sequence long reads Portable device

FIGURE 11.4   Comparisons between Illumina, PacBio, and Oxford Nanopore (https://ge netics.thetech.org/ask-a-geneticist/3rd-generation-sequencing).

TABLE 11.1   Whole Genome Sequencing Platforms

Platform	Sequencing technology	Read length
Illumina	Sequencing by synthesis	Short reads 1 × 36bp–2 × 300bp
Ion Torrent	Sequencing by synthesis	Short reads 200–400bp
PacBio	Single molecule sequencing by synthesis	Long reads Up to 60kb
Oxford Nanopore	Single molecule	Long reads Up to 100kb

11.6 CONCLUSION

The molecular landscapes in food safety investigation with the introduction of high-throughput sequencing platforms like NGS applications have become widely accepted. It is without any doubt that the application of NGS, be it through whole genome sequencing or more elaborate metagenomics approaches, offers a plethora of opportunities in the areas of food safety and microbial risk assessment. Sequencing technology is rapidly evolving as each commercially available DNA sequencing platform and their new versions are released. The older platforms, as well as their released new versions, have similarities and differences relative to the others depending on the performance and mechanism of sequencing/operation principles. They all have their distinct characteristics. Depending on the application, one technology may be better than others. The technology adopted relies upon the utilization of sequencing data, the sequencing performance, and enhancing highly efficient potential can lead to low sample sequencing costs. The technology adopted depends on the type of sample, genome size, and application. For example, metagenomic analysis targeting the detection of all microbial genes in a sample needs much more coverage, restricting the number of specimens that can be used in a single run, and the cost of sequencing per sample is typically increased.

For investigations related to disease persistence, evolution, and outbreaks, many FBPs are well researched with the aid of sequenced genomic data of WGS. The knowledge base repository, National Center for Biotechnology (NCBI) (http:/www.ncbi.nlm.nih.gov/genome/browse/) has the genomic data of prokaryotes and eukaryotes. Ensembl Genomes includes significantly important genomes; the current data set contains 44,048 genomes, Ensembl Fungi contains 1014 genomes, Ensemble metazoan contains 78 genomes for invertebrates and 236 genomes for vertebrate species. Ensembl plant and protists contains 67 and 237 genomes respectively. Bacteriophages are important to bacterial genome size, growth, and virulence (Brüssow et al. 2004). Key insights about the features and the conceivable significance of phages to FBDs can be attributed to shotgun metagenomics (Nieuwenhuijse and Koopmans 2017).

Metagenomics creates sequencing data from the DNA sample, facilitates individual strains to be identified, and allows prediction of microbial communities' encoded functions. This technique permits *in situ* assessment of population diversity (Venter et al. 2004; Baker et al. 2006) and determines specific gene families in an environment (Tyson et al. 2004). The metagenomics platform is used for the precise detection, identification, and characterization of food pathogens (Leonard et al. 2015; Leonard et al. 2016; Aw et al. 2016) and in the food chain environment (Yang et al. 2016). Although bacterial pathogens are documented in food, several hours of culture-dependent enrichment and a high depth of sequencing are required to ensure that the genomic diversity is identified within the sample (Sekse et al. 2017). Metagenomics thus offers an opportunity to examine the diversity and complex abundance of microorganisms within a sample in more unbiased ways than metabarcoding (Forbes et al. 2017). The existence of genetic markers for organisms, serotypes, virulence, and AMR genes, etc. can provide useful, rapid insights through shotgun metagenomics, although these markers cannot currently be assigned, because of the nature of the metagenomic data present in different bacterial genomes (Leonard et al. 2016; Yang et al. 2016). Future developments will have the ability to investigate phylogeny, using metagenomics, metabarcoding, and bioinformatics (Ottesen et al. 2016; Truong et al. 2017). Because of the declining costs, increased resolution, and value-added secondary analysis NGS approaches offer to food safety, these modern approaches will continue to replace traditional molecular subtyping methods and drive improvements in global food safety.

REFERENCES

Aird, D., Ross, M. G., Chen, W. S., et al. 2011. Analyzing and minimizing PCR amplification bias in Illumina sequencing libraries. *Genome Biology* 12(2):R18.

Allard, M. W., Strain, E., Melka, D., Bunning, K., Musser, S. M., Brown, E. W. and Timme, R. 2016. Practical value of food pathogen traceability through building a whole-genome sequencing network and database. *Journal of Clinical Microbiology* 54(8):1975–83.

Ambardar, S., Gupta, R., Trakroo, D., Lal, R. and Vakhlu, J. 2016. High throughput sequencing: An overview of sequencing chemistry. *Indian Journal of Microbiology* 56(4):394–404.

Ansorge, W. J. 2016. Next generation DNA sequencing (II): Techniques, applications. *Journal of Next Generation, Sequence and Application* S1:005.

Ardui, S., Ameur, A., Vermeesch, J. R. and Hestand, M. S. 2018. Single molecule real-time (SMRT) sequencing comes of age: Applications and utilities for medical diagnostics. *Nucleic Acids Research* 46(5):2159–68.

Ashley, E. A., Butte, A. J., Wheeler, M. T., et al. 2010. Clinical assessment incorporating a personal genome. *Lancet* 375(9725):1525–35. doi:10.1016/S0140-6736(10)60452-7.

Ashton, P. M., Nair, S., Peters, T. M., et al. 2016. *Salmonella* whole genome sequencing implementation group., identification of *Salmonella* for public health surveillance using whole genome sequencing. *Peer Journal* 4:e1752. doi:10.7717/peerj.1752.

Aw, T. G., Wengert, S. and Rose, J. B. 2016. Metagenomic analysis of viruses associated with field-grown and retail lettuce identifies human and animal viruses. *International Journal of Food Microbiology* 223:50–6. doi:10.1016/j.ijfoodmicro.2016.02.008.

Bai, Y., Sartor, M. and Cavalcoli, J. 2012. Current status and future perspectives for sequencing livestock genomes. *Journal of Animal Science and Biotechnology* 3(1):8.

Baker, A. C., Goddard, V. J., Davy, J., Scroeder, D. C., Adama, D. G. and Wilson, W. H. 2006. Identification of a diagnostic marker to detect freshwater cyanophages of filamentous cyanobacteria. *Applied and Environmental Microbiology* 72(9):5713–9.

Barba, M., Czosnek, H. and Hadidi, A. 2014. Historical perspective, development and applications of next-generation sequencing in plant virology. *Viruses* 6(1):106–36.

Braslavsky, I., Hebert, B., Kartalov, E. and Quake, S. R. 2003. Sequence information can be obtained from single DNA molecules. *Proceedings of the National Academy of Sciences of the United States of America* 100(7):3960–4.

Brüssow, H., Canchaya, C. and Hardt, W. D. 2004. Phages and the evolution of bacterial pathogens: From genomic rearrangements to lysogenic conversion. *Microbiology and Molecular Biology Reviews: MMBR* 68(3):560–602. doi:10.1128/MMBR.68.3.560-602.2004.

Buehler, B., Hogrefe, H. H., Scott, G., et al. 2010. Rapid quantification of DNA libraries for next-generation sequencing. *Methods* 50(4):S15–8.

Croucher, N. J. and Didelot, X. 2015. The application of genomics to tracing bacterial pathogen transmission. *Current Opinion in Microbiology* 23:62–7.

DeAngelis, M. M., Wang, D. G. and Hawkins, T. L. 1995. Solid-phase reversible immobilization for the isolation of PCR products. *Nucleic Acids Research* 23(22):4742–3.

Derrington, I. M., Butler, T. Z., Collins, M. D., Manrao, E., Pavlenok, M., Niederweis, M. and Gundlach, J. H. 2010. Nanopore DNA sequencing with MspA. *Proceedings of the National Academy of Sciences of the United States of America* 107(37):16060–5. doi:10.1073/pnas.1001831107.

Deurenberg, R. H., Bathoorn, E., Chlebowicz, M. A., et al. 2017. Application of next generation sequencing in clinical microbiology and infection prevention. *Journal of Biotechnology* 243:16–24. doi:10.1016/j.jbiotec.2016.12.022.

DiagenodeBioruptor. 2019. https://www.diagenode.com/en/categories/bioruptor-shearing-device.

Dohm, J. C., Lottaz, C., Borodina, T. and Himmelbauer, H. 2008. Substantial biases in ultra-short read data sets from high-throughput DNA sequencing. *Nucleic Acids Research* 36(16):e105.

Dwivedi, H. P. and Jaykus, L. A. 2011. Detection of pathogens in foods: The current state-of-the-art and future directions. *Critical Reviews in Microbiology* 37(1):40–63. doi:10.3109/1040841X.2010.506430.

EFSA. 2013. Scientific opinion on VTEC-seropathotype and scientific criteria regarding pathogenicity assessment. *EFSA Journal* 11(4):3138–243. doi:10.2903/j.efsa.2013.3138.

Eid, J., Fehr, A., Gray, J., et al. 2009. Real-time DNA sequencing from single polymerase molecules. *Science* 323(5910):133–8.

Fakruddin, R. M. M., Chowdhury, A., Hossain, N., Mahajan, S. and Islam, S. 2013. Pyrosequencing - A next generation sequencing technology. *World Applied Science Journal* 24(12):1558–71.

Forbes, J. D., Knox, N. C., Ronholm, J., Pagotto, F. and Reimer, A. 2017. Metagenomics: The next culture-independent game changer. *Frontiers in Microbiology* 8:1069. doi:10.3389/fmicb.2017.01069.

Franz, E., Gras, L. and Dallman, T. 2016. Significance of whole genome sequencing for surveillance, source attribution and microbial risk assessment of foodborne pathogens. *Current Opinion in Food Science* 8:74–9. doi:10.1016/j.cofs.2016.04.004.

Goodwin, S., McPherson, J. D. and McCombie, W. R. 2016. Coming of age: Ten years of next-generation sequencing technologies. *Nature Reviews Genetics* 17(6):333–51. doi:10.1038/nrg.2016.49.

Hawkins, T. L., Connor-Morin, T. O., Roy, A. and Santillan, C. 1994. DNA purification and isolation using a solid-phase. *Nucleic Acids Research* 22(21):4543–4.

Heather, J. M. and Chain, B. 2016. The sequence of sequencers: The history of sequencing DNA. *Genomics* 107(1):1–8.

Henson, J., Tischler, G. and Ning, Z. 2012. Next-generation sequencing and large genome assemblies. *Pharmacogenomics* 13(8):901–15.

Hickman, S. E., Kingery, N. D., Ohsumi, T., Borowsky, M. L., Wang, L. C., Means, T. K. and El Khoury, J. 2013. The microglial sensome revealed by direct RNA sequencing. *Nature Neuroscience* 16(12):1896–905. doi:10.1038/nn.3554.

Hindson, B. J., Ness, K. D., Masquelier, D. A., et al. 2011. High-throughput droplet digital PCR system for absolute quantitation of DNA copy number. *Analytical Chemistry* 83(22):8604–10.

Hindson, C. M., Chevillet, J. R., Briggs, H. A., et al. 2013. Absolute quantification by droplet digital PCR versus analog real-time PCR. *Nature Methods* 10(10):1003–5.

IlluminaTruSeq DNA Nano. 2019. https://www.illumina.com/products/by-type/sequencing-kits/library-prep-kits/truseq-nano-dna.html.

Jackson, B. R., Tarr, C., Strain, E., et al. 2016. Gerner-smidt implementation of nationwide real-time whole-genome sequencing to enhance *Listeriosis* outbreak detection and investigation. *Clinical Infectious Diseases* 63(3):380–6. doi:10.1093/cid/ciw242.

Jain, M., Koren, S., Quick, J., et al. 2017. Nanopore sequencing and assembly of a human genome with ultra-long reads. *BioRxiv*. doi:10.1101/128835.

Joseph, S. and Forsythe, S. J. 2012. Insights into the emergent bacterial pathogen *Cronobacter* sp., generated by multilocus sequence typing and analysis. *Frontiers in Microbiology* 3:397.

Kchouk, M., Gibrat, J. F. and Elloumi, M. 2017. Generations of sequencing technologies: From first to next generation. *Biology and Medicine* 9(3):395.

Khodakova, D., Wanga, C. and Zhang, D. Y. 2016. Diagnostics based on nucleic acid sequence variant profiling: PCR, hybridization, and NGS approaches. *Advanced Drug Delivery Reviews* 105(A):3–19.

Kotrova, M., Trka, J., Kneba, M. and Brüggemann, M. 2017. Is next-generation sequencing the way to go for residual disease monitoring in acute lymphoblastic leukemia? *Molecular Diagnosis and Therapy* 21(5):481–92.

Laurie, M. T., Bertout, J. A., Taylor, S. D., Burton, J. N., Shendure, J. A. and Bielas, J. H. 2013. Simultaneous digital quantification and fluorescence-based size characterization of massively parallel sequencing libraries. *BioTechniques* 55(2):61–7.

Lee, H., Gurtowski, J., Yoo, S., et al. 2016. Third-generation sequencing and the future of genomics. *BioRxiv*. http://www.ncbi.nlm.nih.gov/pubmed/048603.

Leonard, S. R., Mammel, M. K., Lacher, D. W. and Elkins, C. A. 2015. Application of metagenomic sequencing to food safety: Detection of Shiga toxin-producing *Escherichia coli* on fresh bagged spinach. *Applied and Environmental Microbiology* 81(23):8183–91. doi:10.1128/AEM.02601-15.

Leonard, S. R., Mammel, M. K., Lacher, D. W. and Elkins, C. A. 2016. Strain-level discrimination of Shiga toxin producing *Escherichia coli* in spinach using metagenomic sequencing. *PLoS ONE* 11(12):e0167870.

Levy, S. E. and Myers, R. M. 2016. Advancements in next-generation sequencing. *Annual Review of Genomics and Human Genetics* 17:95–115.

Lindstedt, B. A., Torpdahl, M., Vergnaud, G., et al. 2013. Use of multilocus variable-number tandem repeat analysis (MLVA) in eight European countries, 2012. *Eurosurveillance* 18(4):20385.

Liu, L., Li, Y., Li, S., et al. 2012. Comparison of next-generation sequencing systems. *Journal of Biomedical Biotechnology* 2012: 251364. doi:10.1155/2012/251364.

Loman, N. J. and Pallen, M. J. 2015. Twenty years of bacterial genome sequencing. *Nature Reviews in Microbiology* 13:1–9. doi:10.1038/nrmicro3565.

Loman, N. J. and Watson, M. 2015. Successful test launch for nanopore sequencing. *Nature Methods* 12(4):303–4.

Lu, H., Giordano, F. and Ning, Z. 2016. Oxford nanopore minion sequencing and genome assembly. *Genomics, Proteomics and Bioinformatics* 14(5):265–79.

Magi, A., Semeraro, R., Mingrino, A., Giusti, B. and Aurizio, D.R. 2017. Nanopore sequencing data analysis: State of the art, applications and challenges. *Briefings in Bioinformatics* 110(47):1– 17. doi:10.1093/bib/bbx062.

Mardis, E. and McCombie, W. R. 2017. Library quantification: Fluorometric quantitation of double-stranded or single-stranded DNA samples using the qubit system. *Cold Spring Harbor Protocols* 6. http://www.ncbi.nlm.nih.gov/pubmed/094730.

Maxam, A. M. and Gilbert, W. 1977. A new method for sequencing DNA. *Proceedings of the National Academy of Sciences of the United States of America* 74(2):560–4. doi:10.1073/pnas.74.2.560.

Metzker, M. L. 2010. Sequencing technologies - the next generation. *Nature Reviews Genetics* 11(1):31–46.

Morey, M., Fernandez-Marmiesse, A., Castineiras, D., Fraga, J. M., Couce, M. L. and Cocho, J. A. 2013. A glimpe into past, present, and future DNA sequencing. *Molecular Genetics and Metabolism* 110(1–2):3–24.

Nakano, K., Shiroma, A., Shimoji, M., et al. 2017. Advantages of genome sequencing by long-read sequencer using SMRT technology in medical area. *Human Cell* 30(3):149–61. doi:10.1007/s13577-017-0168-8.

Nieuwenhuijse, D. F. and Koopmans, M. P. G. 2017. Metagenomic sequencing for surveillance of food- and waterborne viral diseases. *Frontiers in Microbiology* 8:230– 9. doi:10.3389/fmicb.2017.00230.

Ottesen, A., Ramachandran, P., Reed, E., et al. 2016. Enrichment dynamics of *Listeria monocytogenes* and the associated microbiome from naturally contaminated ice cream linked to a listeriosis outbreak. *BMC Microbiology* 16(1):275.

Pareek, C. S., Smoczynski, R. and Tretyn, A. 2011. Sequencing technologies and genome sequencing. *Journal of Applied Genetics* 52(4):413–35.

Pillai, S., Gopalan, V. and Lam, A. 2017. Review of sequencing platforms and their applications in phaeochromocytoma and paragangliomas. *Critical Reviews in Oncology/Hematology* 116:58–67.

Pinheiro, L. B., Coleman, V. A., Hindson, C. M., Herrmann, J., Hindson, B. J., Bhat, S. and Emslie, K. R. 2012. Evaluation of a droplet digital polymerase chain reaction format for DNA copy number quantification. *Analytical Chemistry* 84(2):1003–11.

Pushkarev, D., Neff, N. F. and Quake, S. R. 2009. Single-molecule sequencing of an individual human genome. *Nature Biotechnology* 27(9):847–52.

Quail, M. A., Smith, M., Coupland, P., et al. 2012. A tale of three next generation sequencing platforms: Comparison of Ion Torrent, Pacific Biosciences and Illumina MiSeq sequencers. *BMC Genomics* 13:341. doi:10.1186/1471-2164-13-341.

Quainoo, S., Coolen, J. P. M., van Hijum, S. A. F. T., Huynen, M. A., Melchers, W. J. G., van Schaik, W. and Wertheim, H. F. L. 2017. Whole-genome sequencing of bacterial pathogens: The future of nosocomial outbreak analysis. *Clinical Microbiology Reviews* 30(4):1015–63. doi:10.1128/CMR.00016-17.

Reuter, J. A., Spacek, D. V. and Snyder, M. P. 2015. High-throughput sequencing technologies. *Molecular Cell* 58(4):586–97. doi:10.1016/j.molcel.2015.05.004.

Rhee, M. and Burns, M. A. 2006. Nanopore sequencing technology: Research trends and applications. *Trends in Biotechnology* 24(12):580–6.

Roberts, C. H., Last, A., Molina-Gonzalez, S., et al. 2013. Development and evaluation of a next-generation digital PCR diagnostic assay for ocular *Chlamydia trachomatis* infections. *Journal of Clinical Microbiology* 51(7):2195–203. doi:10.1128/JCM.00622-13.

Robin, J. D., Ludlow, A. T., LaRanger, R., Wright, W. E. and Shay, J. W. 2016. Comparison of DNA quantification methods for next generation sequencing. *Scientific Reports* 6:24067.

Rohland, N. and Reich, D. 2012. Cost-effective, high-throughput DNA sequencing libraries for multiplexed target capture. *Genome Research* 22(5):939–46.

Rothberg, J. M., Hinz, W., Rearick, T. M., et al. 2011. An integrated semiconductor device enabling non-optical genome sequencing. *Nature* 475(7356):348–52. doi:10.1038/nature10242.

Salipante, S. J., Scroggins, S. M., Hampel, H. L., Turner, E. H. and Pritchard, C. C. 2014. Microsatellite instability detection by next generation sequencing. *Clinical Chemistry* 60(9):1192.

Sanger, F., Nicklen, S. and Coulson, A. R. 1977. DNA sequencing with chain-terminating inhibitors. *Proceedings of the National Academy of Sciences of the United States of America* 74(12):5463–7.

Schadt, E. E., Turner, S. and Kasarskis, A. 2010. A window into third-generation sequencing. *Human Molecular Genetics* 19(R2):R227–40. doi:10.1093/hmg/ddq416.

Sekse, C., Holst-Jensen, A., Dobrindt, U., Johannessen, G. S., Li, W., Spilsberg, B. and Shi, J. 2017. High throughput sequencing for detection of foodborne pathogens. *Frontiers in Microbiology* 8:2029. doi:10.3389/fmicb.2017.02029.

Shendure, J. and Ji, H. 2008. Next-generation DNA sequencing. *Nature Biotechnology* 26(10):1135–45. doi:10.1038/nbt1486.

Shin, S. C., Ahn, D. H., Su, J. K., Lee, H., Oh, T. J., Lee, J. E. and Park, H. 2013. Advantages of single-molecule real-time sequencing in high-GC content genomes. *PLoS ONE* 8(7):e68824.

Slatko, B. E., Garner, A. F. and Ausubel, F. M. 2018. Overview of next-generation sequencing technologies. *Current Protocols in Molecular Biology* 122(1):e59. doi:10.1002/cpmb.59.

Smith, L. M., Sanders, J. Z., Kaiser, R. J., et al. 1986. Fluorescence detection in automated DNA sequence analysis. *Nature* 321(6071):674–9.

Ståhlberg, A., Krzyzanowski, P. M., Egyud, M., Filges, S., Stein, L. and Godfrey, T. E. 2017. Simple multiplexed PCR-based barcoding of DNA for ultrasensitive mutation detection by next-generation sequencing. *Nature Protocols* 12(4):664–82.

Ståhlberg, A., Krzyzanowski, P. M., Jackson, J. B., Egyud, M., Stein, L. and Godfrey, T. E. 2016. Simple, multiplexed, PCR-based barcoding of DNA enables sensitive mutation detection in liquid biopsies using sequencing. *Nucleic Acids Research* 44(11):11.

Swaminathan, B., Barrett, T. J., Hunter, S. B., Tauxe, R. V. and CDC PulseNet Task Force. 2001. PulseNet: The molecular subtyping network for foodborne bacterial disease surveillance, United States. *Emerging Infectious Disease Journal* 7(3):382–9.

Thompson, J. F. and Steinmann, K. E. 2010. Single molecule sequencing with a HeliScope genetic analysis system. *Current Protocols in Molecular Biology*. doi:10.1002/0471142727.mb0710s92.

Truong, D. T., Tett, A., Pasolli, E., Huttenhower, C. and Segata, N. 2017. Microbial strain-level population structure and genetic diversity from metagenomes. *Genome Research* 27(4):626–38.

Tyson, G. W., Chapman, J., Hugenholtz, P., et al. 2004. Community structure and metabolism through reconstruction of microbial genomes from the environment. *Nature* 428(6978):37–43. doi:10.1038/nature02340.

Valouev, A., Ichikawa, J., Tonthat, T., et al. 2008. A high-resolution, nucleosome position map of C. *elegans* reveals a lack of universal sequence-dictated positioning. *Genome Research* 18(7):105–63.

Venter, J. C., Remington, K., Heidelberg, J. et al. 2004. Environmental genome shotgun sequencing of the Sargasso sea. *Science* 304(5667):66–74. doi:10.1126/science.1093857.

Wang, Y., Wen, Z., Shen, J., et al. 2014. Comparison of the performance of Ion Torrent chips in noninvasive prenatal trisomy detection. *Journal of Human Genetics* 59(7):393–6. doi:10.1038/ jhg.2014.40.

Weile, J. and Knabbe, C. 2009. Current applications and future trends of molecular diagnostics in clinical bacteriology. *Analytical and Bioanalytical Chemistry* 394(3):731–42.

White, R. A., Blainey, P. C., Fan, H. C. and Quake, S. R. 2009. Digital PCR provides sensitive and absolute calibration for high throughput sequencing. *BMC Genomics* 10:116.

Xu, Y., Vaidya, B., Patel, A. B., Ford, S. M., McCarley, R. L. and Soper, S. A. 2003. Solid-phase reversible immobilization in microfluidic chips for the purification of dye-labeled DNA sequencing fragments. *Analytical Chemistry* 75(13):2975–84.

Yang, X., Noyes, N. R., Doster, E., et al. 2016. Use of metagenomic shotgun sequencing technology to detect foodborne pathogens within the microbiome of the beef production chain. *Applied and Environmental Microbiology* 82(8):2433–43. doi:10.1128/AEM.00078-16.

Yuan, Y., Xu, H. and Leung, R. K. 2016. An optimized protocol for generation and analysis of Ion Proton sequencing reads for RNA-Seq. *BMC Genomics* 17:403. doi:10.1186/s12864-016-2745-8.

CHAPTER 12

Microbial Genomes
Potential Systems for WGS and NGS

Debadarshee Das Mohapatra and Smaranika Pattnaik

CONTENTS

12.1 INTRODUCTION

Original genome sequencing seeks to understand the order of DNA nucleotides, or base of a genome, which is an organism's complete set of genetic instructions. Next-generation sequencing (NGS) is a blanket term collectively referring to high-throughput DNA sequencing strategies that can produce large amounts of genomic data in a single reaction by diverse methodologies. NGS is also referred to in the literature as "deep," "high-throughput" sequencing. In this context, microbial whole-genome sequencing (MHGS) is an essential tool for mapping genomes of unique organisms and comparing with known organisms to fish out the genes of interest. WGS promises to be transformative for the practice of clinical microbiology (Koser et al. 2012) because NGS is transforming microbiology. With the increased accessibility and decrease in cost of sequencing and the optimization of the 'Wet laboratory,' components are rapidly revolutionizing clinical and public health. In addition, the multiplex capability and high yield of current day DNA sequencing instruments has made bacterial whole genome sequencing a routine affair (Seeman 2014).

Typing of foodborne pathogens was one of the earliest applications of WGS and proof-of-concept has been demonstrated for the superiority of WGS over traditional typing methods for a range of high priority foodborne pathogens, including *Salmonella enterica*, *Listeria monocytogenes*, *Campylobacter* sp., and STEC strains (Moran-Gilad 2017).

WGS provides physiogenomics of the whole genome of the biosample isolates, thereby determining the genotypes of each sub-strain of microbial strains of clinical relevance. Hence, a genotypic link can be constructed for the microbial candidates involved in disease epidemiology. It is relevant to mention that, traditionally, researchers did not consider genotypic studies for the clinically relevant microorganisms because the clinically relevant microorganisms were characterized through biochemical as well as immunological assays. The genotype studies were not valid for the different strains or serotypes of medically important microorganisms. Nevertheless, the advent of WGS paved the way for identifying the variation of genomes among the strains. WGS can detect single-nucleotide polymorphisms (SNPs), thus inferring the existence of variation in sequences of strains also. Further, development of bioinformatics tools (Van Goethem et al. 2019) for analyzing the gene sequences and annotating with structural and functional characteristics abetted the WGS approaches in clinical setups.

In addition, using massively parallel (or "next-generation" sequencing) DNA sequencing technologies, it is now possible to examine the complete or nearly complete genomes of micro-organisms with an output of WGS, theoretically distinguishing strains which differ at only a single nucleotide (Salipante et al. 2015). Furthermore, WGS analysis has also revolutionized the food safety industry by enabling high-resolution typing of foodborne bacteria. The US Food and Drug Administration (FDA) uses WGS data to assess levels of genetic relatedness among isolates (Pightling et al. 2018). NGS, through high-throughput screening can generate thousands to millions of sequence reads and up to several hundred billion base pairs (bp) of sequence information per sample. As the read length, error rate, number of reads, and sequenced bases vary substantially, selective amplification (targeted) and non-selective, random (shotgun) approaches are in practice. Therefore, the number of high-quality genomes for the most important food pathogens is already high and rapidly growing. It is pertinent to mention here that foodborne pathogens are causing a great number of diseases with significant effects on human health and economy (Bintsis 2017).

WGS of pathogens has become more accessible and affordable as a tool for genotyping. Analysis of the entire pathogen genome via WGS could provide unprecedented resolution in discriminating even highly related lineages of bacteria and revolutionize outbreak analysis in hospitals (Quainoo et al. 2017). The Institute for Genomic Research (TIGR) published the first complete genome sequence for a free-living organism, *Haemophilus influenzae*, in 1995 (Fraser and Ligett 2005), followed by *Saccharomyces cerevisiae* and *Escherichia coli* sequencings. WGS-based omics analytical tools can be used for pandemic, endemic, as well as sporadic nosocomial outbreaks and a comparative study could be deduced. In addition, WGS of bacterial pathogens has shown potential for epidemiological surveillance, outbreak detection, and infection control (Oniciuc et al. 2018), as antimicrobial resistance (AMR) surveillance is a critical step within risk assessment schemes.

12.2 FOODBORNE MICROBIAL STRAINS

Foodborne microbial strains, specifically bacterial strains, play a large role in food pathology. The bacterial strains belonging to various biotypes and serotypes enter into the body, and with use of various virulent factors, are able to colonize and execute the infection. While residing inside the body, they can resist the action of immunogens as well as prescribed antibiotics. The infection may persist for a long period of time, leading to chronic infections, and possibly death. These foodborne bacterial strains (FBBS) pose a serious threat to public and community health. Figure 12.1 depicts the source food and the multiplying bacteria strains involved in typical epidemiological outbreaks. It is interesting to note here that the strains are all pathotypes and serotypes of *Escherichia coli* only. The parent *E. coli* undergoes horizontal as well as vertical evolution, with emergence of various serotypes (Pattnaik 2017). As the serotypes are associated with pathogenicity, they are also called pathotypes. Figure 12.1 illustrates various types of food contaminated with shiga toxin producing *E. coli* (STEC) serotypes. Raw milk, water cress, packed salad, slaw garnish, butcher house meat, lamb feeding events, open farms, and even leaf salad in a restaurant could be prime food sources for the growth and multiplication of STEC 0157:H7 PT21/28, STEC 0157:H7 PT2, STEC 0157:H7 PT8, 0157, 0157:H7 PT21/28, 0157:H7 PT21/28, 0157:H7, 0157:H7 PT8, and PT 54 respectively. This illustrates the ability of *E. coli* bacteria to adapt and survive in various foodstuffs. The organism itself is diversified using 'horizontal transfer of genes' mechanisms. There is a shuttle of antibiotic resistant genes (ARGs) thus causing mutations with the emergence of new bacterial strains. Of course, there is a degree of variation in mutation and adoption in SNP (Schürch et al. 2018). However, the foodborne bacteria always look forward for acquiring 'genome plasticity,' to follow Nature's rule of 'survival of the fittest.' WGS strategy also unravels the orthodox genes responsible for toxin productions.

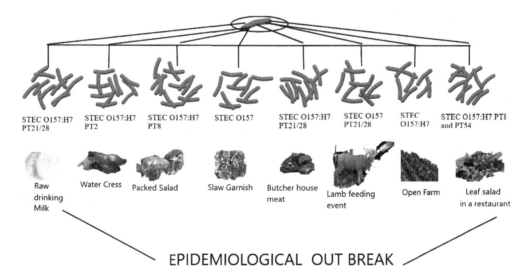

FIGURE 12.1 Foodborne outbreaks of shiga toxin-producing *Escherichia coli* (STEC) derived from WGS data with the evolution of eight numbers of STEC strains from a parent strain of *Escherichia coli*.

This chapter considers the proteins/toxins expressed from the conserved gene domains of bacterial strains, which are predominant in the clinical setup by bringing recurrent foodborne infections. The foodborne bacterial strains (FBBS), namely *Bacillus cereus*, *Listeria monocytogenes*, *Staphylococcus aureus*, *Escherichia coli*, *Salmonella typhi*, *Shigella dysenteriae*, and *Campylobacter jejuni* are briefly introduced and the WGS reports (Table 12.1) are discussed concisely under two headings namely, Gram-positive bacteria and Gram-negative bacteria.

12.2.1 Foodborne Gram-Positive Bacterial Strains as Potential System for WGS and NGS

12.2.1.1 Bacillus cereus

B. cereus being Gram-positive, are long rods. The sizes vary from 2μm to 10μm. They are ubiquitous in the environment, and are opportunistic to cause nosocomial infections in hospital environments. They are the common contaminants of foods of varying pH, as this organism can survive in extreme conditions. The bacteria produce two types of toxins, namely, (a) the emetic (vomiting) and (b) the diarrheal one, thus causing two types of illness. The emetic and the diarrheal syndromes are caused by emetic toxin produced by bacteria during the growth phase in the food and diarrheal toxins in the small intestine, respectively. Although the conventional approaches were practiced to identify the organism, Carolle et al. (2019) had attempted the first WGS-based characterization of isolates linked to an outbreak caused by members of the *B. cereus* group. It was inferred that there was sequence variation due to SNPs.

Isolation and identification of *B. cereus* strains from dried foods, samples of powdered infant formula (PIF), medicated fish feed, and dietary supplements (Carter et al. 2018) and integrated WGS/MLSA analyses identified genetic diversity of 64 strains expressing haemolysin BL, nonhemolytic enterotoxin, cytotoxin K, and enterotoxin FM toxin genes.

12.2.1.2 Staphylococcus aureus

In food microbiology, *Staphylococcus aureus* has a particular organization. With spherical morphology of this bacterium is usually gram-positive. Due to its unique mode of division, they are found in grape-like clusters, hence the Greek name Staphylococci. Staphylococcal gastrointestinal diseases are described as the foremost prevalent foodborne intoxicants worldwide. The strains of food poisoning episodes harbor the *egc* cluster, comprising *seg*, *sei*, *sem*, *sen*, and *seo* genes. Each of these genes is linked to typical symptoms of staphylococcal food poisoning. The enterotoxins can be separated into two groups; (a) the classical (SEA-SEE) and (b) the newer (SEG-SEIY and counting) enterotoxin groups (Fisher et al. 2018).

Attempts have been made in the recent past to sequence the whole genome of *S. aureus* and attribute functional aspects using "annotation" tools. Zerbino and Birney (2008) had used Illumina Next era XT protocol and assembled the contigs using Velvet Optimizer software. Further, the sequences annotated using PROKKA (Seemann 2014), and the core- and pan-genomes were determined using default values. The aligned, ungapped core genome was subjected to SNP. In addition, the phylogenetic analysis had been performed by RAxML with 100 bootstrap support replicates (Stamatakis 2015). The resulting phylogenetic tree and accessory genome (non-unique genes as well as cryptic genes) sequencing results gave a better understanding about the candidate genome.

TABLE 12.1 The Whole Genome Sequencings of Foodborne Pathogens

Names of Bacteria	Source of Food	NGS Strategy of WGS Data	Analysis	Authors
Bacillus cereus	Spices and PIF	MLSA Strategy	Identification of hemolysin BL, Nonhemolytic enterotoxin, Cytotoxin K, and enterotoxin	Carter et al. 2018
		Solexa paired-end sequencing technology	4 exotoxin genes	Qiu et al. 2012
		MLSA	Intragenomic diversity	Carter et al. 2019
Listeria momocytogenes	Human	Core-genome multilocus sequence typing (cgMLST) by SNP	Reflecting unknown disease cluster and variation	Halbedel et al. 2018
			emrC, berABC, qacH-Tn6188 and qacC	Hurley et al. 2019
Staphylococcus aureus	Cheese	Core- and pan- genomes Ilumina Nextera XT	*spa, MLST, PVL* SCC*mec* type	Cunningham et al. 2017
		SNP calling and gMLST	Gene variants (Alleles)	
		RGC and SHGC	egc cluster (seg, sei, sem, sen and seo)	Jøhler et al. 2018
			seg, sei, sem, sen, seo grouped in an exotoxin gene cluster, and *seh* a transposon-associated gene	Macori et al. 2020
			PAI containing enterotoxin C and enterotoxin-like L	Franck et al.2017
Escherichia coli STECO26 STEC	Buffet restaurant Children HUS	WGS typing *In silico* virulome	stx *stx2*	Mossong et al, 2015 Abdalhamid et al.2019 González-Escalona and Kase 2019
Verotoxigenic *Escherichia coli* (VTEC) O157:H7	HUS	An in-house developed single nucleotide variant phylogenomics (SNVPhyl) pipeline and the BioNumerics whole genome Multilocus Sequence Typing (wgMLST) tool	Genetic difference detection among stereotypes	Rumore et al. 2018

(Continued)

TABLE 12.1 (CONTINUED)   The Whole Genome Sequencings of Foodborne Pathogens

Names of Bacteria	Source of Food	NGS Strategy of WGS Data	Analysis	Authors
E. coli	Cilantro, mixed salad and arugula	MLST20	Tetracycline (TET) gene	Reid et al. 2020
Salmonella typhi		WGS of plasmid *Mcr1*	safer, cfa, pefABCD, spvBCF, class 1 integrons carrying resistance determinants (In1582 and In167)	Gupta et al. 2019
			gyrA	Wang et al. 2019
	Human GI tract	SNP analysis and nucleotide difference approach	Pan-genome tree K-mer tree, SNP tree Nucleotide difference tree	Leekitcharoenphon et al. 2014
Shigella dysenteriae		Shiga typer	Geno serotyping	Wu et al. 2019
Shigella sonnie	Foodborne illness		Chromosomal Int2/Tn7 SpA plasmid (SRL), pABC-3, strA dfrA14	Baker et al. 2017
Shigella flexneri	Public health unit	SNP	O' Ag Synthesis	Chattaway et al. 2017
Sh. flexneri	Stool sample	Long-read MinION and short-read Ion Torrent 400	*sull1, tetA, tetR,* aph(6)-1d, *aph(3″)-1b*	Sethuvel et al. 2019
Campylobacter jejuni	Raw milk	GSMLST	Genetic and evolutionary relatedness	Oakeson et al. 2018
	Human, poultry and bovine routinely collected human stools consumed with lettuce	MLST, SNP	Type VI secretion system (T6SS) Gene clusters	Rokney et al. 2018 Joensen et al. 2018
	Diarrheal isolates	Sequence types (STs), serotypes, phylogenetic relatedness and pan-genome	Phylogeny, resistome and virulome	Fiedoruk 2019
Complete microbial WGS	Food	SNP and genomic MLST b	Relatedness	Jagadeesan et al. 2019

Therefore, Frank et al. (2017) had taken interest and identified *spa, MLST, PVL,* and *SCCmec* gene cassettes of said bacteria using WGS approaches. Cunningam et al. (2017) strategically took near-complete genomes, assembled de novo with paired-end and long-mate-pair (8 kb) libraries. By assembling the paired-end data, they could analyze the structural and functional aspect of the genome of interest. The research group had also applied the SNP analysis using the in-house pipeline. The investigators used two assembly strategies to generate core genome multi-locus sequence typing (cgMLST). The strategies were (a) the near-complete genome data, which had been generated with the in-house commercial software and (b) assembling paired-end data. As a consequence, the resolved assemblies were used to perform cgMLST.

12.2.1.3 Listeria monocytogenes

Listeria monocytogenes is also a gram-positive, non-spore-forming, intracellular bacilli, that can withstand extremely temperatures. Hence, the bacterium is able to grow in food stuffs stored in refrigerators. The bacterial strains are prevalent ubiquitously in the surroundings and hence, are designated food contaminants. *L. monocytogenes* is the major causative agent of Listeriosis. Although the disease is uncommon it can be severe due to a higher degree of intensity of the disease. It is reported that poultry meat and products are the main vehicles of pathogenic strains of *L. monocytogenes* for humans (Jamshidi and Zeinali 2019).

The German–Austrian Consultant Laboratory for *L. monocytogenes* implemented whole-genome sequencing (WGS) in 2015 for the improvement of outbreak 'cluster detection.' It was mentioned that, a total of 424 human *L. monocytogenes* isolates were collected from 2007 to 2017 and were subjected to WGS and core-genome multilocus sequence typing (cgMLST). It was added that cgMLST grouped the isolates into 38 complexes, reflecting 4 known and 34 unknown disease clusters confirmed by SNP calling (Halbedel et al. 2018).

Likewise, the whole-genome sequencing approaches (Hurley et al. 2019) were applied as a tool to characterize and track 100 *L. monocytogenes* isolates collected from three food processing environments. The WGS results asserted that the *emrC* gene cluster was present in most of the isolated strains. In addition, *bcrABC, qacH-Tn6188,* and *qacC* genes were identified as potential food poisoning toxin producers. This is to say that WGS is being applied as an identifying tool for a nationwide Listeriosis outbreaks. It was discovered that a protein named 'Internalin' interacts with the E-Cadherin receptor of human cells, making easy its entry into the sensitive host cells. Therefore, WGS had contributed a lot for diagnosing the master genes which express the effector proteins to cause the major outbreaks. However, WGS has rarely been used in smaller local outbreaks. Figure 12.2 depicts the putative proteins of the respective gram-positive bacteria. The toxins/proteins of Gram-negative bacteria eluted from WGS strategies: a: Hemolysin BL-binding component of *Bacillus cereus* (Madegowda et al. 2008); b: Staphylococcal enterotoxin C3 (Chi et al. 2002); c: Internalin of *Listeria monocytogenes* (Schubert et al. 2002) respectively.

The said proteins were retrieved from the RCSB-PDB database (www.rcsb.org). The structural PDB files were visualized as protein 3D structures within the Discovery studio (www.3dsbiovia.com). Further, the protein sequences translated from respective PDB files were analyzed in the BLASTn (www.ncbi.ac.in) portal followed by conserved domain search and assigning respective protein families and protein super families. The precise families and super families indicate the sort and performance of every one of the proteins/toxins related to foodborne illness, detected by WGS approaches.

a   b   c

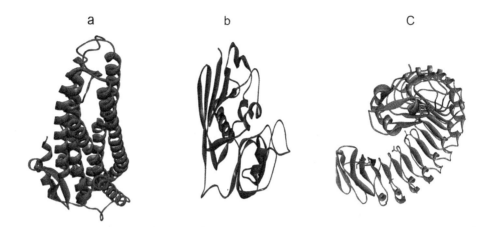

FIGURE 12.2   The toxins/proteins of gram-positive bacteria eluted from WGS strategies: (a) Hemolysing BL-binding component of *Bacillus cereus*; (b) Staphylococcal enterotoxin C3; (c) Internalin of *Listeria monocytogenes*.

12.2.2 Foodborne Gram-Negative Bacterial Strains as Potential System for WGS and NGS

12.2.2.1 Escherichia coli

Escherichia coli is also an emerged contaminant of food and beverages. The strains such as *E. coli* O157 and O104, are frequently associated with foodborne diseases in developing countries, like India, as well as developed countries, like United States or European countries (Yang et al. 2017). *E. coli*, a gram-negative, with hyphen-like morphology is considered the most studied bacteria due to its short generation time and a simplified growth curve. Furthermore, its nutritional requirement is minimal. Therefore, genome sequencing technologies have already been applied to this lab-friendly tiny workhorse. Generally, *E. coli* strains are non-pathogenic and found as normal microbiota in gut and urinary tracts. However, due to acquisition of virulent genes from other bacterial, strains like *Shigella* sp., *Salmonella* sp., etc., the bacteria become pathogenic. Based upon attainment of genes, and pathophysiology, the pathogenic *E. coli* strains are evolved as enterotoxigenic *E. coli* (ETEC), enteropathogenic *E. coli* (EPEC), enterohaemorrhagic *E. coli* (EHEC), enteroaggregative *E. coli* (EAEC), entero invasive *E. coli* (EIEC), shiga toxin producing *E. coli* (STEC) strains. Although the pathotypes are categorized based upon serotypings (Pattnaik 2017), but WGS and NGS sequencing approaches are adopted and phylogenetic relationships are interpreted. Among all the pathotypes, STEC strain is linked with food poisoning due to expression of shiga-like toxin. In a review, by Jenkins et al. (2019), it was stated that WGS-pairwise SNP data could derive genetic relatedness of STEC strains of *Escherichia coli*, originated from different contaminated foods.

Shiga-toxin producing *Escherichia coli* (STEC) O26: H11 is the second most common cause of severe diarrhea and hemolytic uremic syndrome (HUS) worldwide. The implementation of WGS enhances the detection and in-depth characterization of these non-O157 STEC strains. Abdalhamid et al. (2019) have provided a comprehensive profile of virulence factors for each strain, retrieved from the WGS database. It was reported that shiga-like toxins (Fraser et al. 1994) encoded by *stx* gene might be also *stx1*, *stx2* sub types. Their studies have shown that strains encoding *stx2* were highly virulent and commonly associated with more severe diseases like HUS than those encoding *stx*. The WGS could also determine genomic diversities of STEC strains (Lang et al. 2019). Gonzalez-Escalona and Kase (2019) have described the genome sequences of 331 STECs isolated from foods regulated by the FDA 2010–2017, and had determined their genomic identity, serotype, sequence type, virulence potential, and prevalence of antimicrobial resistance.

12.2.2.2 Shigella *spp.*

The genus *Shigella* is notorious for its pathogenicity by contaminating foodstuffs. Street food often harbor *Shigella* bacilli. Morphologically, *Shigella* is very similar with *Escherichia coli*, but the serodiagnosis can differentiate the *Shigella* sp. *Shigella* consists of 4 species (serogroups) and >54 serotypes: *S. dysenteriae* (15 serotypes), *S. flexneri* (18 serotypes), *S. boydii* (20 serotypes), and *S. sonnei* (1 serotype) causing bacillary dysentery. The cells colonize massively using a number of molecular pathways. Invasin is a protein responsible for invasion. This is self-chaperoning of the type III secretion system needle tip protein (Johnson et al. 2007). These serotypes are distinguished solely through the somatic (O) antigen, or lipopolysaccharide, expressed on the bacterial cell surface. This bacterium undergoes frequent phase variation between fimbriated and afimbriated serotypes. The phase variation is the output of gene rearrangement. Hence, WGS is needed for characterization of these two serotypes. But there is a big literature gap in studies relating to *Shigella* genomics. Although WGS-based *in silico* serotyping improves accuracy for *Shigella* identification through a combinatorial 'geno-serotyping,' genetic markers do not include rare serotypes of *Shigella*. The possible reason behind this could be due to onset of convergence evolution in *Shigella* strains (Song et al. 2016).

12.2.2.3 Salmonella *spp.*

Salmonella is also one of the most frequently isolated foodborne pathogens. It is a major worldwide public health concern, accounting for 93.8 million foodborne illnesses and 155,000 deaths per year (Eng et al. 2015). *Salmonella enterica* serovar enteritidis a main contributor to the virulence of gastrointestinal illness. *Salmonella enterica* serotype Typhi and *S. enterica* serotype Paratyphi A, B (tartrate negative), and C cause typhoid and paratyphoid fevers. Stomach ache, nausea, vomiting, fever, and weakness constitute the syndromic symptoms, the 'Salmonellosis.' The bacteria enter the human gut through contaminated food. Here, it neutralizes the action of gastric juices and makes an invasive entry into the mucous of the large and small intestines as well. With sufficient nutrition, the bacterium initiates its replication. Due to invasion into the mucous membranes, there is stimulation of inflammatory reaction, leading synthesis of cytokines and other chemokines, which cause tissue damage, hemolysis, and related events. Typhoid toxin is secreted into the lumen of the Salmonella-containing vacuole by a secretion mechanism strictly dependent on TtsA, a specific muramidase that facilitates toxin transport through the peptidoglycan layer (Geiger et al. 2018).

Eggs and egg-containing foods are frequently identified as food vehicles for the growth and multiplication of *Salmonella* sp. Egg-based sauces and Vietnamese style sandwiches, which often contain pâté and raw egg butter, are potent sources of *Salmonella* outbreak (Ford et al., 2016). Advances in NGS technologies have made it possible to examine WGS as a potential molecular subtyping tool for Salmonella outbreak detection and source tracing. An outbreak of foodborne typhoid fever due to *Salmonella enterica* serotype Typhi in Japan was reported for the first time in 16 years (Kobyashi et al. 2016), and it was asserted that only WGS could provide a retrospective epidemiological study to determine the active strains. The identified, *S. enteritidis* isolates were sequenced, and SNP-based cluster analyses were made. It was understood that four-SNP pairwise diversity was in concordance with epidemiologically defined outbreak typings.

12.2.2.4 Campylobacter jejuni

In industrialized countries, the plausible explanation for bacterial gastroenteritis is *Campylobacter jejuni*. Being zoonotic in origin, this bacterium causes acute gastroenteritis, in extreme cases it enters into blood stream and causes bacteremia. A potential virulent enzyme named 5'-methylthioadenosine/S-adenosylhomocysteine nucleosidase (Parveen and Cornell 2011) causes the *Campylobacter* outbreak. It has been estimated that cases *Campylobacter* gastroenteritis have surpassed *Salmonella* infection in developed countries like the US. This is a gram-negative bacterium with helical morphology. However, it can take on coccal forms when exposed to atmospheric oxygen. They are motile with single polar flagellum. Gram-positive bacteria carry out Peptidoglycan (PG) acetylation, thereby resisting the action of lysozymes. *C. jejuni* is the only gram-negative bacterium that can perform acetylation to its PG layer. The genes encoding putative PG acetyltransferase A (PatA) and B (PatB) are highly conserved in *C. jejuni* (Iwata et al. 2016). However, outbreaks caused by said bacteria are rarely reported, which may reflect limitations of surveillance, for which molecular typing is not routinely performed (Joenson et al. 2020). But Oakeson et al. (2018) demonstrated the superior resolution of genetic and evolutionary relatedness generated by WGS data analysis, compared to pulsed-field gel electrophoresis (PFGE). Their study retrospectively applied WGS and a reference-free bioinformatic analysis of *Campylobacter jejuni* associated with raw milk. Rokney et al. (2018) had analyzed the phylogeny, diversity, and prevalence of virulence factors using WGS of a national sample of *C. jejuni* clinical, food, and animal isolates collected over a 10-year period (2003–2012). Fiedoruk et al. (2019) have designed a study comprising phylogeny, resistome, and virulome in order to assess similarities between genomes of *C. jejuni* strains, isolated from children suffering from acute diarrhea in north eastern Poland, in comparison to *C. jejuni* genomes stored in public databases. The WGS analysis revealed notable diversity among clinically relevant strains. The 3D structures of proteins/toxins/enzymes [shiga-like toxin of *Escherichia coli* (Boisen et al. 2014), Invasin of *Shigella flexneri* (Johnson et al. 2007), TtsA of *Salmonella typhi* (Geiger et al. 2018), and 5'-methylthioadenosine/S-adeno sylhomocysteine nucleosidase of *Campylobacter jejuni* (Ducati et al. 2018)] of gram-negative bacteria are given in Figures 12.3 and 12.4.

12.3 WGS AND NGS APPROACHES TO FOODBORNE FUNGAL STRAINS

In microbial the consortium, fungal as well as viral strains have the utmost importance in food pathophysiology. The fungi are considered to be the most opportunistic food pathogens. Being dimorphic in nature (both filamentous and yeast-like) they survive in decaying organic matter of food. The spores of fillamentose fungi and cells of yeast like fungi can enter the gut and produce toxins. Once the fungal strains are established inside the body, then it is very difficult to eradicate them from the affected tissues. But, approx. 40% of researchers do not take interest in fungal strains because of their longer period of growth and multiplication time, in contrast to bacteria which take only 20–50 minutes of generation time. Moreover, the eukaryotic genome size of fungal strains is very large and composite. It is observed that strains of *Mucor* sp., *Penicillium* sp., *Fusarium* sp., *Aspergillus* sp., and *Candida* sp. are associated with foods and vegetables and cause

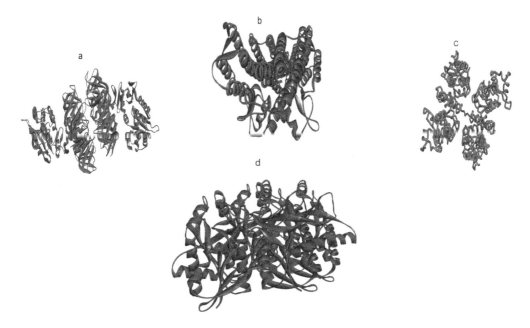

FIGURE 12.3   The toxins/proteins of gram-negative bacteria eluted from WGS strategies: (a) Shiga-like toxin of *Escherichia coli*, (b) Invasin of *Shigella flexneri*; (c) *TtsA* of *Salmonella typhi*, (d) 5'-methylthioadenosine/S-adenosylhomocysteine nucleosidase of *Campylobacter jejuni*.

diseases in immunocompromised patients. The cheese industry suffers most with fungal contaminants belonging to strains *Acremonium, Alternaria, Aspergillus, Aureobasidium, Botrytis, Cladosporium, Epicoccum, Eurotium, Exophiala, Fusarium, Gliocladium, Lecanicillium, Mucor, Penicillium, Rhizopus,* and *Wallemia* (Kure and Eskaar 2019). Sekse et al. (2017) have reported that very few genome assemblies are available in public fungal genomic databases. Hence, the WGS portfolio of fungal strains seems to be inconspicuous because the pathogenomics of candidate fungal strains are underestimated and also understudied.

12.4   WGS AND NGS APPROACHES TO FOODBORNE VIRAL STRAINS

WGS data analysis on viruses that originate from food sources is scarce, although the primary genome, that of RNA bacteriophage MS2, was sequenced in 1976, during a truly heroic feat of direct determination of an RNA sequence trailed by the genome of bacteriophage φX174. On the other hand, bacteriophages have a key role in gut microbiota. Earlier Arnold et al. (2016) had given some insights on the emerging tools to detect gut bacteriophages. Pacheco (2019) had administered physiological and genomic characterization of six virulent bacteriophages of shiga toxin-producing *Escherichia coli* O157: H7 f. Also, Waller et al. (2014) had identified 20 viral taxa in 252 human gut metagenomic samples, using taxon-specific marker genes. Additionally, an in-depth bacteriophage family was also discovered by Yutin et al. (2018). And, the Illumina platform (Rihtman et al. 2016) and CRISPR-map (Gogleva et al. 2014) have been utilized to study the gut bacteriophage genes.

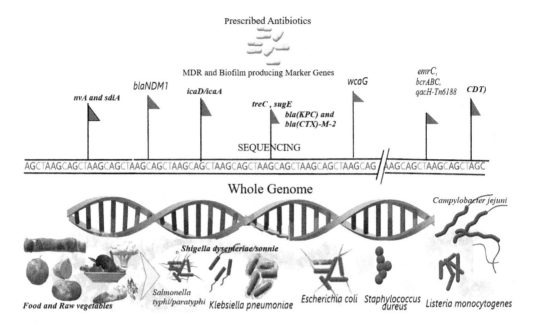

FIGURE 12.4   A model depicting the elution of some marker drug resistance genes from the NGS data retrieved from WGS of respective bacterial strains. SdiA, a transcription factor of the LuxR family that regulates genes in response to N-acyl homoserine lactones (AHLs) produced by other species of bacteria, NDM-1 (New Delhi metallo-beta-lactamase), ica D and ica A (intercellular adhesion biofilm genes D and A), Trehalose-6-phosphate hydrolase (tre C) can hydrolyse trehalose-6-phosphate to glucose and glucose 6-phosphate, sugE, a multidrug efflux system protein, blaKPC, β-latcamase producing carabapenem resistant *Klebsiella pneumoniae*, blaCTX-M-2, is a class of β lactamases. The CTX-M-type enzymes are a group of molecular class A extended-spectrum β-lactamases (ESBLs) that exhibit an overall preference for cefotaxime (CTX; hence the CTX-M name) and ceftriaxone, wcaG (GDP-L-fucose synthase), emrC (Outer membrane protein), ABC-BCR–The symbol ABL1 is derived from Abelson, the name of a leukemia virus which carries a similar protein. The symbol BCR is derived from breakpoint cluster region, a gene which encodes a protein that acts as a guanine nucleotide exchange factor for Rho GTPase proteins. qacH-Tn6188, a multidrug efflux protein associated with Transposon detected in *Listeria monocytogenes*, CDT (cytolethal distending toxin).

12.5 CONCLUSION

From this brief overview on the foodborne organisms causing ailment among people, it's observed that bacterial strains having a place in both gram-positive and gram-negative categories are involving inside the beat position of the board. WGS provided information beyond the identity and relationship of strains; it promotes public health by improving the security, quality, and shelf life of foods. In addition, compared to any other related tools, WGS has advantages in identifying pathogenicity, virulence factors, adaptation, survival, resistance to biocides, metals and antimicrobial drugs, and genome plasticity dynamism (hgt effect) and in addition, the conserved domains of putative proteins/toxins of foodborne bacteria genes or gene cassettes. However, the domain of fungal strains is deprived of WGS approaches. The WGS strategies could have been a right approach for metataxonomic studies for related fungal spp. like *Mucor* and *Aspergillus*. Viral strains belonging to bacteriophages group are predominant in foods. However, because of random and frequent mutations of bacteriophages, the WGS strategies are not yet ready to vanquish viral strains and their putative genes/proteins.

REFERENCES

Abdalhamid, B., Mccutchen, E. L., Bouska, A. C., Weiwei, Z., Loeck, B., Hinrichs, S. H. and Iwen, P. C. 2019. Whole genome sequencing to characterize Shiga toxin-producing *Escherichia coli* O26 in a public health setting. *Journal of Infection and Public Health* 12(6):884–9.

Arnold, J. W., Roach, J. and Azcarate-Peril, M. A. 2016. Emerging technologies for gut microbiome research. *Trends in Microbiology* 24(11):887–901.

Baker, K. S., Campos, J., Pichel, M., et al. 2017. Whole genome sequencing of *Shigella sonnei* through PulseNet Latin America and Caribbean: Advancing global surveillance of foodborne illnesses. *Clinical Microbiology and Infection* 23(11):845–53.

Bintsis, T. 2017. Foodborne pathogens. *AIMS Microbiology* 3(3):529.

Boisen, N., Hansen, A. M., Melton-Celsa, A. R., Zangari, T., Mortensen, N. P., Kaper, J. B., et al. 2014. The presence of the pAA plasmid in the German O104: H4 Shiga toxin type 2a (Stx2a)–producing enteroaggregative *Escherichia coli* strain promotes the translocation of Stx2a across an epithelial cell monolayer. *The Journal of Infectious Diseases* 210(12):1909–19.

Carroll, L. M., Wiedmann, M., Mukherjee, M., Nicholas, D. C., Mingle, L. A., Dumas, N. B., Cole, J. A. and Kovac, J. 2019. Characterization of emetic and diarrheal *Bacillus cereus* strains from a 2016 foodborne outbreak using whole-genome sequencing: Addressing the microbiological, epidemiological, and bioinformatic challenges. *Frontiers in Microbiology* 10:144.

Carter, L., Chase, H. R., Gieseker, C. M., Hasbrouck, N. R., Stine, C. B., Khan, A., Ewing-Peeples, L. J., Tall, B. D. and Gopinath, G. R. 2018. Analysis of enterotoxigenic *Bacillus cereus* strains from dried foods using whole genome sequencing, multi-locus sequence analysis and toxin gene prevalence and distribution using endpoint PCR analysis. *International Journal of Food Microbiology* 284:31–9.

Carter, J. C., Van Wijk, M. and Rowsell, M. 2019. Symptoms of 'food addiction' in binge eating disorder using the Yale Food Addiction Scale version 2.0. *Appetite* 133:362–9.

Chattaway, M. A., Greig, D. R., Gentle, A., Hartman, H. B., Dallman, T. J. and Jenkins, C. 2017. Whole-genome sequencing for national surveillance of *Shigella flexneri*. *Frontiers in Microbiology* 8:1700.

Chi, Y. I., Sadler, I., Jablonski, L. M., Callantine, S. D., Deobald, C. F., Stauffacher, C. V. and Bohach, G. A. 2002. Zinc-mediated dimerization and its effect on activity and conformation of staphylococcal enterotoxin type C. *Journal of Biological Chemistry* 277(25):22839–46.

Cunningham, S. A., Chia, N., Jeraldo, P. R., et al. 2017. Comparison of whole-genome sequencing methods for analysis of three methicillin-resistant *Staphylococcus aureus* outbreaks. *Journal of Clinical Microbiology* 55(6):1946–53.

Ducati, R. G., Harijan, R. K., Cameron, S. A., Tyler, P. C., Evans, G. B. and Schramm, V. L. 2018. Transition-state analogues of *Campylobacter jejuni* 5′-methylthioadenosine nucleosidase. *ACS Chemical Biology* 13(11):3173–83.

Eng, S. K., Pusparajah, P., Mutalib, N.-S. A., Ser, H., Chan, K. and Lee, L.-H. 2015. *Salmonella*: A review on pathogenesis, epidemiology and antibiotic resistance. *Frontiers in Life Science* 8:284–93.

Fiedoruk, K., Daniluk, T., Rozkiewicz, D., Oldak, E., Prasad, S. and Swiecicka, I. 2019. Whole-genome comparative analysis of *Campylobacter jejuni* strains isolated from patients with diarrhoea in north eastern Poland. *Gut Pathogens* 11(1):32.

Fisher, E. L., Otto, M. and Cheung, G. Y. 2018. Basis of virulence in enterotoxin-mediated staphylococcal food poisoning. *Frontiers in Microbiology* 9:436.

Franck, K. T., Gumpert, H., Olesen, B., Larsen, A. R., Petersen, A., Bangsborg, J., Albertsen, P., Westh, H. and Bartels, M. D. 2017. *Staphylococcal aureus* enterotoxin C and enterotoxin-like L associated with post-partum mastitis. *Frontiers in Microbiology* 8:173.

Fraser, M. E., Chernaia, M. M., Kozlov, Y. V. and James, M. N. 1994. Crystal structure of the holotoxino from *Shigella dysenteriae* at 2.5 Å resolution. *Nature Structural Biology* 1(1):59–64.

Fraser-Liggett, C. M. 2005. Insights on biology and evolution from microbial genome sequencing. *Genome Research* 15(12):1603–10.

Ford, L., Moffatt, C. R., Fearnley, E., Miller, M., Gregory, J., Sloan-Gardner, T. S., et al. 2018. The epidemiology of *Salmonella enterica* outbreaks in Australia, 2001–2016. *Frontiers in Sustainable Food Systems* 2(86):1–8. doi:10.3389/fsufs.2018.00086

Geiger, T., Pazos, M., Lara-Tejero, M., Vollmer, W. and Galán, J. E. 2018. Peptidoglycan editing by a specific LD-transpeptidase controls the muramidase-dependent secretion of typhoid toxin. *Nature Microbiology* 3(11):1243–54.

Gogleva, A. A., Gelfand, M. S. and Artamonova, I. I. 2014. Comparative analysis of CRISPR cassettes from the human gut metagenomic contigs. *BMC Genomics* 15(1):202.

Gonzalez-Escalona, N. and Kase, J. A. 2019. Virulence gene profiles and phylogeny of Shiga toxin-positive *Escherichia coli* strains isolated from FDA regulated foods during 2010–2017. *PLoS ONE* 14(4):e0214620.

Gupta, S. K., Sharma, P., McMillan, E. A., Jackson, C. R., Hiott, L. M., Woodley, T., Humayoun, S. B., Barrett, J. B., Frye, J. G. and McClelland, M. 2019. Genomic comparison of diverse *Salmonella* serovars isolated from swine. *PLoS ONE* 14(11):e0224518.

Halbedel, S., Prager, R., Fuchs, S., Trost, E., Werner, G. and Flieger, A. 2018. Whole-genome sequencing of recent *Listeria monocytogenes* isolates from Germany reveals population structure and disease clusters. *Journal of Clinical Microbiology* 56(6):e00119-18. doi:10.1128/JCM.00119-18

Hurley, D., Luque-Sastre, L., Parker, C. T., et al. 2019. Whole-genome sequencing-based characterization of 100 *Listeria monocytogenes* isolates collected from food processing environments over a four-year period. *mSphere* 4(4):e00252-19.

Iwata, T., Watanabe, A., Kusumoto, M. and Akiba, M. 2016. Peptidoglycan acetylation of *Campylobacter jejuni* is essential for maintaining cell wall integrity and colonization in chicken intestines. *Applied and Environmental Microbiology* 82(20):6284–90.

Jagadeesan, B., Gerner-Smidt, P., Allard, M. W., et al. 2019. The use of next generation sequencing for improving food safety: Translation into practice. *Food Microbiology* 79:96–115.

Jamshidi, A. and Zeinali, T. 2019. Significance and characteristics of *Listeria monocytogenes* in poultry products. *International Journal of Food Science* 2019:1–7. doi:10.1155/2019/7835253

Jenkins, C., Dallman, T. J. and Grant, K. A. 2019. Impact of whole genome sequencing on the investigation of food-borne outbreaks of Shiga toxin-producing *Escherichia coli* serogroup O157: H7, England, 2013 to 2017. *Eurosurveillance* 24(4). http://www.ncbi.nlm.nih.gov/pubmed/1800346.

Joensen, K. G., Kiil, K., Gantzhorn, M. R., Nauerby, B., Engberg, J., Holt, H. M., et al. 2020. Whole-genome sequencing to detect numerous *Campylobacter jejuni* outbreaks and match patient isolates to sources, Denmark, 2015–2017. *Emerging Infectious Diseases* 26(3):523.

Joensen, K. G., Kuhn, K. G., Müller, L., Björkman, J. T., Torpdahl, M., Engberg, J., Holt, H. M., Nielsen, H. L., Petersen, A. M., Ethelberg, S. and Nielsen, E. M. 2018. Whole-genome sequencing of *Campylobacter jejuni* isolated from Danish routine human stool samples reveals surprising degree of clustering. *Clinical Microbiology and Infection* 24(2):201–5.

Johler, S., Macori, G., Bellio, A., Acutis, P. L., Gallina, S. and Decastelli, L. 2018. Characterization of *Staphylococcus aureus* isolated along the raw milk cheese production process in artisan dairies in Italy. *Journal of Dairy Science* 101(4):2915–20.

Johnson, S., Roversi, P., Espina, M., et al. 2007. Self-chaperoning of the type III secretion system needle tip proteins iPad and BipD. *Journal of Biological Chemistry* 282(6):4035–44.

Kobayashi, T., Kutsuna, S., Hayakawa, K., Kato, Y., Ohmagari, N., Uryu, H., Takei, R., et al. 2016. An outbreak of food-borne typhoid fever due to *Salmonella enterica* serotype Typhi in Japan reported for the first time in 16 Years. *The American Journal of Tropical Medicine and Hygiene* 94(2):289–91.

Köser, C. U., Ellington, M. J., Cartwright, E. J., Gillespie, S. H., Brown, N. M., Farrington, M., Holden, M. T., Dougan, G., Bentley, S. D., Parkhill, J. and Peacock, S. J. 2012. Routine use of microbial whole genome sequencing in diagnostic and public health microbiology. *PLoS Pathogens* 8(8). doi:10.1371/journal.ppat.1002824

Kure, C. F. and Skaar, I. 2019. The fungal problem in cheese industry. *Current Opinion in Food Science* 29:14–9.

Lang, C., Hiller, M., Konrad, R., Fruth, A. and Flieger, A. 2019. Whole genome-based public health surveillance of less common STEC serovars and untypable strains identifies four novel O genotypes. *Journal of Clinical Microbiology* 54(10):1–12.

Leekitcharoenphon, P., Nielsen, E. M., Kaas, R. S., Lund, O. and Aarestrup, F. M. 2014. Evaluation of whole genome sequencing for outbreak detection of *Salmonella enterica*. *PLoS ONE* 9(2):e87991. doi:10.1371/journal.pone.0087991

Macori, G., Bellio, A., Bianchi, D. M., et al. 2020. Genome-wide profiling of enterotoxigenic *Staphylococcus aureus* strains used for the production of naturally contaminated cheeses. *Genes* 11(1):33.

Madegowda, M., Eswaramoorthy, S., Burley, S. K. and Swaminathan, S. 2008. X-ray crystal structure of the B component of Hemolysin BL from *Bacillus cereus*. *Proteins:Structure, Function, and Bioinformatics* 71(2):534–40.

Moran-Gilad, J. 2017. Whole genome sequencing (WGS) for food-borne pathogen surveillance and control–taking the pulse. *Eurosurveillance* 22(23):30547. doi:10.2807/1560-7917.ES.2017.22.23.30547.

Mossong, J., Decruyenaere, F., Moris, G., Ragimbeau, C., Olinger, C. M., Johler, S., Perrin, M., Hau, P. and Weicherding, P. 2015. Investigation of a staphylococcal food poisoning outbreak combining case–control, traditional typing and whole genome sequencing methods, Luxembourg, June 2014. *Eurosurveillance* 20(45):30059.

Oakeson, K. F., Wagner, J. M., Rohrwasser, A. and Atkinson-Dunn, R. 2018. Whole-genome sequencing and bioinformatic analysis of isolates from foodborne illness outbreaks of *Campylobacter jejuni* and *Salmonella enterica*. *Journal of Clinical Microbiology* 56(11):e00161-18. doi:10.1128/JCM.00161-18

Oniciuc, E. A., Likotrafiti, E., Alvarez-Molina, A., Prieto, M., Santos, J. A. and Alvarez-Ordóñez, A. 2018. The present and future of whole genome sequencing (WGS) and whole metagenome sequencing (WMS) for surveillance of antimicrobial resistant microorganisms and antimicrobial resistance genes across the food chain. *Genes* 9(5):268.

Pacheco, B. I. C. 2019. *Physiological and Genomic Characterization of Six Virulent Bacteriophages of Shiga Toxin-Producing Escherichia coli O157: H7 for Biocontrol and Detection Applications*. Thesis, University of Alberta. doi:10.7939/r3-v0vd-bc72

Parveen, N. and Cornell, K. A. 2011. Methylthioadenosine/S-adenosylhomocysteine nucleosidase, a critical enzyme for bacterial metabolism. *Molecular Microbiology* 79(1):7–20.

Pattnaik, S. 2017. *Diagnostic bacteria genomics*, Mumbai: Himalaya Publishing House.

Pightling, A. W., Pettengill, J. B., Luo, Y., Baugher, J. D., Rand, H. and Strain, E. 2018. Interpreting whole-genome sequence analyses of foodborne bacteria for regulatory applications and outbreak investigations. *Frontiers in Microbiology* 9:1–14. doi:10.3389/fmicb.2018.01482

Quainoo, S., Coolen, J. P., van Hijum, S. A., Huynen, M. A., Melchers, W. J., van Schaik, W. and Wertheim, H. F. 2017. Whole-genome sequencing of bacterial pathogens: The future of nosocomial outbreak analysis. *Clinical Microbiology Reviews* 30(4):1015–63.

Qiu, J., Li, H., Su, H., Dong, J., Luo, M., Wang, J., et al. 2012. Chemical composition of fennel essential oil and its impact on *Staphylococcus aureus* exotoxin production. *World Journal of Microbiology and Biotechnology* 28(4):1399–405.

Reid, C. J., Blau, K., Jechalke, S., Smalla, K. and Djordjevic, S. P. 2020. Whole genome sequencing of *Escherichia coli* from store-bought produce. *Frontiers in Microbiology* 10:3050.

Rihtman, B., Meaden, S., Clokie, M. R., Koskella, B. and Millard, A. D. 2016. Assessing Illumina technology for the high-throughput sequencing of bacteriophage genomes. *PeerJournal* 4:e2055.

Rokney, A., Valinsky, L., Moran-Gilad, J., Vranckx, K., Agmon, V. and Weinberger, M. 2018. Genomic epidemiology of *Campylobacter jejuni* transmission in Israel. *Frontiers in Microbiology* 9:2432.

Rumore, J., Tschetter, L., Kearney, A., Kandar, R., McCormick, R., Walker, M., Peterson, C. L., Reimer, A. and Nadon, C. 2018. Evaluation of whole-genome sequencing for outbreak detection of verotoxigenic *Escherichia coli* O157: H7 from the Canadian perspective. *BMC Genomics* 19(1):870.

Salipante, S. J., SenGupta, D. J., Cummings, L. A., Land, T. A., Hoogestraat, D. R. and Cookson, B. T. 2015. Application of whole-genome sequencing for bacterial strain typing in molecular epidemiology. *Journal of Clinical Microbiology* 53(4):1072–9.

Schubert, W. D., Urbanke, C., Ziehm, T., Beier, V., Machner, M. P., Domann, E., Wehland, J., Chakraborty, T. and Heinz, D. W. 2002. Structure of internalin, a major invasion protein of *Listeria monocytogenes*, in complex with its human receptor E-cadherin. *Cell* 111(6):825–36.

Schürch, A. C., Arredondo-Alonso, S., Willems, R. J. L. and Goering, R. V. 2018. Whole genome sequencing options for bacterial strain typing and epidemiologic analysis based on single nucleotide polymorphism versus gene-by-gene–based approaches. *Clinical Microbiology and Infection* 24(4):350–4.

Seemann, T. 2014. Prokka: Rapid prokaryotic genome annotation. *Bioinformatics* 30(14):2068–9.

Sekse, C., Holst-Jensen, A., Dobrindt, U., Johannessen, G. S., Li, W., Spilsberg, B., et al. 2017. High throughput sequencing for detection of foodborne pathogens. *Frontiers in Microbiology* 8:2029.

Sethuvel, D. P. M., Anandan, S., Ragupathi, N. K. D., Gajendiran, R., Kuroda, M., Shibayama, K. and Veeraraghavan, B. 2019. IncFII plasmid carrying antimicrobial resistance genes in *Shigella flexneri*: Vehicle for dissemination. *Journal of Global Antimicrobial Resistance* 16:215–9.

Song, C., Liu, C., Wu, S., Li, H., Guo, H., Yang, B., et al. 2016. Development of a lateral flow colloidal gold immunoassay strip for the simultaneous detection of *Shigella boydii* and *Escherichia coli* O157: H7 in bread, milk and jelly samples. *Food Control* 59:345–51.

Stamatakis, A. 2015. Using RAxML to infer phylogenies. *Current Protocols in Bioinformatics* 51(1):6–14.

Van Goethem, N., Descamps, T., Devleesschauwer, B., Roosens, N. H., Boon, N. A., Van Oyen, H. and Robert, A. 2019. Status and potential of bacterial genomics for public health practice: A scoping review. *Implementation Science: IS* 14(1):1–16.

Waller, A. S., Yamada, T., Kristensen, D. M., Kultima, J. R., Sunagawa, S., Koonin, E. V. and Bork, P. 2014. Classification and quantification of bacteriophage taxa in human gut metagenomes. *The ISME Journal* 8(7):1391–402.

Wang, H., Cai, L., Hu, H., Xu, X. and Zhou, G. 2019. Complete genome sequence of *Salmonella enterica* serovar Enteritidis NCM 61, with high potential for biofilm formation, isolated from meat-related sources. *Microbiology Resource Announcements* 8(2):e01434-18. doi:10.1128/MRA.01434-18

Wu, Y., Lau, H. K., Lee, T., Lau, D. K. and Payne, J. 2019. *In silico* serotyping based on whole-genome sequencing improves the accuracy of *Shigella* identification. *Applied and Environmental Microbiology* 85(7):e00165-19. doi:10.1128/AEM.00165-19

Yang, S. C., Lin, C. H., Aljuffali, I. A. and Fang, J. Y. 2017. Current pathogenic *Escherichia coli* foodborne outbreak cases and therapy development. *Archives of Microbiology* 199(6):811–25.

Yutin, N., Makarova, K. S., Gussow, A. B., Krupovic, M., Segall, A., Edwards, R. A. and Koonin, E. V. 2018. Discovery of an expansive bacteriophage family that includes the most abundant viruses from the human gut. *Nature Microbiology* 3(1):38–46.

Zerbino, D. R. and Birney, E. 2008. Velvet: Algorithms for de novo short read assembly using de Bruijn graphs. *Genome Research* 18(5):821–9.

CHAPTER 13

Single Cell Sequencing of Microbial Species

Charles Oluwaseun Adetunji, Wadzani Dauda Palnam,
Juliana Bunmi Adetunji, Benjamin Ewa Ubi,
Ayodele Eugene Ayeni, Devarajan Thangadurai,
Muhammad Akram, Fahad Said Khan,
Olugbenga Samuel Michael, Olugbemi Tope Olaniyan,
Ajayi Kolawole Temidayo, Saher Islam,
Arowora Kayode Adebisi, and Daniel Ingo Hefft

CONTENTS

13.1 INTRODUCTION

For 16S rRNA-based studies, sequencing of single-cell genomes has helped in providing quick access to a library of microbial community genes that could facilitate quick analyses of genomes of microbial strain variants. This might be linked to the fact that traditional techniques can detect only a relatively few of the species present in such microbial community (Blainey 2013; Lasken and McLean 2014). Also, comparisons among differing microbial species at genome level single-cell genomics have been identified as one of the easiest methods that could be utilized for effective identification (Blainey 2013; Tyml et al. 2019).

Single-cell genome sequencing has been identified as a unique technique among all the current genomic approaches, because it provides quick access to datasets without any level of uncertainty in the classification of reads in line with the strains of origin and finely resolving phase variation existing at the genome level of complex microbial populations (Blainey 2013). Based on the need to gain insights into the complex and diverse microbial ecosystem, great advances in culture-independent methods by genome sequencing technologies have evolved that play a crucial role toward introducing solutions in this regard. The application of culture-independent methods for analyses of environmental microbial taxa has gained higher status especially toward the advancement and its broad applications in metagenomic sequencing studies and single-cell genome sequencing as compared to single-gene polymerase chain reaction (PCR) based methods (Blainey 2013; Ahrendt et al. 2018).

Metagenomics has been identified as a technique that analyzes the sequences of the entire DNA derived from an environmental bulk sample without subjecting the samples to the culturing technique. This technique has achieved numerous breakthroughs in accessing the diversity of uncultivated microorganisms. However, metagenomics has several limitations, which entail its ability to capture fine-scale strain heterogeneity within closely related microbial populations during assembly of short-read sequence data and inability to resolve the link between the genome and any additional cellular DNA like plasmids or phages. Moreover, the sequencing of the genomes of single cells has several advantages that could help in resolving some of these aforementioned limitations regarding metagenomics which has increased its application for some years (Blainey 2013; Lasken and McLean 2014; Ahrendt et al. 2018) in microbial diversity studies. The utilization of metagenomics and single-cell genomic approaches have become a boon for novel discoveries by bringing transformation to the world of microbial ecology, diversity, evolutionary history, metabolism, agriculture, biotechnology, and human health (Ahrendt et al. 2018). Genomic DNA sequencing of individual cells provides single-cell amplified genomes (SAGs), which enable it to: 1) function as reference for genomes that could facilitate quick assembly of sequenced data for novel uncultivated microbial species, which have close relatedness and 2) elucidate strain variations within species by associating the core genes with the individual strains and providing crucial insights for adequate prediction of biological function and adaptation. Single-cell sequencing also aids in the detection of pathogens with a low abundance level and their transmission between the host and their environment. Single cell genomics method is a technically challenging process if high quality single-cell sequencing data need to be obtained (Lasken and McLean 2014).

Single cell genomic analysis workflows have the following steps: cell isolation, cell lysis or permeabilization, whole-genome amplification which normally occurs through a multiple displacement amplification method, whole-genome amplification product screening, barcoded library construction, sequencing and informatics, de novo genome assembly, contaminant identification and removal, genome quality check and categorization as well as database submission (Blainey 2013). Therefore, this chapter intends to provide detailed information on the utilization of single cell sequencing for the identification of microbial species.

13.2 SINGLE CELL SEQUENCING

With the use of new and powerful genomic technologies, DNA sequencing from single cells has brought fresh opportunities to microbiology, allowing easy understanding of microbial species without any conventional culture conditions. Thus, bacterial species can be sequenced directly from environmental, food, or clinical samples. Moreover, with the technical improvements in bioinformatics technologies, single-cell sequencing, metagenomic analysis, and genome assembly, the analysis of complex bacterial communities is now possible for even previously uncharacterized microbial species (Lasken 2012). This section explores the genomic sequencing of bacteria and yeast single cells (Figure 13.1).

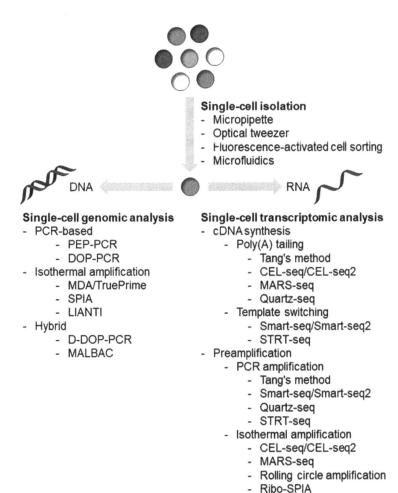

FIGURE 13.1 Overview of microbial single-cell analysis (Adapted from Chen et al. 2017, Creative Commons Attribution 4.0 International).

13.2.1 Single Cells of Bacteria

Kuchina et al. (2019) reported a single-cell RNA sequencing method that is effective for both gram-negative and gram-positive bacteria and able to reveal transcriptional states that remain unknown at a population level. The authors also provide adequate in-depth gene expression profiles associated with certain physiologic functions such as competence, prophage induction in PBSX, and unforeseen activation of metabolic pathway in a cell subpopulation.

Chijiiwa et al. (2020) reported a study of microbial responders to dietary fiber in the mouse gut microbiota using single-cell genomics. According to their study, it was discovered that mouse feed with inulin leads to several changes in their fecal microbiome composition and increases the population of *Bacteroides* spp. Moreover, the increase in the population of *Bacteroides* spp. in the intestine simultaneously resulted in higher levels of succinate generation in the mouse intestine. The parallel single-cell genome sequencing technique employed, named SAG-gel platform, produced 346 SAGs from mouse gut microbes before and after dietary inulin supplementation. Further classification revealed 267 bacteria, cutting across 2 phyla, 4 classes, 7 orders, and 14 families. The genomes of the predominant inulin-responders, *Bacteroides* spp., were successfully identified, as well as their polysaccharide utilization loci and the specific metabolic pathways required for their production of succinate. The technique employed in the study has the potential for application in human and animal health practices. It provided an understanding of uncultivable bacteria in the intestinal microbiota, allowed an in-depth isolation and identification of the specific functions of uncultured bacteria within the gut, as well as an estimation of the metabolic lineages involved in the fermentation of dietary fiber by bacteria.

Although single-cell sequencing does not involve the process of microorganism cultivation, it has the prospects of contributing many novel genomes most especially from previously uncharacterized strains. In addition to this, the method has the capability to reveal subtle differences across species by allowing more insight into species-distinguishing genes and by expanding the pan-genome, the total identified gene diversity within species. Biological function and adaptation of species are revealed by the variations within the genome, as well as features such as pathogen infectivity and virulence, transmission, and onset of resistance to antibiotics, which are very important in studying foodborne pathogens (Lasken and McLean 2014).

Hwang et al. (2018) revealed that substantial progress has been seen in the evolution and advancement in the application of genomic techniques for analysis of food related samples using next-generation sequencing (NGS) technologies within the past few years. This has led to the massive knowledge output in the area of pathophysiology mechanisms underlying some disease conditions, particularly cancer biology and microbial diversification. Studies have shown that single-cell analyses will enable researchers to unravel the genomic profiling of the bulk of microbial populations utilizing NGS-based technologies. It is known that single-cell RNA sequencing has great potential to uncover an array of microbial populations which could help in effective prediction of disease outbreaks.

Li and Clevers (2010) revealed that adequate understanding of cell and molecular biology has introduced tremendous progress toward the application of single-cell RNA sequencing analysis. Studies have shown that single-cell RNA sequencing applications can be channeled toward unraveling the features of tumorigenesis which cannot be determined originally with the conventional studies. This technique can be applied for proper understanding of the various pathways that are involved in cancer treatment, phylogenetic connections in cells, liquid biopsy, lineage tracing or phylogeny, gene mutations, stem cell controlling mechanisms in disease and healthy conditions, and molecular diagnosis.

13.2.2 Single Cells of Yeast

Zhang et al. (2010) described a rapid procedure for the extraction of genomic DNA from a large group of fungal strains. In this method, termed thermolysis, cells were rinsed with water to get rid of potential PCR inhibitors and were incubated at 85°C in a lysis buffer to cause cell walls and membranes to break down. Regions under study from high and single-copy number genes were then successfully amplified from the DNA samples that were extracted. The method allowed the gene samples to be preserved at 20°C for up to one year or more and required small samples of starting fungi or yeast DNA material.

The potential of single-cell sequencing is manifest in the capability to discover unknown or uncultivable fungal species. When coupled with other tools, such as fluorescence in situ hybridization (FISH), it can be much more efficient. Previously unstudied niches should be explored and new techniques should be applied to discover and characterize these unknown species. Hence, the importance of unmapped habitats and uncommon techniques is apparent (Wu et al. 2019). Researchers should consider these options for the identification of yet undiscovered fungal species. It is thus clear that when single-cell sequencing is complemented with other tools in the near future, it will play important roles in revealing the genomes of uncultivated microorganisms and the underlying understanding behind the functions of microbial community and other aspects of microbial ecology.

The 16S ribosomal RNA gene sequence has been the main identification signature for the majority of bacterial phyla. Single-cell sequencing platforms have allowed for an increased understanding of the rudimentary biology, taxonomy, and evolution of microbial species and their diverse functions in human health and environmental ecosystems (Lasken and Mclean 2014).

Saint et al. (2019) revealed heterogeneity of gene expression in yeast during periods of mitotic division and stress adaptations. The study, along with single-cell RNA sequencing also employed single-cell imaging and Bayesian true count recovery. The results showed the groups of genes that were highly variable during rapid cell division of yeasts in constant culture conditions. Before the stationary phase, there was a gradual, coordinated adaptation in gene regulation, and then subsequently, high variations in gene expression when growth decreased. Their study showed gene expression heterogeneity determined different physiological conditions.

Khatri et al. (2017) reported a comparative study of the yeast *Saccharomyces boulardii* with *Saccharomyces cerevisiae* was performed to reveal proper understanding on the similarities at the genome-level and differences between them. Genomic single-cell sequencing of two strains of *S. boulardii* was carried out, and then compared to 145 strains of *S. cerevisiae*. Ty1, Ty3, Ty4, the Ty elements, and associated long-terminal repeat (LTR) were present in *S. cerevisiae,* but were absent in *S. boulardii*, except for Ty2 and Ty5 elements. The genes encoding HXT11 and HXT9 (the hexose transporters) as well as those for asparagine utilization (ASP3-1, ASP3-2, ASP3-3, and ASP4-4) were absent in the probiotic yeast strains. There were also differences in repeat periods and copy number of repeats for the flocculin gene in the yeast.

Nadal-Ribelles et al. (2019) reported a study that investigates correlations between clonal yeast populations with respect to their RNA transcripts. The authors developed a procedure that could digitally count transcript start sites in a strand- and isoform-specific manner. A negative correlation was observed in the expression of sense–antisense pairs, while paralogs and transcripts that showed divergence were co-expressed. When index sorting was coupled with single-cell RNA sequencing, there was a linear relationship between the size of cells and RNA content. The results suggest that there is a selective advantage in individual cells brought about by functional transcript diversity within yeast populations.

Woolfit et al. (2007) reported detailed information on the single-cell sequencing of the wine-spoilage yeast *Dekkera (Brettanomyces) bruxellensis*. The organism's genome was compared to that of *Candida albicans* and the baker's yeast *Saccharomyces cerevisiae*. The results show that the rate of chromosomal rearrangement estimated in *D. bruxellensis* was slower than that calculated for *C. albicans*. The result obtained from the proteome analysis of *D. bruxellensis* is enriched with transporters and revealed that the frequency of amino acid evolution was higher than that of *C. albicans*. Moreover, it was also established that *D. bruxellensis* has a number of genes involved in the metabolism of nitrogen and lipid, a property thought to reflect the adaptation of the organism to a low nutrient and high ethanol niche. The results provided brief information on the evolution of *D. bruxellensis* as well as the genetic justifications for its metabolic capabilities.

13.3 SINGLE-CELL SEQUENCING: CASE STUDIES

Tasic et al. (2016) revealed the importance of single-cell sequencing in molecular biology by studying the complex population of cells in diverse fields of science, such as neurobiology, immunology, microbiology, tissue mosaicism, biochemistry, physiology, development, and cancer. Lake et al. (2016) revealed that biological diversity is an important area in which single-cell sequencing can be applied to study genomic profiling of rare cells, delineating population diversity, classifying cell types, and tracing cell lineages. The authors showed that many areas still remained unexplored particularly the area of virology. In human physiology, single-cell sequencing has been utilized to study transcriptional mechanisms due to excitation in special senses and as a biomarker identification in embryogenic cells. Future benefits in single-cell sequencing utilization should be directed toward mutation diversity. The authors also suggested that *in situ* single-cell sequencing should be applied to single cell genomic data measurement and the preservation of spatial orientation in cells thereby linking the single cell genotypes and phenotypes (Figure 13.2).

Dey et al. (2015) made tremendous progress by utilizing single cells to measure both the RNA gene expression and copy number status. Also, Fan et al. (2015) was able to measure several single cells using DNA beads containing barcodes and microwells in parallel. Vera-Rodriguez et al. (2016) revealed the importance of single cell sequencing technology in biological heterogeneity. The authors suggested that these high-resolution properties of single cell approaches are possible due to ability to perform amplification, isolation of transcriptome, epigenome plus genome, together with NGS. The authors noted that the utilization of single cell sequencing approaches has expanded the fields of immunology, cancer biology, physiology, microbiology, and embryology and perinatal medicine.

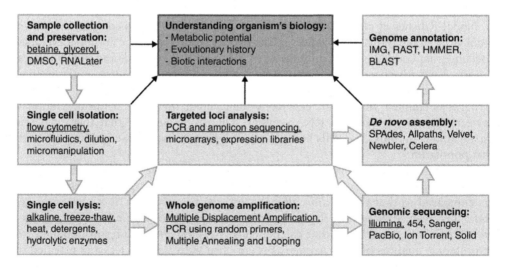

FIGURE 13.2 Methods and workflow of microbial single cell genomics (Reproduced with permission from Stepanauskas 2012, Copyright © Elsevier 2012).

Court et al. (2016) revealed that advancement in NGS together with single cell genomic study has been able to unravel the secrets behind biological diversity or heterogeneity. The authors also highlighted several shortcomings of single cell sequencing due to high cost and difficulty in holistic sequence of single cell epigenome, genome, or transcriptome. Again, to acquire a deeper understanding of temporal and spatial molecular measurement of single cells, real-time, *in vivo* or *in situ* analysis plus sequencing of the RNA or DNA must be obtained; hence, single-cell sequencing has become a powerful tool to unravel the complexity in nature and microbial physiology.

Rinke et al. (2014) revealed that without the advancement in single cell sequencing, it will be practically impossible to uncover most of the hidden microbial community as numerous microbial species have remained unidentified and uncultured until recent times. The advancement in single cell sequencing has helped to broaden the microbial phylogenetic tree resulting in a better understanding of taxonomy, evolution, and their basic physiology in the diverse environment and ecology in relation to human health and wellness (Figure 13.3).

Dean et al. (2002) showed that the expansion of single cell sequencing has enabled scientists to add reference genome to taxonomic groups across different microbial species. The authors revealed additionally that genomic polymorphism in different strains of human pathogenic microbes have been uncovered by whole-genome comparison using single-cell sequencing. Many designs of DNA amplification methods for bacteria pathogenic study are now been adapted for diploid cell analysis in human disease and health. Recently, novel whole genome amplification approaches in single cell are adapted to understand microbial eukaryotes plus multicellular or unicellular plants designed to improve understanding in many uncultured bacteria through assembly of DNA sequences in microbial ecology, evolution, plus physiology.

Zeng et al. (2018) highlighted that single cell sequencing is very important in the field of physiology for transcriptome classification of neuronal cell types resulting in the study of heterogeneity and the mechanism involved in neural function and development in health and diseases. Concerning the neurophysiology function using single cell sequencing, most of the study requires cells not in their natural environment so much of the spatial information is lost making detection ineffective. Nonetheless, the author noted that *in situ* single cell sequencing would resolve this challenge and will serve as a classical tool to cellular and molecular approaches.

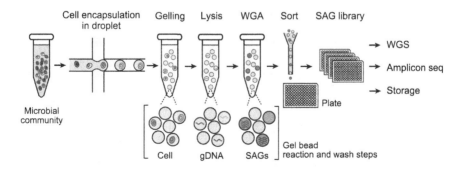

FIGURE 13.3 Workflow for SAG-gel-based single-cell genome sequencing of bacterial cells in a complex microbial community (Adapted with slight modification from Chijiiwa et al. 2020, Creative Commons Attribution 4.0 International).

Wang and Song (2017) revealed the role and capacity of single cell sequencing in the study of microbial or biological heterogeneity. Using high resolution, many advanced technologies such as NGS transcriptome, epigenome, and genome can be adapted together with single cell isolation to study the physiological properties of a single cell from a large ecological community. Basic medical research has utilized single cell sequencing to add knowledge and value to clinical applications of single cell sequencing. Coskun et al. (2016) showed that biological heterogeneity can be studied by utilizing advanced techniques, such as single cell genomic, epigenomic sequencing, and NGS in basic science. The author also noted that the high cost limits the availability of single cell sequencing, hence biomedical engineers should urgently develop low-cost effective single cell sequencing.

McLean and Lasken (2014) revealed that many diverse microbial species are unknown until recently when advancement in genomic study and molecular biology technologies became available to unravel the bacteria phyla and identify basic genetic constituents linked with the phenotypic and genotypic expression. Single-cell sequencing has been suggested to have enormous potential to significantly expand the knowledge in microbial phylogenetic family. Many studies have revealed that large scale analysis of the single-cell studies have gained significant insight into the human microbiome and taxonomic groups. This approach also ensures adequate comparison between genomic polymorphism across different strains and assembly with great opportunity to uncover multicellular or unicellular plants or microbial eukaryotes of notable pathogen.

McLean and Lasken (2014) demonstrated that single cell sequencing is an important tool in the study of infectious disease. Many new discoveries have been made concerning unknown species of pathogens in the environment through genomic analysis. The authors revealed that microbial genomics study using DNA sequence from single cells have provided numerous answers to many of the unknown gaps in research. Until recently, it was almost impossible to sequence genomic content from a single cell, but now many microbial phyla have been identified through gene sequencing and have been registered. The process involves isolation of cells, amplification of the copy number and subsequent whole-genome sequencing. With these, many pathogen genomes can be sequenced from biological samples without necessarily culturing them. Kathju et al. (2010) described how progress in single-cell genomics application has broadened and uncovered data from pathogenic microbes through direct sampling. Using single cell sequencing, pathogenic genomes in environment or biofilm can be uncovered.

13.4 CONCLUSION

This chapter delivers proper insights on some recent advancements in single cell sequencing of microbial species. This chapter also highlighted that single-cell sequencing could enable researchers to unravel the genomic profiling of the bulk of microbial populations by utilizing NGS based technologies. It is known that single-cell sequencing has great potential to uncover an array of microbial population and predict the direction of disease outbreak. Therefore, this chapter has provided detailed information on applications of recent advancements in single cell sequencing of microbial species. Specific examples were also given on applications of sequencing for adequate identification of microbial species from the single cells of fungi and bacteria.

REFERENCES

Ahrendt, S. R., Quandt, C. A., Ciobanu, D., et al. 2018. Leveraging single-cell genomics to expand the fungal tree of life. *Nature Microbiology* 3:1417–28.

Blainey, P. C. 2013. The future is now: Single-cell genomics of bacteria and archaea. *FEMS Microbiology Reviews* 37(3):1–29.

Chen, Z., Chen, L. and Zhang, W. 2017. Tools for genomic and transcriptomic analysis of microbes at single cell level. *Frontiers Microbiology* 8:1831. doi:10.3389/fmicb.2017.01831.

Chijiiwa, R., Hosokawa, M., Kogawa, M., et al. 2020. Single-cell genomics of uncultured bacteria reveals dietary fiber responders in the mouse gut microbiota. *Microbiome* 8:5. doi:10.1186/s40168-019-0779-2.

Coskun, A. F., Eser, U. and Islam, S. 2016. Cellular identity at the single-cell level. *Molecular Biosystems* 12:2965–79.

Court, C. M., Ankeny, J. S., Sho, S., et al. 2016. Reality of single circulating tumor cell sequencing for molecular diagnostics in pancreatic cancer. *Journal of Molecular Diagnostics* 18:688–96.

Dean, F. B., Seiyu, H., Linhua, F., et al. 2002. Comprehensive human genome amplification using multiple displacement amplification. *Proceedings of the National Academy of Sciences USA* 99:5261–6.

Dey, S. S., Kester, L., Spanjaard, B., Bienko, M. and van Oudenaarden, A. 2015. Integrated genome and transcriptome sequencing of the same cell. *Nature Biotechnology* 33:285–9.

Fan, H. C., Fu, G. K. and Fodor, S. P. 2015. Combinatorial labeling of single cells for gene expression cytometry. *Science* 347(6222):1258367. doi:10.1126/science.1258367.

Hwang, B., Lee, J. H. and Bang, D. 2018. Single-cell RNA sequencing technologies and bioinformatics pipelines. *Experimental and Molecular Medicine* 50:96. doi:10.1038/s12276-018-0071-8.

Kathju, S., Lasken, R. S., Satish, L., et al. 2010. Multiple displacement amplification as an adjunct to PCR-based detection of *Staphylococcus aureus* in synovial fluid. *BMC Research Notes* 3:259. doi:10.1186/1756-0500-3-259.

Khatri, I., Tomar, R., Ganesan, K., Prasad, G. S. and Subramanian, S. 2017. Complete genome sequence and comparative genomics of the probiotic yeast *Saccharomyces boulardii*. *Scientific Reports* 7:371. doi:10.1038/s41598-017-00414-2.

Kuchina, A., Brettner, L. M., Paleologu, L., et al. 2019. Microbial single-cell RNA sequencing by split-pool barcoding. *Bio-Archive* 869248. doi:10.1101/869248.

Lake, B. B., Ai, R., Kaeser, G. E., et al. 2016. Neuronal subtypes and diversity revealed by single-nucleus RNA sequencing of the human brain. *Science* 352:1586–90.

Lasken, R. S. 2012. Genomic sequencing of uncultured microorganisms from single cells. *Nature Reviews Microbiology* 10(9):631–40.

Lasken, R. S. and McLean, J. S. 2014. Recent advances in genomic DNA sequencing of microbial species from single cells. *Nature Reviews Genetics* 15(9):577–84.

Li, L. and Clevers, H. 2010. Coexistence of quiescent and active adult stem cells in mammals. *Science* 327: 542–5.

McLean, J. S. and Lasken, R. S. 2014. Single cell genomics of bacterial pathogens: Outlook for infectious disease research. *Genome Medicine* 6:108. doi:10.1186/s13073-014-0108-0.

Nadal-Ribelles, M., Islam, S., Wei, W., et al. 2019. Sensitive high-throughput single-cell RNA-seq reveals within-clonal transcript correlations in yeast populations. *Nature Microbiology* 4(4):683–92.

Rinke, C., Lee, J., Nathet, N., et al. 2014. Obtaining genomes from uncultivated environmental microorganisms using FACS-based single-cell genomics. *Nature Protocols* 9(5):1038–48.

Saint, M., Bertaux, F., Tang, W., et al. 2019. Single-cell imaging and RNA sequencing reveal patterns of gene expression heterogeneity during fission yeast growth and adaptation. *Nature Microbiology* 4(3):480–91.

Stepanauskas, R. 2012. Single cell genomics: An individual look at microbes. *Current Opinion in Microbiology* 15(5):613–20.

Tasic, B., Menon, V., Nguyen, T. N., et al. 2016. Adult mouse cortical cell taxonomy revealed by single cell transcriptomics. *Nature Neuroscience* 19:335–46.

Tyml, T., Date, S. V. and Woyke, T. 2019. A single-cell genome perspective on studying intracellular associations in unicellular eukaryotes. *Philosophical Transactions of the Royal Society B Biological Sciences* 374(1786):20190082. doi:10.1098/rstb.2019.0082.

Vera-Rodriguez, M., Michel, C. E., Mercader, A., et al. 2016. Distribution patterns of segmental aneuploidies in human blastocysts identified by next-generation sequencing. *Fertility and Sterility* 105(4):1047–55.

Wang, J. and Song, Y. 2017. Single cell sequencing: A distinct new field. *Clinical and Translational Medicine* 6:10. doi:10.1186/s40169-017-0139-4.

Woolfit, M., Rozpędowska, E., Piškur, J. and Wolfe, K. H. 2007. Genome survey sequencing of the wine spoilage yeast *Dekkera (Brettanomyces) Bruxellensis*. *Eukaryotic Cell* 6:721–33.

Wu, B., Hussain, M., Zhang, W., Stadler, M., Liu, X. and Xiang, M. 2019. Current insights into fungal species diversity and perspective on naming the environmental DNA sequences of fungi. *Mycology* 10:127–40.

Zeng, Z., Miao, N. and Sun, T. 2018. Revealing cellular and molecular complexity of the central nervous system using single cell sequencing. *Stem Cell Research and Therapy* 9:234. doi:10.1186/s13287-018-0985-z.

Zhang, Y. J., Zhang, S., Liu, X. Z., Wen, H. A. and Wang, M. 2010. A simple method of genomic DNA extraction suitable for analysis of bulk fungal strains. *Letters in Applied Microbiology* 51:114–8.

PART **III**

High-Throughput Sequencing Technology Applications for Food Safety and Quality

Current Trends in the Detection of Foodborne Pathogens and Maintenance of Microbial Food Safety and Quality Based on Ribosomal RNA Approaches

*Charles Oluwaseun Adetunji, Ayodele Eugene Ayeni,
Olugbemi Tope Olaniyan, Muhammad Akram,
Benjamin Ewa Ubi, Saher Islam, Umme Laila,
Mehvish Ajaz, Wadzani Dauda Palnam,
Devarajan Thangadurai, Olugbenga Samuel Michael,
Juliana Bunmi Adetunji, Fahad Said Khan, and Daniel Ingo Hefft*

CONTENTS

14.1 INTRODUCTION

Current trends indicate that foodborne pathogens are still a major public health concern with the emergence or re-emergence of novel chemical or biological pathogens. Most of these illnesses are generated from viruses, protozoa, and bacteria, and other food biohazards like mycotoxins, veterinary drugs, prions, environmental contaminants, and pesticide residues (Motarjemi and Käferstein 1997). It has been observed that *Shigella* spp. is found in the intestinal tract of human beings. *Streptococcus pyogenes* is normally carried in the throat of asymptomatic carriers or diseased persons. *Staphylococcus aureus* is normally found on the skin and in the nose of human beings. It has been discovered that foodborne viruses have been recognized as a serious public health concern that show up in the human intestine. Also, the majority of the food that favors the multiplication of *Salmonella typhi* are normally found in food that requires no additional cooking. Numerous outbreaks of foodborne diseases could be linked to the presence of viruses and *Shigella* spp., while the majority of staphylococcal foodborne illness are normally associated with salty, high-protein, and moist foods (Ollinger-Snyder and Matthews 1996).

A few RNAs that are not encrypted, especially small RNAs (sRNAs) have emerged as the main controllers for many cell forms in multicellular eukaryotes (Garzon et al. 2009). Moreover, microRNA (miRNA) profiles have become another diagnostic tool (Dalmay 2008; Negrini et al. 2009). Although profiles of miRNA possess less information than mRNA, the potential value can be higher due to the regulatory function of miRNAs (Dalmay 2008). sRNAs have been tested for other gram-negative microorganisms, including *Salmonella typhimurium* and *Pseudomonas aeruginosa* (Sharma and Vogel 2009). Moreover, some other sRNAs are involved in stress reactions, iron homeostasis, the outer layer protein biosphere, and sugar digestion and play a fundamental and central role in the detection of many microorganisms. Therefore, this chapter intends to provide comprehensive information and current trends in the detection of foodborne pathogen and maintenance of microbial food safety and quality based on ribosomal RNA approaches.

14.2 FOODBORNE PATHOGENS

Motarjemi and Käferstein (1997) revealed that many parameters significantly contribute to the emergence of foodborne pathogens, including alterations of pathogens, urbanization, development, new lifestyles, health systems, beliefs and practices, existing knowledge, demographic changes, trade in food, travel, migration, animal feed, poverty, and pollution. Studies have shown that the spread of these pathogens may likely continue in the coming years if efforts are not unified by all countries to stem or control foodborne pathogens (Käferstein 1997). Aznar and Alarcón (2002) described the main species of bacteria pathogens responsible for seafood spoilage. They used molecular biology techniques for a faster, more reliable, and reproducible detection of different pathogenic microbes. The author pointed out that studies of the genus *Vibrio* spp. have found it to be responsible for the outbreak of human pathogenic illness from the consumption of improperly prepared shellfish. Their study indicated that the recent advancements in molecular biology techniques have progressively changed and enhanced surveillance monitoring in public health and also facilitated the gross reduction in seafood-borne pathogens.

Tam et al. (2012) showed that traditional approaches utilized in the evaluation of foodborne pathogenic outbreaks are generally not efficient and effective. Recently, many emerging techniques and technologies for the evaluation and understanding of foodborne disease outbreaks have emerged onto the scene. Polymerase chain reaction (PCR) has proven to be one of the modern and sophisticated ways of analysis in real-time outbreak situations. This has been applied as a frontline tool and significantly improved the outbreak investigation process. Almeida et al. (2010) revealed that the next-generation sequencing approach in combination with computational biology, such as bioinformatics, has led to significant developments in the identification of foodborne pathogens, particularly in meat products. It has been affirmed that the majority of the results documented from previous studies in the past during the assessment and identification of pathogenic microbes was monitored utilizing the classic microbiological techniques such as culturing, which could only analyze cultivable microorganisms; hence, a large portion of microbiota went undetected. Today, the application of molecular microbiology techniques has played a role in numerous advancements toward the detection of foodborne pathogens. Specific examples of researchers/authors that have utilized the 16S ribosomal RNA approach for the identification of food and pathogens are highlighted in the following sections.

14.2.1 *Listeria monocytogenes*

L. monocytogenes can develop serious infections in animals and humans (but mostly animals) (Hain et al. 2007; Graves et al. 2009). As far as the population is concerned, most *Listeria* cases were documented through complex feeding. The main clinical indicators that appear in the human body include meningitis, sepsis, premature birth, prenatal illness, and gastroenteritis. Pregnancy was considered in approximately 20%–30% of patients with listeriosis, regardless of appropriate antimicrobial therapy (Hof et al. 2007). Numerous scientists have applied ribosomal RNA approaches for the detection of different species of *L. monocytogenes* (Johansson et al. 2002; Wong et al. 2004; Shen and Higgins 2005; Stritzker et al. 2005; Hain et al. 2007; Cossart and Toledo-Arana 2008). Moreover, some other authors have also proposed the untranslated RNA proximity for surface proteins involved in the disruption of monocytogenes and suggested the post-description of long RNA groups in pathogens (Loh et al. 2006).

14.2.2 *Staphylococcus aureus*

This microorganism is one of the leading drivers of disease in treatment clinics around the world. Due to the large number of clinical crises and diseases caused by this microbe, a rigorous, new, and effective approach to treatment is required. Several authors have applied ribosomal RNA approaches for the detection of *Staphylococcus aureus* in food samples (Guvakova et al. 1995; Morfeldt et al. 1995; Dunman et al. 2001; Huntzinger et al. 2005; Pichon and Felden 2005; Boisset et al. 2007).

14.2.3 *Streptococcus pyogenes*

Streptococcus pyogenes, or Group A *streptococcus* (GAS) is another bacterial organism important only to humans. Most gram-positive species of bacteria of note can cause death and other grievous effects (Courtney et al. 2002; Carapetis et al. 2005). Numerous streptococcal contaminations of food can be linked to improper handling and storage of cold foods, most especially salads. *Streptococcus pyogenes* seems to pass from hand lesions and the pharynx of the food handlers. Some symptoms associated with *Streptococcus pyogenes* include submandibular lymphadenopathy, sore throat, enlarged tonsils, and pharyngeal erythema. In view of the aforementioned, Katzenell et al. (2001), conducted work that focused on the streptococcal contamination of food such as salad–especially those containing eggs. The authors stated that a limited incubation period and a higher level of attack rate (51%–90%) than in transmission by droplets was observed. It was documented that 7 out of 17 reports showed an M-untypeable serotype of virulent nature. The authors also suggested that food handlers should ensure that they follow all guidelines for the prevention of food poisoning associated with foodborne streptococcal pharyngitis as well as the necessary rules for food storage and preparation; most especially, cold salads containing eggs should not be stored overnight before serving.

14.2.4 *Enterococcus faecalis*

These organisms have been identified as one of the major pathogens responsible for foodborne hazards (Dulebohn et al. 2007). Although is it naturally found in the human intestine, it has been cited as a post-nosocomial pathogen due to high antibiotic resistance (Ogier and Serror 2008). It is also responsible for several disorders–most especially in young people–which include endocarditis, gonorrhea, pneumonia, peritonitis, abdominal pain, urinary incontinence, and sepsis (Eaton and Gasson 2001; Gilmore et al. 2002; Paulsen et al. 2003; Verneuil et al. 2004; Verneuil et al. 2005; Riboulet-Bisson et al. 2008).

14.2.5 *Clostridium difficile*

Clostridium difficile has been identified as one of the microorganisms that affects humans and animals as a result of contamination. Typical symptoms caused by this strain include pseudo-membrane colitis and abdominal pain (Rupnik et al. 2009).

14.3 16S RIBOSOMAL RNA APPROACHES FOR THE IDENTIFICATION OF FOODBORNE PATHOGENS

The 16S ribosomal RNA (16S rRNA) genes in bacteria are predominantly utilized for the evaluation of the phylogenetic relatedness and identification of bacterial species identification; it is a standard method widely used to identify and classify microorganisms. The 16S rRNA sequence analysis has the advantage of being simple compared to culture methods and does not require special conditions for each bacterial species; it is usually used in characterization studies and targeted in the molecular surveillance of microbial communities. *Shigella* species, *Escherichia coli*, and serotypes of *Salmonella enterica* are examples that may be identified using 16S rRNA. The 16S rRNA contains highly conserved and hypervariable regions. The hypervariable regions are mostly chosen for assessment of taxonomic archaeal and bacterial diversity and are usually assessed using electrophoretic methods to identify the distinct bands of individual species (Kakinuma et al. 2003; Lin et al. 2004; Mayo et al. 2014; Lee et al. 2019).

Axelsson et al. (2012) revealed that meat and cereal-based fermented food products are processed by local microflora in the environment. The authors noted that recent advancements in molecular biology techniques have brought tremendous progress to the identification of many microbial pathogens in meat and cereal products. Some of these techniques include PCR, 16S rDNA sequencing, repetitive element PCR, denaturing gradient gel electrophoresis, restriction fragment length polymorphism, pulsed-field gel electrophoresis (PFGE), matrix-assisted laser desorption/ionization-time of flight mass spectrometry plus restriction endonuclease analysis–PFGE. El-Hadedy and El-Nour (2012) studied ready-made vegetable salads and utilized molecular approaches to identify the presence of microorganisms such as *Staphylococcus aureus* and *Escherichia coli*.

Hellberg et al. (2013) reported 16S rRNA partial sequencing for the identification of *Listeria* at the species level. The authors built a 16S rRNA partial gene reference library for *Listeria* spp. differentiation and molecular subtyping. The exclusivity and spiking tests proved effective for their purposes, as the spiking results allowed species differentiation. Microbial spoilage in food can be monitored during storage periods by monitoring the metabolic activities of microorganisms during those times, as well as monitoring the microbial diversity of foods during food processing. Abalunan et al. (2013) suggested that it is pertinent to adopt ribosomal DNA analysis produced by PCR-denaturing gradient gel electrophoresis to detect microbial populations inherent to these fermented foods and proffer possible solutions. This emerging area of biotechnology utilizes barcodes imprinted in fermented foods for possible tracing of pathogens or any unwanted food hazards.

Cauchie et al. (2016) used a culture-dependent and culture-independent metagenetic approach based on high-throughput 16S rRNA gene sequencing to examine the genetic diversity of white pudding over different refrigeration storage periods. Similarly, Lin et al. (2004) conducted a study using 16S rRNA gene to identify *Salmonella* serovars. Meat obtained from poultry is one of the most popular meat products worldwide, hence there is a need for quality assessment and food safety checks. Multiple pathogens using a 16S rRNA-derived geno-biochip were used by Shin et al. (2016) to simultaneously identify, *Bacillus cereus*, *Campylobacter jejuni*, *Clostridium perfringens*, *Escherichia coli* O157:H7, *Listeria monocytogenes*, *Salmonella enterica*, *Shigella dysenteriae*, *Staphylococcus aureus*, *Vibrio cholerae*, and *Yersinia enterocolitica*. Heperkan et al. (2016) carried out another study on powdered infant formula (PIF) and follow-up formula using PCR with partial 16S rRNA gene sequencing to identify *Cronobacter sakazakii*. Although other pathogenic bacteria were absent in the samples, *Cedecea lapagei*, an opportunistic pathogen was identified in the PIF. The authors suggested that results obtained from their study necessitate more effective control measures in the production of infant foods which can prevent all these foodborne pathogens.

Alshaikh et al. (2017) reported the isolation of pathogenic bacteria present in baby food using biochemical tests and 16S rRNA analysis. The analysis showed the presence of three strains of bacteria identified as pathogens using biochemical tests in the five food samples tested. The pathogens were identified to the species level as *Bacillus cereus*, which can be a threat to infant health. Alikunhi et al. (2017) reported the evaluation and identification of bacteria in 13 different fish from three coastal sites and a local fish market in Jeddah, Saudi Arabia. The authors approached the study using culture-dependent and 16S rRNA identification. Based on 16S rRNA sequencing, *Rahnella aquatilis* and *Photobacterium damselae*, known as human pathogens capable of causing foodborne illness and severe antibiotic resistance, were among the dominant species from fish muscle. Others, such as *Hafnia* sp. and *Pseudomonas stutzeri*, also occurred in the fish muscle samples tested.

Tilahun et al. (2018) also reported the utilization of 16S rRNA gene in the identification of lactic acid bacteria. Teff is used for the production of injera, a soft fermented local food among the Ethiopians. The molecular investigation was carried out on 97 teff dough samples collected from different sources at different fermentation stages. Acid production potentials were used as the basis for the selection of bacteria for molecular characterization. Karatuğ et al. (2018) reported the characterization of *Salmonella* isolates, a very common bacterial pathogen on food. Four molecular characterization approaches were employed including 16S rRNA sequencing, PCR restriction fragment length polymorphism, PFGE, and invasion A (*invA*)–*Salmonella* plasmid virulence C (*spvC*) gene PCR. The 38 food isolates of *Salmonella* employed in this study had been characterized by biochemical tests prior to the study. Surprisingly, 16S rRNA sequencing results showed that four of the 38 isolates were *Escherichia coli*, *Proteus mirabilis*, *Citrobacter murliniae*, and the others were *Salmonella enterica*. All but two of the isolates (94%) had the *invA* gene, and 11 % had the *spvC* gene. Both *invA* and *spvC* are target genes utilized for the molecular identification of *Salmonella* spp., but *spvC* is plasmid-borne and *invA* is found in the *Salmonella* chromosome. Interestingly, the *E. coli*, *P. mirabilis,* and *C. murliniae* isolates also displayed the *invA* gene, revealing evidence of a horizontal kind of gene transfer among *Enterobacteriaceae* family members. This study illustrates the superior features of molecular methods of identification as opposed to biochemical assay, showing properties in bacteria, which would otherwise be unknown.

Yu et al. (2019) assessed the incubation impact at 30°C and 7°C on overall aerobic bacterial diversity and count and evaluated the performance of ISO methods for microbiological quality checks, using culture combined matrix assisted laser desorption/ionization-time of flight mass spectroscopy (MALDI-TOF MS) detection and 16S rRNA sequencing. Sixteen chicken carcasses were identified and no substantial impact of incubation temperature on total aerobic bacterial diversity and level was detected. For the 16S rRNA sequencing, nucleic acid extraction was carried out, and then the V3-V4 region of 16S rRNA gene was amplified. A high bacterial diversity was revealed by the 16S rRNA sequencing and MALDI-TOF MS, whereas culture revealed the presence of *Escherichia coli, Staphylococcus,* and bacteria phenotypically similar to *Pseudomonas*. Chukwu et al. (2019) reported the use of 16S rRNA gene sequencing, culture-dependent, and biochemical methods to identify foodborne pathogens from products sold in Lagos, Nigeria. The authors isolated 30 isolates which included 29 that were subjected to 16S rRNA gene sequencing to establish their genus level (96.7%), while 26 isolates were identified to the species level, and 1 was unclassified due to a large number of undetermined nucleotides. The study validated the accuracy of 16S rRNA as an important gene in identifying foodborne pathogens.

Most recently, Câmara et al. (2020) reported a study identifying Pico cheese as a reservoir for *Enterococcus*, a species that possess antibiotic resistance and virulence properties. The study was performed via genotypic and phenotypic testing for the virulence property using *cylA, esp, asa1, gelE, tdc, odc, efaA, ace, hdc1,* and *hdc2* as gene markers. Other physiological tests included the production of DNase, gelatinase, and biogenic amine as well as hemolysis and antibiotic resistance. All *Enterococci* tests displayed antibiotic resistance to aminoglycosides, and two of the isolates showed vancomycin resistance that is not attributable to *vanB* and *vanA*. One of the isolates was β-hemolytic, none produced DNase, and 54 % produced gelatinase. The most abundant virulence genes were *gelE, efaA, tdc,* and *ace*. Important to note was that no isolate lacked virulence factors, thereby making them inappropriate to use in adjunct cultures for Pico cheese. The results highlighted the importance of assessing *E. faecalis* isolates from the artisanal cheese for food safety implications. These studies revealed the importance of 16S rRNA in the molecular identification and characterization of microorganisms for food safety and quality assessment, with particular emphasis on foodborne pathogens, and then others that contribute to food enhancement, such as lactic acid bacteria.

14.4 SMALLER RNAs (sRNAs) FOR MICROBIAL DETERMINATION

Newly discovered RNA arrangements that are linked to the destructive components of various gram-positive microorganisms should not be used exclusively as potential anti-effect drugs, and furthermore, they are misused as focal point guides. In this regard, sRNAs can be of benefit, in contrast to mRNA or oligonucleotides DNA, given their supposed administrative capacity. The potential for this set of mixtures as molecular markers is currently a dynamic area of research. The text offers many different configurations for rapid demonstration tests that rely on corrosive nuclear links (Waters and Storz 2009).

14.4.1 Hfq-Binding sRNA: Recent Discoveries

It has been highlighted that increases in the level of expression of sRNA in microorganisms results from the synthesis of basic and clear mRNAs, which enhance or inhibit protein synthesis in experiments (Waters and Storz 2009). Cis-bag sRNAs are all connected to their mRNA, although trans-bag sRNAs are only half-connected to their complex mRNAs. In the gram-negative microorganisms, the interaction between permanent sRNA targets is usually dependent on RNA, for example, Hfq, the sRNA-mRNA duplex develops. But in most gram-positive microorganisms, the role of Hfq in the regulation of sRNA interference is less clear (Bouloc and Repoila 2016). However, Hfq increases resistance and weight loss on gram-positive *L. monocytogenes* (Christiansen et al. 2004) and gram-negative *Salmonella* spp. (Sittka et al. 2007). At the beginning of their study, Rf-binding properties of Hfq were effectively used as a target to detect sRNA (Christiansen et al. 2006; Sittka et al. 2008). Although LhrA and GcvB have different causes, quantities, and nucleotide agents, the two depend on Hfq for potency and regulatory control (Sharma et al. 2007; Nielsen et al. 2010; Nielsen et al. 2011; Dugar et al. 2013; Taveirne et al. 2013). Subsequently, several bioinformatics analyses were conducted to evaluate the expression of sRNAs in cells including *E. coli*, *L. monocytogenes*, *Bacillus subtilis*, and *Pseudomonas aeruginosa*, while some of these techniques have been used for the identification of more than 200 sRNAs (Livny and Waldor 2007). In similar cases, experimental studies have already been conducted that include cDNA sequencing and RNA inactivation, in some cases, leading to the identification of new transcription (Sharma and Vogel 2009; Sharma et al. 2010). Micro-matrix insertion is a great way to adapt to genome-wide sRNA detection. Thus, surface and next RNA have been applied in the detection of foodborne pathogenic foods (Landt et al. 2008; Perez et al. 2009; Toledo-Arana et al. 2009; Kumar et al. 2010).

14.5 CURRENT TRENDS IN MICROBIAL FOOD SAFETY

Noticeable progress has been achieved in the past two decades in the area of detection and monitoring, but many areas still need to be addressed to achieve a total reduction of foodborne pathogens. It has been established that traditional approaches do not compare with conventional methods in terms of rapid, efficient, accurate, and reliable detection of pathogens in food samples. The application of genomic techniques entails the use of PCR, microarray technology, and next-generation assays. Eley et al. (1992) asserted that public safety in terms of foodborne pathogens requires rapid detection and monitoring. In recent times nucleic acid-based PCR combined with reverse-transcriptase PCR, based on messenger RNA has been utilized for effective detection. These techniques are culture-independent approaches employed in the detection of pathogenic organisms. Moreover, the authors demonstrated that phage-based methods show more promise due to high speed, cost, and sensitivity for food testing. Eley (1992) showed that a reduction in the amount of time needed for the detection of pathogens in food is linked to the nucleic acid probe technology, which enhances specificity and cost effectiveness. The author revealed that the utilization of the ELISA technique will not give the desired result due to low specificity, but the amplification of DNA sequencing using PCR would considerably ensure a higher speed in the level of bacteria detection in food samples.

Callon et al. (2006) revealed that molecular biology techniques have led to many advancements in biological and medical science research through the utilization of modern techniques for effective detection and control of foodborne pathogens in dairy products. These new techniques have helped the food industry to control the food production chain and increase the level of safety in food samples. RNA-based RT-PCR techniques have provided numerous data, especially on the level of metabolic activity of the human microbiota and ecosystem, utilizing culture-dependent or independent molecular approaches. Gene sequencing, metagenomics, and transcriptomics are promising tools utilized for microbial diversity research in dairy food products. Chen and Knabel (2007) described how the implementation of food policies and regulations could be deployed in the food industry that would result in the reduction of foodborne pathogens. Gilbert et al. (2008) revealed that in the field of metagenomics, sequencing is not the challenge, but data storage and computation together with the high cost of running metagenomic sequences is. The authors assert that the depth of coverage relies on the type of sequencing technology and the fact that, for many years, Sanger technology has brought about advancements in metagenomics analysis.

Ludmila (2009) provided detailed information on the application of metagenomics that could provide adequate knowledge on microbial communities, environmental microbes, and their physiological processes. This tool enables scientists to use high-resolution genomic analysis to quickly connect specific phyla to particular functions in the environment. Al-Haggar et al. (2013) discussed how genomic investigations using high-throughput sequencing could generate large amounts of data within a short time. The authors agreed that gene expression study, epigenetics, transcriptomics together with genomics involves the use of next-generation DNA sequencing techniques in a short period. Some merits of this technique include the reduction in the cost of analyzing samples. RNA sequencing has made a tremendous impact on gene expression projects globally with many methods of standardizing RNA sequencing data. Valiollahi et al. (2014) described how real-time monitoring and nanopore sequencing as the latest advancement in third-generation sequencing tools is being utilized for fluorescent resonant energy transfer due to simplicity, scalability, increased DNA polymerase products, reduced miscalculation, and better practicability and efficiency.

Manisha et al. (2015) indicated that generally across the globe, food biosafety has increasingly become an area of concern due to infection caused by the consumption of foodborne pathogens. The authors assert that the primary goal of a scientist is to reduce the time for the detection and monitoring of pathogens across the food production chain. Available evidence suggests a few methods, such as PCR, enzyme immunoassay, microarrays, plus flow cytometry. Alcaine et al. (2015) commented on the search for more rapid, sensitive, and viable ways of analyzing foodborne pathogens compared to the less effective and more laborious methods involved in traditional culture-based approaches. The authors explained further that mRNA-based tests utilizing reverse transcriptase PCR combined with phage amplification and lysis, enzyme assay endpoint detection, and immunoassay method seem to be the solution to analyzing food microbes successfully, even though this method is highly susceptible to mRNA fragility. Nguyen et al. (2016) reported a multiplex PCR procedure for simultaneous detection of *E. coli* O157:H7, *Salmonella* spp., and *L. monocytogenes* in food. In addition to the PCR, the authors also used conventional culture and biochemical methods to identify the pathogens. The virulence genes *invA*, *stx*, and *hlyA* were respectively used as targets for the three pathogens. Because of the efficient identification of the three pathogens, the multiplex PCR assay has shown potential for use in routine diagnostic laboratories and as a rapid screening tool in food testing. Pervaiz et al. (2017) revealed that, due to advancements in DNA sequencing and evaluation of data sets in real-time scenarios, several classifications, mutation detections, drug selections, and molecular diagnoses have been done on many healthy and diseased cells.

14.6 CONCLUSION

This chapter has provided detailed information on current trends in the detection of foodborne pathogens and maintenance of microbial food safety and quality based on ribosomal RNA approaches. In addition, numerous challenges surrounding the application of conventional culture-based techniques have given way to more current quick, sensitive, and reliable tools, such as molecular techniques. Screening food products for the presence of pathogens is currently moving toward culture-independent approaches to utilize nucleic acids of the organisms. The amplification of the 16S rRNA in collected samples has made the identification of pathogens a straightforward event. Although several genes are used for identification, the 16S rRNA has thus far proven to be the most reliable due to its advantage of identifying and differentiating microbial communities in food products. Moreover, there is a need to apply significant techniques, such as hazard analysis and critical control points systems, that could minimize the occurrence of foodborne disease attributed to food handlers, through training, education programs, and certification of foodservice managers.

REFERENCES

Abalunan, A. J. F., Teves, F. G. and Madamba, M. R. S. B. 2013. Isolation of fungal species and aflatoxin detection in fermented products. *International Research Journal of Biological Sciences* 2(4):51–4.

Alcaine, S. D., Tilton, L., Serrano, M. A., Wang, M., Vachet, R. W. and Nugen, S. R. 2015. Phage-protease-peptide: A novel trifecta enabling multiplex detection of viable bacterial pathogens. *Applied Microbiology and Biotechnology* 99(19):8177–85.

Al-Haggar, M., Khair-Allaha, B., Islam, M. and Mohamed, A. 2013. Bioinformatics in high throughput sequencing: Application in evolving genetic diseases. *Journal of Data Mining in Genomics and Proteomics* 4(3). doi:10.4172/2153-0602.1000131.

Alikunhi, N. M., Batang, Z. B., AlJahdali, H. A., Aziz, M. A. M. and Al-Suwailem, A. M. 2017. Culture-dependent bacteria in commercial fishes: Qualitative assessment and molecular identification using 16S rRNA gene sequencing. *Saudi Journal of Biological Sciences* 24(6):1105–16.

Almeida, C., Azevedo, N. F., Fernandes, R. M., Keevil, C. W. and Vieira, M. J. 2010. Fluorescence in situ hybridization method using a peptide nucleic acid probe for identification of *Salmonella* spp. in a broad spectrum of samples. *Applied and Environmental Microbiology* 76(13):4476–85.

Alshaikh, K. S., Mutwakil, M. H. and Ahmed, M. M. M. 2017. Isolation and identification of *Bacillus cereus* from infant food using 16S rRNA sequence. *Advances in Biological Research* 11(1):1–5.

Axelsson, L., Rud, I., Naterstad, K., et al. 2012. Genome sequence of the naturally plasmid-free *Lactobacillus plantarum* strain NC8 (CCUG 61730). *Journal of Bacteriology* 194(9):2391–2.

Aznar, R. and Alarcón, B. 2002. On the specificity of PCR detection of *Listeria monocytogenes* in food: A comparison of published primers. *Systematic and Applied Microbiology* 25(1):109–19.

Boisset, S., Geissmann, T., Huntzinger, E., et al. 2007. *Staphylococcus aureus* RNAIII coordinately represses the synthesis of virulence factors and the transcription regulator Rot by an antisense mechanism. *Genes and Development* 21(11):1353–66.

Bouloc, P. and Repoila, F. 2016. Fresh layers of RNA-mediated regulation in gram-positive bacteria. *Current Opinion in Microbiology* 30:30–5.

Callon, C., Delbès, C., Duthoit, F. and Montel, M. C. 2006. Application of SSCP-PCR fingerprinting to profile the yeast community in raw milk salers cheeses. *Systematic and Applied Microbiology* 29(2):172–80.

Câmara, S. P. A., Dapkevicius, A., Silva, C. C. G., Malcata, F. X. and Dapkevicius, M. L. N. E. 2020. Artisanal Pico cheese as reservoir of *Enterococcus* species possessing virulence and antibiotic resistance properties: Implications for food safety. *Food Biotechnology* 34(1):25–41.

Carapetis, J. R., Steer, A. C., Mulholland, E. K. and Weber, M. 2005. The global burden of group A streptococcal diseases. *The Lancet Infectious Diseases* 5(11):685–94.

Cauchie, E., Gand, M., Kergourlay, G., et al. 2016. The use of 16S rRNA gene metagenetic monitoring of refrigerated food products for understanding the kinetics of microbial subpopulations at different storage temperatures: The example of white pudding. *International Journal of Food Microbiology* 247:70–8.

Chen, Y. and Knabel, S. J. 2007. Multiplex PCR for simultaneous detection of bacteria of the genus *Listeria*, *Listeria monocytogenes*, and major serotypes and epidemic clones of *L. monocytogenes*. *Applied and Environmental Microbiology* 73(19):6299–304.

Christiansen, J. K., Larsen, M. H., Ingmer, H., Sogaard-Andersen, L. and Kallipolitis, B. H. 2004. The RNA-binding protein Hfq of *Listeria monocytogenes*: Role in stress tolerance and virulence. *Journal of Bacteriology* 186(11):3355–62.

Christiansen, J. K., Nielsen, J. S., Ebersbach, T., Valentin-Hansen, P., Sogaard-Andersen, L. and Kallipolitis, B. H. 2006. Identification of small Hfq-binding RNAs in *Listeria monocytogenes*. *RNA* 12(7):1383–96.

Chukwu, E. E., Nwaokorie, F. O., Coker, A. O., Avila-Campos, M. J. and Ogunsola, F. T. 2019. 16S rRNA gene sequencing: A practical approach to confirming the identity of food borne bacteria. *IFE Journal of Science* 21(3):13–24.

Cossart, P. and Toledo-Arana, A. 2008. *Listeria monocytogenes*, a unique model in infection biology: An overview. *Microbes and Infection* 10(9):1041–50.

Courtney, H. S., Hasty, D. L. and Dale, J. B. 2002. Molecular mechanisms of adhesion, colonization, and invasion of group A streptococci. *Annals of Medicine* 34(2):77–87.

Dalmay, T. 2008. MicroRNAs and cancer. *Journal of Internal Medicine* 263(4):366–375.

Dugar, G., Herbig, A., Forstner, K. U., et al. 2013. High-resolution transcriptome maps reveal strain-specific regulatory features of multiple *Campylobacter jejuni* isolates. *PLoS Genetics* 9(5):e1003495. doi:10.1371/journal.pgen.1003495.

Dulebohn, D., Choy, J., Sundermeier, T., Okan, N. and Karzai, A. W. 2007. Trans-translation: The tmRNA-mediated surveillance mechanism for ribosome rescue, directed protein degradation, and nonstop mRNA decay. *Biochemistry* 46(16):4681–93.

Dunman, P. M., Murphy, E., Haney, S., et al. 2001. Transcription profiling-based identification of *Staphylococcus aureus* genes regulated by the *agr* and/or *sarA* loci. *Journal of Bacteriology* 183(24):7341–53.

Eaton, T. J. and Gasson, M. J. 2001. Molecular screening of *Enterococcus* virulence determinants and potential for genetic exchange between food and medical isolates. *Applied and Environmental Microbiology* 67(4):1628–35.

Eley, A. 1992. *Microbial food poisoning*, 1st ed. New York: Springer.

Eley, A., Oxley, K. M., Spencer, R. C., Kinghorn, G. R., Ben-Ahmeida, E. T. and Potter, C. W. 1992. Detection of *Chlamydia trachomatis* by the polymerase chain reaction in young patients with acute epididymitis. *European Journal of Clinical Microbiology and Infectious Diseases* 11(7):620–3.

El-Hadedy, D. and El-Nour, S. A. 2012. Identification of *Staphylococcus aureus* and *Escherichia coli* isolated from Egyptian food by conventional and molecular methods. *Journal of Genetic Engineering and Biotechnology* 10(1):129–35.

Garzon, R., Calin, G. A. and Croce, C. M. 2009. MicroRNAs in cancer. *Annual Review of Medicine* 60:167–79.

Gilbert, J. A., Field, D., Huang, Y., et al. 2008. Detection of large numbers of novel sequences in the metatranscriptomes of complex marine microbial communities. *PLoS ONE* 3(8):e3042. doi:10.1371/journal.pone.0003042.

Gilmore, M. S., Coburn, P. S., Nallapareddy, R. S. and Murray, B. E. 2002. Enterococcal virulence. In: *The enterococci: Pathogenesis, molecular biology, and antibiotic resistance*, eds. M. S. Gilmore and D. B. Clewell, 301–54. Washington, DC: ASM Press.

Graves, L. M., Helsel, L. O., Steigerwalt, A. G., et al. 2009. *Listeria marthii* sp. nov., isolated from the natural environment, Finger Lakes National Forest. *International Journal of Systematic and Evolutionary Microbiology* 60(6):1280–8.

Guvakova, M. A., Yakubov, L. A., Vlodavsky, I., Tonkinson, J. L. and Stein, C. A. 1995. Phosphorothioate oligodeoxy nucleotides bind to basic fibroblast growth factor, inhibit its binding to cell surface receptors, and remove it from low affinity binding sites on extracellular matrix. *The Journal of Biological Chemistry* 270(6):2620–7.

Hain, T., Chatterjee, S. S., Ghai, R., et al. 2007. Pathogenomics of *Listeria* spp. *International Journal of Medical Microbiology* 297(7–8):541–57.

Hellberg, R. S., Martin, K. G., Keys, A. L., Haney, C. J., Shen, Y. and Smiley, R. D. 2013. 16S rRNA partial gene sequencing for the differentiation and molecular subtyping of *Listeria* species. *Food Microbiology* 36(2):231–40.

Heperkan, D., Dalkilic-Kaya, G. and Juneja, V. K. 2016. *Cronobacter sakazakii* in baby foods and baby food ingredients of dairy origin and microbiological profile of positive samples. *LWT - Food Science and Technology* 75:402–7.

Hof, H., Szabo, K. and Becker, B. 2007. Epidemiology of listeriosis in Germany: A changing but ignored pattern. *Deutsche Medizinische Wochenschrift* 132(24):1343–8.

Huntzinger, E., Boisset, S., Saveanu, C., et al. 2005. *Staphylococcus aureus* RNAIII and the endoribonuclease III coordinately regulate *spa* gene expression. *The EMBO Journal* 24(4):824–35.

Johansson, J., Mandin, P., Renzoni, A., Chiaruttini, C., Springer, M. and Cossart, P. 2002. An RNA thermosensor controls expression of virulence genes in *Listeria monocytogenes*. *Cell* 110(5):551–61.

Käferstein, F. K. 1997. Food safety: A commonly underestimated public health issue. *World Health Statistics Quarterly. Rapport Trimestriel de Statistiques Sanitaires Mondiales* 50(1–2):3–4.

Kakinuma, K., Fukushima, M. and Kawaguchi, R. 2003. Detection and identification of *Escherichia coli*, *Shigella*, and *Salmonella* by microarrays using the *gyrB* gene. *Biotechnology and Bioengineering* 83(6):721–8.

Karatuğ, N. T., Yüksel, F. N., Akçelik, N. and Akçelik, M. 2018. Genetic diversity of food originated *Salmonella* isolates. *Biotechnology and Biotechnological Equipment* 32(3):638–45.

Katzenell, U., Shemer, J. and Bar-Dayan, Y. 2001. Streptococcal contamination of food: An unusual cause of epidemic pharyngitis. *Epidemiology and Infection* 127(2):179–84.

Kumar, R., Shah, P., Swiatlo, E., Burgess, S. C., Lawrence, M. L. and Nanduri, B. 2010. Identification of novel non-coding small RNAs from *Streptococcus pneumoniae* TIGR4 using high-resolution genome tiling arrays. *BMC Genomics* 11:350. doi:10.1186/1471-2164-11-350.

Landt, S. G., Abeliuk, E., McGrath, P. T., Lesley, J. A., McAdams, H. H. and Shapiro, L. 2008. Small non-coding RNAs in *Caulobacter crescentus*. *Molecular Microbiology* 68(3):600–14.

Lee, S. I., Kim, S. A., Park, S. H. and Ricke, S. C. 2019. Molecular and new-generation techniques for rapid detection of foodborne pathogens and characterization of microbial communities in poultry meat. In: *Food safety in poultry meat production, food microbiology and food safety nature*, eds. K. Venkitanarayanan, S. Thakur and S. Ricke, 235–60. Cham: Springer.

Lin, C.-K., Hung, C. -L., Hsu, S. -C., et al. 2004. An improved PCR primer pair based on 16S rDNA for the specific detection of *Salmonella* serovars in food samples. *Journal of Food Protection* 67(7):1335–43.

Livny, J. and Waldor, M. K. 2007. Identification of small RNAs in diverse bacterial species. *Current Opinion in Microbiology* 10(2):96–101.

Loh, E., Gripenland, J. and Johansson, J. 2006. Control of *Listeria monocytogenes* virulence by 5′-untranslated RNA. *Trends in Microbiology* 14(7):294–8.

Ludmila, C. 2009. Functional metagenomics: Recent advances and future challenges. *Biotechnology and Genetic Engineering Reviews* 26(1):335–52.

Manisha, M., Sangita, B., Satish, K. S. and Ram, K. G. 2015. Molecular detection of foodborne pathogens: A rapid and accurate answer to food safety. *Critical Reviews in Food Science and Nutrition* 56(9):1568–84.

Mayo, B., Rachid, C. T. C., Alegría, A., Leite, A. M. O., Peixoto, R. S. and Delgado, S. 2014. Impact of next generation sequencing techniques in food microbiology. *Current Genomics* 15(4):293–309.

Morfeldt, E., Taylor, D., von Gabain, A. and Arvidson, S. 1995. Activation of alpha-toxin translation in *Staphylococcus aureus* by the trans-encoded antisense RNA, RNAIII. *The EMBO Journal* 14(18):4569–77.

Motarjemi, Y. and Käferstein, F. K. 1997. Global estimation of foodborne diseases. *World Health Statistics Quarterly. Rapport Trimestriel de Statistiques Sanitaires Mondiales* 50(1–2):5–11.

Negrini, M., Nicoloso, M. S. and Calin, G. A. 2009. MicroRNAs and cancer – new paradigms in molecular oncology. *Current Opinion in Cell Biology* 21(3):470–9.

Nguyen, T. T., Giau, V. V. and Vo, T. K. 2016. Multiplex PCR for simultaneous identification of *E. coli* O157:H7, *Salmonella* spp. and *L. monocytogenes* in food. *3 Biotech* 6(2):205. doi:10.1007/s13205-016-0523-6.

Nielsen, J. S., Larsen, M. H., Lillebaek, E. M. et al. 2011. A small RNA controls expression of the chitinase ChiA in *Listeria monocytogenes*. *PLoS ONE* 6(4):e19019. doi:10.1371/journal.pone.0019019.

Nielsen, J. S., Lei, L. K., Ebersbach, T., et al. 2010. Defining a role for Hfq in Gram-positive bacteria: Evidence for Hfq-dependent antisense regulation in *Listeria monocytogenes*. *Nucleic Acids Research* 38(3):907–19.

Ogier, J. C. and Serror, P. 2008. Safety assessment of dairy microorganisms: The *Enterococcus* genus. *International Journal of Food Microbiology* 126(3):291–301.

Ollinger-Snyder, P. and Matthews, M. E. 1996. Food safety: Review and implications for dietitians and dietetic technicians. *Journal of the American Dietetic Association* 96(2):163–170.

Paulsen, I. T., Banerjei, L., Myers, G. S., et al. 2003. Role of mobile DNA in the evolution of vancomycin-resistant *Enterococcus faecalis*. *Science* 299(5615):2071–4.

Perez, N., Trevino, J., Liu, Z., Ho, S. C., Babitzke, P. and Sumby, P. 2009. A genome-wide analysis of small regulatory RNAs in the human pathogen group A *Streptococcus*. *PLoS ONE* 4(11):e7668. doi:10.1371/journal.pone.0007668.

Pervaiz, T., Lotfi, A., Haider, M. S., Haifang, J. and Fang, J. 2017. High throughput sequencing advances and future challenges. *Journal of Plant Biochemistry and Physiology* 5(2):188. doi:10.4172/2329-9029.1000188.

Pichon, C. and Felden, B. 2005. Small RNA genes expressed from *Staphylococcus aureus* genomic and pathogenicity islands with specific expression among pathogenic strains. *Proceedings of the National Academy of Sciences of the United States of America* 102(40):14249–54.

Riboulet-Bisson, E., Sanguinetti, M., Budin-Verneuil, A., Auffray, Y., Hartke, A. and Giard, J. C. 2008. Characterization of the Ers regulon of *Enterococcus faecalis*. *Infection and Immunity* 76(7):3064–74.

Rupnik, M., Wilcox, M. H. and Gerding, D. N. 2009. *Clostridium difficile* infection: New developments in epidemiology and pathogenesis. *Nature Reviews in Microbiology* 7(7):526–36.

Sharma, C. M., Darfeuille, F., Plantinga, T. H. and Vogel, J. 2007. A small RNA regulates multiple ABC transporter mRNAs by targeting C/A-rich elements inside and upstream of ribosome-binding sites. *Genes and Development* 21(21):2804–17.

Sharma, C. M., Hoffmann, S., Darfeuille, F., et al. 2010. The primary transcriptome of the major human pathogen *Helicobacter pylori*. *Nature* 464(7286):250–5.

Sharma, C. M. and Vogel, J. 2009. Experimental approaches for the discovery and characterization of regulatory small RNA. *Current Opinion in Microbiology* 12(5):536–46.

Shen, A. and Higgins, D. E. 2005. The 5′ untranslated region-mediated enhancement of intracellular listeriolysin O production is required for *Listeria monocytogenes* pathogenicity. *Molecular Microbiology* 57(5):1460–73.

Shin, H. H., Hwang, B. H. and Cha, H. J. 2016. Multiplex 16S rRNA-derived geno-bio-chip for detection of 16 bacterial pathogens from contaminated foods. *Biotechnology Journal* 11(11):1405–14.

Sittka, A., Lucchini, S., Papenfort, K., et al. 2008. Deep sequencing analysis of small non-coding RNA and mRNA targets of the global posttranscriptional regulator, Hfq. *PLoS Genetics* 4(8):e1000163. doi:10.1371/journal.pgen.1000163.

Sittka, A., Pfeiffer, V., Tedin, K. and Vogel, J. 2007. The RNA chaperone Hfq is essential for the virulence of *Salmonella typhimurium*. *Molecular Microbiology* 63(1):193–217.

Stritzker, J., Schoen, C. and Goebel, W. 2005. Enhanced synthesis of internalin A in *aro* mutants of *Listeria monocytogenes* indicates posttranscriptional control of the *inlAB* mRNA. *Journal of Bacteriology* 187(8):2836–45.

Tam, C. C., O'Brien, S. J., Tompkins, D. S. et al. 2012. Changes in causes of acute gastroenteritis in the United Kingdom over 15 years: Microbiologic findings from 2 prospective, population-based studies of infectious intestinal disease. *Clinical Infectious Diseases* 54(9):1275–86.

Taveirne, M. E., Theriot, C. M., Livny, J. and DiRita, V. J. 2013. The complete *Campylobacter jejuni* transcriptome during colonization of a natural host determined by RNAseq. *PLoS ONE* 8(8):e73586. doi:10.1371/journal.pone.0073586.

Tilahun, B., Tesfaye, A., Muleta, D., Bahiru, A., Terefework, Z. and Wessel, G. 2018. Isolation and molecular identification of lactic acid bacteria using 16S rRNA genes from fermented *teff* (*Eragrostis tef* (Zucc.)) dough. *International Journal of Food Science* 2018:8510620. doi:10.1155/2018/8510620.

Toledo-Arana, A., Dussurget, O., Nikitas, G., et al. 2009. The *Listeria* transcriptional landscape from saprophytism to virulence. *Nature* 459(7249):950–6.

Valiollahi, E., Farsi, M., Fevereiro, P. and Kakhki, A. M. 2014. Bioinformatic characterization and expression analysis of miRNAs in *Solanum lycopersicum*. *Plant Omics* 7(2):108–16.

Verneuil, N., Rince, A., Sanguinetti, M., et al. 2005. Contribution of a PerR-like regulator to the oxidative-stress response and virulence of *Enterococcus faecalis*. *Microbiology* 151(12):3997–4004.

Verneuil, N., Sanguinetti, M., Le Breton, Y., et al. 2004. Effects of the *Enterococcus faecalis hypR* gene encoding a new transcriptional regulator on oxidative stress response and intracellular survival within macrophages. *Infection and Immunity* 72(8):4424–31.

Waters, L. S. and Storz, G. 2009. Regulatory RNAs in bacteria. *Cell* 136(4):615–628.

Wong, K. K., Bouwer, H. G. and Freitag, N. E. 2004. Evidence implicating the 5′ untranslated region of *Listeria monocytogenes* actA in the regulation of bacterial actin-based motility. *Cellular Microbiology* 6(2):155–66.

Yu, Z., Peruzya, M. F., Dumolin, C., Joossens, M. and Houf, K. 2019. Assessment of food microbiological indicators applied on poultry carcasses by culture combined MALDI-TOF MS identification and 16S rRNA amplicon sequencing. *Food Microbiology* 82:53–61.

CHAPTER 15

Application of WGS Technologies in Disease Diagnostics, Surveillance, Transmission, and Outbreak Investigation in the Food Sector

Nayana Aluparambil Radhakrishnan, Sreejith Sreekumaran,
Akshaya Chekkara Thandayan Santhosh, Sunil Pareek,
and Radhakrishnan Edayileveettil Krishnankutty

CONTENTS

15.1 INTRODUCTION

Food safety is one of the major human health concerns throughout the world, hence, rapid detection of foodborne pathogens is essential. Enter whole-genome sequencing (WGS)-based technologies, which enable the rapid identification, control, and prevention of microbial hazards in foods (Cook and Nightingale 2018). It helps public health authorities by providing better surveillance and rapid detection of foodborne diseases (FBDs). WGS also provides the high level of bacterial strain discrimination necessary for the rapid investigation and source attribution of pathogens (Brown et al. 2019).

Conceptually speaking, WGS is a quite simple technology, where deoxyribonucleic acid (DNA) is purified, processed, and sequenced, and the results are analyzed and visualized using bioinformatics tools. It reveals the complete DNA makeup of an organism. This technical simplicity provides advantages over other conventional molecular methods which rely upon different features. Due to the same, application of WGS in food safety management, can contribute to greater consumer protection, trade facilitation, and food/nutrition security (Franz et al. 2016).

The resolution efficiency offered by WGS for subtyping different bacterial, viral, fungal, and parasitic pathogens is also very high (Ronholm 2018). Data generated by WGS can also have application in secondary data analyses, such as virulence gene detection, antibiotic resistance gene profiling, mobile genetic element identification, and geographic origin attribution (Koutsoumanis et al. 2019). This allows scientists to conduct successful epidemiological investigations of foodborne outbreaks that would not have been possible with traditional techniques.

The epidemiological discrimination efficiency provided by WGS is far ahead compared with traditional typing methods (Ronholm 2018). Hence, the currently used phenotypic and genotypic reference methods are likely to be replaced by WGS in the coming years. WGS also enhances cluster detection and improves accuracy and resolution in compared with pulsed-field gel electrophoresis (PFGE) and Multiple Locus Variable-number Tandem Repeat Analysis (MLVA) as revealed with studies on *Salmonella* (Deng et al. 2015) and *Listeria* (Schmid et al. 2014). Based on the progress made so far, WGS is becoming an effective tool for food safety management. Extensive studies are required to further the applications of WGS in food security; potential applications of WGS for food safety and quality are discussed in the following sections (Figure 15.1).

15.2 SUBTYPING TECHNIQUES

Whole-genome sequencing of food associated microorganisms involves the identification of both known and unknown spoilage/pathogenic organisms present in the food. They may be culturable or non-culturable organisms. Culture-dependent methods are generally used to isolate microorganisms from food products, production facilities, and affected individuals. In the case of non-culturable organisms, the methods used are molecular or phage typing, serotyping, etc. Genome sequencing of both culturable and non-culturable organisms facilitate its strain-level identification and may enable tracing of the source of the pathogen. Even though various molecular subtyping techniques based on amplification and restriction digestion have been used for microbial identification, DNA sequencing is remarkable for its accuracy. However, the length of the fragment to be sequenced and the correctness and quality of nucleotide data obtained will depend on the sequencing platform used (Figure 15.2) (Paramithiotis et al. 2018).

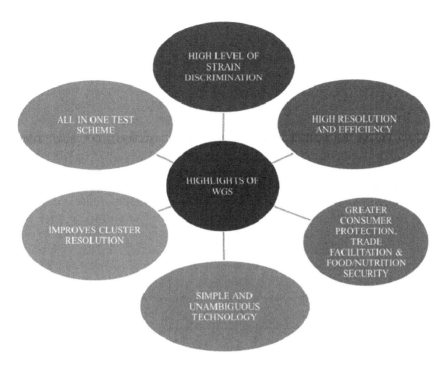

FIGURE 15.1   Highlights of whole genome sequencing.

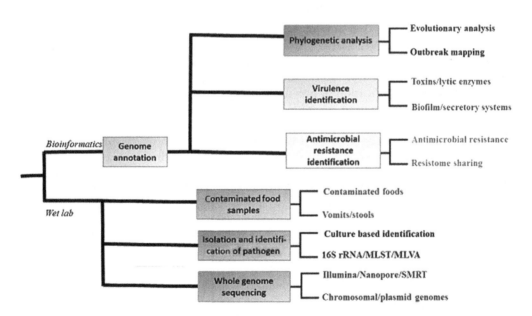

FIGURE 15.2   Pectoral representation of the sample processing and bioinformatics analysis for a foodborne outbreak investigation.

15.2.1 Classical Subtyping Methods

Classical methods, such as serotyping and phage typing, have been used worldwide for more than 50 years by public health agencies for the identification of human foodborne outbreaks and foodborne disease surveillance (Miljković-Selimović et al. 2009). Serotyping is a grouping method based on the agglutination reaction between the given antisera and the cell surface antigen of the pathogen. The somatic (O), flagellar (H), and capsular (Vi) antigens of the pathogens are the generally targeted antigens of most bacterial species (Olsen et al. 1993). A number of polymeric chain reaction (PCR)-based identification methods have further been introduced such as real-time PCR and multiplex PCR for the detection of foodborne bacterial pathogens (Salazar et al. 2015). PCR-based methods have the advantage over traditional methods, as they provide concise results from heterogeneous food samples.

Phage typing is mainly used for the detection of single strains of bacteria and to trace the source of outbreaks. It has been used to detect foodborne pathogens such as *Salmonella*, *Campylobacter*, *Listeria*, and *Shigella* (Olsen et al. 1993). In one study, a panel of lytic phages was inoculated on a lawn of inoculum of isolate under investigation, and the pattern of lysis over the bacterial lawn was analyzed to determine the phage type. For several decades, this method was used as an epidemiological tool for differentiating *Salmonella* and *Escherichia coli* O157:H7 (Cowley et al. 2015). But poor resolving power, high cost, and lack of technical expertise limited its wider use.

15.2.2 Gene Amplification-Based Techniques

Gene amplification-based techniques mainly include PCR-based techniques, variable-number tandem repeat (VNTR), and amplified fragment length polymorphism (AFLP) (Ronholm 2018). PCR-based amplification techniques can also be used to identify the antimicrobial resistance genes and virulent strains of foodborne organisms.

Arbitrarily primed PCR (AP-PCR) has been used for the amplification of unknown genomic regions by using arbitrary primers. A number of studies have also reported success with random amplification of polymorphic DNA (RAPD)-based methods to distinguish strains of *A. baumannii*, *C. albicans*, *H. pylori*, *P. mirabilis*, *H. capsulatum*, *H. somnus*, *Leptospira* sp., *S. aureus*, and *L. Pneumophilia* (Lukinmaa et al. 2004). But it is less reproducible compared with REP-PCR, where the primers complementary to the interspersed repetitive consensus sequences are used for the DNA amplification (Li et al. 2009). REP-PCR has also proven beneficial for typing bacteria of epidemiological importance such as *Acinetobacter* spp. (Pasanen et al. 2014), *N. meningitides* (Versalovic et al. 1994), *S. pneumoniae* (Koeuth et al. 1995), *C. difficile* (Polage et al. 2015), and *L. monocytogenes* (Soni et al. 2014). The enterobacterial repetitive intergenic consensus (ERIC) obtained from *E. coli* and *S. typhimurium* were also successfully used for the DNA typing (Wilson and Sharp 2006). Genotyping has also been carried out by VNTR/MLVA (multiple locus VNTR analysis). VNTR is a location in the genome where a short nucleotide sequence is organized as tandem repeat. The number of tandem repeats per locus may vary between strains within a given species. MLVA is a PCR-based genotyping method based on the polymorphic analysis of multiple VNTR loci (Lukinmaa et al. 2004). AFLP is another method which has been applied for bacterial typing. AFLP uses the combination of both restriction enzyme digestion and PCR amplification. AFLP offers great flexibility in the number of loci that can be amplified simultaneously in one PCR reaction and has the same discriminatory power as PFGE (Li et al. 2009).

15.2.3 Restriction Digestion-Based Technique

The restriction digest is a procedure that includes the restriction fragment length polymorphism (RFLP) and PFGE. Food Safety and Public Health use PFGE for tracking the spread of various bacteria and fungi that cause food poisoning and is known as the "gold standard" of molecular typing methods. In PFGE, the entire bacterial genome is cut with one or two restriction enzymes. The length of the pieces of DNA resulting from this digestion will provide a pattern that could differentiate bacterial strains. PFGE is able to determine the length of the DNA fragments in relation to the other samples used (Olsen et al. 1993). PulseNet, the PFGE database, is further used for tracking foodborne bacteria, such as *Escherichia coli* O157:H7, nontyphoidal *Salmonella, Shigella*, and *Listeria monocytogenes* (Li et al. 2009). However, replacing the PFGE with WGS could provide better discriminatory power and precision (Kubota et al. 2019). In the case of RFLP, the polymorphic nature of the locations of restriction enzyme sites within the defined genetic regions has been exploited. Only the genomic DNA fragments which could hybridize to the probes are visible in RFLP analysis, which greatly simplifies the analysis. The length of the restriction fragment will differ from strain to strain, and the similarity of the generated pattern of restriction fragments can be used to differentiate strains and to analyze the genetic relatedness (Li et al. 2009).

15.2.4 Sequencing-Based Techniques

By DNA sequencing, original genetic information of an organism will be obtained and the sequence data of individual gene and/or the whole genome can be used for bacterial subtyping. DNA sequencing-based genotyping has contributed to the identification of single nucleotide polymorphisms (SNPs), sequence deletions or insertions, and genes under positive selection. This is an efficient way to identify genetic changes and helps to phylogenetically classify bacterial strains (Li et al. 2009). It could also be used for the detection of multiple SNPs within a small region of DNA. In multilocus sequence typing (MLST), housekeeping genes are used where nucleotide sequence variation rate is relatively slow. Here, the analysis requires the detailed interpretation of nucleotide sequence data of selected genes (Brown et al. 2019).

Currently three different types of MLST analysis have been employed for the identification of phylogenetic relationship. These are the housekeeping genes-based MLST analysis, the core genome-based MLST analysis, and the whole genome-based MLST analysis. In housekeeping genes-based MLST analysis, 7 different housekeeping genes which are present in all the selected genomes are used for the variation analysis to observe sequence types and clonal complexes. In the core genome MLST analysis, the core genes are conserved in the query and the reference genomes are used for the variation analysis. This method gives better resolution for the phylogenetic analysis than the previous method. Whole genome-based MLST analysis is done by comparing the genome of the isolate, and it helps to distinguish even the closely related genome sequences. However, the phylogenetic analysis will be ambiguous, as it considers the accessory genomes, such as genes present in mobile genetic elements (Brown et al. 2019). After the different sequences present within a bacterial species are assigned with different allele numbers, the alleles at each of the loci will be used to define the allelic profile and subsequent sequence type (ST). The resulting nucleotide data are then compared with a database of previously identified allelic types (Cook and Nightingale 2018). Established databases, like pubMLST/BIGSdb, are used to determine the individual alleles and multilocus ST of the isolate. Bacterial

Isolate Genome Sequence database (BIGSdb, published in 2010), enables the PubMLST to include all the levels of sequence data, from single gene sequences up to and including complete, finished genomes (Jolley et al. 2018). It is now cheaper to sequence an entire genome by massive parallel sequencing than to sequence the MLST genes individually. The generated STs can further be calculated using the platform provided by the Center for Genomic Epidemiology. It has been used for the characterization of a large number of organisms, including *Streptococcus pneumoniae*, *Staphylococcus aureus*, *Campylobacter jejuni*, *Streptococcus pyogenes*, and *Haemophilus influenza* as well as *Candida albicans* (Greenwood et al. 2012) which also include potential foodborne pathogens.

15.3 METHODS BASED ON WHOLE GENOME SEQUENCING

WGS is the process of complete gene sequencing of an organism. From the WGS data of bacteria, larger genetic scaffolds are assembled and compared with reference databases for accurate identification (Taboada et al. 2017). There are clone-by-clone and whole-genome shotgun methods for DNA sequencing. In the clone-by-clone method, smaller sections of the genome are cloned into bacteria. The produced clones are further processed to produce overlapping base pairs. From the sequence data of the clones, the overlapping portions are identified and used to reassemble the clone. Whereas in the whole-genome shotgun sequencing method, the isolated genomic DNA is randomly sheared, cloned into vector, and sequenced. Random shotguns are reassembled based on overlapping sequences to generate the final sequence data. This method is labor intensive, is only good for shorter reads (Versalovic et al. 1994), and has seen tremendous changes with the introduction of other technologies.

15.3.1 First-Generation Sequencing

Sanger and Maxam-Gilbert sequencing technologies initiated the field of DNA sequencing. The first genome in the Human Genome Project and the first bacterial genome were sequenced using Sanger sequencing. This method is also referred to as dideoxy sequencing or chain termination DNA sequencing. Here, chain terminating dideoxynucleotides are incorporated by the DNA polymerase during the DNA amplification to interrupt the elongation of DNA. Initially, radiolabeled dideoxynucleotides were used for the detection of consensus sequences. Later, it was simplified with fluorescent base labeling, which allowed the automation of the process and higher throughput (Kulski 2016). The first full sequence of the virus phiX174 (Smith et al. 2003) and the first bacterial genome sequence of *Haemophilus influenzae* (Fleischmann et al. 1995) were also sequenced by the Sanger method. In Maxam-Gilbert sequencing, chemical degradation is used for cleaving nucleotides. The cleaved nucleotide bases (C, T+C, G, A+G) can be separated according to their sizes by electrophoresis. Sanger sequencing is favored over Maxam-Gilbert, as it is user-friendly and simpler (Kchouk et al. 2017).

15.3.2 Next-Generation Sequencing (NGS)

The emergence of second-generation sequencing broke through the limitations of first-generation sequencing. This high-throughput sequencing method was first introduced by

Roche's 454 technology in 2005 and commercialized at lower cost (Kulski 2016). The NGS platforms available to most microbiology laboratories include Illumina (Illumina), 454 (Roche), Ion Torrent (Life Technologies, now Thermo Fisher Scientific), and ABI/SOLID systems. These can help in the comparative epidemiological investigations and studies of massive microbial sequences. The basic characteristics of these techniques include high throughput, lower cost, high speed, and direct detection of the sequencing output without electrophoresis. However, each NGS platform has a variation in the approach for DNA preparation, sequencing, imaging, and comparative analysis.

The Roche/454 pyrosequencer uses the pyrosequencing technique and detects the pyrophosphate which is released when a nucleotide is added to a new DNA strand. The Roche GS-FLX can generate high sequence output with long reads. The longer reads produced by them helps in the mapping of repetitive regions. However, when compared with the competitors, they have high error rates and generates less data per run (Kchouk et al. 2017).

The most widely used NGS platform Ilumina/Solexa Genome Analyzer (GA) produces shorter reads (300 bp). This also uses the sequencing by synthesis approach and the reads can be increased up to 600 bp by paired end sequencing (Kulski 2016). The Ion Torrent semiconductor-based technique is able to produce reads with lengths of 400 bp but is less accurate than others (Ronholm et al. 2016). Another NGS sequencer, the Supported Oligonucleotide Ligation and Detection (SOLID) sequencing was acquired by ABI in 2007. ABI/SOLID works on sequencing by ligation, also producing short reads up to 75 bp with an output up to 30 Gb/run (Mayo et al. 2014).

The short DNA sequences (reads) produced by these second-generation sequencers require *de novo* assembly before analysis. Burrows-Wheeler Aligner, Novoalign, MOSAIK, and SMALT are some examples of reference-guided assemblers (Pightling et al. 2015).

15.3.3 Third-Generation Sequencing

Third-generation sequencing was introduced as a solution for the problems caused by second-generation sequencing. It offers fast, low cost sequencing with easy sample preparation and processing. In addition, TGS can produce long reads of several kilobases, thereby solving the assembly problems. The single molecule real-time sequencing approach (SMRT) was developed by Quake laboratory; and Illumina (Moleculo) is a third-generation sequencer capable of producing long reads exceeding 1 kb (Mayo et al. 2014). The most widely used TGS platforms are SMRT sequencers (Pacific Biosciences) and MinION sequencers (Oxford Nanopore Technology) (Kchouk et al. 2017). MinION sequencing allows routine WGS analysis anywhere with minimal reagents and sequencing equipment (Ronholm et al. 2016). It was used in the West African 2014–2015 Ebola outbreak to support the monitoring and surveillance of the transmission and the evolution of the virus (Quick et al. 2016). However, in comparison with second-generation sequencing, the TGS technique has a high error rate and lack of accuracy. PacBio SMRT platforms have an error rate of about 13% by insertion and deletion errors, and these errors are randomly distributed along the long read. MinION produces a high error rate of approximately 12% dominated by mismatches, insertions, and deletion (Kchouk et al. 2017).

These second and third-generation sequencing methods are always used as complementary for routine studies and surveillance. So, hybrid sequencing is employed where

TGS is used to generate a closed-genome scaffold and a deep coverage by second-generation techniques for accurate SNP analysis. PBcR is one example of such assemblers built to work with hybrid data (Ronholm et al. 2016).

15.4 DATA ANALYSIS

Whole-genome sequence datasets are analyzed through bioinformatics tools and processes. Bioinformatics analyses are often performed in a multistep manner, and multi-software analysis (pipelines) are set up to run automatically. These pipelines aggregate the isolates into clusters based on genetic differences and are visualized with phylogenetic trees that can be used to identify closely related isolates. The genetic variants analysis pipeline consists of 8 steps: quality control of raw reads, determination of reference strain, mapping of the pre-processed reads to the reference strain, post-alignment processing, quality control of the mapped reads, genome annotation, phylogenetic tree construction, and phylogenetic analysis (Oakeson et al. 2017). Genome Analysis Toolkit (GATK) is a widely used tool to analyze the second-generation WGS data (Hwang et al. 2019). Using the field programmable gate array technology (FPGA), the Dynamic Read Analysis for GENomics (DRAGEN) Bio-IT platform (Illumina) can conduct analysis within minutes (Lightbody et al. 2019). Wtdbg25 and CANU6 are the softwares used for the assembly of third-generation sequencing of WGS data (Ruan and Li 2020). This has been further supported with help from cloud computing for massive analysis of WGS data (Wang et al. 2018).

15.5 WGS IN FOOD MICROBIOLOGY

WGS has revolutionized the analysis of microbial pathogens of food. The technology has immense application in the epidemiological study of foodborne diseases.

15.5.1 Foodborne Bacteria

WGS has recently been used for the analysis of many of the food pathogens such as *L. monocytogenes*, *Salmonella*, *E. coli*, *Campylobacter*, *V. cholerae*, etc. *L. monocytogenes* is the causative agent of listeriosis which causes severe infections of the central nervous system, septicemia, gastroenteritis, and abortion (Zhu et al. 2017). Investigation of listeriosis is difficult, as the incubation period is long. PFGE had been used as the gold standard for typing listeriosis, but there are challenges, as the mobile genetic element present in *L. monocytogenes* changes frequently. Because of its superior discriminatory power, WGS replaced the traditional methods, and specific analyses have been developed for outbreak detection (Pietzka et al. 2019). The study by Pietzka et al. (2019) compared the potential of WGS to identify the source of a listeriosis outbreak with the PFGE analysis. This showed WGS-based typing to have superior discriminatory power and yields, better data accuracy, greater laboratory throughput, and lower cost when compared with the PFGE. WGS, combined with SNP analysis, has also been used to differentiate persistent *L. monocytogenes* from sporadic strains in multiple independent food-associated environments (Stasiewicz et al. 2015). The discriminatory power of WGS has also been used to identify outbreaks caused by more than one strain of *Listeria* (Tham et al. 2002).

WGS was instrumental in the investigation of a cluster of seven cases of listeriosis in Austria and Germany during 2011–2013 (Schmid et al. 2014). As the 7 isolates shared a common serotype, analysis with PFGE and AFLP was found to be difficult; this further indicates the potential of WGS analysis. Core Genome Multi Locus Sequence Typing (cgMLST)-based data analyses are also used for the appropriate subtype differentiation of *L. monocytogenes* (Orsi et al. 2019). The clustered regularly interspaced palindromic repeat (CRISPR) associated locus subtype sequences and the sequences of mobile genetic elements can also be used to supplement cgMLST analysis to interpret WGS. This interpretation, along with epidemiological data, has been found to be effective to describe *L. monocytogenes* strains during outbreak investigations (Ronholm et al. 2016).

Salmonella is one of the major causes of foodborne disease. The genomic homogeneity of clonal serovars of *Salmonella* subspecies such as *S. enteritidis*, *S. typhimurium*, and *S. montevideo*, makes WGS the most appropriate option as the traditional typing methods are inadequate for cluster resolution (Den et al. 2014). Whole-genome sequencing is a viable platform for the evaluation and bench-marking of molecular subtyping methods, such as cgMLST, MLVA, CRISPR-MVLST, and PFGE, for tracing major lineages and ecological origins of *S. enterica* serotype Enteritidis (Deng et al. 2015). On the basis of WGS data, several packages, such as SeqSero (Zhang et al. 2015) and Salmonella In Silico Typing Resource (SISTR) (Yoshida et al. 2016), have been developed for predicting the *Salmonella* serotypes. In the case of *Campylobacter,* MLST, PFGE, and flaA typing methods have been replaced with WGS, as it can provide better genetic attribution with respect to the source (Cody et al. 2013). Data from WGS are also now providing insights into the epidemiology and surveillance of *V. cholerae* infections (Rahaman et al. 2015).

15.5.2 Foodborne Viruses

WGS methods have also been used for subtyping and detection of foodborne viruses. It can be used for the detection of virus-related outbreaks and to track virus transmission routes (Ronholm et al. 2016). For example, WGS is used to investigate the NoV transmission events and interhost dynamics and to analyze genomic variations (Kundu et al. 2013). Foodborne outbreak of Hepatitis A virus (HAV) is also a serious threat across the globe; WGS is a good option to track this virus (Chiapponi et al. 2014). In addition to this, WGS is also used for the improved detection and identification of many other viruses (Ogorzaly et al. 2015; Peretti et al. 2015).

15.5.3 Detecting Virulence and Toxins

Virulence factors of microorganisms enable them to survive under adverse conditions and to adapt to new environmental niches. The virulence genes encode membrane to secretory proteins (toxins, exo-enzymes, type I to VI secretion systems), biofilm-forming proteins, siderophores, as well as the polysaccharides that form the capsules. The pivotal steps in initiating infection by pathogen generally involve adhesion, colonization, invasion, and/or resistance to host defense mechanisms, and antimicrobials. These functions are provided by the virulence factors and are often transferred by horizontal gene transfer (Hacker et al. 2000). WGS analysis enables the identification of complete virulence factors of a pathogen. The virulence factors from the genome data may be identified by homology search with known virulence genes, by comparing with strains having various levels of virulence, or by predicting the horizontally acquired genes. The virulence determinants associated with phage-driven exchange and the antibiotic resistance could be accurately predicted from the chromosome and plasmid by whole-genome analysis.

WGS has already been used for the rapid and reliable detection of *E. coli* virulence genes and to trace the *E. coli* outbreaks (Manning et al. 2008). User-friendly websites, including the Virulence Finder (Joensen et al. 2014) and SuperPhy (Ruan and Li 2020), are available for the automatic detection of virulence genes from WGS data. Web tools (Serotype Finder) are also available for serotyping of *E. coli* and to help in outbreak investigations (Fratamico et al. 2016). In addition, several databases are also available to identify the pathogenicity islands which are the specific genomic loci where the virulence factors are localized. Such databases include the Pathogenicity Island Database (PAIDB) (Yoon et al. 2015), Horizontal Gene Transfer Database (HGT-DB) (Garcia-Vallve et al. 2003), IslandViewer database (Dhillon et al. 2015), and the PredictBias database (Pundhir et al. 2008).

15.5.4 Detecting Antimicrobial Resistance

WGS-based analysis is a valuable method to predict the antibiotic resistance of pathogens using bioinformatics tools. Several databases dedicated to the antibiotic resistance markers are currently available. Antibiotic Resistance Gene (ARG) databases are generally online platforms that offer antimicrobial resistance-related reference data along with web tools for sequence analysis and annotation of ARGs. There are two types of public databases for the study of antimicrobial resistance. They are generalized antimicrobial resistance (AR) databases and specialized AR databases (Boolchandani et al. 2019). Generalized AR databases incorporate a wide range of ARGs and mechanistic information, while specialized AR databases are created to meet specific needs and/or to provide extensive information on specific gene families. Generalized databases include ARDB (Liu and Pop 2009), ARG-ANNOT (both currently archived), and CARD (Liu and Pop 2009). Specialized antimicrobial resistance databases include the Lactamase Engineering Database (LacED) (Thai et al. 2009), β-Lactamase Database (BLDB) (Naas et al. 2017), and the Comprehensive β-lactamase Molecular Annotation Resource (CBMAR) (Srivastava et al. 2014). ARBD created by Liu and Pop is a manually curated database. It was aimed at centralizing all the information about antimicrobial resistance, identifying and characterizing new genes, and providing consistent annotation of resistance information in newly sequenced organisms (Liu and Pop 2009). The ARDB database contains resistance information for 23,137 genes from 1737 bacterial species. McArthur et al. (2013) developed a database called Comprehensive Antibiotic Resistance Database (CARD) in 2013. It is a tool combining a large antibiotic resistance gene database of 4221 genes. The Antibiotic Resistance Ontology tool and the Resistance Gene Identifier (RGI) enable the identification of antibiotic resistance genes from partial or whole genome sequencing data.

Another rapid and simple bioinformatic tool for the detection of known and putative antibiotic resistance genes in bacterial genomes is the ARG-ANNOT database created by Gupta et al. (2014). Here, a local BLAST program enables the sequence analysis, and the database includes 1689 genes causing resistance to different antibiotics. Information about all classes of β-lactamases enzymes is gathered in BLDB. Detailed biochemical and molecular information about known and novel β-lactamases are provided by Comprehensive B-Lactamase Molecular Annotation Resource (CBMAR). Lactamase

Engineering Database (LacED) is a specialized resource that contains mutational and structural data and offers integrated tools for sequence analysis (Thai et al. 2009).

15.6 APPLICATION OF WGS IN FOOD MICROBIOLOGY

15.6.1 Foodborne Outbreak Detection and Traceback Investigation Using WGS

Surveillance and outbreak investigation are generally carried out by subtyping of pathogens. As WGS offers high resolution for subtyping different bacterial, viral, fungal, and parasitic pathogens, it can be used for retrospective comparison of microorganisms associated with epidemiologically suspected outbreaks or for prospective laboratory surveillance of high-burden diseases, such as listeriosis and salmonellosis.

Retrospective comparisons can only be done with the coordinated efforts guided by public health professionals involving epidemiologists, (molecular) microbiologists, bioinformaticians, and clinicians (Koutsoumanis et al. 2019). WGS combined epidemiological investigation can enhance the discriminatory power to recognize the low intensity and extended time period of outbreaks allowing them to be linked with food products (Lassen et al. 2016).

Integrated foodborne disease surveillance combines data collected from different parts of the food chain (farm to fork) and provides comprehensive information for identifying and confirming outbreaks, monitoring disease trends, identifying risk factors and populations, and improving food production and public health practices. Its goals are to identify the sources and patterns of endemic and emerging disease and to support an efficient and coordinated multi-agency response to health risks along the food chain (Galanis et al. 2012). This helps to improve the ability of rapid source tracing of the contaminant and to make an estimate of the relative contribution of different food sources toward human FBD (Pires et al. 2014).

15.6.2 Source Attribution Using WGS

Source attribution refers to the understanding of different sources of foodborne pathogens which cause a particular foodborne disease (Mughini-Gras et al. 2018). The sources can be both reservoirs (environment, animals) and vehicles (food items). Different points of attribution contribute to the development of disease, like the point of reservoir or the point of exposure (Pires et al. 2018). Microbial subtyping is found to be more reliable for source attribution. It is based on comparing subtypes found among strains isolated from human cases with the subtypes found in a potential pathogen source. Frequency matching models and population genetic models are the two types of microbial subtyping (Pires et al. 2018). Frequency matching model is used with traditional subtyping, like with serotyping results, while the population genetic model is used with genotyping results. WGS can have significant applications on the same.

15.6.3 Microbial Risk Assessment Using WGS

Risk assessment refers to the scientific evaluation of known or potential adverse health effects resulting from human exposure to foodborne hazards. It consists of hazard identification, hazard characterization, exposure assessment, and risk characterization (Koutsoumanis et al. 2019). Hazard identification involves the identification of agents capable of causing adverse health effects and which may be present in a particular food or group of foods. It can be assessed by WGS by looking into the pathogenic characteristics (Wright et al. 2016). Hazard characterization involves the qualitative and/or quantitative evaluation of the nature of the adverse health effects associated with the hazards present in the food. For hazard characterization, a dose–response assessment should be performed if data are available. WGS can be employed here in combination with phenotypic data (Pielaat et al. 2016) to revise the current dose–response models for more targeted pathogen–human interactions (Kovac et al. 2017). Exposure assessment involves the qualitative and/or quantitative evaluation of the likely intake of the hazards via food as well as through exposures from other sources, if relevant. WGS can be used to predict the ability of a microorganism to grow or survive within the host or the food, as well as during processing, storage, and distribution of foods (den Besten et al. 2018; Fritsch et al. 2019). Risk characterization involves the qualitative and/or quantitative estimation of the probability of occurrence and severity of known or potential adverse health effects in a given population based on hazard identification, hazard characterization, and exposure assessment. Risk characterization is expected to benefit from WGS as a consequence of its implementation in the previous risk assessment steps–hazard identification, hazard characterization, and exposure assessment.

15.7 CONCLUSION

WGS represents a new approach for tracking foodborne pathogens, and thereby it can be applied in FBD outbreak investigations, source attribution, and microbial risk assessment, which prove its promising roles in FBD surveillance and response. Thus, WGS has an influential role in food safety and quality control, as it has the ability to detect, resolve, and even prevent FBD outbreaks rapidly. A combined approach of WGS and epidemiological investigations could be more noteworthy for the rapid delineation of an outbreak. There is no doubt that in the coming years, this multipurpose technology, WGS, will become used even more widely than today. More research and developments need to be conducted for implementing this technology as an effective tool in maintaining food safety and quality.

REFERENCES

Boolchandani, M., D'Souza, A. W. and Dantas, G. 2019. Sequencing-based methods and resources to study antimicrobial resistance. *Nature Reviews Genetics* 20(6):356–70.

Brown, E., Dessai, U., McGarry, S. and Gerner-Smidt, P. 2019. Use of whole-genome sequencing for food safety and public health in the United States. *Foodborne Pathogens and Disease* 16(7):441–50.

Chiapponi, C., Pavoni, E., Bertasi, B., et al. 2014. Isolation and genomic sequence of hepatitis A virus from mixed frozen berries in Italy. *Food and Environmental Virology* 6(3):202–6.

Cody, A. J., McCarthy, N. D., van Rensburg, M. J., et al. 2013. Real-time genomic epidemiological evaluation of human *Campylobacter* isolates by use of whole-genome multilocus sequence typing. *Journal of Clinical Microbiology* 51(8):2526–34.

Cook, P. W. and Nightingale, K. K. 2018. Use of omics methods for the advancement of food quality and food safety. *Animal Frontiers : The Review Magazine of Animal Agriculture* 8(4):33–41.

Cowley, L. A., Beckett, S. J., Chase-Topping, M., Perry, N., Dallman, T. J., Gally, D. L. and Jenkins, C. 2015. Analysis of whole genome sequencing for the *Escherichia coli* O157: H7 typing phages. *BMC Genomics* 16(1):271.

Den Bakker, H. C., Allard, M. W., Bopp, D., et al. 2014. Rapid whole-genome sequencing for surveillance of *Salmonella enterica* serovar enteritidis. *Emerging Infectious Diseases* 20(8):1306.

Den Besten, H. M., Amézquita, A., Bover-Cid, S., et al. 2018. Next generation of microbiological risk assessment: Potential of omics data for exposure assessment. *International Journal of Food Microbiology* 287:18–27.

Deng, X., Shariat, N., Driebe, E. M., et al. 2015. Comparative analysis of subtyping methods against a whole-genome-sequencing standard for *Salmonella enterica* serotype enteritidis. *Journal of Clinical Microbiology* 53(1):212–8.

Dhillon, B. K., Laird, M. R., Shay, J. A., et al. 2015. IslandViewer 3: More flexible, interactive genomic island discovery, visualization and analysis. *Nucleic Acids Research* 43(W1):W104–8.

Fleischmann, R. D., Adams, M. D., White, O., et al. 1995. Whole-genome random sequencing and assembly of *Haemophilus influenzae* Rd. *Science* 269(5223):496–512.

Franz, E., Gras, L. M. and Dallman, T. 2016. Significance of whole genome sequencing for surveillance, source attribution and microbial risk assessment of foodborne pathogens. *Current Opinion in Food Science* 8:74–9.

Fratamico, P. M., Deb Roy, C., Liu, Y., Needleman, D. S., Baranzoni, G. M. and Feng, P. 2016. Advances in molecular serotyping and subtyping of *Escherichia coli*. *Frontiers in Microbiology* 7:644.

Fritsch, L., Felten, A., Palma, F., et al. 2019. Insights from genome-wide approaches to identify variants associated to phenotypes at pan-genome scale: Application to *L. monocytogenes'* ability to grow in cold conditions. *International Journal of Food Microbiology* 291:181–8.

Galanis, E., Parmley, J. and De With, N. and British Columbia Integrated Surveillance of Foodborne Pathogens Working Group. 2012. Integrated surveillance of Salmonella along the food chain using existing data and resources in British Columbia, Canada. *Food Research International* 45(2):795–801.

Garcia-Vallve, S., Guzmán, E., Montero, M. A. and Romeu, A. 2003. HGT-DB: A database of putative horizontally transferred genes in prokaryotic complete genomes. *Nucleic Acids Research* 31(1):187–9.

Greenwood, D., Slack, R. C., Barer, M. R. and Irving, W. L. 2012. *Medical microbiology: A guide to microbial infections: Pathogenesis, immunity, laboratory diagnosis and control*. Elsevier Health Sciences.

Gupta, S. K., Padmanabhan, B. R., Diene, S. M., Lopez-Rojas, R., Kempf, M., Landraud, L. and Rolain, J. M. 2014. ARG-ANNOT, a new bioinformatic tool to discover antibiotic resistance genes in bacterial genomes. *Antimicrobial Agents and Chemotherapy* 58(1):212–20.

Hacker, J. and Kaper, J. B. 2000. Pathogenicity islands and the evolution of microbes. *Annual Reviews in Microbiology* 54(1):641–79.

Hwang, K. B., Lee, I. H., Li, H., Won, D. G., Hernandez-Ferrer, C., Negron, J. A. and Kong, S. W. 2019. Comparative analysis of whole-genome sequencing pipelines to minimize false negative findings. *Scientific Reports* 9(1):1–10.

Joensen, K. G., Scheutz, F., Lund, O., Hasman, H., Kaas, R. S., Nielsen, E. M. and Aarestrup, F. M. 2014. Real-time whole-genome sequencing for routine typing, surveillance, and outbreak detection of verotoxigenic *Escherichia coli*. *Journal of Clinical Microbiology* 52(5):1501–10.

Jolley, K. A., Bray, J. E. and Maiden, M. C. 2018. Open-access bacterial population genomics: BIGSdb software, the PubMLST.org website and their applications. *Wellcome Open Research* 3.

Kchouk, M., Gibrat, J. F. and Elloumi, M. 2017. Generations of sequencing technologies: From first to next generation. *Biology and Medicine* 93(2):105–11. doi: 10.1016/j.ygeno.2008.10.003.

Koeuth, T., Versalovic, J. and Lupski, J. R. 1995. Differential subsequence conservation of interspersed repetitive *Streptococcus pneumoniae* BOX elements in diverse bacteria. *Genome Research* 5(4):408–18.

Koutsoumanis, K., Allende, A., Alvarez-Ordóñez, A., et al. 2019. Whole genome sequencing and metagenomics for outbreak investigation, source attribution and risk assessment of food-borne microorganisms. *EFSA Journal,* 17(12):e05898.

Kovac, J., den Bakker, H., Carroll, L. M. and Wiedmann, M. 2017. Precision food safety: A systems approach to food safety facilitated by genomics tools. *TrAC Trends in Analytical Chemistry* 96:52–61.

Kubota, K. A., Wolfgang, W. J. and Baker, D. J. et al. 2019. PulseNet and the changing paradigm of laboratory-based surveillance for foodborne diseases. *Public Health Reports* 134(2S):22S–8S.

Kulski, J. K. 2016. Next-generation sequencing - An overview of the history, tools, and "Omic" applications. In: *Next generation sequencing - Advances, applications and challenges*, eds. J. K. Kulski, pp. 3–60. Springer Nature, Singapore.

Kundu, S., Lockwood, J., Depledge, D. P., et al. 2013. Next-generation whole genome sequencing identifies the direction of Norovirus transmission in linked patients. *Clinical Infectious Diseases* 57(3):407–14.

Lassen, S. G., Ethelberg, S., Björkman, J. T., et al. 2016. Two *Listeria* outbreaks caused by smoked fish consumption - Using whole-genome sequencing for outbreak investigations. *Clinical Microbiology and Infection* 22(7):620–4.

Li, W., Raoult, D. and Fournier, P. E. 2009. Bacterial strain typing in the genomic era. *FEMS Microbiology Reviews* 33(5):892–916.

Lightbody, G., Haberland, V., Browne, F., Taggart, L., Zheng, H., Parkes, E. and Blayney, J. K. 2019. Review of applications of high-throughput sequencing in personalized medicine: Barriers and facilitators of future progress in research and clinical application. *Briefings in Bioinformatics* 20(5):1795–811.

Liu, B. and Pop, M. 2009. ARDB - Antibiotic resistance genes database. *Nucleic Acids Research* 37(S1):D443–7.

Lukinmaa, S., Nakari, U. M., Eklund, M. and Siitonen, A. 2004. Application of molecular genetic methods in diagnostics and epidemiology of food-borne bacterial pathogens. *Acta Pathologica, Microbiologica et Immunologica Scandinavica (Apmis)* 112(11-12):908–29.

Manning, S. D., Motiwala, A. S., Springman, A. C., et al. 2008. Variation in virulence among clades of *Escherichia coli* O157:rH7 associated with disease outbreaks. *Proceedings of the National Academy of Sciences of the United States of America* 105(12):4868–73.

Mayo Pérez, B., Rachid, C. T., Alegría, Á., Leite, A. M., Peixoto, R. S. and Delgado, S. 2014. Impact of next generation sequencing techniques in food microbiology. *Current Genomics* 15(4):293–309.

McArthur, A. G., Waglechner, N., Nizam, F., et al. 2013. The comprehensive antibiotic resistance database. *Antimicrobial Agents and Chemotherapy* 57(7):3348–57.

Miljković-Selimović, B., Kocić, B., Babić, T. and Ristić, L. 2009. Bacterial typing methods. *Acta Facultatis Medicae Naissensis* 26(4):225–33.

Mughini-Gras, L., Franz, E. and van Pelt, W. 2018. New paradigms for *Salmonella* source attribution based on microbial subtyping. *Food Microbiology* 71:60–7. doi:10.1016/j.fm.2017.03.002.

Naas, T., Oueslati, S., Bonnin, R. A., et al. 2017. Beta-lactamase database (BLDB)– structure and function. *Journal of Enzyme Inhibition and Medicinal Chemistry* 32(1):917–9.

Oakeson, K. F., Wagner, J. M., Mendenhall, M., Rohrwasser, A. and Atkinson-Dunn, R. 2017. Bioinformatic analyses of whole-genome sequence data in a public health laboratory. *Emerging Infectious Diseases* 23(9):1441.

Ogorzaly, L., Walczak, C., Galloux, M., Etienne, S., Gassilloud, B. and Cauchie, H. M. 2015. Human adenovirus diversity in water samples using a next-generation amplicon sequencing approach. *Food and Environmental Virology* 7(2):112–21.

Olsen, J. E., Brown, D. J., Skov, M. N. and Christensen, J. P. 1993. Bacterial typing methods suitable for epidemiological analysis. Applications in investigations of salmonellosis among livestock. *Veterinary Quarterly* 15(4):125–35.

Orsi, R. H., Jagadeesan, B., Baert, L. and Wiedmann, M. 2019. Comparative analysis of tools and approaches for source tracking *Listeria monocytogenes* in a food facility using whole-genome sequence data. *Frontiers in Microbiology* 10:947.

Paramithiotis, S., Hadjilouka, A. and Drosinos, E. H. 2018. Molecular typing of major foodborne pathogens. In: *Foodborne diseases*, eds. A. Grumezescu and A. M. Holban, pp. 421–72. Academic Press.

Pasanen, T., Koskela, S., Mero, S., Tarkka, E., Tissari, P., Vaara, M. and Kirveskari, J. 2014. Rapid molecular characterization of *Acinetobacter baumannii* clones with rep-PCR and evaluation of carbapenemase genes by new multiplex PCR in hospital district of Helsinki and Uusimaa. *PLoS ONE* 9(1):e85854.

Peretti, A., FitzGerald, P. C., Bliskovsky, V., Buck, C. B. and Pastrana, D. V. 2015. Hamburger polyomaviruses. *The Journal of General Virology* 96(4):833.

Pielaat, A., Kuijpers, A., Delfgou-van Asch, E., van Pelt, W. and Wijnands, L. 2016. Phenotypic behavior of 35 *Salmonella enterica* serovars compared to epidemiological and genomic data. *Procedia Food Science* 7:53–8.

Pietzka, A., Allerberger, F., Murer, A., et al. 2019. Whole genome sequencing based surveillance of *L. monocytogenes* in food for early detection and investigations of Listeriosis outbreaks. *Frontiers in Public Health* 7:139.

Pightling, A. W., Petronella, N. and Pagotto, F. 2015. Choice of reference-guided sequence assembler and SNP caller for analysis of *Listeria monocytogenes* short-read sequence data greatly influences rates of error. *BMC Research Notes* 8(1):748.

Pires, S. M., Duarte, A. S. and Hald, T. 2018. Source attribution and risk assessment of antimicrobial resistance. In: *Antimicrobial resistance in bacteria from livestock and companion animals*, eds. S. Schwarz, L. M. Cavaco J. Shen, pp. 619–35. Washington, DC: American Society for Microbiology.

Pires, S. M., Vieira, A. R., Hald, T. and Cole, D. 2014. Source attribution of human salmonellosis: An overview of methods and estimates. *Foodborne Pathogens and Disease* 11(9):667–76.

Polage, C. R., Gyorke, C. E., Kennedy, M. A., et al. 2015. Overdiagnosis of *Clostridium difficile* infection in the molecular test era. *JAMA Internal Medicine* 175(11):1792–801.

Pundhir, S., Vijayvargiya, H. and Kumar, A. 2008. PredictBias: A server for the identification of genomic and pathogenicity islands in prokaryotes. *In Silico Biology* 8(3,4):223–34.

Quick, J., Loman, N. J., Duraffour, S., et al. 2016. Real-time, portable genome sequencing for Ebola surveillance. *Nature* 530(7589):228–32.

Rahaman, M. H., Islam, T., Colwell, R. R. and Alam, M. 2015. Molecular tools in understanding the evolution of *Vibrio cholerae*. *Frontiers in Microbiology* 6:1040.

Ronholm, J. 2018. Game changer-next generation sequencing and its impact on food microbiology. *Frontiers in Microbiology* 9:363.

Ronholm, J., Nasheri, N., Petronella, N. and Pagotto, F. 2016. Navigating microbiological food safety in the era of whole-genome sequencing. *Clinical Microbiology Reviews* 29(4):837–57.

Ruan, J. and Li, H. 2020. Fast and accurate long-read assembly with wtdbg2. *Nature Methods* 17(2):155–8.

Salazar, J. K., Wang, Y., Yu, S., Wang, H. and Zhang, W. 2015. Polymerase chain reaction-based serotyping of pathogenic bacteria in food. *Journal of Microbiological Methods* 110:18–26.

Schmid, D., Allerberger, F., Huhulescu, S., et al. 2014. Whole genome sequencing as a tool to investigate a cluster of seven cases of listeriosis in Austria and Germany, 2011–2013. *Clinical Microbiology and Infection* 20(5):431–6.

Smith, H. O., Hutchison, C. A., Pfannkoch, C. and Venter, J. C. 2003. Generating a synthetic genome by whole genome assembly: φX174 bacteriophage from synthetic oligonucleotides. *Proceedings of the National Academy of Sciences of the United States of America* 100(26):15440–5.

Soni, D. K., Singh, M., Singh, D. V. and Dubey, S. K. 2014. Virulence and genotypic characterization of *Listeria monocytogenes* isolated from vegetable and soil samples. *BMC Microbiology* 14(1):241.

Srivastava, A., Singhal, N., Goel, M., Virdi, J. S. and Kumar, M. 2014. CBMAR: A comprehensive β-lactamase molecular annotation resource. *Database* 2014:bau111.

Stasiewicz, M. J., Oliver, H. F., Wiedmann, M. and den Bakker, H. C. 2015. Whole-genome sequencing allows for improved identification of persistent *Listeria monocytogenes* in food-associated environments. *Applied and Environmental Microbiology* 81(17):6024–37.

Taboada, E. N., Graham, M. R., Carriço, J. A. and Van Domselaar, G. 2017. Food safety in the age of next generation sequencing, bioinformatics, and open data access. *Frontiers in Microbiology* 8:909.

Thai, Q. K., Bös, F. and Pleiss, J. 2009. The Lactamase Engineering Database: A critical survey of TEM sequences in public databases. *BMC Genomics* 10(1):390.

Tham, W., Aldén, J., Ericsson, H., et al. 2002. A listeriosis patient infected with two different *Listeria monocytogenes* strains. *Epidemiology and Infection* 128(1):105–6.

Versalovic, J., Schneider, M., De Bruijn, F. J. and Lupski, J. R. 1994. Genomic fingerprinting of bacteria using repetitive sequence-based polymerase chain reaction. *Methods in Molecular and Cellular Biology* 5(1):25–40.

Wang, Y., Li, G., Ma, M., He, F., Song, Z., Zhang, W. and Wu, C. 2018. GT-WGS: An efficient and economic tool for large-scale WGS analyses based on the AWS cloud service. *BMC Genomics* 19(1):959.

Wilson, L. A. and Sharp, P. M. 2006. Enterobacterial repetitive intergenic consensus (ERIC) sequences in *Escherichia coli*: Evolution and implications for ERIC-PCR. *Molecular Biology and Evolution* 23(6):1156–68.

Wright, A., Ginn, A. and Luo, Z. 2016. Molecular tools for monitoring and source-tracking *Salmonella* in wildlife and the environment. In: *Food safety risks from wildlife*, eds. M. Jay-Russell and M. Doyle, pp. 131–50. Cham: Springer.

Yoon, S. H., Park, Y. K. and Kim, J. F. 2015. PAIDB v2. 0: Exploration and analysis of pathogenicity and resistance islands. *Nucleic Acids Research* 43(D1):D624–30.

Yoshida, C. E., Kruczkiewicz, P., Laing, C. R., Lingohr, E. J., Gannon, V. P., Nash, J. H. and Taboada, E. N. 2016. The Salmonella in silico typing resource (SISTR): An open web-accessible tool for rapidly typing and subtyping draft Salmonella genome assemblies. *PLoS ONE* 11(1):e0147101.

Zhang, S., Yin, Y., Jones, M. B., et al. 2015. Salmonella serotype determination utilizing high-throughput genome sequencing data. *Journal of Clinical Microbiology* 53(5):1685–92.

Zhu, Q., Gooneratne, R. and Hussain, M. A. 2017. *Listeria monocytogenes* in fresh produce: Outbreaks, prevalence and contamination levels. *Foods* 6(3):21. doi:10.3390/foods6030021.

CHAPTER 16

High-Throughput Sequencing for Detection of Foodborne Pathogens in Food Safety

Mahesh Pattabhiramaiah and Shanthala Mallikarjunaiah

CONTENTS

16.1 INTRODUCTION

Food, which is an essential part of daily life, passes through several manufacturing phases before reaching the customer. Foodborne diseases are usually infectious or toxic and are caused by toxins entering the body through the ingestion of food or water. Many individuals may be at risk for foodborne diseases (WHO 2007a). Foodborne diseases most commonly elicit gastrointestinal symptoms, but these diseases can also cause neurological, gynecological, immunological, and other symptoms, which can lead to multi-organ failure including cancer. In this way, foodborne diseases pose a major threat of injury, morbidity, and mortality in human beings especially in immunocompromised individuals and infants (Stein et al. 2007; WHO 2008; Tauxe et al. 2010). Foodborne pathogens trigger foodborne diseases (FBDs) either specifically (by pathogens) or implicitly by bacterial mycotoxins and endotoxins (Martinovic et al. 2016), and may have catastrophic health and financial implications both in developed and developing countries (Pires et al. 2012; Henao et al. 2015). Diarrheal diseases are a significant fraction of FBDs, with a relatively high effect on children (Pires et al. 2015). FBDs are typically caused by foodborne pathogens (FBPs), such as bacteria, viruses, parasites, and a few fungi.

Factors vital to the development of pathogens include human hosts, animal hosts, and their interactions with humans, the pathogen itself, and the environment, including how food is produced, processed, and stored. For instance, changes in the vulnerability of the host due to malnutrition, age, and other conditions can lead to new infections in vulnerable populations. Genetic exchange or mutations in the organisms can create new strains with the potential to cause disease. Exposure to emerging pathogens through variations in eating habits, climate, mass production, food processing, and intensified globalization of food supply may trigger pathogens to evolve in new demographics or new geographical regions.

Bacteria, viruses, and parasites are accountable for many of the foodborne diseases (Scallan et al. 2011). Therefore, in many nations, sporadic infections or eruptions are rigorously investigated. Technological advancements have also led to an increase in the transmission of foodborne pathogens across international borders, adversely affecting trade and food safety (Frank et al. 2011; Bernard et al. 2014) leading to the establishment of intricate restrictions by individual governments and international organizations to strengthen food safety. Consequently, distinct food production systems are burdened with providing food fit for human consumption.

Food should not contain any infectious microorganisms (bacteria, viruses, amoebae, fungi, algae, yeast, or Giardia), parasites (e.g., worm eggs, larvae, or cysts), toxins, poisons, or toxic chemicals, such as pesticides. Poor hygiene practices and viruses are also considered major risk factors for foodborne diseases (Rolando 2011). In food contamination, bacteria growing on the food before its consumption can produce toxins that may cause illness. The common foodborne diseases are outlined in Table 16.1 which describes the key forms of food poisoning.

16.2 THE EFFECTS OF FOODBORNE PATHOGENS

Microbiota plays an important role in every food matrix, ranging from fermentation, to contamination and spoilage. The pathogenic properties of foodborne microorganisms are strain-dependent. Besides, the epidemic potential of a foodborne strain within its population can vary in function of its genetic makeup and ecological items (EFSA 2013). Foodborne pathogens are major causes of morbidity and mortality throughout the world, and the ability to conduct epidemiological investigations and intervene in foodborne illnesses is a critical part of the existing public health infrastructure. Illness caused by foodborne pathogens represents an important economic and public health burden worldwide. More than 200 diseases are caused by ingesting physical hazards (nails, stones, feather, bone, etc.), chemical compounds (pesticides, heavy metals, etc.) and biological hazards like bacteria, viruses, fungi, protozoa, etc., from contaminated food (Figures 16.1–16.3). This rising issue of public health has major socioeconomic impacts due to a loss of profitability in medical systems and damage to hospitality and commerce. The global incidence of sickness and death is greatly influenced by these diseases. Foodborne diseases are caused by food contamination and can exist in the food production, supply, and consumption chain at any point. They are the result of numerous sources of contamination in the environment, including water, soil or air pollution, and inadequate food storage and processing.

TABLE 16.1   Pathogens and Foodborne Illness

Pathogen	Signs and Symptoms	Food Involved
Anisakis simplex (infection)	Abdominal pain, vomiting, coughing	Saltwater fish
Bacillus cereus (toxico-infection)	Nausea, abdominal pain, diarrhea, vomiting	Cereal products, rice, custards and sauces, meatloaf
Campylobacter jejuni (infection)	Diarrhea (sometimes bloody), severe abdominal pain, fever, anorexia, headache, vomiting	Raw milk, poultry, beef liver, raw clams, contaminated water
Clostridium botulinum (intoxication)	Vertigo, double vision, difficulty swallowing, speaking, and breathing; weak muscles, respiratory paralysis, frequently fatal	Home-canned low-acid food, garlic and oil mixtures, vacuum-packed fish, fermented fish eggs, fish, marine mammals
Clostridium perfringens (toxico-infection)	Abdominal pain, diarrhea	Cooked meat, poultry, gravy, sauces, soups
Cryptosporidium (infection)	Severe diarrhea, low-grade fever, and severe intestinal distress	Any food product that comes into contact with a contaminated person or contaminated water
Escherichia coli O157:h7 (*E. coli*) (toxico-infection)	Severe abdominal pain, diarrhea (sometimes bloody), nausea, vomiting, fever, chills, headache, muscular pain, bloody urine	Soft unpasteurized cheese, contaminated water, any undercooked animal source foods, especially hamburger
Giardiasis lamblia (infection)	Abdominal pain, diarrhea, fever, cramps	Water, raw vegetables, and fruits
Hepatitis A (infection)	Fever, anorexia, nausea, abdominal pain, jaundice	Shellfish, contaminated water, any food contaminated by feces, urine or blood of infected humans and other primates
Listeria monocytogenes (infection)	Nausea, vomiting, stomach cramps, diarrhea, headache, constipation, fever	Unpasteurized milk, soft cheeses, undercooked poultry, prepared meats, unwashed raw vegetables
Norovirus (infection)	Nausea, vomiting, diarrhea, abdominal pain	Contaminated water, food, or food contact surfaces

Source: http://www.foodsafe.ca/docs/Foodborne_Illness_Chart.pdf

Hundreds of outbreaks of foodborne pathogens occur around the world every year. In the European Union (EU), a total of 5,648 foodborne outbreaks were reported in 2011, affecting 69,553 people (EFSA 2013). Most foodborne pathogens are transmitted from the environment or animals through food products to humans. To minimize the occurrence of FBPs in the food production chain, and thereby increase food safety, knowledge of the behavior of the pathogens and detection methods are needed. Foodborne pathogens are a public health threat and their control is a challenge to the food industry. The occurrence and transcription of genes, which encode specialized functions, play a vital role in the survival of pathogens, their growth under harsh conditions and in food, their ability to produce toxins and virulence, and their pathogenicity and antimicrobial resistance.

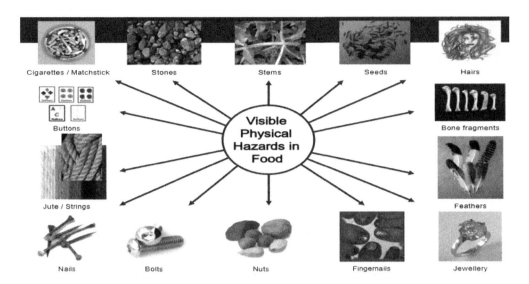

FIGURE 16.1   Physical hazards in foods. (Source: https://www.pmg.engineering/2019/08/07/foreign-material-control/.)

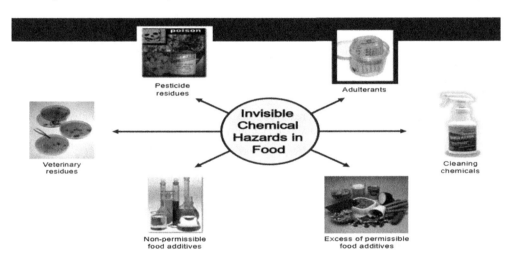

FIGURE 16.2   Chemical hazards. (Source: https://www.pmg.engineering.)

The risks associated with foodborne pathogenic microorganisms have been documented for decades. Similarly, the application of heat to low acid foods in insulated containers handled botulism problems. In all countries, foodborne illness is a growing cause of morbidity and the list of possible foodborne microbial pathogens continues to expand despite significant advances in food science and technology. In addition, foodborne disease in most countries is a leading cause of preventable death and financial burden. Regrettably, there are limited data on the impacts of foodborne diseases in most countries.

Malnutrition and illness comprise the majority of child mortality in developing countries. In addition, undernourished babies and children are far more vulnerable to foodborne threats and the dangers of diarrheal disease. The high incidence of diarrheal

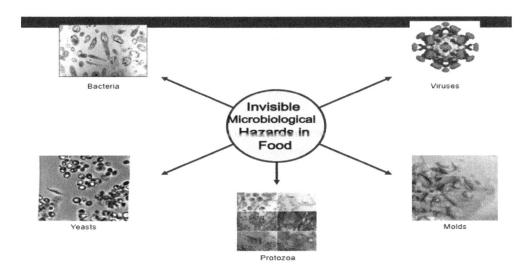

FIGURE 16.3 Biological/microbial hazards. (Souce: https://www.pmg.engineering.)

disease in under-developed countries again emphasizes the significance of food safety issues (WHO 2007a). In contrast, serious diseases frequently associated with non-hygienic services, often lead to infant mortality. Such outbreaks, in effect, worsen malnutrition, contributing to a negative feedback loop of illness and mortality (Prüss-Ustün et al. 2011). Reasonably high mortality and low quality of life for children, in several parts of the world, are significant concerns for the public health community, particularly nutritionists. The leading causes of infant mortality in underdeveloped countries can be attributed to pneumonia and diarrhea, both entirely avoidable and curable diseases. Diarrhea is, therefore, a significant public health issue. The leading causes of diarrhea are inaccessibility of drinking water, intake of polluted water, lack of hygiene facilities, denial of access to sanitation, congested living conditions, consumption of tainted food due to inadequate food storage, and consumption of raw fruit (Hénock and Dovie 2007).

Even though not well established, developing countries are assumed to have the greatest influence on the foodborne disease issue due to the prevalence of a wide variety of foodborne diseases, as well as those triggered by parasites (WHO 2008). But the high prevalence of diarrheal diseases in many developed countries reveals significant fundamental food safety concerns in those localities as well (WHO 2007b). Worldwide, potable water coverage rose from 77% in 1990 to 87% in 2008, with just 84% of the people in underdeveloped countries having better access to water supplies, in contrast to almost 100% in developed nations. Nevertheless, only 60% percent of the population in sub-Saharan Africa (SSA) has access to safe, potable water. Among infants under 5 years of age, 70% of fatalities are often attributed to biologically contaminated food. Such contaminants are more widely triggered by foodborne parasites, even though mycotoxin contamination is more common.

Foodborne diseases encompass a wide range of illnesses from diarrhea to cancers. Most present as gastrointestinal issues, though they can also produce neurological, gynecological, and immunological symptoms. Diseases causing diarrhea are a major problem in all countries of the world, though the burden is carried disproportionately by low- and middle-income countries and by children less than 5 years of age.

Nearly one person in ten throughout the globe falls ill annually, contributing to more than 420,000 deaths after consuming tainted food. Children suffer tremendously and 125,000 fatalities in people below the age of five are recorded annually. Diarrheal disorders are the majority of these instances. Foodborne diseases have severe effects on kidney and liver damage, brain and nerve diseases, reactive arthritis, cancer, and death. Foodborne diseases in low-and middle-income countries may be closely linked to poverty, but they represent a growing worldwide public health concern. The risks of food contamination and the transport of contaminated foodstuffs across national borders are increased by increased foreign trade and increasingly complex food chains. Such problems are amplified and linked to rising towns, climate change, migration, and increasing international travel.

Most cases of food poisoning typically occur from bacteria, viruses, fungi, protozoa, and parasites due to insufficient food handling. Pathogens, when present in negligible amounts, are not harmful to most healthy adults because of the immune system. When food is not handled properly, harmful pathogens multiply and get transmitted. Contaminated foods don't look, taste, or smell differently compared to uncontaminated food. A simple method of microbial contaminant detection is needed for low cell numbers, since the presence in humans or foods of a single cell of a pathogen may be adequate for an infectious dose. *Campylobacter*, *Clostridium perfringens*, *E. coli*, *Listeria*, *Salmonella*, *Bacillus cereus*, Botulism, Hepatitis A, Norovirus, *Shigella*, *Staphylococcus aureus*, and *Vibrio* species causing vibriosis are the bacteria and viruses that cause the most illnesses.

FBD monitoring seeks to eliminate the disease burden caused by food contamination. Surveillance aims at tracking disease patterns, estimating the burden of the disease, identifying vulnerable populations, assessing contamination sources and routes of transmission, and detecting and managing outbreaks. The monitoring system feedback is used to advise policies and enhance techniques for prevention. The surveillance is usually laboratory-based and ultimately integrated with food chain surveillance.

Previously, the standard surveillance methods for many foodborne identifying bacterial pathogens were antigen testing, pulsed-field gel electrophoresis (PFGE) and multilocus variable number tandem repeat analysis (MLVA). However, the rates of strain discrimination in these approaches are lower than whole-genome sequencing (WGS). WGS allows real-time detection of numerous potential pathogens that would have gone undetected by conventional methods.

16.3 HIGH-THROUGHPUT SEQUENCING OF FOODBORNE PATHOGENS

Food, feed and environmental clinical samples harbor complex and diverse microbial communities. Over the last few decades, detection of foodborne pathogens has relied on the culture and isolation of unknown bacterial pathogens. However, long incubation periods are required to culture the most genera and species. Today, modern techniques that characterize pathogens rely on 16S rRNA gene amplicon sequencing and metagenomics which permit detecting, identifying, and monitoring of FBPs. Fast, robust, and low-cost methods of detecting foodborne pathogens include the culture-independent high-throughput sequencing (HTS) of 16S rRNA genes amplified by PCR from food samples. To identify and characterize food spoilage associated with pathogenic and beneficial organisms, shotgun metagenomic sequencing of the microbiomes and 16S rRNA-based sequencing studies have been employed. HTS can yield thousands to millions of sequence reads and up to several hundred billion base pairs (bp) of sequence information per sample. There is great variation in the read length, error, and number of reads and sequenced bases. The approaches include selective amplification (targeted) and non-selective, random (shotgun). For the most significant food pathogens, the number of high-quality genomes is now large and increasing rapidly, gaining in part from the relatively small genome sizes of most microorganisms (around 100 Mbp). In scenarios where there is no sequenced reference genome, *de novo* sequencing and genome assembly is required. *De novo* assembly in order to acquire a draft sample genome based on short-read sequence data (<250 bp) of high-quality single isolates cultivated is complex but can be (semi) automated (Emond-Rheault et al. 2017).

16.3.1 Sequencing of Foodborne Bacteria (FBB)

Salmonellosis is an infectious disease caused by *Salmonella* and causes diarrhea and deaths. *Salmonella* is more prevalent in immune-compromised people and is caused by consuming uncooked eggs, raw poultry and meat, infected raw vegetables and fruits, as well as raw milk and milk products made of unpasteurized milk. In the United States, gastrointestinal illness is mainly caused by *Salmonella enterica* serovar *enteritidis* cannot be detected by conventional methods. To investigate outbreak detection and to trace back a source, next-generation sequencing (NGS) technologies are used to investigate WGS as a potential molecular subtyping tool. The *S. enteritidis* isolate analyses of seven epidemiologically documented foodborne outbreaks and intermittent isolates were used to assess the utility of WGS for detecting outbreaks (Angela et al. 2015).

NGS approaches have been used in the identification of the genetic and evolutionary diversity and the epidemiological origins of outbreak strains of the significant *Salmonella enterica* serovars (Holt et al. 2008; Chin et al. 2011). In contrast to PFGE and MLVA in *Salmonella*, STEC, and *Listeria*, WGS enhances cluster identification and improves resolution and accuracy (Dallman et al. 2015; Morganti et al. 2018; Waldram et al. 2018; Reimer et al. 2019; Ung et al. 2019). Traditional microbial typing methods lack the ability to discriminate different strains and the epidemiological; however, the genome was successfully identified using WGS (Pearce et al. 2018).

For years, *Salmonella* serotyping has become the benchmark for the surveillance of *Salmonella* and enables monitoring of changes in the prevalence of certain serotypes in specific regions which are significant evidence for established clusters (Grimont and Weill 2007; Ashton et al. 2016; Yoshida et al. 2016). Until now, the detection of *Salmonella* disease outbreaks was performed with numerous typing methods, such as phage typing, MLVA, or PFGE (Ribot et al. 2006;Wuyts et al. 2013; Barco et al. 2014; Inns et al. 2015). With the advancement of next generation sequencing technology, these classic typing methods have become less popular, and today's sequencing approaches provide high-resolution isolate characterization and outbreak detection based on single nucleotide polymorphism and whole-genome MLST (wgMLST)-based gene-by-gene analysis (Nadon et al. 2017; Palma et al. 2018; Pearce et al. 2018).

Clostridium perfringens is a significant causal agent of foodborne and non-foodborne gastroenteritis, characterized as a gram-positive, spore-forming, anaerobic, rod-shaped bacterium (Grass et al. 2013). *Clostridium perfringens* is the fourth most widespread causative agent of foodborne outbreaks in France and Europe and an omnipresent ecological bacterium. *C. perfringens* is often alluded to as a "buffet germ" since it develops most rapidly in large food quantities, such as casseroles, stews, and gravies that are often exposed in unsafe areas at room temperature. This bacterium's ability to form resistant spores leads to its persistence in many harsh niches, including water, sewage, food, and human and animal intestinal microbiota (Xiao 2014; Li et al. 2016). Gastroenteritis, enterotoxemia, and necrotic enteritis are caused by *C. perfringens* in animals (Keyburn et al. 2008; Uzal et al. 2015). It can multiply quite rapidly, and the most vulnerable are young children and the elderly. Disease typically occurs when certain foods are consumed that are infected by these bacteria which create adequate toxin to induce abdominal cramping and diarrhea (Scallan et al. 2011). Food such as meat, poultry products, as well as vegetables and crops are contaminated by *C. perfringens*. The *C. perfringens* entero-toxin (CPE) encoded by the *cpe* gene causes these foodborne outbreaks.

Studies relying on WGS were conducted to identify the genomic epidemiology and virulence gene content of *C. perfringens* strains associated with outbreaks. Between 2011 and 2017, phylogenomic research was carried out on 109 *C. perfringens* isolates (human and food) acquired from clusters of disease in England and Wales using a standard Illumina library preparation protocol prior to sequencing on Illumina MiSeq. Preliminary results demonstrated WGS 'improved discriminating power in profiling outbreak *C. per-fringens* strains as contrasted to the existing laboratory technique of fluorescent amplified fragment length polymorphism (FAFLP) investigation by Public Health England. Further study reported that, over a 5-year period, isogenic *C. perfringens* strains were affiliated with nine distinctive care home-related outbreaks, suggesting a possible common source associated to such outbreaks or propagation across space and time (Raymond et al. 2019).

The *Mycoplasma genitalia* genome was sequenced using Roche/454 sequencing plat-form and 580,069 bp was sequenced in a single run for the first time with 99.96% accu-racy and coverage of 96% of the genome. This method was first to sequence a complete human genome with a maximum reading length of 400 Mb per cycle, which then grew to approximately 700 bp (Berglund et al. 2011).

Campylobacter jejuni causes bacterial gastroenteritis worldwide; however, this pathogen's epidemiology is only poorly established, and it is challenging to shed new light on this area since most cases are sporadic and go unreported (Kaakoush et al. 2015). The infection caused by *Campylobacter* bacteria is often correlated with consuming raw or undercooked poultry and meat or cross-contamination with other foods. Freezing decreases the amount of *Campylobacter* bacteria in raw meat but does not fully elimi-nate it; therefore, it is necessary to properly cook the meat. In summer, campylobacte-riosis manifests itself frequently and is most widespread in children and adolescents. Until recently, vital molecular typing resources for studying *C. jejuni* infection epidemi-ology included PFGE and conventional seven-locus multilocus sequence typing (MLST) approaches that have contributed significantly to scientific knowledge of the existence of *C. jejuni* infections in patients and potential reservoirs (Carrillo et al. 2012).

These strategies, however, were hard to integrate in the framework of systematic sur-veillance; furthermore, shortcomings of these approaches have been revealed, indicating that such strategies may be inadequate as the only subtyping methods used to determine epidemiological links between *C. jejuni* isolates (Biggs et al. 2011; Revez et al. 2014; Llarena et al. 2016). The emergence of a new generation with effective tools for epide-miological monitoring of *C. jejuni* is indeed absolutely important as a serious solution to the shortcomings of the previous typing methods (Carrillo et al. 2012). WGS is poised to evolve as the preferred typing tool for *C. jejuni* outbreak investigations and has the scope for regular monitoring. WGS has the potential to enhance monitoring and identification of outbreaks for *C. jejuni* infections and to understand the evolutionary and epidemiological dynamics of *C. jejuni*-causing campylobacteriosis (Ann et al. 2017).

Yahara et al. (2017) examined the impact of various stages of the poultry production chain on *Campylobacter* populations using WGS and genome wide association studies (GWAS). Disease-related SNPs were differentiated in complexes ST-21 and ST-45, and analysis of the function of genes containing related elements indicated roles for format metabolism, aerobic survival, oxidative respiration, and nucleotide salvage, enabling possible linkages to environmental robustness and virulence.

Highly discriminatory subtyping of isolates is required to achieve a national surveillance network that can diagnose ongoing *Campylobacter* outbreaks in real time. Therefore, by utilizing WGS-based typing, the incidence of *Campylobacter* outbreaks among people in Denmark was more comprehensively assessed (Joensen et al. 2020). In order to correlate the clinical isolates with their origins, they were compared with food and animal isolates which covered the key putative origins of human *Campylobacter* infections (i.e., contact or ingestion of animals or animal products, mainly contaminated poultry meat). While infections with *Campylobacter* are mainly foodborne, a subsequent case–control survey in Denmark revealed that association between animals and the environment could account for a significant proportion of domestic infections (Kuhn et al. 2018). Many other causes of contamination recorded involve potable water, bathing water, unpasteurized milk, fruit, and vegetables (Verhoeff et al. 2011; Domingues et al. 2012; Mughini et al. 2012; Kaakoush et al. 2015).

For public health surveillance of several foodborne pathogens including *Salmonella*, *Listeria*, and Shiga toxin–producing *E. coli*, WGS offers high-resolution discriminatory subtyping and has been successfully implemented (Joensen et al. 2014; Kvistholm et al. 2016; Kuhn et al. 2018). Previous studies have revealed that WGS is appropriate for *Campylobacter* outbreak investigations (Kovanen et al. 2014; Revez et al. 2014; Clark et al. 2016; Lahti et al. 2017), and subsequently it was shown that WGS can track clinical infections directly back to chicken slaughter loads (Kovanen et al. 2016).

Staphylococcus aureus is typically found in the bodies, throats, and noses of healthy people and animals. However, if it gets into food products, it can replicate and develop harmful toxins, becoming a common human pathogen and a significant contributor of infection. Symptoms of staphylococcal food poisoning include nausea, stomach cramps, vomiting, and diarrhea. Cooking can kill staphylococcal bacteria, but their toxins are heat-resistant. Immunocompromised persons are likely to be infected from *S. aureus*. WGS provides an overview of the genome structure and gene content and has been widely implemented in disease investigations and genomic correlations. In one study, a blend of genetic, phenotypical, and computational methods was used to classify six enterotoxigenic *S. aureus* strains. The six strains were used to produce naturally infected cheeses that were screened for staphylococcal enterotoxins using the European screening process. The findings derived from the evaluation of toxins released in cheese, coupled with the genomic characteristics, provided a snapshot of the strains which can be used as reference material for the development of staphylococcal enterotoxin-positive cheese (Macori et al. 2019). *Escherichia coli*, also known as *E. coli*, are a large group of bacteria. Although most strains of *E. coli* are harmless, some can make you very sick. *E. coli* O157: H7 (STEC) causes food poisoning, and its consequences can be dangerous.

Application of the Ion Torrent personal genome machine sequencer has culminated in the discovery and genomic characterization of the entero-hemorrhagic *E. coli* O104:H4 strain liable for the May 2011 outbreak in Germany (Mellmann et al. 2011). Changes in the expression of microbe genes associated with different stressors–genetic and environmental, also acid tolerance or stress response mechanism–could also be identified using HT-NGS methods. The generated population genome data could be used to evaluate the susceptibility of antimicrobials or to detect changes in resistance. This method may be useful in verifying the existence of such mutations or polymorphisms in genes that may be correlated with antimicrobial resistance (Joseph and Read 2010) that could be difficult to ascertain using other techniques. Awareness of the whole genome sequence, in conjunction with other genetic approaches (MLVA and PCR), could contribute to a plethora of information for discovering new targets and help to establish more successful strategies for managing transmission routes of pathogenic microbes, or for combatting multidrug resistance (Sukhnanand et al. 2005; Fitzgerald et al. 2007; McQuiston et al. 2008; Xi et al. 2008; Wise et al. 2009).

16.3.2 Sequencing of Foodborne Viruses (FBVs)

Viruses are the greatest reservoirs of genetic diversity on the planet (Suttle 2005), and they are constantly evolving under strong selection pressure (Sironi et al. 2015). Second and third-generation sequencing can be employed to sequence the whole genome of viruses, providing information about the entire viral populaces amongst infected organisms or ecological samples, identifying the direction and transmission source in the event of an outbreak, and thereby assisting in intervention and forensic investigation. This was made abundantly clear with the 2014–2015 Ebola outbreak (Ladner et al. 2014; Check 2015; Quick et al. 2016).

Viruses play a multiple role in pathogenetic contamination of food. Some viruses, such as bacteriophages, can influence the population structure and virulence of microbes, while other viruses can be FBPs themselves (Newell et al. 2010; EFSA 2011; Hayes et al. 2017). To be transmittable via food, a virus must be environmentally stable and contagious in a food matrix for a certain period. Since viruses cannot replicate inside of a food matrix, the infectious dose is minimal. FBVs contaminate foods primarily via human fecal material (Rodríguez-Lazaro et al. 2012) and are denatured by routine cooking or frying activities. Crude foods, such as soft berries, herbs, shellfish, ready-to-eat foods, and cold-served foods are typically the sources of infection (Halliday et al. 1991; Hedberg and Osterholm 1993; de Wit et al. 2003; Fiore 2004; EFSA 2011, 2015).

Most viruses infect through person-to-person contact route or the foodborne route. The vast majority of outbreaks spread concurrently through individuals and foodborne organisms. The spread of zoonotic viruses is also suspected to be associated with food and water transmission which disseminates the manifestation of zoonotic events, e.g., through bushmeat handling (Nieuwenhuijse and Koopmans 2017). The reporting and surveillance of foodborne viruses is limited, hence, sequencing the genome of FBVs will shed light onto real impact and diversity of foodborne viruses (Nieuwenhuijse and Koopmans 2017).

The most prominent FBV positive-sense, single-stranded, non-enveloped RNA viruses are Norovirus (NoV), Hepatitis A virus (HAV) and Hepatitis E virus (HEV), and the double-stranded RNA Rotavirus (RV) (Newell et al. 2010; EFSA 2011). Examples of other (zoonotic) RNA viruses suspected of communicating through food are the severe acute respiratory syndrome (SARS), Middle East respiratory syndrome (MERS), and Ebola viruses (Newell et al. 2010; Nieuwenhuijse and Koopmans 2017). In comparison to the alarmingly large majority of bacteriophages which are double-stranded DNA viruses, most FBP viruses are RNA viruses and require specific sample processing or nucleic acid extraction methods.

The novel HTS systems provide useful information for major viral outbreaks (Quick et al. 2016). In one study, assembly of RNAseq metagenomics and shotgun read (mean contig size 680 bp) were used for mapping viral genomes on samples from field grown and retail lettuce into a reference database (Aw et al. 2016). The common infection in humans corresponded to RV and a small fraction of the reads were attributed to other viruses.

RNAseq and mapping technique was used to investigate the viral contamination of fresh parsley plants that were irrigated with fecally tainted river water (Fernandez-Cassi et al. 2017). HEV and NoV were associated with FBPs in a small fraction (<1%) of the reads. Large outbreaks of gastroenteritis are triggered by extremely contagious NoV which is most responsible for the diarrheal disease and mortalities worldwide (de Wit et al. 2003; Pires et al. 2015) and is the foremost source of foodborne illness in the USA (Scallan et al. 2011; Moore et al. 2015; Imamura et al. 2016). A combination of RNAseq on virus suspensions and PCR-enriched targeted sequencing (genotyping approach similar to metataxonomics) was used to characterize NoV diversity in shellfish from two commercial producers in Japan using HTS.

Foodborne outbreaks of the Hepatitis A virus (HAV) are recorded throughout the world (Aggarwal and Goel 2015). HAV, belonging to the *Picornaviridae* family is a single-stranded, positive-sense, non-enveloped RNA virus. A major HAV outbreak in Italy involving 1,202 cases was recorded in 2013 (Rizzo et al. 2013; Tavoschi et al. 2015). A link between beef intake and an increased risk of colorectal cancer was proposed (zur Hausen 2012). The existence in beef products of oncogenic viruses, such as polyomaviruses which survive cooking, is a possible explanation of this relationship (zur Hausen 2012). The sequencing of polyomaviruses in the metagenome was investigated in retail meat products after virion enrichment (Peretti 2015).

16.3.3 Sequencing of Foodborne Fungal Pathogens

Plants are frequently infected by "field fungi" in the field and synthesize toxic compounds; upon ingestion, these plants or their derived products may cause disease (Lee et al. 2015; Stoev 2015). *Fusarium* spp.-infected cereal grains produce zearalenone, fumonisins, and trichothecenes (e.g., toxins such as deoxynivalenol, T-2, and HT-2), and *Penicillium* spp.-infected fruits produce patulin. During postharvest storage, numerous fungi infest a wide range of food and feed products. *Aspergillus* spp. and *Penicillium* spp. are notable examples, developing acute or chronically toxic compounds, such as aflatoxins, ohratoxins, citrinin, and other antimicrobials with indirect health impacts through gut microbiota modulation (Gillings et al. 2015; Stoev 2015). During the processing of food, some mycotoxins are persistent. They can accelerate infection from a wide range of fungal and non-fungal pathogens (Antonissen et al. 2014; Stoev 2015), and it is postulated that the fungal infections are mainly contributed by mycotoxins (Withlow and Hagler 2016).

The NCBI repository indicates that there are approximately 2,515 genome assemblies and 29 complete fungal genomes accessible. However, there are very few cases of high-throughput sequencing (HTS) in epidemiological and fungal FBP virulence studies (Billmyre et al. 2014; Lee et al. 2014; Litvintseva et al. 2014; Vaux et al. 2014; Litvintseva et al. 2015). Few scientific studies concern unique cases of pathogenesis of fungal food (Lee et al. 2014; Vaux et al. 2014). The main characterizations of FBD are metataxonomic and molecular fungal isolates, and mycobiotas generally are limited to PCR amplifications of one or a few genetic loci followed by the Sanger sequencing

(Byrnes et al. 2010; Desnos-Ollivier et al. 2015; Wang et al. 2015c). The importance of fungi as FBPs is probably understudied compared to bacteria and viruses.

Various fungi are opportunistic and contagious FBPs (Clemons et al. 2010; Iriart et al. 2010; Gurgui et al. 2011; Kazan et al. 2011; Benedict et al. 2016). Immunocompromised people are largely infected by hostile fungal infections (Brown et al. 2012; Bitar et al. 2014; Benedict et al. 2016). HTS is widely used to investigate the outbreak of fungal FBP and of numerous sporadic cases of outbreaks of fungus-associated FBD (Benedict et al. 2016).

Commercial yogurts contaminated with fungus, *Mucor circinelloides,* lead to illness with nausea, vomiting, and diarrhea. The lethal fungal infection is triggered by the mucormal fungal infection, mucormycosis, which is steadily on the rise. The strains of *M. circinelloides, F. lusitanicus,* and *F. griseocyanus* were identified as fewer common causes of infection, however, *Mucor circinelloides* and were identified as the most virulent *M. circinelloides* subspecies commonly associated with human infections as detected by the multilocus sequence typing of yogurt. WGS investigation of the yogurt isolate established it as being close to the *M. circinelloides, F. circinelloides* subgroup, with a greater percentage of divergence with the *M. circinelloides, F. lusitanicus* subgroup (Soo Chan Lee et al. 2014).

The US Center for Disease Control and Prevention (CDC) uses WGS to help resolve outbreaks of fungal diseases. WGS offers comprehensive data on fungi and other disease-triggering pathogens. The main purposes of WGS are to track, identify, and prevent the spread of disease-causing fungi, as well as to understand how people become exposed to infection-causing fungi, identify how geographically dispersed pathogen disease-causing fungi are spreading, find the source of outbreaks, and discover how fungi cause disease. *Candida auris,* an evolving fungus that poses a significant global health threat, have been detected throughout the world using whole genomic (Oltean et al. 2019).

16.3.4 Sequencing of Foodborne Protozoan Parasites

Toxoplasma gondii is an intracellular obligatory protozoan parasite, causing serious health problems in individuals who are at high risk for food poisoning: pregnant women, infants, older adults and immunocompromised people. Symptoms can be flu-like and include swollen lymph glands, muscle aches and pains that last months, and eye indications, culminating in diminished or blurry vision or discomfort, like redness or tearing.

Toxoplasmosis is a foodborne infection which is of significant concern and poses the highest risk of illness among all parasitic infections (WHO 2015; Limon et al. 2017; Bouwknegt et al. 2018). Raw pork meat and meat products are considered to be one of the key risk determinants for *T. gondii* infection, since viable *T. gondii* parasites have been isolated from it (van der Giessen et al. 2007; Kijlstra et al. 2009; Limon et al. 2017). ABI PRISM 310 Genetic Analyzer (Applied Biosystems, Foster City, CA) with ABI Prism Large Dye Terminator v. 3.1, Cycle Sequencing Kits and Big Dye XTerminator Purification Kit (Applied Biosystems) were used for the sequencing of the amplicons. Geneious V. 11.1.4. Software (Geneious, Wellington, New Zealand) was used to analyze sequences and compare with sequences in the NCBI database using Blast.

16.3.5 Sequencing of FBP in Plant-Derived/Fermented Products

Vegetables are one of the carriers of pathogenic microbes and have contributed to the outbreaks of foodborne illness (Buchholz et al. 2011; Frank et al. 2011). Irrigation wastewater from farms comprises a substantial number of coliform bacteria reservoirs and cross-contaminates the surface of plants. Several microorganisms colonize on the plant surface located above ground (phyllosphere) (Lindow and Brandl 2003). Recent NGS studies have established how seasonal irrigation and soil type can influence the phyllosphere microbiome. In the romaine lettuce phyllosphere microbiome, the strongest link between the compositions of the microbiome were temperature fluctuations (Williams et al. 2013).

Soil is a central factor influencing vegetable and fruit microbiomes. 16S rDNA sequencing of microbes in leaves, flowers, and fruits had a similar taxon with that of the soil microbiome in which the plants were grown (Zarraonaindia et al. 2015). The irrigation water that contains pathogenic microorganisms may, therefore, contaminate soil, leading to cross-contamination of fruit or vegetables (Van Dyk et al. 2016), leading to outbreaks (Buchholz et al. 2011).

Bacterial communities that are unique are found to be associated on the surface of fresh fruits and vegetables (Leff and Fierer 2013). A high abundance of Enterobacteriaceae was documented in sprouts, spinach, lettuce, tomato, pepper, and strawberries, while Actinobacteria, Bacteroidetes, Firmicutes, and Proteobacteria were dominant in other fruits, like apples, peaches, grapes, and mushrooms. Unless proper cleaning is practiced, all individuals are susceptible to these microbes. NGS is a choice platform for the identification of microorganisms linked with ready-to-eat and raw food products.

Until shotgun metagenomics was used (Ottesen et al. 2013) efforts to detect and isolate *Salmonella* (a common contaminant) from tomatoes were unsuccessful.

In the case of cilantro (coriander leaves), the microbiota was characterized and samples were identified as *Salmonella* using 16S rDNA shotgun metagenomics (Jarvis et al. 2015). Before enrichment, gram-negative Proteobacteria dominated in cilantro samples. However, the composition had transformed mainly to gram-positive Firmicutes after enrichment for 24 hrs (Ottesen et al. 2013). In bagged spinach, STEC was detected using shotgun metagenomics (Leonard et al. 2015). It was proposed that the serotype and virulence genes of contaminating pathogen should be determined by the DNA sequence.

One of the first microbial food ecology studies involving NGS characterized the bacterial group in "meju," a Korean, soybean-derived, fermented food. Fermented soybean pastes are assorted with seasonings, vegetables, and rice in many Asian products, such as "cheonggukjang," "doenjang," and "kochjang" (Nam et al. 2012 a, b). NGS techniques have been widely used to identify the microbial populations of fermented products. Predominant microbes identified using NGS included *Bacillus* spp. (*B. amyloliquefaciens*, *B. licheniformis*, and *B. subtilis*), followed by *Enterococcus* spp. (*E. faecalis*, *E. faecium*), *Lactobacillus* spp., and species of other LAB genera (*Leuconostoc*, *Weissella*, *Tetragenococcus*).

A fermented Korean product, "kimchi," that combines vegetables, fermented marine products ("jeotgal"), and seasonings with a rich, starchy paste (rice, wheat) was the subject in studies utilizing pyrosequencing (Jung et al. 2011; Park et al. 2012). Pyrosequencing revealed that at the end of fermentation *Lactobacillus* spp. were dominant while during the early stages of fermentation, other species dominated, such as *Pseudomonas*, *Enterobacter*, *Leuconostoc* spp., and *Weissella* spp. (Park et al. 2012).

Spanish- and Greek-style fermented table olives were sequenced using the pyrosequencing of rDNA- and cDNA-derived amplicons (Cocolin et al. 2013). A larger number of Enterobacteriaceae, such as *Enterobacter*, *Citrobacter*, *Escherichia*, and *Klebsiella*, were associated with these samples. The microbiome of rye and wheat sourdough fermentation was examined using 16S rDNA pyrosequencing (Ercolini et al. 2013). *Weissella* spp. was the leading bacteria present in rye flour. *Lactobacillus sakei*, *Leuconostoc* spp., *Weissella* spp., and *Lactococcus lactis* were the dominant species present in wheat sourdoughs.

Cultures of kefir grains and kefir beverages of different origin have also been studied using the pyrosequencing of amplicons of rDNA (Dobson et al. 2011; Leite et al. 2012). *Lactobacillus kefiranofaciens, Lactobacillus kefiri, Lactobacillus parakefiri, Lactobacillus buchneri,* and others were the prevalent microbes identified by pyrosequencing. Further, *Acetobacter* and *Lactococcus* were the minor bacterial species present (Dobson et al. 2011; Leite et al. 2012). The research verified that the beverages' microbiota were distinct from that of the grains (Dobson et al. 2011). NGS-based approaches play a major role in revealing complex microbiome changes and characterizing the microbiome of raw food products.

16.3.6 Sequencing of FBP Present in Milk and Fermented Dairy Products

Farm livestock is a major source of milk pathogens (Arqués et al. 2015). In advanced countries, the most prevalent potential pathogens associated with milk or dairy products include *Campylobacter, Escherichia coli* O157:H7, *Listeria monocytogenes, Salmonella* spp., and *Staphylococcus aureus* (Jakobsen et al. 2011); they are also the principal pathogen threats linked to raw milk (Kousta et al. 2010; Yang et al. 2012; Claeys et al. 2013) and raw cheese (Verraes et al. 2015).

Different parameters influence the microbial composition of milk under natural conditions, such as microorganisms present on the teat, animal feed, water supply quality, and equipment hygiene (Quigley et al. 2013). Therefore, it is important to identify pathogens in dairy products to ensure health. The elimination, or at least the reduction, of foodborne diseases (Giacometti et al. 2013) is one of the most critical tasks in food health. Consequently, pathogens in different food products need to be identified in the early stages of development, thereby reducing the number of outbreaks of FBPs. WGS has been vital to the accurate detection and identification of foodborne bacteria in diverse types of dairy products (including pasteurized milk) that are responsible for outbreaks; those efforts would not have been possible without these molecular approaches.

The core microbiota of cheese consists of *Lactobacillus* sp., *Streptococcus* sp., and *Lactococcus* sp. The other detectable genera also identified by sequencing in the milk and cheese, of cow, sheep, and goat include *Enterococcus* sp., *Streptococcus* sp., and *Leuconostoc* sp., *Phylobacterium* sp., *Carnobacterium* sp., *Stenotrophomonas* sp., *Pseudomonas* sp., and *Kurthia* sp. were also detected (Bao et al. 2012).

Total microbial DNA of milk of healthy cows has been pyrosequenced, and substantial variations in their microbial compositions have been identified with clinical or subclinical mastitis (Bhatt et al. 2012; Kuehn et al. 2013). A greater abundance of *Brevundimonas, Burkholderia, Sphingomonas,* and *Stenotrophomonas* were found in the mastitis samples reads, while *Pseudomonas, Psychrobacter,* and *Ralstonia* reads were in larger proportions of healthy samples (Kuehn et al. 2013). The authors suggested that a microbial mixture rather than a single pathogen may cause some cases of mastitis based on these findings. Along with sequences of the host species *S. aureus, E. coli,* and *Enterobacter, Yersinia* species were found. Large numbers of viral sequences were also present, and it was hypothesized that the 'natural resistance' in milk against pathogens might be provided by phages present (Bhatt et al. 2012).

The microbiota of cow milk, particularly for the identification of the microbial species involved in mastitis, was examined by the pyrosequencing of rDNA amplicons.

Escherichia coli, Klebsiella spp., *Staphylococcus aureus, Streptococcus uberis*, and *Trueperella pyogenes* were the most common pathogens causing mastitis which was detected by pyrosequencing (Oikonomou et al. 2012).

Streptococcus thermophilus accounted for approximately 99 percent of the sequences identified compared to raw milk, the majority of which were mesophilic LAB organisms. The reported *Vibrio* spp. contained 4% of raw milk and 0.1% of milk processed, respectively (Delgado et al. 2013). The small amount of DNA sequences of *Geobacillus toebii* and *Methylobacterium populi* was isolated from milk and dairy products that were not previously identified.

The variability and dynamics of the bacterial communities in the production and ripening of conventional cheese, including Danish raw milk cheeses, artisanal Irish cheeses, Latin-style cheeses, the Polish cheese Oscypek, and water buffalo mozzarella were also studied by pyrosequencing (Masoud et al. 2010; Alegría et al. 2012; Quigley et al. 2012; Lusk et al. 2012; Ercolini et al. 2012). In all the studied cheeses and production stages, the specific forms, depending on the cheese technology used and/or the use (or not) of starter cultures, LAB were shown to be dominant.

Pyrosequencing identified sequences of microbes in cheese including *Bifidobacterium* spp. in Oscypek (Alegría et al. 2012), *Faecalibacterium, Prevotella*, and *Helcococcus* in Irish cheeses (Quigley et al. 2012), and *Exiguobacterium* in Latin-style cheeses (Lusk et al. 2012). The new species *Vibrio casei* (Bokulich et al. 2013), which was originally isolated from French washed rind cheese (Blicher et al. 2010), has recently been correlated with the Vibrionaceae read sequences.

Culture-independent molecular techniques such as pyrosequencing, which allow fast insights into the composition, structures, and dynamics of microbial communities, have been used to respond to the diversity of microbial communities in cheese manufacturing and ripening (Alegria et al. 2012). The microbial diversity and phylogeny of buttermilk, a fermented milk product, were assessed using 16S rRNA-based pyrosequencing (Sathyanarayan et al. 2013).

The Illumina MiSeq tool was used to sequence 18 isolates in a milk processing environment. For library preparation, the prepmkit NEBNext® UltraTM II FS library (NEB) was used. The MiSeq reagent kit V3 (Illumina) was used to sequence and isolate the genus *Bacillus*. These sequences were analyzed using bioinformatics tools and for library preparation of obtained sequences, NEBNext® UltraTM II FS library prep kit (NEB) was used, which is available at https://cge.cbs.dtu.dk/services/.

From a dairy powder processing unit, previously characterized *Cronobacter sakazakii* was retrieved via a wgMLST approach and later selected for WGS using Oxford Nanopore technology. For identifying spore-forming bacteria in skimmed powdered milk, partial sequencing of 16S rRNA (a 570-bp part of the V3–V5 region) was performed. For library preparation, a DNA library preparation kit, Nextera® XT (Illumina), was used and a MiSeq reagent kit was chosen for running sequencing. Bioinformatics annotation tools were used for the trimming, assembly, and quality control of the sequences. For identification and characterization purposes, WGS contigs assembler was subsequently used.

The development and maturation areas of two artisan cheese factories were examined for bacterial and fungal populations in a similar study and involving the same technology (Bokulich and Mills 2013). *Debaryomyces* and *Lactococcus* were the predominant eukaryotic and prokaryotic organism-associated microbes present in the fermentation of milk and the establishment of these microbes on processing surfaces played a vital role in their transmission between sequential fermentations. The Illumina technology was used to assess microflora in the cheese-making industry and to study its effect on cheese microbiota (Bokulich and Mills 2013).

The 16S rDNA sequences provided details about the number of genera such as *Streptococcus, Lactococcus, Corynebacterium, Lactobacillus*, and *Kocuria* in human milk samples, and clear evidence of causality in the existing pathogenic strains *Cronobacter* spp. and *Salmonella enterica*.

The dynamics of Bola de Ocosingo's microbial population during the cheese processing were sequenced using pyrosequencing which revealed the dominance of *S. thermophilus, L. lactis, L. helveticus, L. delbrueckii*, and *L. plantarum*, all reported with potential probiotic effect. In the ripened cheese, pyrosequencing has shown very low proportions of *E. coli/S. flexnerii*.

Sequencing of 16S rRNA was conducted on an Illumina MiSeq System with a v2 500-cycles kit. 16S rDNA sequence analysis was performed, which offers identification accuracy and aims to reduce the loss of natural biodiversity, including potentially significant strains. Environmental microbiota is critical for the attainment and preservation of quality and safety in food processing plants.

MLST has been productive in the molecular analysis of epidemiology (Vogler et al. 2011; Revez et al. 2014; Cui et al. 2013; Revez et al. 2014) and for identifying *Lactobacillus* (Diancourt et al. 2007; Picozzi et al. 2010; Tanigawa and Watanabe 2011), *Pediococcus* (Calmin et al. 2008), *Enterococcus* (Burgos et al. 2009), *Oenococcal* (Bilhère et al. 2009), and *Lactococcus* (Passerini et al. 2010; Fernández et al. 2011). In different types of cheese made from cow and buffalo milk, the predominant genera *Streptococcus* and *Lactobacillus* were identified (Bilhère et al. 2009; Passerini et al. 2010; Fernández et al. 2011). The texture, taste, and aroma of cheese products are greatly influenced by *Streptococcus*, which produces bacteriocins that protect the cheese from microbial spoilage (Yu et al. 2011).

16.3.7 Sequencing of FBPs Associated with Meat

The animal genomic research trend emerged after the Human Genome Project began in 2001 (Lander et al. 2001; Venter et al. 2001). To produce a significant volume of sequence data efficiently at low costs (and a low quality), the promise of high-throughput (HT)-NGS based several studies on the sequencing and assembly of the entire genome of animals of food production, including bovine/cattle (Womack 2006), pig (Mote and Rothschild 2006), sheep (Cockett 2006), equine/horse (Chowdhary and Raudsepp 2006), and avian/chicken (Burt 2006). These genome sequence innovations have presented an argument for the advancement of genetic tools within the domestic animal industry and understanding of animal growth (Elsik et al. 2009). The research on the genome of the cattle is a good example of how functions and regulatory mechanisms are established that support improved production of milk and meat. Technological developments and advancement in the field of HT-NGS allow for a wide range of applications of these techniques, including analysis of the development of drug resistance in bacterial pathogens (Joseph and Read 2010).

These existing technologies can therefore also be implemented directly to improve the production of animals and increase food security. Increased control and quality assurance of food in terms of food health have become the persistent reasons for the protection of animal feed, sanitation and the welfare of livestock, and the conservation of animal systems. The advancement of DNA technology has allowed better monitoring and surveillance systems to be developed. Despite this, over the past few decades, foodborne problems have not decreased (Harlizius et al. 2004). In the framework of pre-harvest food safety, HT-NGS tools will play an important role toward the improvement of the productivity of food animals and genetic disease resistance (Harlizius et al. 2004).

Marination is a conventional method of treatment frequently used during food production. NGS-based strategies were used to research the complexity of microbiome and structure transitions during marination and to determine whether such a process would prolong the shelf life of the product. In a common case, when marinated with acetic acid and subsequently packed into a changed atmosphere, the poultry product was easily spoiled. A greater understanding of bacterial food spoilage has facilitated by NGS approaches. These sequence-based surveys identified the spoilage microbes, but also demonstrated the interaction between different bacterial communities to avoid spoilage. 16S rDNA sequence-based identification of the bacterial microflora of meat contributing to spoilage indicated a heterofermentative lactic acid bacteria *Leuconostoc gasicomitatum* as the spoilage organism. The marinating of meat augmented the percentage of spoilage-associated *Leuconostoc gasicomitatum,* which contributed to faster spoilage, thereby minimizing the percentage of useful microbes such as *Brochothrix thermosphacta, Clostridium* species, and Enterobacteriaceae in the sample (Nieminen et al. 2012a).

To evaluate microbial diversity variations throughout storage under varying conditions, community profiling via rDNA pyrosequencing has been used in a meat system (Ercolini et al. 2011). For example, *Photobacterium, Pseudomonas, Brochothrix, Bacillus, Carnobacterium, Enterococcus, Hafnia, Myroides, Rahnella, Staphylococcus, Serratia, Streptococcus, Weissella,* and *Xanthomonas* are microbes in minced pork meat samples. Improved salt-meat treatment with higher salt levels in combination with vacuum packing increased the abundance of Enterobacteriaceae, Enterococcaceae, and Leuconostocaceae subpopulations which was useful in delaying spoilage (Fougy et al. 2016).

Pyrosequencing has also been used to compare the composition and diversity of microbial communities in marinated and unmarinated broiler fillets (Nieminen et al. 2012b). Bacterial diversity and succession were analyzed by the sequencing of rDNA amplicons and then later by shotgun metagenomics. Smaller populations of *Carnobacterium, Vagococcus, Brochothrix, Clostridium,* Enterobacteriaceae, and *Vibrio* were found in marinated meat compared to the unmarinated fillets (Nieminen et al. 2012b). *Vagococcus* and *Vibrio,* the predominant communities in the unmarinated samples, had never before been associated with the shelf life of meat. In addition, increased proportions of LAB species belonging to the Lactobacillaceae and Leuconostocaceae families were encountered in the marinated samples.

Storage conditions affect the meat microbiota. Further, pyrosequencing identified complex changes in the majority of microbial communities and could adversely affect the consistency of the meat. The integration of NGS with microbial metabolites research (metabolomics) can also help identify microorganisms' beneficial and damaging interactions in the food ecosystem.

One way to monitor the poultry carrier state is to use genetic sequencing techniques (Calenge et al. 2011). The most common foodborne pathogens *Salmonella* and *Campylobacter* may both be transmitted to the poultry intestinal tract. Therefore, the inhibition of these pathogens in chickens may reduce foodborne pathogens' transmission into humans. The regulation and management of *Salmonella* in chickens are among the most promising areas of food safety research (Chambers and Gong 2011). The mechanisms involved with colonization of *Salmonella enterica* in the chicken gut were analyzed using HT-NGS.

Campylobacter jejuni is a zoonotic microbe that primarily infects the human gut via the alimentary chains by consuming undercooked raw chicken meat, ready-to-eat raw food, or unpasteurized milk. In developed countries, *C. jejuni* pathogen is mainly responsible for foodborne infections and is a main cause of apprehension in the health community. The main source of campylobacteriosis in humans is by consumption or handling of contaminated poultry meat (Jorgensen et al. 2002; Scallan et al. 2011). Despite existing research efforts to track, classify, and monitor transmission of this foodborne pathogen, control techniques and preventive measures have also been established at the farm level. The lack of a tool to identify strains of Campylobacter makes it more challenging to trace the causes of outbreaks. More recent work has led to the discovery of different *C. jejuni* clonal groups through genome-based approaches (multilocus sequence sort and microarray) (Dingle et al. 2001). However, Campylobacter's high plasticity obviates the use of genotyping technology in favor of better methods, including WGS (Lefébure et al. 2010).

16.4 APPLICATION OF WGS IN FOOD SAFETY

Food quality and food safety are key components for creating secure and economically viable food sources and are crucial for enhancing food security globally (WHO 2015). Preserving and enhancing the microbiological quality and achieving food safety, including security, involve the identification of known and unknown microbes, particularly spoilage and pathogenic organisms that infiltrate the food supply chain. Using omics-based approaches (i.e., genomics, transcriptomics, proteomics, and metabolomics) may expand the reach of screening programs and therefore can improve the identification of a variety of issues, particularly pathogen identification, foodborne disease epidemic detection, microbial source monitoring, niche adaptation, antimicrobial resistance, and product shelf life. By implementing improved screening and subtyping methods both for known and unknown pathogens, the omics strategies can boost food quality, food safety, and, consequently, public health. Effective tools to track particular bacterial strains to their source and our awareness of these pathogens can be optimized to avoid potential contamination incidents, particularly detecting and remediating strains that survive in a given niche, including the processing plant area (Miller et al. 2013; Zhou et al. 2016; Forbes et al. 2017). Indeed, despite improvements, the projected worldwide incidence of foodborne diseases remains alarmingly high: 600 million foodborne diseases, 420,000 deaths attributable, and 31 major food safety hazards in 2010 (WHO 2015) as well as the associated economic and social losses.

Application of NGS in controlling food safety is expected to be a boon for growing global trade. WGS has emerged, offering tremendous promise for the investigation, assessment, and control of future microbiological food safety problems and diseases. It enables the detection and characterization of microorganisms with an unparalleled level of accuracy, thus effectively mitigating most of the ambiguity that compromises our ability to safely and effectively handle microbiological food safety issues.

In the evaluation of bacterial diversity, culture-dependent approaches are limited in sensitivity. Conventional methods for detecting foodborne pathogens are selective, based on the cultivation of microorganisms; but they are time-consuming. Thus, diverse rapid detection methods have been established that are more responsive, accurate, time-efficient, labor-saving and effective for the rapid detection of foodborne pathogens in food products to avoid outbreaks. Platforms for nucleic acid sequencing are rapidly being used for characterizing bacteria and identifying bacterial populations via complex environmental matrices. The common sequencing approaches used to identify microbial communities include sequencing of 16S rDNA, metagenomics, and metatranscriptomics with regard to food. Metagenomics and metatranscriptomics methods have further advanced the performance of a microbiome by offering deep insight into the compositional and functional features of fermented and non-fermented food microbiomes. The high-throughput ability implemented by NGS and the improved efficiency provided for comprehensive genome investigation allows researchers to carry out sequencing and processing of bacterial genomes during the outbreak of a common foodborne disease investigation. Molecular-level sequencing protocols and in silico-level automation enhance technical capabilities for reading hundreds of DNA base pairs in length, generating gigabases of data in one run applied to contaminated food samples.

During the assessment of a contamination incident, reliable source monitoring is one of WGS' key tools in food safety management. Knowing whether the detected pathogen or spoilage agent is the consequence of a transient or repeated contamination incident is vital in determining the underlying cause of the contamination and will promote the implementation or testing of control measures. WGS may be used to enhance manufacturer and raw material control and to maximize efforts of initiatives for the monitoring of environmental pathogens. Enhanced root-cause research would contribute to a profound understanding of transmission routes as well as to finding new contamination sources. WGS offers insights to enhance the 'hazard detection' phase in the framework of microbial risk analysis. Current information about species is most frequently obtained by investigating well characterized laboratory strains that do not necessarily accurately reflect the broader population's phenotypical diversity. Several domains, like computational food microbiology and thermal processing, are likely to gain from phenotypic prediction using WGS data.

The use of WGS for risk evaluation is intended to assess the relative contribution of various livestock, environmental and food sources including specific food commodity and processing sources for the attribution of sporadic foodborne diseases impacting human health (Pires et al. 2009).

16.5 CONCLUSIONS AND FUTURE PERSPECTIVES

Efficient application of WGS in food safety has contributed to impressive developments in foodborne pathogen detection and surveillance. The next phase toward enhancing food safety would be to use a wide range of foodborne pathogen genomic sequencing tools to obtain functional data and to detect biomarkers predicting the pathogenic potential of isolates and other food safety phenotypes. Dynamic forecasting will be important for improving investigative techniques and evaluating food safety threats at a subspecies level and can have a significant effect on public health and the commercial feasibility of food companies.

The systematic use of WGS as component of the screening of pathogenic diseases greatly enhances our ability to identify and evaluate outbreaks and to track infection dynamics. Simultaneously, the overwhelming range of microbial and human interactions that lead to outbreak is emerging like never before. WGS is primarily used to classify and assess microbial genomes with an epidemiological perspective. The use of a wide range of whole genome sequencing strategies would facilitate the revelation of complex interactions within microbial consortia in food environments in the coming years, through our comprehension about how to use precious microbial resources to ensure efficiency levels as well as food quality and health.

Since WGS provides unparalleled precision to identify the extent of relatedness between bacterial and viral isolates, it accentuates established epidemiological tools by enabling transmission routes. WGS can diagnose infections, forecast the presence of undiagnosed instances and intermediates in transmission networks, indicate the possible route of propagation, and classify undetected risk factors for forward transmission. Such information is critical in the bid to circumvent or reduce outbreaks, in the development and assessment of prevention strategies, and in the utilization of government healthcare services.

Eventually, the quintessential expansion of the influence of NGS would be a significant decrease in the expense of the food industry. The price of generating bacterial genomic sequences is still steadily decreasing, and it is anticipated that this trend will continue into the near future.

The analysis of WGS data currently includes expert judgments in the fields of microbiology, epidemiology, and bioinformatics. WGS has not substituted a successful epidemiological investigation and will not replace it, but rather the two can be used as compatible datasets for the efficient demarcation of epidemic incidents. The enhanced clarity provided by WGS offers future epidemiologists with historic data the opportunity to link instances which had been dismissed as sporadic in the intervening years. Data from WGS may also be used for secondary research, such as assessment of the efficacy of contamination control approaches in the harvest-to-fork-to-flush arena and allocation of geographical origin. Several software programs will assist in the genome research. Furthermore, it is crucial that sequencing does not eliminate culture entirely, as acquiring an isolate is of prime integral part of microbiology and secondary evaluation. These modern approaches will continue to replace traditional molecular subtyping methods and drive improvements in global food safety because of the decreasing costs, enhanced resolution, and value-added secondary analysis NGS offers to food safety.

REFERENCES

Aggarwal, R. and Goel, A. 2015. Hepatitis A: Epidemiology in resource-poor countries. *Current Opinion in Infectious Diseases* 28(5):488–96. doi:10.1097/QCO.00000 00000000188.

Alegría, Á., Szczesny, P., Mayo, B., Bardowski, J. and Kowalczyk, M. 2012. Biodiversity in Oscypek, a traditional Polish Cheese, determined by culture-dependent and -independent approaches. *Applied and Environmental Microbiology* 78(6):1890–8. doi:10.1128/AEM.06081-11.

Angela, J., Taylor, V., Lappi, W., et al. 2015. Characterization of foodborne outbreaks of *Salmonella enterica* serovar *enteritidis* with whole-genome sequencing single nucleotide polymorphism-based analysis for surveillance and outbreak detection. *Journal of Clinical Microbiology* 53(10):3334–40. doi:10.1128/JCM.01280-15.

Ann-Katrin, L., Eduardo, T. and Mirko, R. 2017. Whole-genome sequencing in epidemiology of *Campylobacter jejuni* infections. *Journal of Clinical Microbiology* 55(5):1269–75. doi:10.1128/JCM.00017-17.

Antonissen, G., Martel, A., Pasmans, F., et al. 2014. The impact of *Fusarium* mycotoxins on human and animal host susceptibility to infectious diseases. *Toxins* 6(2):430–52.

Arqués, J. L., Rodríguez, E., Langa, S., Landete, J. M. and Medina, M. 2015. Antimicrobial activity of lactic acid bacteria in dairy products and gut: Effect on pathogens. *Biomedical Research International* 2015:1–9.

Ashton, P. M., Nair, S., Peters, T. M., et al. 2016. *Salmonella* whole genome sequencing implementation group., identification of *Salmonella* for public health surveillance using whole genome sequencing. *Peer Journal* 4:e1752. doi:10.7717/peerj.1752.

Aw, T. G., Wengert, S. and Rose, J. B. 2016. Metagenomic analysis of viruses associated with field-grown and retail lettuce identifies human and animal viruses. *International Journal of Food Microbiology* 223:50–6. doi:10.1016/j.ijfoodmicro.2016.02.008.

Bao, Q., Yu, J., Liu, W., et al. 2012. Predominant lactic acid bacteria in traditional fermented yak milk products in the Sichuan Province of China. *Dairy Science and Technology* 92(3):309–19. doi:10.1007/s13594-012-0061-x.

Barco, L., Ramon, E., Cortini, E., et al. 2014. Molecular characterization of *Salmonella enterica* serovar 4,(5),12:i. *DT193 ASSuT Strains from Two Outbreaks in Italy*. *Foodborne Pathogens and Disease* 11(2):138–44.

Benedict, K., Chiller, T. M. and Mody, R. K. 2016. Invasive fungal infections acquired from contaminated food or nutritional supplements: A review of the literature. *Foodborne Pathogens and Disease* 13(7):343–9. doi:10.1089/fpd.2015.2108.

Berglund, E. C., Kiialainen, A. and Syvanen, A. C. 2011. Next-generation sequencing technologies and applications for human genetic history and forensics. *Investigative Genetics* 2(1):1–15.

Bernard, H., Faber, M., Wilking, H., et al. 2014. Outbreak investigation team. *Eurosurveillance* 19(8):20719.

Bhatt, V. D., Ahir, V. B., Koringa, P. G., et al. 2012. Milk microbiome signatures of sub-clinical mastitis-affected cattle analysed by shotgun sequencing. *Journal of Applied Microbiology* 112(4):639–50.

Biggs, P. J., Fearnhead, P., Hotter, G., et al. 2011. Whole-genome comparison of two *Campylobacter jejuni* isolates of the same sequence type reveals multiple loci of different ancestral lineage. *PLoS ONE* 6(11):e27121. doi:10.1371/journal.pone.0027121.

Bilhère, E., Lucas, P. M., Claisse, O. and Lonvaud-Funel, A. 2009. Multilocus sequence typing of *Oenococcus oeni*: Detection of two subpopulations shaped by intergenic recombination. *Applied and Environmental Microbiology* 75(5):1291–300. doi:10.1128/AEM.02563-08.

Billmyre, R. B., Croll, D., Li, W., et al. 2014. Highly recombinant VGII *Cryptococcus gattii* population develops clonal outbreak clusters through both sexual macroevolution and asexual microevolution. *mBio* 5(4):e01494–14. doi:10.1128/mBio.01494-14.

Bitar, D., Lortholary, O., Le Strat, Y., et al. 2014. Population-based analysis of invasive fungal infections, France, 2001–2010. *Emerging Infectious Diseases* 20(7):1149–55. doi:10.3201/eid2007.140087.

Blicher, A., Neuhaus, K. and Scherer, S. 2010. *Vibrio casei* sp. nov. isolated from the surface of two French red smear soft cheeses. *International Journal of Systematic and Evolutionary Microbiology* 60(8):1745–9.

Bokulich, N. A. and Mills, D. A. 2013. Facility-specific "house" microbiome drives microbial landscapes of artisan cheese making plants. *Applied and Environmental Microbiology* 79(17):5214–23. doi:10.1128/AEM.00934-13.

Bokulich, N. A., Ohta, M., Richardson, P. M. and Mills, D. A. 2013. Monitoring seasonal changes in winery-resident microbiota. *PLoS ONE* 8(6):e66437.

Bouwknegt, M., Devleesschauwer, B., Graham, H., Robertson, L. and vander, G. J. 2018. The Euro-FBP workshop participants, Prioritisation of food-borne parasites in Europe, 2016. *Eurosurveilliance* 2018:23. doi:10.2807/1560-7917.ES.2018.23.9.17-00161.

Brown, G. D., Denning, D. W. and Levitz, S. M. 2012. Tackling human fungal infections. *Science* 336(6082):647. doi:10.1126/science.1222236.

Buchholz, U., Bernard, H., Werber, D., et al. 2011. German outbreak of *Escherichia coli* O104:H4 associated with sprouts. *New England Journal of Medicine*:11–23. doi:10.1056/NEJMoa1106482.

Burgos, M. J., Lopez, R. L., Abriouel, H., Omar, N. B. and Galvez, A. 2009. Multilocus sequence typing of *Enterococcus faecalis* from vegetable foods reveals two new sequence types. *Foodborne Pathogens and Disease* 6(3):321–7. doi:10.1089/fpd.2008.0169.

Burt, D. 2006. The chicken genome. In: *Vertebrate genomes*, edited by Volff J.-N., 123–37. Basel, Karger Press.

Byrnes, E. J. III, Li, W., Lewit, Y., et al. 2010. Emergence and pathogenicity of highly virulent *Cryptococcus gattii* genotypes in the northwest United States. *PLoS Pathogens* 6(4):e1000850. doi:10.1371/journal.ppat.1000850.

Calenge, F., Legarra, A. and Beaumont, C. 2011. Genomic selection for carrier-state resistance in chicken commercial lines. *BMC Proceedings* 5:S24.

Calmin, G., Lefort, F. and Belbahri, L. 2008. Multi-loci sequence typing (MLST) for two lacto-acid bacteria (LAB) species: *Pediococcus parvulus* and *P. damnosus*. *Molecular Biotechnology* 40(2):170–9. doi:10.1007/s12033-008-9073-4.

Carrillo, C. D., Kruczkiewicz, P., Mutschall, S., Tudor, A., Clark, C. and Taboada, E. N. 2012. A framework for assessing the concordance of molecular typing methods and the true strain phylogeny of *Campylobacter jejuni* and *C. coli* using draft genome sequence data. *Frontiers in Cellular and Infection Microbiology* 2:57–64.

Chambers, J. R. and Gong, J. 2011. The intestinal microbiota and its modulation for *Salmonella* control in chickens. *Food Research International* 44(10):3149–59.

Check, H.E. 2015. Pint-sized DNA sequencer impresses first users. *Nature* 521(7550):15–16. doi:10.1038/521015a.

Chin, C. S., Sorenson, J., Harris, J. B., et al. 2011. The origin of the Haitian cholera outbreak strain. *New England Journal of Medicine* 364(1):33–42.

Chowdhary, B. and Raudsepp, T. 2006. The horse genome. *Genome Dynamics* 2:97–110.

Claeys, W. L., Cardoen, S., Daube, G. et al. 2013. Raw or heated cow milk consumption: Review of risks and benefits. *Food Control* 31(1):251–62.

Clark, C. G., Berry, C., Walker, M., et al. 2016. Genomic insights from whole genome sequencing of four clonal outbreak *Campylobacter jejuni* assessed within the global *C. jejuni* population. *BMC Genomics* 17(1):990.

Clemons, K. V., Salonen, J. H., Issakainen, J., Nikoskelainen, J., McCullough, M. J., Jorge, J. J. and Stevens, D. A. 2010. Molecular epidemiology of *Saccharomyces cerevisiae* in an immune-compromised host unit. *Diagnostic Microbiology and Infectious Disease* 68(3):220–7. doi:10.1016/j.diagmicrobio.2010.06.010.

Cockett, N. 2006. The sheep genome. *Genome Dynamics* 2:79–85.

Cocolin, L., Alessandria, V., Botta, C., Gorra, R., De Filippis, F., Ercolini, D. and Rantsiou, K. 2013. NaOH-Debittering induces changes in bacterial ecology during table olives fermentation. *PLoS ONE* 8(7):e69074.

Cui, Y., Yu, C., Yan, Y., et al. 2013. Historical variations in mutation rate in an epidemic pathogen, *Yersinia pestis*. *Proceedings of the National Academy of Sciences of the United States of America* 110(2):577–82. doi:10.1073/pnas.1205750110.

Dallman, T. J., Ashton, P. M., Byrne, L., et al. 2015. Applying phylogenomics to understand the emergence of Shiga-toxin-producing *Escherichia coli* O157:H7 strains causing severe human disease in the UK. *Microbial Genomics* 1(3):e000029. doi:10.1099/mgen.0.000029.

de Wit, M. A., Koopmans, M. P. G. and van Duynhoven, Y. 2003. Risk factors for Norovirus, Sapporo-like virus, and group A rotavirus gastroenteritis. *Emerging Infectious Diseases* 9(12):1563–70. doi:10.3201/eid0912.020076.

Delgado, S., Rachid, C. T., Fernández, C. C., Rychlik, T., Alegría, A., Peixoto, R. S. and Mayo, B. 2013. Diversity of thermophilic bacteria in raw pasteurized and selectively-cultured milk as assessed by culturing PCR-DGGE and Pyrosequencing. *Food Microbiology* 36(1):103–11.

Desnos-Ollivier, M., Patel, S., Raoux-Barbot, D., Heitman, J., Dromer, F. and Group, F. C. S. 2015. Cryptococcosis serotypes impact outcome and provide evidence of *Cryptococcus neoformans* speciation. *mBio* 6(3):e00311–15. doi:10.1128/mBio.00311-15.

Diancourt, L., Passet, V., Chervaux, C., Garault, P., Smokvina, T. and Brisse, S. 2007. Multilocus sequence typing of *Lactobacillus casei* reveals a clonal population structure with low levels of homologous recombination. *Applied and Environmental Microbiology* 73(20):6601–11. doi:10.1128/AEM.01095-07.

Dingle, K. E., Colles, F. M., Wareing, D. R. et al. 2001. Multilocus sequence typing system for *Campylobacter jejuni*. *Journal of Clinical Microbiology* 39(1):14–23.

Dobson, A., O'sullivan, O., Cotter, P. D., Ross, P. and Hill, C. 2011. High-throughput sequence-based analysis of the bacterial composition of kefir and an associated kefir grain. *FEMS Microbiology Letters* 320(1):56–62.

Domingues, A. R., Pires, S. M., Halasa, T. and Hald, T. 2012. Source attribution of human campylobacteriosis using a meta-analysis of case-control studies of sporadic infections. *Epidemiology and Infection* 140(6):970–81.

EFSA 2011. Scientific opinion on an update on the present knowledge on the occurrence and control of foodborne viruses. *EFSA Journal* 9(7):2190–285. doi:10.2903/j.efsa.2011.2190.

EFSA 2013. Scientific opinion on VTEC-seropathotype and scientific criteria regarding pathogenicity assessment. *EFSA Journal* 11(4):3138–243. doi:10.2903/j.efsa.2013.3138.

EFSA 2015. The European Union summary report on trends and sources of zoonoses, zoonotic agents and food-borne outbreaks in 2013. *EFSA Journal* 13:165. doi:10.2903/j.efsa.2015.3991.

Elsik, C. G., Tellam, R. L., Worley, K. C., et al. 2009. The genome sequence of taurine cattle: A window to ruminant biology and evolution. *Science* 324(5926):522–8. doi:10.1126/science.1169588.

Emond-Rheault, J. G., Jeukens, J., Freschi, L., et al. 2017. A syst-OMICS approach to ensuring food safety and reducing the economic burden of salmonellosis. *Frontiers in Microbiology* 8:996.

Ercolini, D., De Filippis, F., La Storia, A. and Iacono, M. 2012. "Remake" by high-throughput sequencing of the microbiota involved in the production of water buffalo Mozzarella cheese. *Applied and Environmental Microbiology* 78(22):8142–8145.

Ercolini, D., Ferrocino, I., Nasi, A., et al. 2011. Monitoring of microbial metabolites and bacterial diversity in beef stored under different packaging conditions. *Applied and Environmental Microbiology* 77(20):7372–81.

Ercolini, D., Pontonino, E., de Filippis, F., Minervini, F., La Storia, A., Gobbetti, M. and Di Cagno, R. 2013. Microbial ecology dynamics during rye and wheat sourdough preparation. *Applied and Environmental Microbiology* 79(24):7827–36.

Fernández, E., Alegría, Á., Delgado, S., Martín, M. C. and Mayo, B. 2011. Comparative phenotypic and molecular genetic profiling of wild *Lactococcus lactis* subsp. *lactis* strains of the *L. lactis* subsp. *lactis* and *L. lactis* subsp. *cremoris* genotypes, isolated from starter-free cheeses made of raw milk. *Applied and Environmental Microbiology* 77(15):5324–35. doi:10.1128/AEM.02991-10.

Fernandez-Cassi, X., Timoneda, N., Gonzales-Gustavson, E., Abril, J. F., Bofill-Mas, S. and Girones, R. 2017. A metagenomic assessment of viral contamination on fresh parsley plants irrigated with fecally tainted river water. *International Journal of Food Microbiology* 257:80–90. doi:10.1016/j.ijfoodmicro.2017.06.001.

Fiore, A. E. 2004. Hepatitis A transmitted by food. *Clinical Infectious Diseases : An Official Publication of the Infectious Diseases Society of America* 38(5):705–15. doi:10.1086/381671.

Fitzgerald, C., Collins, M., van Duyne, S., Mikoleit, M., Brown, T. and Fields, P. 2007. Multiplex, bead-base suspension array for molecular determination of common *Salmonella* serogroups. *Journal of Clinical Microbiology* 45(10):3323–34.

Forbes, J. D., Knox, N. C., Ronholm, J., Pagotto, F. and Reimer, A. 2017. Metagenomics: The next culture-independent game changer. *Frontiers in Microbiology* 8:1069. doi:10.3389/fmicb.2017.01069.

Fougy, L., Desmonts, M., Coeuret, G., et al. 2016. Reducing salt in raw pork sausages increases spoilage and correlates with reduced bacterial diversity. *Applied and Environmental Microbiology* 82(13):3928–39.

Frank, C., Faber, M. S., Askar, M., et al. 2011. Large and ongoing outbreak of haemo-lyticuraemic syndrome, Germany. *Eurosurveillance* 16(21):19878.

Giacometti, J., BuretićTomljanović, A. and Josić, D. 2013. Application of proteomics and metabolomics for investigation of food toxins. *Food Research International* 54(1):1042–51.

Gillings, M. R., Paulsen, I. T. and Tetu, S. G. 2015. Ecology and evolution of the human microbiota: Fire, farming and antibiotics. *Genes* 6(3):841–57. doi:10.3390/genes6030841.

Grass, J. E., Gould, L. H. and Mahon, B. E. 2013. Epidemiology of foodborne disease outbreaks caused by *Clostridium perfringens*, United States, 1998–2010. *Foodborne Pathogens and Diseases* 10(2):131–6. doi:10.1089/fpd.2012.1316.

Grimont, P. A. D. and Weill, F. X. 2007. Antigenic formulae of the *Salmonella* serovars. In: WHO Collaborating Centre for Reference and Research on Salmonella WHO, 9th ed. 9:1–166.

Gurgui, M., Sanchez, F., March, F., et al. 2011. Nosocomial outbreak of *Blasto schizomyces capitatus* associated with contaminated milk in a haematological unit. *Journal of Hospital Infection* 78(4):274–8. doi:10.1016/j.jhin.2011.01.027.

Halliday, M. L., Kang, L. Y., Zhou, T. K., et al. 1991. An epidemic of hepatitis-A attributable to the ingestion of raw clams in Shanghai, China. *Journal of Infectious Diseases* 164(5):852–9. doi:10.1093/infdis/164.5.852.

Harlizius, B., van Wijk, R. and Merks, J. M. M. 2004. Genomics for food safety and sustainable animal production. *Journal of Biotechnology* 113(1–3):33–42.

Hayes, S., Mahony, J., Nauta, A. and van Sinderen, D. 2017. Metagenomic approaches to assess bacteriophages in various environmental niches. *Viruses* 9(6):127. doi:10.3390/v9060127.

Hedberg, C. W. and Osterholm, M. T. 1993. Outbreaks of food-borne and waterborne viral gastroenteritis. *Clinical Microbiology Reviews* 6(3):199–210. doi:10.1128/CMR.6.3.199.

Henao, O. L., Jones, T. F., Vugia, D. J., Griffin, P. M. and Network F. D. A. S. 2015. Foodborne diseases active surveillance network - 2 decades of achievements, 1996-2015. *Emerging Infectious Diseases* 21(9):1529–36.

Hénock, B. N. Y. and Dovie, D. B. K. 2007. Diarrheal diseases in the history of public health. *Archives of Medical Research* 38(2):159–163.

Holt, K., Thomson, N., Wain, J., et al. 2008. Pseudogene accumulation in the evolutionary histories of *Salmonella enterica* serovars Paratyphi A and Typhi. *BMC Genomics* 10:36.

Imamura, S., Haruna, M., Goshima, T., Kanezashi, H., Okada, T. and Akimoto, K. 2016. Application of next-generation sequencing to investigation of Norovirus diversity in shellfish collected from two coastal sites in Japan from 2013 to 2014. *Japanese Journal of Veterinary Research* 64(2):113–22. doi:10.14943/jjvr.64.2.113.

Inns, T., Lane, C., Peters, T., et al. 2015. A multi-country *Salmonella enteritidis* phage type 14b outbreak associated with eggs from a German producer: 'near real-time' application of whole genome sequencing and food chain investigations, United Kingdom, May to September 2014. *Eurosurveillance* 20(16):21098.

Iriart, X., Fior, A., Blanchet, D., Berry, A., Neron, P. and Aznar, C. 2010. *Monascus ruber*: Invasive gastric infection caused by dried and salted fish consumption. *Journal of Clinical Microbiology* 48(10):3800–2. doi:10.1128/JCM.01000-10.

Jakobsen, R. A., Heggebø, R., Sunde, E. B. and Skjervheim, M. 2011. *Staphylococcus aureus* and *Listeria monocytogenes* in Norwegian raw milk cheese production. *Food Microbiology* 28(3):492–6.

Jarvis, K. G., White, J. R., Grim, C. J., et al. 2015. Cilantro microbiome before and after nonselective pre-enrichment for *Salmonella* using 16S rRNA and metagenomic sequencing. *BMC Microbiology* 15:160. doi:10.1186/s12866-015-0497-2.

Joensen, K. G., Kiil, K., Gantzhorn, M. R., et al. 2020. Whole-genome sequencing to detect numerous *Campylobacter jejuni* outbreaks and match patient isolates to sources, Denmark, 2015–2017. *Emerging Infectious Diseases* 26(3):523–32.

Joensen, K. G., Scheutz, F., Lund, O., Hasman, H., Kaas, R. S., Nielsen, E. M. and Aarestrup, F. M. 2014. Real-time whole-genome sequencing for routine typing, surveillance, and outbreak detection of verotoxigenic *Escherichia coli*. *Journal of Clinical Microbiology* 52(5):1501–10.

Jorgensen, F., Bailey, R., Williams, S., et al. 2002. Prevalence and numbers of *Salmonella* and *Campylobacter* sp. on raw, whole chickens in relation to sampling methods. *International Journal of Food Microbiology* 76(1–2):151–4.

Joseph, S. J. and Read, T. D. 2010. Bacterial population genomics and infectious disease diagnostics. *Trends in Biotechnology* 28(12):611–8.

Jung, J. Y., Lee, S. H., Kim, J. M., et al. 2011. Metagenomic analysis of "kimchi" a traditional Korean fermented food. *Applied and Environmental Microbiology* 77(7):2264–74.

Kaakoush, N. O., Castaño-Rodríguez, N., Mitchell, H. M. and Man, S. M. 2015. Global epidemiology of Campylobacter infection. *Clinical Microbiology Reviews* 28(3):687–720.

Kazan, E., Maertens, J., Herbrecht, R., et al. 2011. A retrospective series of gut aspergillosis in haematology patients. *Clinical Microbiology and Infection : The Official Publication of the European Society of Clinical Microbiology and Infectious Diseases* 17(4):588–94. doi:10.1111/j.1469-0691.2010.03310.x.

Keyburn, A. L., Boyce, J. D., Vaz, P., et al. 2008. NetB, a new toxin that is associated with avian necrotic enteritis caused by *Clostridium perfringens*. *PLoS Pathogens* 4(2):e26. doi:10.1371/journal.ppat.0040026.

Kijlstra, A., Meerburg, B. G. and Bos, A. P. 2009. Food safety in free-range and organic livestock systems: Risk management and responsibility. *Journal of Food Protection* 12:2448–681.

Kousta, M., Mataragas, M., Skandamis, P. and Drosinos, H. 2010. Prevalence and sources of cheese contamination with pathogens at farm and processing levels. *Food Control* 21(6):805–15.

Kovanen, S., Kivistö, R., Llarena, A. K., et al. 2016. Tracing isolates from domestic human *Campylobacter jejuni* infections to chicken slaughter batches and swimming water using whole-genome multilocus sequence typing. *International Journal of Food Microbiology* 226:53–60.

Kovanen, S. M., Kivistö, R. I., Rossi, M., et al. 2014. Multilocus sequence typing (MLST) and whole-genome MLST of *Campylobacter jejuni* isolates from human infections in three districts during a seasonal peak in Finland. *Journal of Clinical Microbiology* 52(12):4147–54.

Kuehn, J. S., Gorden, P. J. O., Munro, D., et al. 2013. Bacterial community profiling of milk samples as a means to understand culture-netative bovine clinical mastitis. *PLoS ONE* 8(4):e61959.

Kuhn, K. G., Nielsen, E. M., Mølbak, K. and Ethelberg, S. 2018. Determinants of sporadic *Campylobacter* infections in Denmark: A nationwide case-control study among children and young adults. *Clinical Epidemiology* 10:1695–707.

Kvistholm, J. A., Nielsen, E. M., Björkman, J. T., et al. 2016. Whole-genome sequencing used to investigate a nationwide outbreak of listeriosis caused by ready-to-eat delicatessen meat, Denmark, 2014. *Clinical Infectious Diseases: An Official Publication of the Infectious Diseases Society of America* 63(1):64–70.

Ladner, J. T., Beitzel, B., Chain, P. S. G., et al. 2014. Threat characterization consortium, standards for sequencing viral genomes in the era of high-throughput sequencing. *mBio* 5(3):e01360-14. doi:10.1128/mBio.01360-14.

Lahti, E., Rehn, M., Ockborn, G., Hansson, I., Ågren, J., Engvall, E. O. and Jernberg, C. 2017. Outbreak of campylobacteriosis following a dairy farm visit: Confirmation by genotyping. *Foodborne Pathogens and Disease* 14(6):326–32.

Lander, E. S., Linton, L. M., Birren, B., et al. 2001. Initial sequencing and analysis of the human genome. *Nature* 409(6822):860–921. doi:10.1038/35057062.

Lee, H. B., Patriarca, A. and Magan, N. 2015. *Alternaria* in food: Ecophysiology, mycotoxin production and toxicology. *Mycobiology* 43(2):93–106. doi:10.5941/MYCO.2015.43.2.93.

Lee, S. C., Blake, R., Alicia, B. L., et al. 2014. Analysis of a food-borne fungal pathogen outbreak: Virulence and genome of a *Mucor circinelloides* isolate from yogurt. *mBio* 5(4):e01390-14. doi:10.1128/mBio.01390-14.

Lefébure, T., Bitar, P. D., Suzuki, H. and Stanhope, M. J. 2010. Evolutionary dynamics of complete *Campylobacter pangenomes* and the bacterial species concept. *Genome Biology and Evolution* 2:646–55.

Leff, J. W. and Fierer, N. 2013. Bacterial communities associated with the surfaces of fresh fruits and vegetables. *PLoS ONE* 8(3):e59310. doi:10.1371/journal.pone.0059310.

Leite, A. M., Mayo, B., Rachid, C. T., Peixoto, R. S., Silva, J. T., Paschoalin, V. M. and Delgado, S. 2012. Assessment of the microbial diversity of Brazilian kefir grains by PCR-DGGE and Pyrosequencing analysis. *International Journal of Food Microbiology* 31(2):215–21.

Leonard, S. R., Mammel, M. K., Lacher, D. W. and Elkins, C. A. 2015. Application of metagenomic sequencing to food safety: Detection of Shiga Toxin-producing *Escherichia coli* on fresh bagged spinach. *Applied and Environmental Microbiology* 81(23):8183–91. doi:10.1128/AEM.02601-15.

Li, J., Paredes-Sabja, D., Sarker, M. R. and McClane, B. A. 2016. *Clostridium perfringens* sporulation and sporulation-associated toxin production. *Microbiology Spectrum* 4(3):1–27. doi:10.1128/microbiolspec.TBS-0022-2015.

Limon, G., Beauvais, W., Dadios, N., Villena, I., Cockle, C., Blaga, R. and Guitian, J. 2017. Cross-sectional study of *Toxoplasma gondii* infection in pig farms in England. *Foodborne Pathogens and Disease* 14(5):269–81.

Lindow, S. E. and Brandl, M. T. 2003. Microbiology of the phyllosphere. *Applied and Environmental Microbiology* 69(1).1875 83. doi.10.1128/AEM.69.4.1875-1883. 2003.

Litvintseva, A. P., Hurst, S., Gade, L., et al. 2014. Whole-genome analysis of *Exserohilum rostratum* from an outbreak of fungal meningitis and other infections. *Journal of Clinical Microbiology* 52(9):3216–22. doi:10.1128/JCM.00936-14.

Litvintseva, A. P., Marsden-Haug, N., Hurst, S., et al. 2015. Valley fever: Finding new places for an old disease: *Coccidioides immitis* found in Washington State soil associated with recent human infection. *Clinical Infectious Diseases : An Official Publication of the Infectious Diseases Society of America* 60(1):E1–3. doi:10.1093/cid/ciu681.

Llarena, A. K., Zhang, J., Vehkala, M., et al. 2016. Monomorphic genotypes within a generalist lineage of *Campylobacter jejuni* show signs of global dispersion. *Microbial Genomics* 10(10):2–9.

Lusk, T. S., Ottensen, A. R., White, J. R., Allard, M. W., Brown, E. W. and Kase, J. E. 2012. Characterization of microflora in Latin-style cheeses by next-generation sequencing technology. *BMC Microbiology* 12:254.

Macori, G., Bellio, A., Bianchi, D. M., et al. 2019. Genome-wide profiling of enterotoxigenic *Staphylococcus aureus* strains used for the production of naturally contaminated cheeses. *Genes* 11(1):12–27.

Martinovic, T., Andjelkovic, U., Gajdosik, M. S., Resetar, D. and Josic, D. 2016. Foodborne pathogens and their toxins. *Journal of Proteomics* 147:226–35. doi:10.1016/j.jprot.2016.04.029.

Masoud, W., Takamiya, M., Vogensen, F. K., Lillevang, S., Al-Soud, W. A., Sørensen, S. J. and Jakobsen, M. 2010. Characterization of bacterial populations in Danish raw milk cheeses made with different starter cultures by denaturing gradient gel electrophoresis and Pyrosequencing. *International Dairy Journal* 21:142–8.

McQuiston, J. R., Herrera-Leon, S., Wertheim, B. C., Doyle, J., Fields, P. I., Tauxe, R. V. and Logsdon, J. M. 2008. Molecular phylogeny of the Salmonellae: Relationships among *Salmonella* species and subspecies determined from four housekeeping genes and evidence of lateral gene transfer events. *Journal of Bacteriology* 190(21):7060–7.

Mellmann, A., Harmsen, D., Cummings, C. A., et al. 2011. Prospective genomic characterization of the German enterohemorrhagic *Escherichia coli* O104:H4 outbreak by rapid next generation sequencing technology. *PLoS ONE* 6(7):e22751.

Miller, R. R., Montoya, V., Gardy, J. L., Patrick, D. M. and Tang, P. 2013. Metagenomics for pathogen detection in public health. *Genome Medicine* 5(9):81. doi:10.1186/gm485.

Moore, M. D., Goulter, R. M. and Jaykus, L. A. 2015. Human Norovirus as a foodborne pathogen: Challenges and developments. *Annual Review of Food Science and Technology* 6:411–33. doi:10.1146/annurev-food-022814-015643.

Morganti, M., Bolzoni, L., Scaltriti, E., et al. 2018. Rise and fall of outbreak-specific clone inside endemic pulsotype of *Salmonella*. Insights from high-resolution molecular surveillance in Emilia-Romagna, Italy, 2012 to 2015. *Eurosurveillance Weekly* 4(5):12–23.

Mote, B. and Rothschild, M. 2006. Cracking the genomic piggy bank: Identifying secrets of the pig genome. *Genome Dynamics* 2:86–96. doi:10.1159/000095097.

Mughini, G., Smid, L., Wagenaar, J. H., et al. 2012. Risk factors for campylobacteriosis of chicken, ruminant, and environmental origin: A combined case-control and source attribution analysis. *PLoS ONE* 7(8):e42599.

Nadon, C., van Walle, I., Gerner-Smidt, P., et al. 2017. PulseNet International: Vision for the implementation of whole genome sequencing (WGS) for global foodborne disease surveillance. *Eurosurveillance* 22(23):30544. doi:10.2807/1560-7917. ES.2017.22.23.30544.

Nam, Y. D., Lee, S. Y. and Lim, S. I. 2012a. Microbial community analysis of Korean soybean pastes by next-generation sequencing. *International Journal of Food Microbiology* 155(1–2):36–42.

Nam, Y. D., Yi, S. H. and Lim, S. I. 2012b. Bacterial diversity of cheonggukjang a traditional Korean fermented food analyzed by barcoded Pyrosequencing. *Food Control* 28(1):135–42.

Newell, D. G., Koopmans, M., Verhoef, L., et al. 2010. Food-borne diseases - The challenges of 20 years ago still persist while new ones continue to emerge. *International Journal of Food Microbiology* 139:S3–15. doi:10.1016/j.ijfoodmicro.2010.01.021.

Nieminen, T. T., Koskinen, K., Laine, P., et al. 2012a. Comparison of microbial communities in marinated and unmarinated broiler meat by metagenomics. *International Journal of Food Microbiology* 157(2):142–9.

Nieminen, T. T., Välitalo, H., Säde, E., Paloranta, A., Koskinen, K. and Björkroth, J. 2012b. The effect of marination on lactic acid bacteria communities in raw broiler fillet strips. *Frontiers in Microbiology* 3:376.

Nieuwenhuijse, D. F. and Koopmans, M. P. G. 2017. Metagenomic sequencing for surveillance of food- and waterborne viral diseases. *Frontiers in Microbiology* 8:230–9. doi:10.3389/fmicb.2017.00230.

Oikonomou, G., Machado, V. S., Santisteban, C., Schukken, Y. H. and Bicalho, R. C. 2012. Microbial diversity of bovine mastitis milk as described by Pyrosequencing of metagenomic 16S rDNA. *PLoS ONE* 7(10):e47671.

Oltean, H. N., Etienne, K. A., Roe, C. C., Gade, L., McCotter, O. Z., Engelthaler, D. M. and Litvintseva, A. P. 2019. Utility of whole-genome sequencing to ascertain locally acquired cases of coccidioidomycosis, Washington, USA external icon. *Emerging Infectious Diseases* 3(3):501–6. doi:10.3201/eid2503.181155.

Ottesen, A. R., Gonzalez, A., Bell, R., et al. 2013. Co-enriching microflora associated with culture based methods to detect *Salmonella* from tomato phyllosphere. *PLoS ONE* 8(9):e73079. doi:10.1371/journal.pone.0073079.

Palma, F., Manfreda, G., Silva, M., et al. 2018. Genome-wide identification of geographical segregated genetic markers in *Salmonella enterica* serovar Typhimurium variant 4,(5),12:i. *Scientific Reports* 8(1):15251.

Park, E. J., Chun, J., Cha, C. J., Park, W. S., Jeon, C. O. and Bae, J. W. 2012. Bacterial community analysis during fermentation of ten representative kinds of "kimchi" with barcoded Pyrosequencing. *International Journal of Food Microbiology* 30(1):197–204.

Passerini, D., Beltramo, C., Coddeville, M., Quentin, Y., Ritzenthaler, P., Daveran-Mingot, M. L. and Le Bourgeois, P. 2010. Genes but not genomes reveal bacterial domestication of *Lactococcus lactis*. *PLoS ONE* 5(12):1–9. doi:10.1371/journal. pone.0015306.

Pearce, M. E., Alikhan, N. F., Dallman, T. J., Zhou, Z., Grant, K. and Maiden, M. C. J. 2018. Comparative analysis of core genome MLST and SNP typing within a European *Salmonella* serovar enteritidis outbreak. *International Journal of Food Microbiology* 274:1–11.

Peretti, A., FitzGerald, P. C., Bliskovsky, V., Buck, C. B. and Pastrana, D. V. 2015. Hamburger polyomaviruses. *Journal of General Virology* 96(4):833–9. doi:10. 1099/vir.0.000033.

Picozzi, C., Bonacina, G., Vigentini, I. and Foschino, R. 2010. Genetic diversity in Italian *Lactobacillus sanfranciscensis* strains assessed by multilocus sequence typing and pulsed-field gel electrophoresis analyses. *Microbiology* 156(7):2035–45. doi:10.1099/mic.0.037341-0.

Pires, S. M., Evers, E. G., van Pelt, W., et al. 2009. Attributing the human disease burden of foodborne infections to specific sources. *Food Borne Pathogens and Disease* 6(4):417–24. doi:10.1089/fpd.2008.0208.

Pires, S. M., Fischer-Walker, C. L., Lanata, C. F., et al. 2015. Aetiology-specific estimates of the global and regional incidence and mortality of diarrhoeal diseases commonly transmitted through food. *PLoS ONE* 10(12):e0142927. doi:10.1371/ journal.pone.0142927.

Pires, S. M., Vieira, A. R., Perez, E., Wong, D. L. F. and Hald, T. 2012. Attributing human foodborne illness to food sources and water in Latin America and the Caribbean using data from outbreak investigations. *International Journal of Food Microbiology* 152(3):129–38. doi:10.1016/j.ijfoodmicro.2011.04.018.

Prüss-Ustün, A., Vickers, C., Pascal, H. and Roberto, B. 2011. Knowns and unknowns on burden of disease due to chemicals: A systematic review. *Environmental Health* 10:9. doi:10.1186/1476-069X-10-9.

Quick, J., Loman, N. J., Duraffour, S., et al. 2016. Real-time, portable genome sequencing for Ebola surveillance. *Nature* 530(7589):228–32. doi:10.1038/nature16996.

Quigley, L., McCarthy, R. O., O'Sullivan, et al. 2013. The microbial content of raw and pasteurized cow milk as determined by molecular approaches. *Journal of Dairy Science* 96(8):4928–37.

Quigley, L., O'Sullivan, O., Beresford, T. P., Ross, R. P., Fitzgerald, G. F. and Cotter, P. D. 2012. High-throughput sequencing for detection of subpopulations of bacteria not previously associated with artisanal cheeses. *Applied and Environmental Microbiology* 78(16):5717–23.

Raymond, K., Shabhonam, C., Anais, P., et al. 2019. Phylogenomic analysis of gastroenteritis-associated *Clostridium perfringens* in England and Wales over a 7-year period indicates distribution of clonal toxigenic strains in multiple outbreaks and extensive involvement of enterotoxin-encoding (CPE) plasmids. *Microbial Genomics* 5(10):10. doi:10.1099/mgen.0.000297.

Reimer, A., Weedmark, K., Petkau, A., et al. 2019. Shared genome analyses of notable listeriosis outbreaks, highlighting the critical importance of epidemiological evidence, input datasets and interpretation criteria. *Microbial Genomics* 5(1):1–9. doi:10.1099/mgen.0.000237.

Revez, J., Llarena, A. K., Schott, T., et al. 2014. Genome analysis of *Campylobacter jejuni* strains isolated from a waterborne outbreak. *BMC Genomics* 15:768. doi:10.1186/1471-2164-15-768.

Ribot, E. M., Fair, M. A., Gautom, R., Cameron, D. N., Hunter, S. B., Swaminathan, B. and Barrett, T. J. 2006. Standardization of pulsed-field gel electrophoresis protocols for the subtyping of *Escherichia coli* O157:H7, *Salmonella*, and *Shigella* for PulseNet. *Foodborne Pathogens and Disease* 3(1):59–67.

Rizzo, C., Alfonsi, V., Bruni, R., et al. 2013. Ongoing outbreak of hepatitis A in Italy: Preliminary report as of 31 May 2013. *Eurosurveillance* 18(27):20518.

Rodríguez-Lazaro, D., Cook, N., Ruggeri, F. M., et al. 2012. Virus hazards from food, water and other contaminated environments. *FEMS Microbiology Reviews* 36(4):786–814. doi:10.1111/j.1574-6976.2011.00306.x.

Rolando, R. 2011. Hand washing a big deal? *SA Food Review*. http://www.foodreview.co.za/safety-andhygiene/32-hand-washing-a-big-deal.

Sathyanarayanan, J., Muthuirulan, P., Jeyaprakash, R. and Paramasamy, G. 2013. Microbial diversity and phylogeny analysis of buttermilk, a fermented milk product, employing 16S rRNA-based Pyrosequencing. *Food Biotechnology* 27(3):213–21. doi:10.1080/08905436.2013.811084.

Scallan, E., Hoekstra, R. M., Angulo, F. J., et al. 2011. Foodborne illness acquired in the United States - Major pathogens. *Emerging Infectious Diseases* 17(1):7–15.

Sironi, M., Cagliani, R., Forni, D. and Clerici, M. 2015. Evolutionary insights into host-pathogen interactions from mammalian sequence data. *Nature Reviews Genetics* 16(4):224–36. doi:10.1038/nrg3905.

Stein, C., Kuchenmuller, T., Henrickx, S., Prüss-Ustün, A., Wolfson, L., Engels, D. and Schlundt, J. 2007. The global burden of disease assessments-WHO is responsible? *PLoS Neglected Tropical Diseases* 1(3):e161.

Stoev, S. D. 2015. Foodborne mycotoxicoses, risk assessment and underestimated hazard of masked mycotoxins and joint mycotoxin effects or interaction. *Environmental Toxicology and Pharmacology* 39(2):794–809. doi:10.1016/j.etap.2015.01.022.

Sukhnanand, S., Alcaine, S., Warnick, L. D., et al. 2005. DNA sequence-based subtyping and evolutionary analysis of selected *Salmonella enterica* serotypes. *Journal of Clinical Microbiology* 43(8):3688–98.

Suttle, C. A. 2005. Viruses in the sea. *Nature* 437(7057):356–61. doi:10.1038/nature04160.

Tanigawa, K. and Watanabe, K. 2011. Multilocus sequence typing reveals a novel subspeciation of *Lactobacillus delbrueckii*. *Microbiology* 157(3):727–38. doi:10.1099/mic.0.043240-0.

Tauxe, R. V., Doyle, M. P., Kuchenmüller, T., Schlundt, J. and Stein, C. E. 2010. Evolving public health approaches to the global challenge of foodborne infections. *International Journal of Food Microbiology* 139(1):S16–28.

Tavoschi, L., Severi, E., Niskanen, T., et al. 2015. The joint effects of lifestyle factors and comorbidities on the risk of colorectal cancer: A large Chinese retrospective case-control study. *PLoS ONE* 10(12):e0143696. doi:10.1371/journal.pone.0143696.

Ung, A., Baidjoe, A. Y., Van Cauteren, D., et al. 2019. Disentangling a complex nationwide *Salmonella dublin* outbreak associated with raw-milk cheese consumption, France, 2015 to 2016. *Eurosurveillance Weekly* 24(3):1–9.

Uzal, F. A., McClane, B. A., Cheung, J. K., Theoret, J., Garcia, J. P., Moore, R. J. and Rood, J. I. 2015. Animal models to study the pathogenesis of human and animal *Clostridium perfringens* infections. *Veterinary Microbiology* 179(1–2):23–33. doi:10.1016/j.vetmic.2015.02.013.

van der Giessen, J., Fonville, M., Bouwknegt, M., Langelaar, M. and Vollema, A. 2007. Seroprevalence of *Trichinella spiralis* and *Toxoplasma gondii* in pigs from different housing systems in the Netherlands. *Veterinary Parasitology* 148(3–4):371–4.

Van Dyk, B. N., De Bruin, W., du Plessis, E. M. and Korsten, L. 2016. Microbiological food safety status of commercially produced tomatoes from production to marketing. *Journal of Food Protection* 79(3):392–406. doi:10.4315/0362-028X.JFP-15-300.

Vaux, S., Criscuolo, A., Desnos-Ollivier, M., et al. 2014. Multicenter outbreak of infections by *Saprochaete clavata*, an unrecognized opportunistic fungal pathogen. *mBio* 5(6):e02309–14. doi:10.1128/mBio.02309-14.

Venter, J. C., Adams, M. D., Myers, E. W., et al. 2001. The sequence of the human genome. *Science* 291(5507):1304–51. doi:10.1126/science.1058040.

Verhoeff-Bakkenes, L., Jansen, H. A. P. M., Veld, P. H., Beumer, R. R., Zwietering, M. H. and van Leusden, F. M. 2011. Consumption of raw vegetables and fruits: A risk factor for *Campylobacter* infections. *International Journal of Food Microbiology* 144(3):406–12.

Verraes, C., Vlaemynck, G., Van Weyenberg, S., et al. 2015. A review of the microbiological hazards of dairy products made from raw milk. *International Dairy Journal* 50:32–44.

Vogler, A. J., Chan, F., Wagner, D. M., et al. 2011. Phylogeography and molecular epidemiology of *Yersinia pestis* in Madagascar. *PLoS Neglected Tropical Diseases* 5(9):1–9. doi:10.1371/journal.pntd.0001319.

Waldram, A., Dolan, G., Ashton, P. M., Jenkins, C. and Dallman, T. J. 2018. Epidemiological analysis of *Salmonella* clusters identified by whole genome sequencing, England and Wales 2014. *International Journal of Food Microbiology* 71:39–45.

Wang, S. H., Shen, M., Lin, H. C., Sun, P. L., Lo, H. J. and Lu, J. J. 2015. Molecular epidemiology of invasive *Candida albicans* at a tertiary hospital in northern Taiwan from 2003 to 2011. *Medical Mycology* 53(8):828–36. doi:10.1093/mmy/myv065.

WHO. 2007a. Countries urged to be more vigilant about food safety. www.who.int/mediacentre/news/ releases/2007/pr39/en/.

WHO. 2007b. *Food safety and food borne illnesses. Fact sheet* 237. www.who.int/entity/media_centre/ factsheets/fs237/en/.

WHO. 2008. *WHO initiative to estimate the global burden of foodborne diseases. A summary document.* Geneva, WHO.

WHO. 2015. *WHO estimates of the global burden of foodborne diseases.* Technical report. http://www.who.int/foodsafety/publications/foodborne_disease/fergreport/en/. Geneva.

Williams, T. R., Moyne, A. L., Harris, L. J. and Marco, M. L. 2013. Season, irrigation, leaf age, and *Escherichia coli* inoculation influence the bacterial diversity in the lettuce phyllosphere. *PLoS ONE* 8(7):e68642. doi:10.1371/journal.pone.0068642.

Wise, M. G., Siragusa, G. R., Plumblee, J., Healy, M., Cray, P. J. and Seal, B. S. 2009. Predicting *Salmonella enterica* serotypes by repetitive sequence-based PCR. *Journal of Microbiological Methods* 76(1):18–24.

Withlow, L. and Hagler, W. Jr. 2016. Mold and mycotoxin issues in dairy cattle: Effects, *Prevention and Treatment.* http://articles.extension.org/pages/11768/mold-and-mycotoxin-issues-in-dairy-cattle:-effects-prevention-and-treatment.

Womack, J. 2006. The bovine genome. *Genome Dynamics* 2:69–78.

Wuyts, V., Mattheus, W., de Laminne, G., et al. 2013. MLVA as a tool for public health surveillance of human *Salmonella typhimurium*: Prospective study in Belgium and evaluation of MLVA loci stability. *PLoS ONE* 8(12):e84055.

Xi, M., Zheng, J., Zhao, S., Brown, E. W. and Meng, J. 2008. An enhanced discriminatory pulsed-field gel electrophoresis scheme for subtyping *Salmonella* serotypes Heidelberg, Kentucky, SaintPaul, and Hadar. *Journal of Food Protection* 71(10):2067–72.

Xiao, Y. 2014. *Clostridium perfringens* sporulation, germination and outgrowth in food: A functional genomics approach. Ph. D. Thesis. Wageningen, Wageningen University.

Yahara, K., Meric, G., Taylor, A. J., et al. 2017. Genome-wide association of functional traits linked with *Campylobacter jejuni* survival from farm to fork. *Environmental Microbiology* 19(1):361–80.

Yang, B., Shi, Y., Xia, X., Xi, M., Wang, X., Ji, B. and Meng, J. 2012. Inactivation of foodborne pathogens in raw milk using high hydrostatic pressure. *Food Control* 28(2):273–8.

Yoshida, C. E., Kruczkiewicz, P., Laing, C. R., Lingohr, E. J., Gannon, V. P., Nash, J. H. and Taboada, E. N. 2016. The *Salmonella* In Silico Typing Resource (SISTR): An open web-accessible tool for rapidly typing and subtyping draft *Salmonella* genome assemblies. *PLoS ONE* 11(1):e0147101.

Yu, J., Wang, W. H., Menghe, B. L., et al. 2011. Diversity of lactic acid bacteria associated with traditional fermented dairy products in Mongolia. *Journal of Dairy Science* 94(7):3229–41. doi:10.3168/jds.2010-3727.

Zarraonaindia, I., Owens, S., Weisenhorn, P., et al. 2015. The soil microbiome influences grapevine-associated microbiota. *mBio* 6(2):e02527–14. doi:10.1128/mBio.02527-14.

Zhou, Y., Wylie, K. M., Feghaly, R. E., et al. 2016. Metagenomic approach for identification of the pathogens associated with diarrhea in stool specimens. *Journal of Clinical Microbiology* 54(2):368–75. doi:10.1128/JCM.01965-15.

Zur Hausen, H. 2012. Red meat consumption and cancer: Reasons to suspect involvement of bovine infectious factors in colorectal cancer. *International Journal of Cancer* 130(11):2475–83. doi:10.1002/ijc.27413.

CHAPTER 17

Examining Microbial Ecology of Food Using NGS Techniques

Nayana Aluparambil Radhakrishnan, Amisha Mathew,
Sunil Pareek, and Radhakrishnan Edayileveettil Krishnankutty

CONTENTS

17.1 INTRODUCTION

Food is a rich source of various nutrients like proteins, carbohydrates, and lipids and, at the same time, it can contain complex microbial communities composed of viruses, bacteria, fungi, and protozoa. These can be (i) spoilage causing organisms, (ii) pathogens, or (iii) beneficial microorganisms which can cause food spoilage, contamination, or fermentation (Ercolini 2013). Therefore, specific knowledge about the diverse microbial communities present in a particular food sample is very essential to improve the food quality, shelf life, and safety.

Culturing is the traditional method used for the identification of food microbiota, which mainly involves the isolation and cultivation of microorganisms from the given sample for their identification and typing (Mayo et al. 2014). But culturing is not a reliable method for studying the complete microbiology of any ecosystem, including food, as only 1% of the total microbes are culturable (Amann et al. 1995). To solve the same, culture-independent molecular techniques have been developed to study the microbial

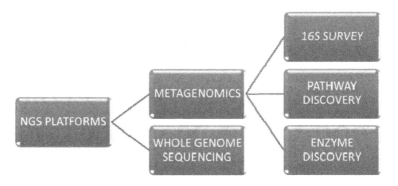

FIGURE 17.1   Next-generation sequencing and its applications in food microbiology.

diversity DNA or RNA. Sanger's dideoxy DNA sequencing formed the basis of the cul-ture-independent method and was time-consuming and expensive. Along with this, dif-ferent amplification techniques mainly targeting the rRNA genes (rDNA) have evolved. Also, techniques like temporal temperature gradient gel electrophoresis (TTGE), denatur-ing gradient gel electrophoresis (DGGE), single-stranded conformation polymorphism (SSCP), terminal restriction fragment length polymorphism (TRFLP), real-time quantita-tive PCR (qPCR), the construction and analysis of 16S rRNA gene libraries, etc. have also been introduced (Mayo et al. 2014). In recent years, there has been tremendous progress in DNA sequencing technologies, collectively called next-generation sequencing (NGS) techniques. These techniques allow the easy identification of microbial taxa, including the ones that are generally uncultivable and the ones that are present in very few num-bers. NGS techniques are now widely used for in-depth understanding of microbial com-position of food and different ecosystems in the food (Giraffa and Neviani 2001).

Food microbiology deals with the study of microorganisms in food having beneficial as well as deleterious effect on the safety and quality of food products. The rapid and cost-effective NGS technologies have reshaped microbial taxonomy and classification and have remolded the face of genome sequencing projects of food-related microbial species (Coenye et al. 2005). Moreover, NGS techniques have helped in profiling food microbial communi-ties and in understanding the molecular mechanisms underlying interesting functionalities in the food ecosystem (Solieri et al. 2013). In food microbiology, NGS approaches can be used in two ways: (1) for determining the whole genome sequence of cultured isolates through "whole-genome sequencing" (WGS) and (2) "metagenomics" where sequences of multiple microbes in the sample are generated using NGS (Figure 17.1). WGS is acquiring popularity as a wonderful tool for surveillance of foodborne illness also (Jagadeesan et al. 2019). Metagenomics has a widespread application in maintaining food safety and qual-ity. It can be used to predict the presence or emergence of spoilage organisms or pathogens in the given food sample based on the analysis of microbial community profile (Jagadeesan et al. 2019). In this chapter, various NGS technologies used to analyze the microbial ecol-ogy of food will be discussed to explore their applications in food microbiology.

17.2 NGS TECHNIQUES

For three decades, first generation sequencing, particularly the Sanger sequencing, ruled the DNA sequencing world. However, the time and cost were its major limitations. By 2005, a new generation of DNA sequencers emerged to overcome the limitations of the available DNA sequencing techniques. Advantages of NGS over first generation techniques include: (1) many millions of short reads are generated in parallel (i.e., high throughput), (2) cost effective, (3) high speed, (4) direct detection of sequencing output without the use of electrophoresis, and (5) discreteness (results are free from noise and signal saturation) (Kchouk et al. 2017).

Short-read sequencing methods can be divided into two main categories: (1) sequencing by synthesis (SBS) and (2) sequencing by ligation (SBL). SBS uses DNA polymerase and ligase enzyme to encircle many DNA strands simultaneously. SBL identifies the position of a particular nucleotide in a given DNA sequence by using the DNA ligase enzyme (Tripathi et al. 2019).

NGS platforms that are commercially used include the Illumina (Illumina), Roche 454, SOLiD, and Ion Torrent (Life Technologies) systems. Third-generation techniques, such as PacBio (Pacific Biosciences), heliscope single molecule sequencing, DNA nanoball sequencing, sequencing with mass spectrometry, nanopore DNA sequencing, tunneling current DNA sequencing, and microscopy-based techniques are now overtaking the second-generation sequencers (Table 17.1).

17.2.1 Roche 454 Pyrosequencer

Rothberg and colleagues launched the first NGS platform, Roche/454 pyrosequencer, in 2005 (Meglecz 2007). This pyrosequencing technique detects the pyrophosphate which is released when a nucleotide is added to a new DNA strand. SBS is the technique behind this technique (Kchouk et al. 2017). DNA samples are fragmented randomly and are attached to a bead carrying primers that are complementary to the fragments. It is then amplified to a million copies using an emulsion PCR (oil-water emulsion). The beads are then transferred to a picotiter plate consisting of approximately one million wells, and within each, sequencing reactions occur independently. The pyrosequencing technique is carried out in a well where a chain of downstream reactions produce light in response to the addition of a nucleotide to the newly synthesized DNA strand. Pyrophosphate released during the nucleotide addition is converted here into adenosine triphosphate, which activates the luciferin-luciferase reaction, releasing light. The intensity of light produced is directly proportional to the number of nucleotides incorporated and the light emitted here is recorded by a camera (Mardis 2008). The latest Roche/454 called GS FLX+ can generate reads with 1,000 bp length with a high sequence output (~1million reads per run). Consensus accuracy of 99.9997% at 15X coverage for the *E. coli* genome has been reported with Roche's FLX+ platform (Mayo et al. 2014).

17.2.2 Illumina Platform

In 2006, a new method for sequencing was developed by Solexa Company. In 2008, the company was acquired by Illumina and came to be known as Illumina/Solexa Genome Analyzer. This uses the SBS approach and is one of the most used techniques (Shendure and Ji 2008; Balasubramanian 2015). Here, randomly fragmented DNA samples are ligated with adapters at both ends, one of which is fixed to a slide on a solid plate. Then using PCR bridge amplification with primers that are complementary to the adaptors and four different reversibly-blocked terminator nucleotides (RT-nucleotides), several copies of the attached DNA sequence is obtained to form a cluster (set of sequence obtained from the same original sequence). DNA polymerase helps in extending the primers. Each

TABLE 17.1 Summary of Commonly Used NGS Platforms

Platform	Chemistry	PCR Amplification	Read Length	Output/ Run (bp)	Error Rate	Run Time	Application	Disadvantage	Type of Instrument	Reference
Sanger sequencing	Asynchronous with base-specific terminator	Standard PCR	700	Few 1000bp	Low	2 hrs	Gene/genome sequencing	PCR biases, low degree of parallelism, expensive	Benchtop	Sanger et al. 1977
Roche 454	SBS Pyro-sequencing	EmPCR	>400	1,000,000	Low	7–10 hrs	*De novo* genome sequencing, RNA-seq, resequencing/ targeted re-sequencing	PCR biases, asynchronous synthesis, homo polymer run, base insertion and deletion errors, EmPCR is cumbersome and technically challenging	Benchtop	Mardis 2008; Metzker 2010
Illumina	Polymerase-based SBS	Bridge Amplificat ion	600	40,000,000	Low	3–4 days	*De novo* genome sequencing, RNA-seq, resequencing/ targeted re-sequencing, metagenomics, ChIP	PCR biases, low multiplexing capability of sample	Benchtop	Mardis 2008; Metzker 2010
SOLiD	SBL	EmPCR	35–40	85,000,000	Low	7 days	Transcript counting, mutation detection, ChIP, RNA-seq	EmPCR is cumbersome, technically challenging PCR, long run time	Benchtop	Mardis 2008; Metzker 2010

(Continued)

TABLE 17.1 (CONTINUED) Summary of Commonly Used NGS Platforms

Platform	Chemistry	PCR Amplification	Read Length	Output/ Run (bp)	Error Rate	Run Time	Application	Disadvantage	Type of Instrument	Reference
Ion Torrent	SBS	EmPCR	200–600	100,000,000	Low	2–4 hrs	*De novo* genome sequencing, RNA-seq, resequencing/ targeted re-sequencing, metagenomics	EmPCR is cumbersome, technically challenging PCR	Benchtop	
PacBio	Phospho-linked fluorescent nucleotides	SMRT	1,000–1,200	100,000,000	High	8 hrs	De novo genome sequencing, RNA-seq, resequencing/ targeted re-sequencing, metagenomics, SNPs and Structural variants detection	High instrument cost, low number of sequence read per run, highest error rates compared with other NGS chemistries	Large scale	Metzker 2010
Oxford Nano-pore	Single molecule	-	30,000–200,000	30,000,000,000	High	1 min–48 hrs	Complete genome of isolates and metagenomics	High error rate	Portable	Mikheyev and Tin 2014

nucleotide is uniquely labeled with a fluorescent dye and an inactive 3' OH group to ensure that only one nucleotide is added. Each incorporated nucleotide is then detected using a charge-coupled device and a computer program by measuring the emitted light released by the nucleotide when excited by a laser. The sequencing cycle continues with the incorporation of a new nucleotide after chemically removing the 3' blocker along with the fluorescent label (Bentley et al. 2008; Reuter et al. 2015).

The Illumina/Solexa platform can currently generate reads of 150–300 bp that could be increased to 300–600 bp using "paired-end" sequencing (same DNA cluster being sequenced from both ends). The biggest advantage of the technology is the reduced price for the high output per run (up to 3,000 Mb) (Mayo et al. 2014).

17.2.3 SOLiD System

Life Technologies marketed another NGS sequencer, Supported Oligonucleotide Ligation and Detection (SOLiD), which was acquired by Applied Biosystems (ABI) in 2007 and developed ABI/SOLiD sequencing technology which works on the SBL approach (Tripathi et al. 2019). The ABI/SOLiD process has multiple rounds of sequencing. DNA fragments, after being attached to a universal adapter, are fixed to a magnetic bead. These fragments are then cloned by emulsion PCR and are placed on a glass slide. A fluorescent-labeled 8-mer is subsequently ligated to the DNA fragments and the color thus emitted by the label is recorded. Once a cycle is completed, the extension product is removed and a new primer which is complementary to the n-1 position of the template is introduced for the second round of ligation (Mardis 2008). This cycle is repeated until each base gets sequenced twice. The sequence of the given DNA fragment can be deduced by decoding the colors to letters of DNA. The main advantage of the ABI/SOLiD platform is its high accuracy (99.99%) because each base is read twice. The main limitation is the relatively short reads (75 bp) and long run time (Mayo et al. 2014).

17.2.4 Ion Torrent

Ion Torrent is a semiconductor-based technique which was commercialized in 2010 by Life Technologies. Unlike other second-generation technologies, Ion Torrent detects the number of hydrogen ions released during the polymerization of the DNA, which is directly proportional to the nucleotides incorporated (Rothberg et al. 2011). Here, modified nucleotides or optics are not used. The release of hydrogen ion will result in a change of pH of the solution, and this change is detected using a sensor which then converts it into voltage signals. The Ion Torrent sequencers can produce reads with lengths of 200, 400, and 600 bp along with throughput that could reach 10 GB for ion proton sequencer. The accuracy is less (98%) compared to other SGS techniques; still it can be used in studies like metagenomics.

17.2.5 PacBio SMRT

Pacific Biosciences developed the first genomic sequencer in 2010 using the Single Molecule Real-Time (SMRT) sequencing approach. SMRT was developed by Quake laboratory and is the most widely used technique in third-generation sequencers (Mayo et al. 2014). Like other sequencing technologies, PacBio SMRT also uses fluorescent-labeled nucleotides, but it proceeds without PCR amplification; instead it records the signal in real-time, light emitted when a nucleotide is incorporated. This uses a microfabricated well called zero-mode waveguides with DNA polymerase attached to it. These wells exploit the property that detects fluorescence produced only at the bottom. During the incorporation of a fluorescent labeled nucleotide by the DNA polymerase, the fluorescent label gets detached. This released luminous signal is then recorded using a sensor (Rhoads and Au 2015). Compared to other techniques, one of the main advantages is that the sample preparation is easy, taking only 4–6 hours rather than days. Large strands of DNA (up to 30 kb) can be sequenced by this technique (Mayo et al. 2014).

17.2.6 Oxford Nanopore Sequencing

The order of nucleotides in a DNA sequence can be determined using the Oxford nanopore sequencing technique. The MinION device released by ONT in 2014 was the first nanopore sequencing device (Bleidorn 2016). These devices are very small and can be attached to the USB of a computer, thus making it ideal for "sequencing on go" (Mikheyev and Tin 2014). The MinION device has 512 channels with nanopores, which can detect nearly 10 bp per second. Strand sequencing is the sequencing method behind this device. The samples are modified before sequencing. Hairpin adapters are ligated to one end, and a motor protein is ligated to the other end. Due to the action of the motor protein, the DNA molecule moves through the nanopore as a single strand and gets translated. When this happens the ionic current varies due to the difference in moving nucleotide through the pore. This variation is recorded on a graphical model and later interpreted to get the sequence (Laver et al. 2015). The direct strand is sequenced first followed by the hairpin adapter and then the inverse strand. The main advantage of the device is its low cost and small size. It can provide long reads up to 150 kbp (Kchouk et al. 2017).

17.3 EXAMINING THE MICROBIAL ECOLOGY OF FOODS BY NGS

The study of the microbial ecology of foods has changed with genomics, transcriptomics, proteomics, and metabolomics and other high-throughput NGS analysis. This provides a more reliable identification and monitoring of food-associated microorganisms than traditional analysis. Microbial analysis mainly depends on the specific food and the target microbes which can be pathogens, spoilage associated, potential starters, and beneficial microorganisms with roles in food contamination, spoilage, and fermentation. The study of the microbial ecology of food by the use of various NGS technologies gives the basic idea of diversity of microbes (Ercolini 2013). The NGS analysis of food samples can be done by a series of steps, including nucleic acid extraction and purification, library construction, sequencing and analysis of the raw sequences using computer software, and the searching of different databases (Mayo et al. 2014) (Figure 17.2).

17.3.1 Nucleic Acid Extraction and Purification

Nucleic acid extraction and purification are the primary steps for sequencing-based studies. Nucleic acid extraction mainly involves four steps: cell disruption, removal of membrane lipids, proteins, and other impurities; nucleic acid purification/binding from bulk;

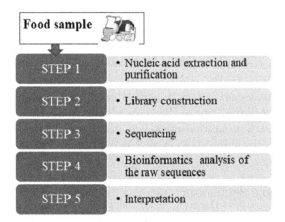

FIGURE 17.2   Steps involved in examining the microbial ecology of foods by NGS.

and nucleic acid concentration (Ali et al. 2017). For the extraction and purification of nucleic acids from the food samples, stringent strategies are required for efficient recovery of nucleic acid and removal of the numerous compounds. Therefore, strategies must be carefully designed for a heterogeneous food system for sufficient homogenization. The well-known method for extraction and purification of nucleic acids involves organic extraction and ethanol precipitation (Tan and Yiap 2009). In addition to this, various kits and methods are also used for the efficient isolation procedures such as Wizard Magnetic DNA Purification for Food (Promega Italia S.r.l., Milano, Italy) and DNeasy Tissue Kit (QIAGEN, Hilden, Germany) (Di et al. 2007). Extraction of genome using silica as an affinity matrix and column-based system are also in use (Ali et al. 2017). Sometimes special consideration should also be given to the lipids and protein components of food, as they can inhibit the PCR amplification and should be removed during the extraction procedure (Bonaiti et al. 2006). A lysis step should include a combination of enzymatic and mechanical treatments for the complete lysis of the cell (Linnarsson 2010). The quality and quantity of the isolated DNA is also a main concern when it is sent for further analysis. Fluorometers (Nakayama et al. 2016), spectrophotometers (Bonaiti et al. 2006), and nanodrop (Binladen et al. 2007) are available for the accurate quantification procedures. Metatranscriptomic studies mainly focus on the mRNA, micro-RNA, tRNA, and other non-coding RNAs. So, for the purification of high-quality non-ribosomal fraction, removal of 90% of rRNA from the total RNA is required (Zhao et al. 2018). Also, cDNA-based techniques can distinguish between viable, active, and inactive cells because they focus on active microbial populations (Cangelosi and Meschke 2014).

17.3.2 Amplification of Target

After the extraction and purification of DNA, the gene of interest can be amplified by PCR. But, based on the particular sequencing platform used, there can be variation in this step. The majority of the diversity analysis of uncultured microorganisms is based on the sequence of PCR-amplified rRNA genes (Hugerth and Andersson 2017). For this, universal primers are used for the PCR amplification to cover most of the microorganisms present in the sample. Genes, such as that of 16S rRNA, help to differentiate the prokaryotic and eukaryotic organelle sequences and 18S, 28S, and fungal internal transcribed spacer (ITS) are used to track the fungal communities (Rivas et al. 2004). For the study of tomato microbiome, Ottesen et al. (2013) have used the V2 region of 16S rRNA genes for Roche pyrosequencing. In cases where the target sequences are known, specific primers can be designed to amplify that particular gene and give a successful PCR outcome (Bonaiti et al. 2006). By applying degenerative primers, the percentage of taxa being targeted by the primers can be increased, but it also increases the issues with amplifying other DNA regions, making it difficult to predict the actual diversity (Mayo et al. 2014). Species diversity and abundance based on rDNA are biased due to the variation in the number of copies of rRNA genes present. Hence, single-copy target genes, such as *rpo*B, *rec*A, *rad*A, *rpo*A, and *gyr*B are also used as phylogenetic markers (Renouf et al. 2006). There are also many computer-based programs such as Primer Prospector and Dege Prime for the design of broad taxonomic primers (Hugerth and Andersson 2017). A major issue with the PCR amplification is the formation of chimeras; this arises due to the sequence similarity between amplicons and the nature of the chain reaction (Nikolaki and Tsiamis 2013). Also, substitutions, insertions, and deletions are the other problems associated with PCR-based amplification (Ercolini 2013). The use of state-of-the-art, high-fidelity polymerases may reduce these error rates.

17.3.3 Library Construction

High-quality library preparation is crucial for the successful NGS experiment. It involves generating a collection of DNA fragments for sequencing (Mayo et al. 2014). NGS library construction is meant to prepare a genomic target, either DNA or RNA, into a form compatible with the sequencing system to be used. The steps involved for preparing RNA or DNA for NGS analysis are: fragmenting and/or sizing the target sequences (RNA or DNA) to a desired length; converting target to double-stranded DNA; attaching oligonucleotide adapters to the ends of target fragments–either the surface of the flow-cell (Illumina) or beads (Ion Torrent). The next step involves clonal amplification of the library, by either cluster generation for Illumina or microemulsion PCR for Ion Torrent (Head et al. 2015). The high quality of sequencing libraries depends on factors, such as quantity and physical characteristics of the RNA or DNA, source material, as well as the desired application (Hugerth and Andersson 2017). Library preparations can be susceptible to bias by genomes that contain unusually high or low guanine-cytosine (GC) content. This can be avoided by the careful selection of polymerases for PCR amplification, thermocycling, conditions, and buffers (Aird et al. 2011).

Numerous kits for making sequencing libraries from DNA are available commercially from a variety of vendors. High-quality reagents with barcoded adapters and PCR primers are also readily available. DNA library preparation using a transposase-based method (Nextera) developed by Illumina (Nextera DNA Sample Prep Kit) is also available (Kulski 2016). To reduce the bias of sequencing the genomes of single cells, multiple displacement amplification by adding a quasi-linear pre-amplification step is also available for the library construction (Head et al. 2015).

17.3.4 Data Analysis

The initial step of data analysis is to assess the quality of raw sequencing data. After that, data is processed based on read quality scores and by the removal of non-biological sequences. Commonly used tools for this task are Fastx, Trim Galore, and Cutadapt (Hugerth and Andersson 2017). Contaminants such as low-quality bases, adapters, primer dimers, and reads that are too short are removed before mapping. The next step is operational taxonomic units (OTU) picking; OTU clustering of sequences will be based on its similarity (Jagadeesan et al. 2019). This step eliminates the sequencing errors formed by PCR, and sequence diversity is reduced to true biological diversity. Popular software packages for OTU picking are Usearch, Qiime (which runs Uclust in the background), and Vsearch, an open-source alternative to Usearch (Tan et al. 2015). By taxonomic assignment (composition approach method), classifying the reads at different taxonomic levels (phylum, class, order, family, or genus) based on the similarities of the sequences with those in a database, it gives information about the relationship of the reads with known microbial groups and helps in ecological and/or functional interpretations (Mayo et al. 2014). Additionally, it allows for comparisons between different studies, even when sequences come from different regions of the marker. For the analysis of 16S, 18S, and large 23S, 28S rDNA sequence databases, such as ribosomal database (RDP), Silva and Greengenes projects, and, for fungus, ITS UNITE database are available (Santamaria et al. 2012). Generally, rDNA sequences sharing 95% of identity are considered to belong to the same genus, and those with 97% or higher identity are considered to belong to the same species (Beye et al. 2018). For the interpretation of genomic and metagenomic data, web tools, such as RAST and MG-RAST can be used (Keegan et al. 2016). Other databases, such as FunGene, are designed to help interpret the phylogenetic markers (like *rpo*B, *gyr*G, *rec*A, etc.) and functional genes (such as those for antibiotic resistance, biodegradation activities, etc.) (Fish et al. 2013). The most popular software programs, Mothur and QIIME, are used to analyze the massive sequencing data both of which are freely distributed and come with manuals and data analysis examples (Hugerth and Andersson 2017). In the IMG database and the PICRUSt software, large numbers of genome and metagenomic datasets are available to infer the number of genes shared by different microbial taxa and provide information on the phenotypic relationships within a microbial community (Langille et al. 2013).

17.4 NGS IN FOOD MICROBIOLOGY

Examining the microbial ecology by the use of the NGS technique is revolutionizing the food industry. It highlights the fate of microbes populating a specific food and helps to differentiate beneficial microbes from contaminants. WGS, SNP, and MLST allow the detailed comparison of individual strains and phylogenetic clustering. Metabarcoding and metagenomic analysis are also helpful for the complete identification of microbial populations of the food (Jagadeesan et al. 2019). Metagenomics coupled with transcriptomics or proteomics helps the identification of functional capacity and biochemical activity of microbial populations (Vieites et al. 2009). DNA and RNA sequence information collected from the food samples will be used to describe a microbial baseline representing the beneficial microbial communities which can be applied to trace the source of contamination and food authentication (Jagadeesan et al. 2019). The genomic data that is obtained by these processes can also be used to detect the presence of antibiotic resistance genes, virulence factors, or other genetic markers (Ronholm 2018). Pyrosequencing of tagged 16S rRNA gene amplicons has been reported to be used for the rapid detection of the microbiomes of fermented plant-related foods such as pearl millet slurries (Humblot and Guyot 2009), meju (Kim et al. 2011), rye (Ehrmann and

Vogel 2005), kim chi (Park et al. 2012) and olives (Cocolin et al. 2013). Pyrosequencing was also used in the USA to detect the *Salmonella* associated outbreak in peanut butter and peanut paste products (Liu 2010).

Pyrosequencing and quantitative real time (qRT)-PCR analysis have also been used for the detection of indigenous microbiota, starter cultures, *Escherichia coli, Listeria,* and *Staphylococcus aureus* from Danish raw milk and cheeses (Masoud et al. 2012). It is also used for the detection of mastitis-causing pathogens such as *Escherichia coli, Klebsiella* spp., *Trueperella pyogenes, Streptococcus uberis,* and *Staphylococcus aureus* (Oikonomou et al. 2012). Both dominant microbes (for example *Lactobacillus kefiranofaciens, L. kefiri, L. parakefiri, L. buchneri,* etc. in the fermented dairy products) and minor bacterial constituents (such as species of *Acetobacter* and *Lactococcus*) can be detected by the pyrosequencing technique (Plessas et al. 2017). It can also help to differentiate microbial communities present in the grain from the fermented final products and from beverages (Leite et al. 2012). It has also been used to study the diversity and ecology of bacterial populations present during the manufacture and ripening of traditional cheeses, artisanal cheeses, and their variations by studying all stages of their production (Fuka et al. 2013). By pyrosequencing of rDNA amplicons, it is easy to characterize the diversity of thermophilic bacteria present in pasteurized milk (Delgado et al. 2013).

Certain processing parameters, like the selection of raw materials and microbial combinations used as starter culture and fermentation conditions, play an important role in shaping the microbial population of food. These parameters can be monitored efficiently by different 16S tagged NGS approaches (Solieri et al. 2013). A study conducted in fen liquor (traditional Chinese fragranced liquor) using 16S pyrosequencing along with qPCR showed that only bacteria grew during the fermentation process while the fungal population remained relatively stable (Li et al. 2011). This study also suggested that variation in biochemical composition during the food fermentation process can be traced by quantification of bacterial and fungal population.

Changes in microbiota and its secreted metabolites (acetoin and butanoic acid) under different storage conditions of meat has also been reported by using NGS-assisted approaches, like pyrosequencing and GC/MS. The results provided insight to estimate the perfect storage condition for meat (Ercolini et al. 2011). Also bacterial WGS has been applied for the typing of foodborne pathogens, like *Listeria monocytogenes* (Gilmour et al. 2010), non-typhoidal *Salmonella* spp., *E. coli* (Brzuszkiewicz et al. 2006), and methicillin-resistant *Staphylococcus aureus.*

The selection of ideal starter culture is essential for fastening the fermentation process and to boost sensory properties. Genome sequencing along with computational techniques can be used both to confirm the identity of the organism and to engineer microbial strains with desired functions for industrial and food production processes (Solieri et al. 2013). The annotated genome sequence of *S. cerevisiae* indicated its functional basis to serve as a starter culture for wine fermentation.

A whole genome-assisted choice of starter culture can also ensure health benefits to the consumer, for example, a well-designed and genome annotated probiotic culture can help in maintaining the balance of microbial community and thus provide good health to the individual (Van Hylckama Vlieg et al. 2011). Oral metagenomic analysis by pyrosequencing revealed that there are certain microbes which have an important role against cariogenic microbes. This indicated their promise to be incorporated as probiotics to fight against dental caries and decay (Belda-Ferre et al. 2012). Likewise, pyrosequencing can be used to trace the genetic basis for the improvement of colitis inflammation when fermented milk product with *Bifidobacterium animalis* subsp. *lactis* is consumed (Veiga et al. 2010).

The combination of NGS with metabolomics would further contribute to characterize beneficial and detrimental activities of the microorganisms present in the food ecosystems. The Illumina technology has also been used to track the bacteria and fungi colonizing the surfaces of winery plants, ripening areas of cheese, and processing surfaces of milk (Walsh et al. 2017). These organisms present on the surface can have beneficial or detrimental role in the final product. NGS techniques will have a revolutionary impact on food microbiology as it can be exploited for pathogen detection, microbial profiling, determining the fate of starters and pathogens over food manufacturing and ripening, and also for predicting product shelf life.

17.5 CONCLUSION

NGS has the potential to become a powerful tool for the culture-independent microbial ecological studies of food. NGS techniques could fuel microbial ecology of food by systematic exploration of the capabilities of the complex microbial consortia in foods and unexpected contaminants including the pathogenic ones. The NGS analysis can also improve the food quality through a detailed understanding of its microbial succession and also has challenges with the extent of taxonomic identification of food microbiota.

REFERENCES

Aird, D., Ross, M. G., Chen, W.-S., et al. 2011. Analyzing and minimizing PCR amplification bias in Illumina sequencing libraries. *Genome Biology* 12(2):1–14. doi:10.1186/gb-2011-12-2-r18.

Ali, N., Rampazzo, R. D. C. P., Costa, A. D. T. and Krieger, M. A. 2017. Current nucleic acid extraction methods and their implications to point-of-care diagnostics. *Biomedical Research International* 2017:9306564.

Amann, R. I., Ludwig, W. and Schleifer, K. H. 1995. Phylogenetic identification and in situ detection of individual microbial cells without cultivation. *Microbiology and Molecular Biolgy Reviews* 59(1):143–69.

Balasubramanian, S. 2015. Solexa sequencing: Decoding genomes on a population scale. *Clinical Chemistry* 61(1):21–4. doi:10.1373/clinchem.2014.221747.

Belda-Ferre, P., Alcaraz, L. D., Cabrera-Rubio, R., Romero, H., Simón-Soro, A., Pignatelli, M. and Mira, A. 2012. The oral metagenome in health and disease. *International Society for Microbial Ecology Journal* 6(1):46–56. doi:10.1038/ismej.2011.85.

Bentley, D. R., Balasubramanian, S., Swerdlow, H. P., et al. 2008. Accurate whole human genome sequencing using reversible terminator chemistry. *Nature* 456(7218):53–9.

Beye, M., Fahsi, N., Raoult, D. and Foumier, P. E. 2018. Careful use of 16S rRNA gene sequence similarity values for the identification of *Mycobacterium* species. *New Microbes and New Infections* 22:24–9. doi:10.1016/j.nmni.2017.12.009.

Binladen, J., Gilbert, M. T. P., Bollback, J. P., Panitz, F., Bendixen, C., Nielsen, R. and Willerslev, E. 2007. The use of coded PCR primers enables high-throughput sequencing of multiple homolog amplification products by 454 parallel sequencing. *PLoS ONE* 2(2):1–9. doi:10.1371/journal.pone.0000197.

Bleidorn, C. 2016. Third generation sequencing: Technology and its potential impact on evolutionary biodiversity research. *Systematics and Biodiversity* 14(1):1–8.

Bonaiti, C., Parayre, S. and Irlinger, F. 2006. Novel extraction strategy of ribosomal RNA and genomic DNA from cheese for PCR-based investigations. *International Journal of Food Microbiology* 107(2):171–9. doi:10.1016/j.ijfoodmicro.2005.08.028.

Brzuszkiewicz, E., Brüggemann, H., Liesegang, H., et al. 2006. How to become a uropathogen: Comparative genomic analysis of extraintestinal pathogenic *Escherichia coli* strains. *Proceedings of National Academy of Sciences of the United States of America* 103(34):12879–84. doi:10.1073/pnas.0603038103.

Cangelosi, G. A. and Meschke, J. S. 2014. Dead or alive: Molecular assessment of microbial viability. *Applied and Environmental Microbiology* 80(19):5884–91. doi:10.1128/AEM.01763-14.

Cocolin, L., Alessandria, V., Dolci, P., Gorra, R. and Rantsiou, K. 2013. Culture independent methods to assess the diversity and dynamics of microbiota during food fermentation. *International Journal of Food Microbiology* 167(1):29–43. doi:10.1016/j.ijfoodmicro.2013.05.008.

Coenye, T., Gevers, D., de Peer, Y. V., Vandamme, P. and Swings, J. 2005. Towards a prokaryotic genomic taxonomy. *FEMS Microbiology Reviews* 29(2):147–67. doi:10.1016/j.femsre.2004.11.004.

Delgado, S., Rachid, C. T. C. C., Fernandez, E., Rychlik, T., Alegría, A., Peixoto, R. S. and Mayo, B. 2013. Diversity of thermophilic bacteria in raw, pasteurized and selectively-cultured milk, as assessed by culturing, PCR-DGGE and Pyrosequencing. *Food Microbiology* 36(1):103–11. doi:10.1016/j.fm.2013.04.015.

Di Pinto, A., Forte, V., Guastadisegni, M. C., Martino, C., Schena, F. P. and Tantillo, G. 2007. A comparison of DNA extraction methods for food analysis. *Food Control* 18(1):76–80.

Ehrmann, M. A. and Vogel, R. F. 2005. Molecular taxonomy and genetics of sourdough lactic acid bacteria. *Trends in Food Science and Technology* 16(1–3):31–42.

Ercolini, D. 2013. High-throughput sequencing and metagenomics: Moving forward in the culture-independent analysis of food microbial ecology. *Applied and Environmental Microbiology* 79(10):3148–55.

Ercolini, D., Ferrocino, I., Nasi, A., et al. 2011. Monitoring of microbial metabolites and bacterial diversity in beef stored under different packaging conditions. *Applied and Environmental Microbiology* 77(20):7372–81.

Fish, J. A., Chai, B., Wang, Q., Sun, Y., Brown, C. T., Tiedje, J. M. and Cole, J. R. 2013. FunGene: The functional gene pipeline and repository. *Frontiers in Microbiology* 4:291. doi:10.3389/fmicb.2013.00291.

Fuka, M. M., Wallisch, S., Engel, M., Welzl, G., Havranek, J. and Schloter, M. 2013. Dynamics of bacterial communities during the ripening process of different Croatian cheese types derived from raw ewe's milk cheeses. *PLoS ONE* 8(11):e80734. doi:10.1371/journal.pone.0080734.

Gilmour, M. W., Graham, M., Van Domselaar, G., et al. 2010. High-throughput genome sequencing of two *Listeria monocytogenes* clinical isolates during a large foodborne outbreak. *BMC Genomics* 11:120. doi:10.1186/1471-2164-11-120.

Giraffa, G. and Neviani, E. 2001. DNA-based, culture-independent strategies for evaluating microbial communities in food-associated ecosystems. *International Journal of Food Microbiology* 67(1–2):19–34.

Head, S. R., Komori, H. K., LaMere, S. A., Whisenant, T., Van Nieuwerburgh, F., Salomon, D. R. and Ordoukhanian, P. 2015. Library construction for next-generation sequencing: Overviews and challenges. *BioTechniques* 56(2):61. doi:10.2144/000114133.

Hugerth, L. W. and Andersson, A. F. 2017. Analysing microbial community composition through amplicon sequencing: From sampling to hypothesis testing. *Frontiers in Microbiology* 1561:1–22. doi:10.3389/fmicb.2017.01561.

Humblot, C. and Guyot, J. P. 2009. Pyrosequencing of tagged 16S rRNA gene amplicons for rapid deciphering of the microbiomes of fermented foods such as pearl millet slurries. *Applied and Environmental Microbiology* 75(13). doi:10.1128/AEM.00451-09.

Jagadeesan, B., Gerner-Smidt, P., Allard, M. W., et al. 2019. The use of next generation sequencing for improving food safety: Translation into practice. *Food Microbiology* 79:96–115.

Kchouk, M., Gibrat, J. F. and Elloumi, M. 2017. Generations of sequencing technologies: From first to next generation. *Biology and Medicine* 9(3). doi:10.4172/0974-8369.1000395.

Keegan, K. P., Glass, E. M. and Meyer, F. 2016. MG-RAST, a metagenomics service for analysis of microbial community structure and function. In: *Microbial environmental genomics*, eds. F. Martin, and S. Uroz, 207–33. New York: Springer. doi:10.1007/978-1-4939-3369-3.

Kulski, J. K. 2016. Next-generation sequencing - An overview of the history, tools, and "Omic" applications. In: *Next generation sequencing - Advances, applications and challenges*, eds. J. K. Kulski, 3–60. London: IntechOpen.

Langille, M. G., Zaneveld, J., Caporaso, J. G., et al. 2013. Predictive functional profiling of microbial communities using 16S rRNA marker gene sequences. *Nature Biotechnology* 31(9):814.

Laver, T., Harrison, J., O'neill, P. A., Moore, K., Farbos, A., Paszkiewicz, K. and Studholme, D. J. 2015. Assessing the performance of the oxford nanopore technologies minion. *Biomolecular Detection and Quantification* 3:1–8.

Leite, A. M., Mayo, B., Rachid, C. T., Peixoto, R. S., Silva, J. T., Paschoalin, V. M. and Delgado, S. 2012. Assessment of the microbial diversity of Brazilian kefir grains by PCR-DGGE and Pyrosequencing analysis. *Food Microbiology* 31(2):215–21.

Li, X. R., Ma, E. B., Yan, L. Z., Meng, H., Du, X. W., Zhang, S. W. and Quan, Z. X. 2011. Bacterial and fungal diversity in the traditional Chinese liquor fermentation process. *International Journal of Food Microbiology* 146(1):31–7. doi:10.1016/j.ijfoodmicro.2011.01.030.

Linnarsson, S. 2010. Recent advances in DNA sequencing methods – General principles of sample preparation. *Experimental Cell Research* 316(8):1339–43. doi:10.1016/j.yexcr.2010.02.036.

Liu, G. E. 2010. Applications and case studies of the next-generation sequencing technologies in food, nutrition and agriculture. *Recent Patents on Food, Nutrition and Agriculture* 1(1):75–9. doi:10.2174/1876142910901010075.

Mardis, E. R. 2008. The impact of next-generation sequencing technology on genetics. *Trends in Genetics* 24(3):133–41.

Masoud, W., Vogensen, F. K., Lillevang, S., Al-Soud, W. A., Sørensen, S. J. and Jakobsen, M. 2012. The fate of indigenous microbiota, starter cultures, *Escherichia coli*, *Listeria innocua* and *Staphylococcus aureus* in Danish raw milk and cheeses determined by Pyrosequencing and quantitative real time (qRT)-PCR. *International Journal of Food Microbiology* 153(1–2):192–202.

Mayo Pérez, B., Rachid, C. T., Alegría, Á., Leite, A. M., Peixoto, R. S. and Delgado, S. 2014. Impact of next generation sequencing techniques in food microbiology. *Current Genomics* 15(4):293–309.

Meglecz, E. 2007. MICROFAMILY (version 1): A computer program for detecting flanking-region similarities among different microsatellite loci. *Molecular Ecology Notes* 7(1):18–20.

Metzker, M. L. 2010. Sequencing technologies the next generation. *Nature Reviews Genetics* 11(1):31–46. doi:10.1038/nrg2626.

Mikheyev, A. S. and Tin, M. M. 2014. A first look at the Oxford nanopore MinION sequencer. *Molecular Ecology Resources* 14(6):1097–102.

Nakayama, Y., Yamaguchi, H., Einaga, N. and Esumi, M. 2016. Pitfalls of DNA quantification using DNA-binding fluorescent dyes and suggested solutions. *PLoS ONE* 11(3):e0150528.

Nikolaki, S. and Tsiamis, G. 2013. Microbial diversity in the era of omic technologies. *BioMed Research International* 958719.

Oikonomou, G., Machado, V. S., Santisteban, C., Schukken, Y. H. and Bicalho, R. C. 2012. Microbial diversity of bovine mastitic milk as described by Pyrosequencing of metagenomic 16S rDNA. *PLoS ONE* 7(10): e47671.

Ottesen, A. R., Peña, A. G., White, J. R., et al. 2013. Baseline survey of the anatomical microbial ecology of an important food plant: *Solanum lycopersicum* (tomato). *BMC Microbiology* 13(1):114.

Park, E. J., Chun, J., Cha, C. J., Park, W. S., Jeon, C. O. and Bae, J. W. 2012. Bacterial community analysis during fermentation of ten representative kinds of kimchi with barcoded Pyrosequencing. *Food Microbiology* 30(1):197–204.

Plessas, S., Nouska, C., Mantzourani, I., Kourkoutas, Y., Alexopoulos, A. and Bezirtzoglou, E. 2017. Microbiological exploration of different types of kefir grains. *Fermentation* 3(1):1.

Renouf, V., Claisse, O., Miot-Sertier, C. and Lonvaud-Funel, A. 2006. Lactic acid bacteria evolution during winemaking : Use of rpoB gene as a target for PCR-DGGE analysis. *Food Microbiology* 23(2):136–45. doi:10.1016/j.fm.2005.01.019.

Reuter, J. A., Spacek, D. V. and Snyder, M. P. 2015. High-throughput sequencing technologies. *Molecular Cell* 58(4):586–97.

Rhoads, A. and Au, K. F. 2015. PacBio sequencing and its applications. *Genomics, Proteomics and Bioinformatics* 13(5):278–89.

Rivas, R., Velázquez, E., Zurdo-Piñeiro, J. L., Mateos, P. F. and Molina, E. M. 2004. Identification of microorganisms by PCR amplification and sequencing of a universal amplified ribosomal region present in both prokaryotes and eukaryotes. *Journal of Microbiological Methods* 56(3):413–26.

Ronholm, J. 2018. Editorial: Game Changer - Next generation sequencing and its impact on food microbiology. *Frontiers in Microbiology* 9:363. doi:10.3389/fmicb.2018.00363.

Rothberg, J. M., Hinz, W., Rearick, T. M., et al. 2011. An integrated semiconductor device enabling non-optical genome sequencing. *Nature* 475(7356):348–52.

Sanger, F., Nicklen, S. and Coulson, A. R. 1977. DNA sequencing with chain-terminating inhibitors. *Proceedings of the National Academy of Sciences of the United States of America* 74(12):5463–7. doi:10.1073/pnas.74.12.5463.

Santamaria, M., Fosso, B., Consiglio, A., et al. 2012. Reference databases for taxonomic assignment in metagenomics. *Briefings in Bioinformatics* 13(6):682–95.

Seul Kim, Y., Kim, M. -C., Kwon, S. -W., Kim, S. -J., Park, I. C., Ka, J. -O. and Weon, H. -Y. 2011. Analyses of bacterial communities in meju, a Korean traditional fermented soybean bricks, by cultivation-based and Pyrosequencing methods. *The Journal of Microbiology* 49(3):340–8. doi:10.1007/s12275-011-0302-3.

Shendure, J. and Ji, H. 2008. Next-generation DNA sequencing. *Nature Biotechnology* 26(10):1135.

Solieri, L., Dakal, T. C. and Giudici, P. 2013. Next-generation sequencing and its potential impact on food microbial genomics. *Annals of Microbiology* 63(1):21–37. doi:10.1007/s13213-012-0478-8.

Tan, B., Ng, C. M., Nshimyimana, J. P., Loh, L. L., Gin, K. Y. H. and Thompson, J. R. 2015. Next-generation sequencing (NGS) for assessment of microbial water quality: Current progress, challenges, and future opportunities. *Frontiers in Microbiology* 6:1027.

Tan, S. C. and Yiap, B. C. 2009. DNA, RNA, and protein extraction: The past and the present. *BioMed Research International* 2009:574398. doi:10.1155/2009/574398.

Tripathi, V., Kumar, P., Tripathi, P., Singh, A. and Kamle, M. 2019. *Microbial genomics in sustainable agroecosystems*, Vol. 2. Singapore: Springer.

Van Hylckama Vlieg, J. E., Veiga, P., Zhang, C., Derrien, M. and Zhao, L. 2011. Impact of microbial transformation of food on health - from fermented foods to fermentation in the gastro-intestinal tract. *Current Opinion in Biotechnology* 22(2):211–9.

Veiga, P., Gallini, C. A., Beal, C., et al. 2010. *Bifidobacterium animalis* subsp. *lactis* fermented milk product reduces inflammation by altering a niche for colitogenic microbes. *Proceedings of the National Academy of Sciences of the United States of America* 107(42):18132–7. doi:10.1073/pnas.1011737107.

Vieites, M., Guazzaroni, M. E., Beloqui, A., Golyshin, P. N. and Ferrer, M. 2009. Metagenomics approaches in systems microbiology. *FEMS Microbiology Reviews* 33(1):236–55. doi:10.1111/j.1574-6976.2008.00152.x.

Walsh, A. M., Crispie, F., Claesson, M. J. and Cotter, P. D. 2017. Translating omics to food microbiology. *Annual Review of Food Science and Technology* 8:113–34.

Zhao, S., Zhang, Y., Gamini, R., Zhang, B. and von Schack, D. 2018. Evaluation of two main RNA-seq approaches for gene quantification in clinical RNA sequencing: PolyA+ selection versus rRNA depletion. *Scientific Reports* 8(1):1–12.

CHAPTER 18

Application of Next-Generation Sequencing Techniques in Food-Related Microbiome Studies

*Ramachandran Chelliah, Eric Banan-Mwine Daliri,
Fazle Elahi, Imran Khan, Shuai Wei, Su-Jung Yeon,
Kandasamy Saravanakumar, Inamul Hasan Madar,
Sumaira Miskeen, Ghazala Sultan, Marie Arockianathan,
Shanmugarathinam Alagarsamy, Thirumalai Vasan,
Myeong-Hyeon Wang, Usha Antony,
Devarajan Thangadurai, and Deog Hwan Oh*

CONTENTS

18.1 INTRODUCTION

It is widely known that foodborne diseases cause considerable morbidity and mortality in humans, particularly in immunosuppressed individuals and in young children (Stark et al. 2009). Many foodborne diseases are caused by bacteria, viruses, and parasites (Newell et al. 2010). Consequently, sporadic infections or outbreaks are reported regularly in many countries. In addition to these, increased globalization has resulted in the transmission of foodborne pathogens across international borders, severely impacting trade and food security (Jennings et al. 2016; King et al. 2017). This has led several national governments and international bodies to establish elaborate controls to improve food safety. Therefore, there is a burden on different food manufacturing systems to provide food that is safe for consumption.

Food, which is an indispensable part of everyday life, undergoes many processing steps before reaching the consumer. The total population of all the microorganisms (microbiomes) plays quite an important role in any food matrix ranging from fermentation to contamination and spoilage. Deeper understanding of the microbial taxonomy is required both to enhance desired food processes, like fermentation, and to alleviate detrimental occurrences like contamination and spoilage. Historically, traditional techniques including the classical gram stain along with individual biochemical characteristics have been used for the isolation, identification, and characterization of bacteria from clinical, food, or environmental origins. Even though they are considered the "gold standard," culture-dependent techniques can only detect 0.1% of a complex community, such as that found in the human intestinal microbiota. In order to expand the understanding of an ecological niche, such as food, techniques are needed to identify or characterize microorganisms and predict the functional interactions of different microbiological communities present in the sample.

During the last decade, advances in sequencing technologies have rapidly changed the analysis of microbial ecology, as next-generation methods are capable of sequencing millions of fragments (Kozich et al. 2013). Based on these high-throughput molecular technologies, even complex environments, like the intestinal microbiota, can be better characterized (Shokralla et al. 2012), thus even visualizing dietary influences in the microbiota. Traditionally, most next-generation sequencing (NGS) related food microbiome studies have largely focused on fermented foods, such as cheese, kim chi, and sausages (Cao et al. 2017; Kok and Hutkins 2018). Different studies have not only enabled the characterization of the microbial composition of fermented foods but have also identified the changes in microbial structure overtime, along with changes in the gene expression patterns related to different fermentation stages (Cao et al. 2017; Tshikantwa et al. 2018). Compared to these fermented foods, studies to identify and characterize the microbiome of non-fermented foods are scarce. This chapter presents the basic principles of the currently available sequencing techniques and how some of these strategies play a key role in understanding microbiomes associated with food, with special emphasis on non-fermented foods.

18.2 NEXT GENERATION SEQUENCING

The first generation of sequencing prevailed for three decades, in particular Sanger sequencing. Be that as it may, the expense and time was a significant hindrance. 2005 and subsequent years have seen the rise of another age of sequencers that have broken through the confinements of first generation sequencing. NGS was initiated in 2007 by Applied Biosystems (Kchouk et al. 2017). Applied Biosystems created a sequencer called SOLiD (Sequencing by Oligonucleotide Ligation and Detection). NGS comprises different types of approaches: sequencing by ligation, sequencing by synthesis, and Ion-proton sequencing (Buermans and Den 2014). These fast, low-cost, and high-producing, parallel-sequencing techniques are able to solve the major drawbacks of previous methods. NGS has revolutionized the fields of genomics and microbiology, allowing us to sequence nucleotides much more quickly and cheaply than before (Heather and Chain 2016).

The Human Genome Project (HGP) aimed to determine the sequences that make up human DNA. It was completed in 2003 ahead of schedule and within budget, and continues to inspire the growth and development of technologies that enable faster and more economical sequencing of genomes (McCombie et al. 2019). While the HGP was very slow and expensive, NGS platforms perform massively parallel sequencing of millions of DNA fragments from a single sample using high-throughput sequencing (Chen et al. 2020). We distinguish the second-generation sequencing technologies which refer to the newest sequencing technologies developed in the NGS environment from the first generation. NGS technologies can sequencing a single molecule without amplification libraries and are capable of generating longer reads at much lower costs and in a shorter time (Rhoads and Au 2015; van Dijk et al. 2018). Next generation sequencing technology describes a number of different approaches including ROCHE 454 sequencing, Ion Torrent, Illumina sequencing, and ABI SOLiD sequencing.

18.2.1 Roche 454 Sequencing

454 Life Sciences brought 454 sequencing to market in 2005. 454 was acquired by Roche in 2007 (Luo et al. 2012). In this technology, multiple fragments of DNA can be sequenced, equivalent to one billion bases; the technology was far cheaper and faster than the methods used in the Human Genome Project. It was the first non-Sanger sequencing to sequence the human genome with new approaches for gene sequencing and small RNA identification, started the era of personalized medicine (Shrestha et al. 2014). 454 sequencing technology uses high-throughput sequencing technology based on large-scale pyrosequencing. The Roche sequencing principle is based on detection of pyrophosphate released during nucleotide integration in a cluster of enzymatic reactions; visible light is generated that is proportional to the amount of amalgamated nucleotides. Roche sequencing generated long reads which are easier to map to the reference genome. Errors found in this technique are insertions and deletions due to homo-polymerization (Luo et al. 2012).

18.2.2 Ion Torrent Sequence

The Ion Torrent sequencing technology was developed by Life Technologies in 2010 (Rusk 2011). It is a core technology using semiconductor technology in chemical and digital information to establish a direct link. The principle idea is to detect the chemical luminescence given off when a proton (H^+) is released and is detected by a semiconductor installed in each well of the chip. The DNA chain is fixed in the semiconductor chip micro holes and is followed by the incorporation of ATGC nucleotides (Kajiyama et al. 2002). With the incorporation of these nucleotides, each base of H^+ ions is released and detected at the bottom of each hole. Through detection of H^+ this sequence is simpler and faster and more flexible (Shendure et al. 2011). Each time the chip was flooded with one nucleotide after other; if it is not the correct nucleotide, no voltage will be found. Ion Torrent has marketed their machine as a rapid, compact, and economical sequencer that can be utilized in a large number of laboratories as a benchtop machine (Kchouk et al. 2017).

18.2.3 Illumina Sequencing

Illumina was an American company founded in 1998. Initially they entered the biological market through the integrated systems for the analysis of genetic variation, biological function, and single nucleotide polymorphism genotyping services (Thomson 2014). Solexa was a genomics company, they had launched an instrument called Genome Analyser in 2006, which is able to sequence 1GB of data in one run, which was a record at that time. In 2009, Illumina started offering a new service of personal human whole-genome sequencing of 30X coverage for $48,000 per genome (Keebler 2010).

18.3 NEXT-GENERATION SEQUENCING IN GUT MICROBIOME RESEARCH

The genome sequencing concept was invented in 1975 (Sanger and Coulson 1975) and was focused on the process of DNA polymerase labeling chain-terminating ddNTPs during DNA replication in vitro. It was costly, time-consuming, and arduous for high-throughput research. NGS strategies are built on the idea of massive parallel sequencing, widely discussed elsewhere (Caboche et al. 2014; Nguyen et al. 2017). Recent advancements in NGS technology have facilitated gut microbiome analysis and allowed genetic and biological variability analysis of unculturable gut microbial ecosystems with economical prices and adequate productivity. Amplicon-based profiling is one of the most commonly used methods of characterizing intestinal microbiome diversity. Here, a taxonomically sensitive gene promoter (usually 16S rRNA for bacteria and archaea) is selected and amplified by PCR from the total DNA. The subsequent amplicons are transcribed, and subsequent analyses of bioinformatics are conducted to measure specific taxonomic distribution and abundance in the sample (Figure 18.1). Comparative analysis of genome map variants through samples clarifies how microbial diversity is correlated with relationships between host and microbe (Ley et al. 2006), or different environmental factors, such as diet (Zhang et al. 2015; Org et al. 2017), medications (Minter et al. 2016), or probiotics (Gaulke et al. 2018). Amplicon sequencing, however, usually only addresses gut microbiome's taxonomic composition. Conclusive evidence of biological roles correlated with the gut microbial culture is unlikely. Finally, adopted computational methods, such as implemented in genetic code (Fadiji and Babalola 2020), were actively promoted to infer the functional potential of the population by merging knowledge on 16S rRNA genes with coding sequences.

Shotgun gene sequencing is an alternative method to classifying the intestinal flora. Instead of amplifying against a particular gene marker, the complete DNA is decoded and evaluated. The computational research typically includes gene interpretations and genetic advances besides taxonomic decoding (Figure 18.1), which offers a more global way of simultaneously exploring gut microbiome's taxonomic classification and functional ability. These taxonomic and functional characteristics may be used to examine the relationships between intestinal microbiota and infection or diets. The metagenomic-wide cluster analysis of 145 European women with type-2 diabetes (T2D) reduced glucose homeostasis, or control subjects recently also revealed strong associations of different gut microbes (e.g., *Roseburia* species and *Faecalibacterium prausnitzii*) and their T2D genes (Ji and Nielsen 2015). An observational framework based on this cohort's metagenomic profiling allowed women to be classified in the pre-T2D cohort who also have elevated blood plasma markers correlated with T2D. In addition, intestinal microbiota taxonomic markers were established to differentiate colorectal carcinoma patients from tumor-free controls that provided non-invasive fecal readings for accurate colorectal cancer detection (Wirbel et al. 2019). The use of such biomarkers acts as a diagnostic imaging environment, which is ideal for individualized medicine.

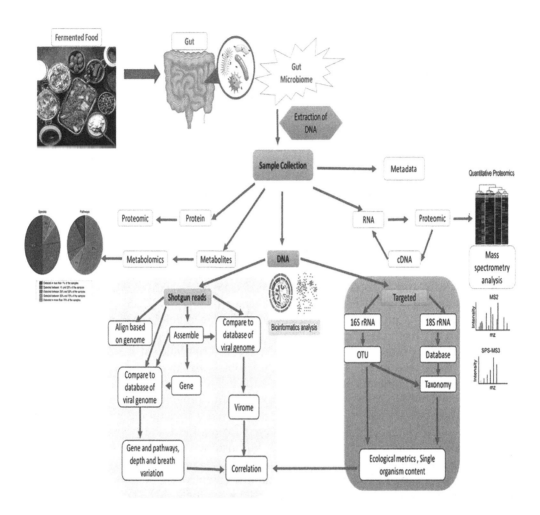

FIGURE 18.1 Workflow of the amplicon metagenomics and metagenomics approaches in food and food-oriented gut microbiome analysis.

While NGS-based profiling has vastly expanded our awareness of the gut microbiome, existing culture-independent metagenomics produce mixed data representing community-level attributes rather than species-specific features. Correspondingly, the belief in high-throughput culture methods were revived (Figure 18.1; Amin and Khan 2016). Subsequent anaerobic culture research on a rich medium has shown that the cultivated samples can recognize as many as 50% of species (Goodman et al. 2011). Using 212 various conditions of culture using mass spectrometry strategies and NGS methodology, as many as 32,500 distinctive colonies were recovered from three stool samples (Cornille et al. 2012). The recognized gut microbiota included 174 species not previously defined in the human intestine and sequenced 31 new species and genera, generating about 10,000 previously unrecognized genes (Ji and Nielsen 2015). In addition, the constituent gut microbiota species with antibiotic resistance were successfully developed from fecal samples by incorporating existing circumstances with rapid phenotypic profiling (Lozupone et al. 2012). Single cell genomics have been presented to evaluate unknown species from a wide range of ecosystems for non-cultural microorganisms (Figure 18.1;

Edet et al. 2017). Such a report is an attempt to accelerate the discovery of new organisms without previous cultivation as single-cell genomics requires a step to amplify the genome from a single cell. Development of single-cell sequence analysis of two major intestinal microbial species, *Gilliamella apicola*, and *Snodgrassella alvi*, revealed significant differences in intra-specific protein-encoding genetic divergence (Koch et al. 2013). Various transcriptomics, proteomics, metabolomics, and phenomics data have become accessible for gut microbiome studies far beyond metagenome and single-cell genomes (Figure 18.1).

18.4 APPLICATION OF NGS TECHNIQUES IN SELECTED MICROBIOME STUDIES

Based on the high-throughput sequencing systems, NGS implementations have already become widespread (Li et al. 2019). There is no question that applying NGS, whether by whole genome sequencing or even more complex metagenomics methods, provides plenty of possibilities in food safety and microbial risk evaluation. The key difference in both methods is the difference in the target used, i.e., for WGS, the nucleic acids are obtained from a pure culture isolate, and, in metagenomics, total nucleic acids are obtained directly from the microbiota present in the investigated area. It can be inferred with whole transcriptome sequencing and meta-transcriptomics at RNA level (Li et al. 2019). For WGS, a culture-based approach, microbial growth is required using selective culture medium and circumstances to retrieve the desired organism. The performance of WGS is the overall DNA content of a microorganism, providing information on genes through diversified selection (e.g., antigen genes), genes undergoing stabilizing preference (e.g., household genes) and genes of interest from a food safety context (e.g., infectivity genes, AMR genes, toxin production), but also non-coding regions and potential episomal regions (Orellana et al. 2019). Therefore, WGS has the ability to identify the highest genetic detail of a bacterial strain, enabling complete characterization of a strain. WGS is also generally applicable, compared to, for example, MLST or PFGE, focusing on the taxonomic community or pathogen under examination. Nevertheless, this would entail standardization in the way data are produced and analyzed, ideally globally, at an inter-laboratory level (Ward et al. 2018). There are programs to encourage this harmonization, such as www.globalmicrobialidentifier.com, and attempts have been made to standardize certain consistency criteria and recommendations on how to effectively use WGS data (Chun et al. 2018). Furthermore, when thoroughly identified, the knowledge can be applied to identify and/or investigate foodborne outbreaks (e.g., path and source monitoring, cross-contamination events), attribution studies, and evaluation of risk of virulence or disease potential, with the goal of incorporating of all these data into risk management assessments, including prevention and control approach studies. Given these implementations, it is necessary to be mindful of some of the problems inherent in culture-dependent approaches. They cannot be deemed empirical because they may underestimate or overstate the microbial diversity present (Shakaya and Chain 2019). Furthermore, a microorganism's genetic content cannot currently be seen separately from its phenotypic features (e.g., its pathogenicity) in relation to host and environmental influences.

Metagenomics methods target the nucleic acids of microorganisms in the atmosphere or food matrix and have the advantage that there is no need for prior cultivation measures to examine the whole population; it is a so-called "culture-independent technique." The sophistication of these analyses makes using the produced data in a food safety and risk assessment sense much more challenging. A solution is to identify related biomarkers that can predict microbial activity (Brul et al. 2012). Here, too, it is necessary to create a link between hosts and environmental factors and biomarker expression level on the one hand and phenotypic behavior on the other. The ongoing improvement in other proteomics and metabolomics studies data would definitely add to this method. Eventually, according to several other methods of detection and typing in which the tool of selection relies on the pathogen and the problem to be discussed, for NGS, the sort of approach and the form and degree of data applied will rely on the intent (Franz et al. 2016).

The curriculum itself aimed to familiarize the NGS in the sense of food welfare and hazard evaluation, using a "learning by doing" concept. First, wet-laboratory knowledge was acquired for two NGS techniques in addition to the requisite theoretical insights: high-throughput sequencing of short reads utilizing Illumina technology with a MiSeq instrument and long-read sequencing based on nanopore technology. The fellow also acquired bioinformatics skills using Unix-based and other tools required to analyze (performance, cutting, construction) the provided sequence data (Franz et al. 2016). Considering the gained practical and theoretical perspectives, further review will help explain how these methods and data can be applied in a microbial threat assessment system.

18.5 APPLICATION OF NGS-BASED STRATEGIES TO RAW MATERIALS AND PRODUCTION ENVIRONMENT SURVEILLANCE

As far as microbiome associated studies are concerned, NGS methods have proved to be by far the most efficient. Yet, there are still challenges and limitations to be considered when applying these techniques to particular cases on food and food-related environments. The following are the conditions to be maintained in order to enhance the NGS studies of food microbiomes.

The quality of obtained nucleic acids–naturally obtained environment samples, such as soil and water, along with stool, saliva, and other, fermented food samples, such as cheese and kimchi, are known to contain high numbers of microorganisms. Subsequently, these samples can offer sufficient template nucleic acid for further analysis. On the contrary, sanitary control and maintenance of strict food production environment standards have made isolation of total DNA from these environments quite challenging (Anvarian et al. 2016). Frequent exposure to detergents from regular cleaning has led to injury to microorganisms colonizing this niche, thereby compromising the ability to recover sufficient template nucleic acid for analysis.

Usage of sufficient controls during sequencing–internal controls during sequencing are absolutely necessary in multi-omic strategies (Brul et al. 2012). However, these are often neglected. A sequencing control should be brought about to contain DNA sequences from known bacterial species. These can be processed in parallel with other samples during sequencing, which could finally yield an estimation of the sequencing errors during downstream bioinformatics analysis. In order to better understand the applications of next-generation sequencing methods, a literature search was performed on the current NCBI PubMed database. Between April 2011 and June 2017, a total of over 126 papers were published describing the various characterizations of different food microbiomes. Out of all these publications which made use of different NGS techniques in the analysis of food microbiome, more than half (63%) used 16S rDNA sequencing, thus suggesting that this strategy has been widely applied for the analysis of food/food production-related studies.

NGS-based methods have been more frequently applied to characterize fermented (63%) types of food matrices compared to non-fermented foods (Figure 18.2c,d). Since 2012, metagenomic and metatranscriptomic approaches have been more widely used for food microbiome characterizations, and the number of studies using these latter approaches has gradually increased over the years. Also, a deeper understanding of next-generation sequencing techniques has been made possible through the outlets that publish studies on the environmental microbiome along the commercial food production chain (Table 18.1).

18.6 METABOLIC MODELING OF THE HUMAN GUT MICROBIOME

While our comprehension of the intestinal flora has advanced rapidly with NGS, an assessment that relies on genomic sequencing is not enough to decode the frameworks of how the microbiome affects human welfare. The metabolic processes of the gut microbiota must be inferred, and the metabolic relationships between intestinal bacteria and the interaction between the host quantified, which will then provide understanding of the molecular mechanisms of the gut microbiome which contribute to human health. A modeling-based methodology in this sense would be a successful way of researching gut microbial metabolic processes at system level (Henson 2015; Kumar et al. 2019).

Genome-scale molecular models (GEMs) are theoretical gene expression-level descriptions of cell function that act as important bioinformatics tools and are commonly used to research microbial metabolism and health impacts (Henson 2015). Historically, the first GEM was created to evaluate microbial activity in 1999 (Schilling et al. 1999). Over 120 GEMs have since been recovered, but modeled species have minimal phylogenetic scope (Gu et al. 2019). GEM restoration and ensuing theoretical research were thoroughly studied elsewhere (Feist and Palsson 2008). Briefly, metabolic reconstructions are focused primarily on genotype–protein-reaction interactions interpreted from genomic variants and related ontological knowledge that link specific genes to regulatory regions and link the genotype–phenotype chart. Usually, draft configurations are assembled by incorporating existing literature knowledge, and the rebuilding metabolic network is transformed into a stoichiometric matrix, where rows reflect metabolism and column reactions and thermodynamic and/or physiological limitations can be implemented to restrict the feasible scope of metabolic activity. Flux balance analysis mimics metabolite flow throughout the metabolic network, allowing the use of GEMs to predict genotype–phenotype interactions (Figure 18.2).

The gut flora focuses on human biology through its metabolic processes, which include energy production, bile acid secretion, choline transformation, and the production of short-chain fatty acids (SCFAs), vitamins, and amino acids (Rowland et al. 2018). Metabolic mapping of gut microbiota based on GEM reconstructions of a few specific species from the predominant phyla in the gut ecosystem and flux balance analysis (FBA) can be implemented to examine key gut microbial species' metabolic capabilities. There are numerous gut microbes, such as *Bacteroides thetaiotaomicron*, *Eubacterium rectale*, *Methanobrevibacter smithii* (Shoaie et al. 2013), and *Bifidobacterium adolescentis* (La-Ongkham et al. 2015). The two most abundant phyla in the human gut biodiversity were reported as Bacteroidetes and Firmicutes, respectively (Sokol et al. 2008; Shoaie et al. 2013). The operational metabolic maps and growth specifications of gut bacteria were explored extensively with GEMs modeling.

Short-chain fatty acids and biosynthesis of organic acid (acetate, butyrate, or propionate) in these species were confirmed by in silico modeling (Mariat et al. 2009; Shoaie et al. 2013).

TABLE 18.1 A Comprehensive List of Publications Using Next-Generations Sequencing Approaches to Study the Environmental Microbiome along the Food Production Chain

Research Target	Country	Methods	Sequencing Platform	Conclusions	References
Sausage processing environment and product	Finland	16S rDNA sequencing	Roche 454 Titanium FLX	Abundant mesophilic psychrotrophs were prevalent throughout sausage production chain microbiomes and with different characteristic patterns of contamination for different genera	Hulman et al. 2015
Ready-to-eat meal plant environment and product	-	16S rDNA sequencing (V1-V3)	Roche 454 GS Junior	*L. gelidum* was identified to be dominant in ready-to-eat meal samples at the end of shelf life, its spoilage characteristic and ability of growing under cold storage should raise industry concern	Pothakos et al. 2015
Cheese factory environment and cheese product	Italy	16S rDNA sequencing (V1-V3, for bacteria); 26S rDNA sequencing (D1-D2, for fungi)	Roche 454 GS Junior	Coexistence of lactic acid bacteria and possible spoilage-associated bacteria was found in core microbiota of cheese factory environment and cheese samples	Stellato et al. 2015
Environment samples alone beef production chain	USA	Shotgun metagenomics sequencing	Illumina HiSeq 2000	No antimicrobial resistant determinants (ARD) were identified in final beef products, indicating slaughter interventions may reduce ARD transmission risk	Noyes 2016
Dairy farm agroecosystems	USA	Shotgun metagenomics sequencing	Ion Torrent Personal Genome Machine	The most abundant antimicrobial resistant genes in dairy agroecosystems were grouped under multidrug transporters	Pitta et al. 2016
Butchery meat and environment samples	Italy	16S rDNA sequencing (V1-V3)	Roche 454 GS Junior platform	The type of retail (large- or small-scale distribution) had no apparent effect on initial fresh meat contamination	Stellato et al. 2017

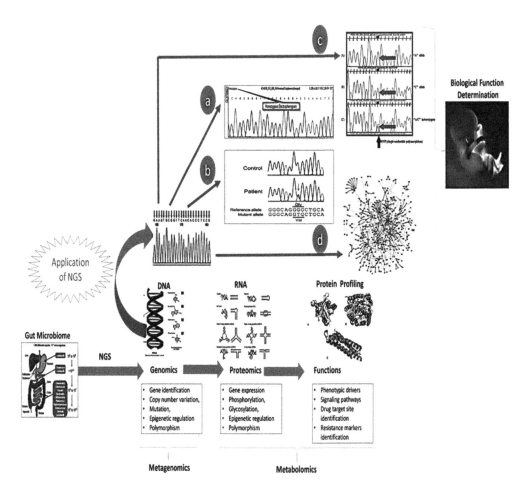

FIGURE 18.2 Schematic overview of the different high-throughput sequencing approaches applicable to food microbiology and suggestions for the sequencing platforms most suitable for each approach (Deurenberg et al. 2017). Application of NGS techniques in food-related microbiome studies: (a) Epigenetic regulation, (b) Mutation analysis, (c) Polymorphism analysis, and (d) Signaling pathways in metabolomics.

Further, *B. thetaiotaomicron* can synthesize all essential human amino acids from inorganic ammonia (Iakiviak 2017). Metabolomic analysis indicates the greater fluxes through acetyl-CoA and anaplerotic oxaloacetate fabrication under amino acid dysfunction were related to propionate overproduction, again coupled with increased flux through the TCA process and decreased NAD^+ reanimation through lactate synthesis (Mitchell et al. 2018). GEMs of some probiotic bacteria were also recovered, such as *Lactococcus lactis* (Ramirez et al. 2010) and *Lactobacillus plantarum* (Koduru 2017). These GEMs help to identify metabolic pathways that the bacteria excreted, which assisted probiotic strain evaluation and standardization.

The gut microbiome is a network of microbial organisms whose metabolism interacts closely with one another and the host's (Wade et al. 2016). Metabolic analysis is thus not only restricted to studying gut flora at the mono-species level but also applies to studying metabolic activity within this multi-species network. Typically, modeling of multi-species

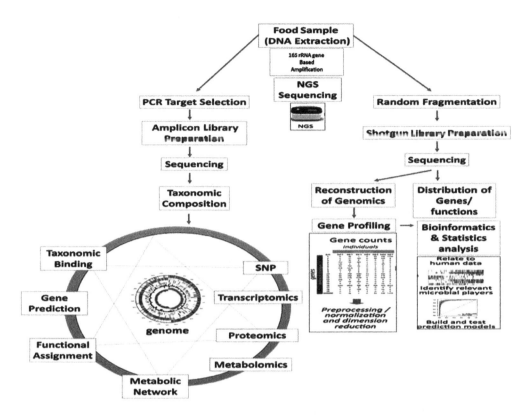

FIGURE 18.3   Gene/genome-centered approach to gut microbiome. The two key metage-nomics methods for gut microbiome studies are typically 16S-rRNA-based amplicon sequencing and whole shotgun sequencing. The taxonomic compositions and functional groups of gut microbial communities that may be correlated with health or disease state can be inferred from metagenome data. In addition, the synthesis of culturomics and NGS methods offers finer knowledge of different gut microbial species' functional fea-tures. Certain available "omics" data (transcriptomics, proteomics, metabolomics, and phenomics) offer much greater insight into the functional role of gut microbes in human health and disease. Integrating these data with metagenomic data, especially metabolic models reconstructed from metagenomics studies, will provide a detailed overview of the metabolic interactions between microbes and host.

microorganism metabolisms in the population is performed by integrating single-species models into a silicon-based joint where nutrients are supplied (Figure 18.3; Prats et al. 2010). This method, therefore, considers each organism in the system as an individual compartment. A common compartment is added to act as a metabolic interaction path-way between various organisms. Based on related experimental results, the resulting group model can be restricted and further applied to derive relationships among constit-uents and analyze the entire system's phenotype under different nutritional constraints. This multi-component approach was applied to infer metabolic interactions between gut microbes (Ji and Nielsen 2015). Acetate is a primary metabolite shared between *B. adolescentis*, and *F. prausnitzii* for growth and butyrate development. *F. prausnitzii* was found to rely on acetate production. Similarly, *E. rectale* also acts as an acetate recipient (Shoaie et al. 2013). A three-species model (*B. thetaiotaomicron*, *E. rectale*, and *M. smithii*) found rivalry for acetate and H_2S production. More integrative analyses of GEM transcriptomics data showed that *E. rectale* changed from polysaccharide to amino acid usage, especially glutamine, in the presence of *B. taiotaomicron*. This shows how GEMs can be used to gain insight into the relationships between organisms, how they cross-feed, and organism commensalism within more complex microbial communities. Beyond the representative gut microbiota species, applying automated GEM reconstructions starting from thousands of annotated gut microbial genome sequences can help to expose the microbe–microbe metabolic interaction landscape (Figure 18.3). Through integrating available GEMs with different environmental factors, the distinct cross-species connections (neutral, commensal, or mutual) and phenotypic properties can be examined, offering insight into the pattern of metabolic activity, population metabolic capacity, and intestinal flora nutritional scenarios.

Intestinal microbiota varies based on the essential dietary nutrient absorption and power sectors; further, the host–microbe relationships play significant roles in immune function (Nicholson et al. 2012). The host–microbe interface model structure is similar to the aforementioned model for predicting microbial connections, which includes combining host metabolism with microbial metabolism. Generic human GEMs (Recon 2, HMR 2.0) and tissue/cell-specific GEMs (liver, muscle, adipocytes, etc.) were used for host–microbe modeling (Nicholson et al. 2012). The authors inferred interactions between *Mycobacterium tuberculosis* and human alveolar macrophages by assembling cytosolic compartments. The interconnected host–microbe GEM allowed the visualization of metabolic differences during infection states, which could also serve as scaffolds to anticipate potential therapeutic targets. In addition, *Plasmodium falciparum* antimicrobial drug targets were examined by combining malaria GEM with human erythrocyte or adipocyte GEM. This interdisciplinary GEM research has been used to classify metabolic interactions among prominent gut microbial species *B. thetaiotaomicron* with different carbohydrate, fat, and protein diets (Cameron et al. 2012). FBA disclosed *B. thetaiotaomicron* offers numerous metabolites changes, which include essential amino acids, nucleotides, and SCFA (such as acetate and propionate) for optimal development.

Though this performance that uses GEMs to restore metabolic interactions suggests these models' utility for intestinal flora studies, some key conceptual obstacles still have to be resolved. Host–microbe GEMs become more complicated, particularly if the overall scenario considers a large number of bacterial communities. Moreover, inter-species or host–microbe systems based on internal metabolic interactions fail to comprehend, for example, how differences in species composition influence microbiome metabolic capacity. Thus, multi-species modeling requires specific methods to address multi-component inward specificity. Additionally, a systematic metabolic modeling approach was modified to study the metabolic function of the gut microbiota as a whole (Figure 18.3; Oberhardt et al. 2008). The community metabolic system can normally be recreated immediately from shotgun metagenomics documentation by disregarding network uses and exchanging metabolites between species (Frioux et al. 2020). In an early study (Greenblum et al. 2012), by incorporating such a biochemical platform at the microbial community level with corresponding gene relative abundance, topological variations at both gene level and network level were established as correlated with obesity and inflammatory bowel disease (IBD). Overall, these community-based methods disregard the boundaries between organisms and sequestration of specific metabolites, offering important insights into the microbiome's metabolic capacity and biological heterogeneity in a complete network.

18.7 GUT MICROBIOME MODELING IN HEALTHCARE AND MEDICINE

Nutrition is an important determining factor of gut flora composition and metabolism (Scott et al. 2013). Complex carbs, vitamins, and minerals are the primary macronutrients that have a significant impact on the formation of intestinal microbiota and metabolism of the host. Monosaccharides (i.e., glucose, galactose) are widely consumed by the gastrointestinal intestinal epithelium, while various dietary polysaccharides, such as soluble fiber, oligosaccharides, and natural fibers, can be metabolized by microbes in the gut but not by the human host (Makki et al. 2018). Proteins are the major energy source in the distal colon after glucose fermentation (Jha and Berrocoso 2016). In silico study of interdisciplinary bacteria–host GEMs on 5 distinct dietary habits vary widely in fat, carbohydrate, and protein content (Chowdhury and Fong 2020). High-carbohydrate consumption offers a strong carbon source with *B. thetaiotaomicron* as an effective dietary natural polymer. Although high-protein diet does not help *B. thetaiotaomicron* production, in accordance with Bacteroides' known in capacity to use proteins as the sole source of carbon and their low proteolytic capacity (Arntzen et al. 2017). In respondents with a high-protein/low-carbohydrate diet, diminished proliferation of bacteroids was also reported in the intestinal flora (Lopez-Legarrea et al. 2014). Thus, the nutrition that influences the gut microbiota can be designed regarding the food nutritive value (Figure 18.3), which significantly improves microbiota development and perks metabolic rate. Therefore, configurations of large-scale GEM rearrangements of gut microbes and prediction reveal potential interaction patterns among gut microbes and focus attention on nutrient prediction and balanced dietary lifestyle.

Attempting to manipulate the gut microbiota with supplements of synbiotics has been shown to influence host metabolism (i.e., homeostasis of glucose; Barratt et al. 2017). Probiotic therapy with *Lactobacillus* strains was characterized more for possible antibacterial activities on enteric pathogens. Consequently, probiotics and their metabolic pathways, called postbiotics, were suggested as dietary supplements for healthy and nutritious cellular homeostasis and as therapeutic aids for IBD treatment (Chong et al. 2019). One major goal in implementing efficient bioactive compounds is limiting towards modification on healthy gut flora. While NGS-based metagenomics presents various gene and species configuration characteristics, changes in metabolite levels remain unresolved. To resolve this problem, the GEM-based model is a desirable solution due to its microbial metabolism predictive capacity (Figure 18.3). The biosynthetic pathway of active postbiotics can be extensively examined with GEMs, leading to the design and analysis of potential probiotic strains with enhanced postbiotic development using metabolic engineering. The composition of gut microbiota in the gastric mucosa is heterogeneous, leading to various prebiotic behaviors at various gut positions (Sorrenti et al. 2020). As shown with polyphenol quercetin, a probiotic strain may not generate a beneficial effect without cooperating with other strains (Barathikannan et al. 2019). Therefore, asserting correlations among gut microbiome species through metabolic frameworks would make it easier to explore probiotic combinations and design algorithms with suitable reaction mechanisms to turn prebiotics into appropriate post-biotics.

Fecal microbiota transfection (FMT) has progressively become a therapeutic tool alternative to traditional therapy, allowing the incorporation of intestinal flora information into clinical use (Wong 2018). FMT proved effective in treating *Clostridium difficile* infections (Vigvári et al. 2019). Typically, FMT includes transplanting fecal bacteria from a healthy person into a host, which is replaced by a stable microbial community. Nonetheless, concerns about pathogen transmission, patient acceptance, and care optimization exist (Freifeld et al. 2011), and, thus, FMT is primarily used for patients without appropriate alternative treatments. Difficult infections that are non-responsive to traditional therapy are treated by transplantation of a synthetic microbiota consisting of 33 bacterial cultures extracted from healthy donor feces (Allegretti et al. 2019). Consequently, using gut microbiome information to enhance public welfare by developing ideal synthetic microbiota for FMT will be the next step (Bircher et al. 2018). Therefore, a bottom-up approach combining single-species GEMs with predetermined roles would involve the development of multi-species activities and interactions and result in potential species configurations with appropriate metabolic processes (Figure 18.3). The relationship between two microbial species may be hostile, commensal, or reciprocal (Schnupf et al. 2017). Modeling all specific inter-microbe relationships and defining all potential metabolite exchanges between organisms will drive complex microbiota assembly by optimizing correlated cooperation. Moreover, by applying metabolic tools and technology and logical microbiota architecture, it will be possible to simplify the dynamics of synthetic microbiota and provide basic information, which can be applied to derive essential pathways of how microbiome affects human health.

18.8 CONCLUSION AND FUTURE PERSPECTIVES

The advancement of NGS-based metagenomics and associated computational methods has brought about wider awareness of the intestinal bacteria and its influence on health infection and nutrition. These sequencing-based surveys concentrate on gut flora's taxonomic or biological configurations, but also provide insight on metabolic interfaces. Therefore, to expose the associated metabolic concept of host–microbe relationships or microbiota–microbe interrelations, it is important to go beyond merely categorizing the intestinal bacteria configuration toward systematic gut microbiome modeling and analysis (Maslowski 2019). Consequently, applying metabolic methodological approaches to intestinal metabolic interactions is a crucial subsequent phase in intestinal microbiota research. Such metabolic redesign at the network level offers a comprehensive interpretation of the biochemical capabilities in the intestinal microbiota species and population and correlates metabolic changes with disease or healthy states. Furthermore, the GEM-based modeling framework is effective for comprehensive gut flora research by quantitatively combining transcriptomics, proteomics, and metabolomics data with environment and biodiversity. Eventually, GEMs-centered in silico growth or projection of metabolite development can be conveniently compared with theoretical evidence to provide possible justification for reported metabolic phenotypes. Intestinal microflora's increased variability and diversity, a simplistic reason for studying metabolic interactions, microbiota profiling using metagenomics (Medlock et al. 2018). Therefore, the research and design structure for simulated microbial diversity must be established further. GEMs have been built up to industrial inter-feeding microbial populations (Minj et al. 2020). Syntrophic growth is achieved in these engineered microbial communities by sharing inter-feeding pathways through organisms. Likewise, symbiotic architecture can be used to grow formulated probiotics with improved nutritional oxidative metabolism or postbiotic biogenesis (Minj et al. 2020). In addition, applying simulation model, which includes systematic modification of gut microbiota in well-designed *in vitro/in vivo* experiments for verification and validation in *in silico* models (Christley et al. 2015). In addition, diet, climate, and hereditary profile modifications may be incorporated into these in vivo/in vitro experimental frameworks to mimic host–microbiota or microorganism–microbe interactions (Bestion et al. 2017; Obadia et al. 2018). Eventually, metabolic modeling coupled with experimental expertise and evidence can significantly enhance our understanding between metabolomics and metagenomics or between microbes and host, offering insight into the clinical application of gut microbiota in diagnoses and therapies.

REFERENCES

Allegretti, J. R., Mullish, B. H., Kelly, C. and Fischer, M. 2019. The evolution of the use of faecal microbiota transplantation and emerging therapeutic indications. *The Lancet* 394(10196):420–31.

Amin, S. and Khan, H. 2016. Revival of natural products: Utilization of modern technologies. *Current Bioactive Compounds* 12(2):103–6.

Anvarian, A. H. P., Cao, Y., Srikumar, S., Fanning, S. and Jordan, K. 2016. Flow cytometric and 16S sequencing methodologies for monitoring the physiological status of the microbiome in powdered infant formula production. *Frontiers in Microbiology* 7:968.

Arntzen, M. Ø., Várnai, A., Mackie, R. I., Eijsink, V. G. and Pope, P. B. 2017. Outer membrane vesicles from *Fibrobacter succinogenes* S85 contain an array of carbohydrate-active enzymes with versatile polysaccharide-degrading capacity. *Environmental Microbiology* 19(7):2701–14.

Barathikannan, K., Chelliah, R., Rubab, M., et al. 2019. Gut microbiome modulation based on probiotic application for anti-obesity: A review on efficacy and validation. *Microorganisms* 7(10):456.

Barratt, M. J., Lebrilla, C., Shapiro, H. Y. and Gordon, J. I. 2017. The gut microbiota, food science, and human nutrition: A timely marriage. *Cell Host and Microbe* 22(2):134–41.

Bestion, E., Jacob, S., Zinger, L., et al. 2017. Climate warming reduces gut microbiota diversity in a vertebrate ectotherm. *Nature Ecology and Evolution* 1(6):1–3.

Bircher, L., Schwab, C., Geirnaert, A. and Lacroix, C. 2018. Cryopreservation of artificial gut microbiota produced with in vitro fermentation technology. *Microbial Biotechnology* 11(1):163–75.

Brul, S., Bassett, J., Cook, P., et al. 2012. 'Omics' technologies in quantitative microbial risk assessment. *Trends in Food Science and Technology* 27(1):12–24.

Buermans, H. P. J. and Den Dunnen, J. T. 2014. Next generation sequencing technology: Advances and applications. *Biochimica et Biophysica Acta (BBA) - Molecular Basis of Disease* 1842(10):1932–41.

Caboche, S., Audebert, C. and Hot, D. 2014. High-throughput sequencing, a versatile weapon to support genome-based diagnosis in infectious diseases: Applications to clinical bacteriology. *Pathogens* 3(2):258–79.

Cameron, E. A., Maynard, M. A., Smith, C. J., Smith, T. J., Koropatkin, N. M. and Martens, E. C. 2012. Multidomain carbohydrate-binding proteins involved in *Bacteroides thetaiotaomicron* starch metabolism. *Journal of Biological Chemistry* 287(41):34614–25.

Cao, Y., Fanning, S., Proos, S., Jordan, K. and Srikumar, S. 2017. A review on the applications of next generation sequencing technologies as applied to food-related microbiome studies. *Frontiers in Microbiology* 8:1829.

Chen, Z., Pham, L., Wu, T. C., et al. 2020. Ultralow-input single-tube linked-read library method enables short-read second-generation sequencing systems to routinely generate highly accurate and economical long-range sequencing information. *Genome Research* 30(6):898–909.

Chong, P. P., Chin, V. K., Looi, C. Y., Wong, W. F., Madhavan, P. and Yong, V. C. 2019. The microbiome and irritable bowel syndrome – A review on the pathophysiology, current research and future therapy. *Frontiers in Microbiology* 10:1136.

Chowdhury, S. and Fong, S. S. 2020. Computational modeling of the human microbiome. *Microorganisms* 8(2):197.

Christley, S., Cockrell, C. and An, G. 2015. Computational studies of the intestinal host-microbiota interactome. *Computation* 3(1):2–28.

Chun, J., Oren, A., Ventosa, A., et al. 2018. Proposed minimal standards for the use of genome data for the taxonomy of prokaryotes. *International Journal of Systematic and Evolutionary Microbiology* 68(1):461–6.

Cornille, A., Pierre, G. L., Marinus, J. M. S., et al. 2012. New insight into the history of domesticated apple: Secondary contribution of the European wild apple to the genome of cultivated varieties. *PLoS Genetics* 8(5):e1002703.

Deurenberg, R. H., Bathoorn, E., Chlebowicz, M. A., Couto, N., Ferdous, M., García-Cobos, S. and Zhou, K. 2017. Application of next generation sequencing in clinical microbiology and infection prevention. *Journal of Biotechnology* 243:16–24.

Edet, U. O., Antai, S. P., Brooks, A. A., Asitok, A. D., Enya, O. and Japhet, F. 2017. An overview of cultural, molecular and metagenomic techniques in description of microbial diversity. *Journal of Advances in Microbiology* 7(2):1–19.

Fadiji, A. E. and Babalola, O. O. 2020. Metagenomics methods for the study of plant-associated microbial communities: A review. *Journal of Microbiological Methods* 170:105860.

Feist, A. M. and Palsson, B. Ø. 2008. The growing scope of applications of genome-scale metabolic reconstructions using *Escherichia coli*. *Nature Biotechnology* 26(6):659–67.

Franz, R., Clavero, C., Kolbeck, J. and Anders, A. 2016. Influence of ionisation zone motion in high power impulse magnetron sputtering on angular ion flux and NbO_x film growth. *Plasma Sources Science and Technology* 25(1):015022.

Freifeld, A. G., Bow, E. J., Sepkowitz, K. A., et al. 2011. Clinical practice guideline for the use of antimicrobial agents in neutropenic patients with cancer: 2010 update by the infectious diseases society of America. *Clinical Infectious Diseases* 52(4):e56–93.

Frioux, C., Singh, D., Korcsmaros, T. and Hildebrand, F. 2020. From bag-of-genes to bag-of-genomes: Metabolic modelling of communities in the era of metagenome-assembled genomes. *Computational and Structural Biotechnology Journal* 18:1722–34.

Gaulke, C. A., Arnold, H. K., Humphreys, I. R., Kembel, S. W., O'Dwyer, J. P. and Sharpton, T. J. 2018. Ecophylogenetics clarifies the evolutionary association between mammals and their gut microbiota. *mBio* 9(5):01348–18.

Goodman, A. L., Kallstrom, G., Faith, J. J., Reyes, A., Moore, A., Dantas, G. and Gordon, J. I. 2011. Extensive personal human gut microbiota culture collections characterized and manipulated in gnotobiotic mice. *Proceedings of the National Academy of Sciences USA* 108(15):6252–7.

Greenblum, S., Turnbaugh, P. J. and Borenstein, E. 2012. Metagenomic systems biology of the human gut microbiome reveals topological shifts associated with obesity and inflammatory bowel disease. *Proceedings of the National Academy of Sciences of the United States of America* 109(2):594–9.

Gu, C., Kim, G. B., Kim, W. J., Kim, H. U. and Lee, S. Y. 2019. Current status and applications of genome-scale metabolic models. *Genome Biology* 20(1):121.

Heather, J. M. and Chain, B. 2016. The sequence of sequencers: The history of sequencing DNA. *Genomics* 107(1):1–8.

Henson, M. A. 2015. Genome scale modelling of microbial metabolism with temporal and spatial resolution. *Biochemical Society Transactions* 43(6):1164–71.

Hultman, J., Waldrop, M. P., Mackelprang, R., et al. 2015. Multi-omics of permafrost, active layer and thermokarst bog soil microbiomes. *Nature* 521(7551):208–12.

Iakiviak, M. 2017. Analysis of the ammonium assimilation pathways of the human colonic bacterium, *Bacteroides thetaiotaomicron*. Doctoral Dissertation, University of Illinois at Urbana-Champaign.

Jennings, S., Stentiford, G. D., Leocadio, A. M., et al. 2016. Aquatic food security: Insights into challenges and solutions from an analysis of interactions between fisheries, aquaculture, food safety, human health, fish and human welfare, economy and environment. *Fish and Fisheries* 17(4):893–938.

Jha, R. and Berrocoso, J. F. 2016. Dietary fiber and protein fermentation in the intestine of swine and their interactive effects on gut health and on the environment: A review. *Animal Feed Science and Technology* 212:18–26.

Ji, B. and Nielsen, J. 2015. From next generation sequencing to systematic modeling of the gut microbiome. *Frontiers in Genetics* 6:219.

Kajiyama, T., Miyahara, Y. and Murakawa, K. 2002. U.S. Patent No. 6,346,383. Washington, DC: U.S. Patent and Trademark Office.

Kchouk, M., Gibrat, J. F. and Elloumi, M. 2017. Generations of sequencing technologies: From first to next generation. *Biology and Medicine* 9(3):1–8.

Keebler, J. E. M. 2010. *Spontaneous Mutation Discovery via High-Throughput Sequencing of Pedigrees*. PhD Dissertation, North Carolina State University, Raleigh, NC.

King, T., Cole, M., Farber, J. M., et al. 2017. Food safety for food security: Relationship between global megatrends and developments in food safety. *Trends in Food Science and Technology* 68:160–75.

Koch, H., Abrol, D. P., Li, J. and Schmid-Hempel, P. 2013. Diversity and evolutionary patterns of bacterial gut associates of corbiculate bees. *Molecular Ecology* 22(7):2028–44.

Koduru, L. 2017. *Investigation of lactic acid bacterial metabolism for probiotic and cell factory application using systems biology approaches*. PhD Thesis, National University of Singapore, Singapore.

Kok, C. R. and Hutkins, R. 2018. Yogurt and other fermented foods as sources of health-promoting bacteria. *Nutrition Reviews* 76(S1):4–15.

Kozich, J. J., Westcott, S. L., Baxter, N. T., Highlander, S. K. and Schloss, P. D. 2013. Development of a dual-index sequencing strategy and curation pipeline for analyzing amplicon sequence data on the MiSeq Illumina sequencing platform. *Applied and Environmental Microbiology* 79(17):5112–20.

Kumar, M., Ji, B., Zengler, K. and Nielsen, J. 2019. Modelling approaches for studying the microbiome. *Nature Microbiology* 4(8):1253–67.

La-Ongkham, O., Nakphaichit, M., Leelavatcharamas, V., Keawsompong, S. and Nitisinprasert, S. 2015. Distinct gut microbiota of healthy children from two different geographic regions of Thailand. *Archives of Microbiology* 197(4):561–73.

Ley, R. E., Peterson, D. A. and Gordon, J. I. 2006. Ecological and evolutionary forces shaping microbial diversity in the human intestine. *Cell* 124(4):837–48.

Li, F., Hitch, T. C., Chen, Y., Creevey, C. J. and Guan, L. L. 2019. Comparative metage-nomic and metatranscriptomic analyses reveal the breed effect on the rumen micro-biome and its associations with feed efficiency in beef cattle. *Microbiome* 7(1):6.

Lopez-Legarrea, P., Fuller, N. R., Angeles Zulet, M., Martinez, J. A. and Caterson, I. D. 2014. The influence of Mediterranean, carbohydrate and high protein diets on gut microbiota composition in the treatment of obesity and associated inflammatory state. *Asia Pacific Journal of Clinical Nutrition* 23(3):360–8.

Lozupone, C. A., Stombaugh, J. I., Gordon, J. I., Jansson, J. K. and Knight, R. 2012. Diversity, stability and resilience of the human gut microbiota. *Nature* 489(7415): 220–30.

Luo, C., Tsementzi, D., Kyrpides, N., Read, T. and Konstantinidis, K. T. 2012. Direct comparisons of Illumina vs. Roche 454 sequencing technologies on the same micro-bial community DNA sample. *PLoS ONE* 7(2):e30087.

Makki, K., Deehan, E. C., Walter, J. and Bäckhed, F. 2018. The impact of dietary fiber on gut microbiota in host health and disease. *Cell, Host and Microbe* 23(6):705–15.

Mariat, D., Firmesse, O., Levenez, F., et al. 2009. The Firmicutes/Bacteroidetes ratio of the human microbiota changes with age. *BMC Microbiology* 9(1):123.

Maslowski, K. M. 2019. Metabolism at the centre of the host–microbe relationship. *Clinical and Experimental Immunology* 197(2):193–204.

McCombie, W. R., McPherson, J. D. and Mardis, E. R. 2019. Next-generation sequenc-ing technologies. *Cold Spring Harbor Perspectives in Medicine* 9(11):a036798.

Medlock, G. L., Carey, M. A., McDuffie, D. G., et al. 2018. Inferring metabolic mecha-nisms of interaction within a defined gut microbiota. *Cell Systems* 7(3):245–57.

Minj, J., Chandra, P., Paul, C. and Sharma, R. K. 2020. Bio-functional properties of probiotic *Lactobacillus*: Current applications and research perspectives. *Critical Reviews in Food Science and Nutrition* 10:1–18.

Minter, M. R., Zhang, C., Leone, V., et al. 2016. Antibiotic-induced perturbations in gut microbial diversity influences neuro-inflammation and amyloidosis in a murine model of Alzheimer's disease. *Scientific Reports* 6:30028.

Mitchell, S. J., Bernier, M., Aon, M. A., et al. 2018. Nicotinamide improves aspects of healthspan, but not lifespan, in mice. *Cell Metabolism* 27(3):667–76.

Newell, D. G., Koopmans, M., Verhoef, L., et al. 2010. Food-borne diseases - The chal-lenges of 20 years ago still persist while new ones continue to emerge. *International Journal of Food Microbiology* 139:S3–15.

Nguyen, M. Q., Wu, Y., Bonilla, L. S., von Buchholtz, L. J. and Ryba, N. J. 2017. Diversity amongst trigeminal neurons revealed by high throughput single cell sequencing. *PLoS ONE* 12(9):e0185543.

Nicholson, J. K., Holmes, E., Kinross, J., et al. 2012. Host-gut microbiota metabolic interactions. *Science* 336(6086):1262–7.

Noyes, C. 2016. *Efficient blockchain-driven multiparty computation markets at scale.* Technical report.

Obadia, B., Keebaugh, E. S., Yamada, R., Ludington, W. B. and William, W. J. 2018. Diet influences host–microbiota associations in *Drosophila*. *Proceedings of the National Academy of Sciences of the United States of America* 115(20):E4547–8.

Oberhardt, M. A., Puchałka, J., Fryer, K. E., Dos Santos, V. A. M. and Papin, J. A. 2008. Genome-scale metabolic network analysis of the opportunistic pathogen *Pseudomonas aeruginosa* PAO1. *Journal of Bacteriology* 190(8):2790–803.

Orellana, L. H., Hatt, J. K., Iyer, R., et al. 2019. Comparing DNA, RNA and protein levels for measuring microbial dynamics in soil microcosms amended with nitrogen fertilizer. *Scientific Reports* 9(1):1–11.

Org, E., Blum, Y., Kasela, S., et al. 2017. Relationships between gut microbiota, plasma metabolites, and metabolic syndrome traits in the METSIM cohort. *Genome Biology* 18(1):1–14.

Pitta, D. W., Pinchak, W. E., Indugu, N., Vecchiarelli, B., Sinha, R. and Fulford, J. D. 2016. Metagenomic analysis of the rumen microbiome of steers with wheat-induced frothy bloat. *Frontiers in Microbiology* 7:689.

Pothakos, V., Devlieghere, F., Villani, F., Björkroth, J. and Ercolini, D. 2015. Lactic acid bacteria and their controversial role in fresh meat spoilage. *Meat Science* 109:66–74.

Prats, C., Ferrer, J., Gras, A. and Ginovart, M. 2010. Individual-based modelling and simulation of microbial processes: Yeast fermentation and multi-species composting. *Mathematical and Computer Modelling of Dynamical Systems* 16(6): 489–510.

Ramirez, K., Ditamo, Y., Rodriguez, L., et al. 2010. Neonatal mucosal immunization with a non-living, non-genetically modified *Lactococcus lactis* vaccine carrier induces systemic and local Th1-type immunity and protects against lethal bacterial infection. *Mucosal Immunology* 3(2):159–71.

Rhoads, A. and Au, K. F. 2015. PacBio sequencing and its applications. *Genomics, Proteomics and Bioinformatics* 13(5):278–89.

Rowland, I., Gibson, G., Heinken, A., et al. 2018. Gut microbiota functions: Metabolism of nutrients and other food components. *European Journal of Nutrition* 57(1):1–24.

Rusk, N. 2011. Torrents of sequence. *Nature Methods* 8(1):44.

Sanger, F. and Coulson, A. R. 1975. A rapid method for determining sequences in DNA by primed synthesis with DNA polymerase. *Journal of Molecular Biology* 94(3):441–8.

Schilling, C. H., Edwards, J. S. and Palsson, B. O. 1999. Toward metabolic phenomics: Analysis of genomic data using flux balances. *Biotechnology Progress* 15(3):288–95.

Schnupf, P., Gaboriau-Routhiau, V., Sansonetti, P. J. and Cerf-Bensussan, N. 2017. Segmented filamentous bacteria, Th17 inducers and helpers in a hostile world. *Current Opinion in Microbiology* 35:100–9.

Scott, K. P., Gratz, S. W., Sheridan, P. O., Flint, H. J. and Duncan, S. H. 2013. The influence of diet on the gut microbiota. *Pharmacological Research* 69(1):52–60.

Shakya, M., Lo, C. C. and Chain, P. S. 2019. Advances and challenges in metatranscriptomic analysis. *Frontiers in Genetics* 10:904.

Shendure, J. A., Porreca, G. J., Church, G. M., et al. 2011. Overview of DNA sequencing strategies. *Current Protocols in Molecular Biology* 96(1):7.

Shoaie, S., Karlsson, F., Mardinoglu, A., Nookaew, I., Bordel, S. and Nielsen, J. 2013. Understanding the interactions between bacteria in the human gut through metabolic modeling. *Scientific Reports* 3:2532.

Shokralla, S., Spall, J. L., Gibson, J. F. and Hajibabaei, M. 2012. Next-generation sequencing technologies for environmental DNA research. *Molecular Ecology* 21(8):1794–805.

Shrestha, R. K., Lubinsky, B., Bansode, V. B., Moinz, M. B., McCormack, G. P. and Travers, S. A. 2014. QTrim: A novel tool for the quality trimming of sequence reads generated using the Roche/454 sequencing platform. *BMC Bioinformatics* 15(1):1–6.

Sokol, H., Pigneur, B., Watterlot, L., Lakhdari, O., Bermúdez-Humarán, L. G., Gratadoux, J. J. and Grangette, C. 2008. *Faecalibacterium prausnitzii* is an anti-inflammatory commensal bacterium identified by gut microbiota analysis of Crohn disease patients. *Proceedings of the National Academy of Sciences USA* 105(43):16731–6.

Sorrenti, V., Ali, S., Mancin, L., Davinelli, S., Paoli, A. and Scapagnini, G. 2020. Cocoa polyphenols and gut microbiota interplay: Bioavailability, prebiotic effect, and impact on human health. *Nutrients* 12(7):1908.

Stark, D., Barratt, J. L. N., Van Hal, S., Marriott, D., Harkness, J. and Ellis, J. T. 2009. Clinical significance of enteric protozoa in the immunosuppressed human population. *Clinical Microbiology Reviews* 22(4):634–50.

Stellato, A., Rajbhandari, S., Turbati, A., et al. 2015. VocBench: A web application for collaborative development of multilingual thesauri. In: *European semantic web conference*, 38–53. Springer, Cham.

Stellato, G., Utter, D. R., Voorhis, A., De Angelis, M., Eren, A. M. and Ercolini, D. 2017. A few *Pseudomonas* oligotypes dominate in the meat and dairy processing environment. *Frontiers in Microbiology* 8:264.

Thomson, M. J. 2014. High-throughput SNP genotyping to accelerate crop improvement. *Plant Breeding and Biotechnology* 2(3):195–212.

Tshikantwa, T. S., Ullah, M. W., He, F. and Yang, G. 2018. Current trends and potential applications of microbial interactions for human welfare. *Frontiers in Microbiology* 9:1156.

van Dijk, E. L., Jaszczyszyn, Y., Naquin, D. and Thermes, C. 2018. The third revolution in sequencing technology. *Trends in Genetics* 34(9):666–81.

Vigvári, S., Vincze, Á., Solt, J., et al. 2019. Experiences with fecal microbiota transplantation in *Clostridium difficile* infections via upper gastrointestinal tract. *Acta Microbiologica et Immunologica Hungarica* 66(2):179–88.

Wade, M. J., Harmand, J., Benyahia, B., et al. 2016. Perspectives in mathematical modelling for microbial ecology. *Ecological Modelling* 321:64–74.

Ward, C. M., To, H. and Pederson, S. M. 2018. ngsReports: An R package for managing FastQC reports and other NGS related log files. *BioRxiv*. http://www.ncbi.nlm.nih.gov/pubmed/313148.

Wirbel, J., Pyl, P. T., Kartal, E., et al. 2019. Meta-analysis of fecal metagenomes reveals global microbial signatures that are specific for colorectal cancer. *Nature Medicine* 25(4):679–89.

Wong, N. 2018. *Developing Nanoparticulate Oligonucleotides as a Target-Specific Antimicrobial for Modulation of Complex Gut Microbiota*. Doctoral Dissertation, University of East Anglia.

Zhang, C., Yin, A., Li, H., et al. 2015. Dietary modulation of gut microbiota contributes to alleviation of both genetic and simple obesity in children. *EBioMedicine* 2(8):968–84.

Index

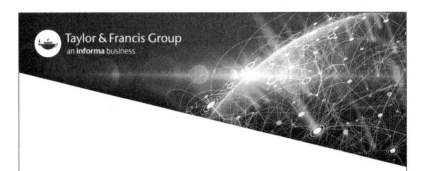

Printed in the United States
By Bookmasters